Introduction to
Oceanography

Introduction to
Oceanography

DAVID A. ROSS
Woods Hole Oceanographic Institution

HarperCollinsCollegePublishers

Acquisitions Editor: Susan McLaughlin
Developmental Editor: Fred Schroyer
Project Editors: Paula Soloway; Rachel Youngman, Hockett Editorial
   Service
Text and Cover Designer: Wendy Ann Fredericks
Cover Photograph: Front cover: Chris Harvey, Tony Stone Images;
   Back cover: Yu Amano, Urbane USA Inc., Photonica.
Art Studio: BurMar
Electronic Production Manager: Valerie A. Sawyer
Desktop Administrator: Sarah Johnson
Manufacturing Manager: Helene Landers
Electronic Page Makeup: RR Donnelley Barbados
Printer and Binder: RR Donnelley & Sons Company
Cover Printer: The Lehigh Press, Inc.

Introduction to Oceanography

Library of Congress Cataloging-in-Publication Data

Ross, David A., 1936–
   Introduction to oceanography / David A. Ross.
      p.   cm.
   Includes bibliographical references and index.
   ISBN 0-673-46938-7
   1. Oceanography.   I. Title
   GC 16.R6   1995   95-1954
   551.46—dc20   CIP

95 96 97 98 9 8 7 6 5 4 3 2 1

Brief Contents

Detailed Contents vi

Preface xv

CHAPTER 1
The Science of the Ocean 1

CHAPTER 2
The Origin and Evolution
of Earth and Its Oceans 21

CHAPTER 3
Reshaping the Ocean: Plate Tectonics 40

CHAPTER 4
Characteristics of the Seafloor 70

CHAPTER 5
Marine Sediments 102

CHAPTER 6
Properties of Water 132

CHAPTER 7
Chemical Processes in the Ocean 153

CHAPTER 8
Physical Aspects of the Ocean 178

CHAPTER 9
Ocean Circulation 198

CHAPTER 10
Waves and Tides 221

CHAPTER 11
Life Forms of the Ocean 244

CHAPTER 12
The Marine Biological Environment 280

CHAPTER 13
Climate, the Ocean, and Global Change 311

CHAPTER 14
Coastal and Estuarine Environments 337

CHAPTER 15
Resources of the Ocean 363

CHAPTER 16
Marine Pollution 395

CHAPTER 17
Marine Archeology 420

CHAPTER 18
Future Uses of the Ocean 438

CHAPTER 19
The Law of the Sea 462

APPENDIX
Terms and Statistics 477

Glossary 479

References 487

Index 489

CHAPTER 11

Detailed Contents

PREFACE   xv

CHAPTER 1
The Science of the Ocean   1

Why Study the Ocean?   2
    The Ocean and the Global Environment   2
    The Ocean and Its Resources   2
    The Ocean and Its Secrets   3

The Early History of Ocean Exploration   3

The Beginning of Oceanography as a Science   4
    Charting and Sounding   4

Box 1-1 The Voyage of Columbus   5

    Darwin, Forbes, and Huxley   6
    Matthew Fontaine Maury   6
    The Challenger Expedition   8
    Nansen and the Poles   9

Modern Oceanography   9

Oceanography Today   11

The Scope of Oceanography   13
    Marine Geology and Geophysics   13
    Chemical Oceanography   14
    Physical Oceanography   14
    Biological Oceanography   14
    Ocean Engineering   14
    Marine Policy   15
    Marine Archeology   15
    Oceanography: The Unified View   15

How to Use This Book   15

Oceanography as a Career   15

Box 1-2 Is a Career in Science a Good Opportunity?   17

Oceanography for the Nonoceanographer   18

Summary   18
Questions   18
Key Terms   19
Further Reading   19

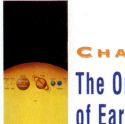

CHAPTER 2
The Origin and Evolution of Earth and Its Oceans   21

The Setting   22

Box 2-1 The Power of 10   22

Origin and Evolution of Our Solar System   23
    The Condensation Hypothesis   23

Box 2-2 Do Other Stars Have Planets?   24

The Early Earth   25
    Origin of Earth's Atmosphere   26
    Origin of the Ocean   27

Further Development of Earth and Its Ocean   28
    Climatic Changes, Glaciers, and Sea Level
        Changes   28

Box 2-3 Early Earth and Favorable Conditions for Life   29

    Extinctions and the Greenhouse Effect   31

Box 2-4 Why Mass Extinctions?   32

The General Character of Earth's Present Ocean   33

Box 2-5 Some Units of Measure   34

    Earth's Ocean and Earth's Oceans   34

Summary   38
Questions   38
Key Terms   38
Further Reading   39

CHAPTER 3

Reshaping the Ocean: Plate Tectonics 40

Early Understanding of Earth's Structure 41

Continental Drift: A Hypothesis to Explain Earth's Crust 41

Box 3-1 Earthquakes and the Internal Structure of Earth 42

   The Continental Drift Hypothesis 43

Box 3-2 Isostasy 44

Seafloor Spreading and Plate Tectonics: A Theory to Explain Earth's Crust 46

Box 3-3 How We Know the Continents Were Once Joined Together 47

   Plenty of Evidence 48

Box 3-4 A Little Geology 49

   Plate Tectonics 51
   Dating the Seafloor 53
   Plate Tectonics: The Motions 54
   Plate Tectonics and Earthquakes 54
   Hot Spots 55
   Ocean Basins and Mountains 56

A Tour Around One Plate 57

Plate Tectonics: Economics, Life-Styles, and Environment 60

The Ridge Program and Some Unanswered Questions 63

Recent Research 66

Summary 68
Questions 68
Key Terms 68
Further Reading 69

CHAPTER 4

Characteristics of the Seafloor 70

Introduction 71

Tools for Studying the Structure of the Seafloor 71

Measuring Earth's Gravity 71
Measuring Earth's Magnetic Field 71

Box 4-1 Sounding Methods, from Rope to Sea Beam 72

Box 4-2 Satellites, Bathymetry, and Gravity 75

   Measuring Ocean Bottom Layers and Structure 76
   Measuring Movement 77

Crustal Structure 79
   Mohorovičić Discontinuity 81
   Density and Isostasy 81

The Continental Margin 80
   The Coastal Region 81
   Continental Shelves 81

Box 4-3 Glaciers on the Continental Shelf 83

   A Model: The U.S. Continental Margin 84
   Continental Slopes 87
   Continental Rises 87
   Submarine Canyons 88
   Turbidity Currents 89
   Marginal Seas 92

The Ocean Basin 93
   The Atlantic Ocean 93
   The Pacific Ocean 95
   The Indian Ocean 96
   The Arctic Ocean 99
   The "Southern Ocean" 99

Summary 99
Questions 100
Key Terms 100
Further Reading 101

CHAPTER 5

Marine Sediments 102

Introduction 103

Box 5-1 Some Tools for Studying the Seafloor 103

Techniques for Sampling and Observing the Seafloor 107
   Drilling Ships 108
   Research Submersibles 109

Box 5-2 The Art of Navigation 111

The Missing Sediments 116

Box 5-3 Walking on the Seafloor 118

Marine Sediments 119
 Terrigenous Sediments 119
 Sediments of the Deep Sea: The Pelagic
  Environment 122
 Biogenous Sediments 122
 Pelagic Clays 125
 Hydrogenous Sediments 125
 Cosmogenous Sediments 125
 Rocks on the Seafloor 125
 Areas of Nondeposition 125

Box 5-4 Seafloor Spreading and Deep Sea Sedimentation 126

 Sediments on the United States Continental
  Margin 128

Paleoceanography 128
 Dating by Magnetic Reversals 128
 Dating by Biogenous Sediment 128
 Dating Problems 129
 The Paleoceanographic Picture 130

Summary 130
Questions 131
Key Terms 131
Further Reading 131

CHAPTER 6
Properties of Water 132

Introduction 133

The Structure of Water 133

Box 6-1 Chemistry Review 134

Some Properties of Water 137
 Dissolving Power 137
 Phase 137
 Density Due to Temperature 138
 Heat Capacity and Calories 138
 Heat Versus Temperature 139
 Pressure 140
 Density Due to Salinity 140
 If Water Were Different... 140

Box 6-2 Some Other Properties of Water 141

Box 6-3 Light in the Ocean 141

Early Studies on the Chemistry of Seawater 143

Box 6-4 How Do Salts Get Into and Out of the Ocean? 145

How Saltwater Is Different from Freshwater 146

Ice in the Ocean 147
 Ice, Glaciers, and Climate Change 147
 Sea Ice 147
 Glaciers and Icebergs 149

Summary 151
Questions 151
Key Terms 151
Further Reading 152

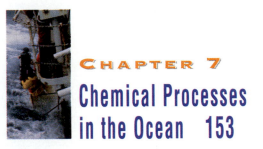

CHAPTER 7
Chemical Processes in the Ocean 153

Recent Research in the Chemical Sea 154

Box 7-1 pH, Alkalinity, Acidity, and Buffering 154

Tools of Chemical Oceanography 156

Box 7-2 Measuring Salinity 158

What's in Seawater? 158
 Dissolved Inorganic Matter 159
 Dissolved Gases 160
 Dissolved Organic Matter 163
 Particulate Matter 163

Box 7-3 Carbon, Carbon Dioxide, and the Ocean 164

Factors Influencing the Ocean's Chemical
 Composition 166
 Chemical Reactions and Residence Time 166
 Ocean Vents and Recycled Seawater 168

Box 7-4 Vents and More Vents 169

 Life and the Ocean: Biochemical Reactions 170
 Physical Processes 171
 Human Input 172

Isotopes and Radioactivity in the Ocean 172
 Stable Isotopes 173
 Radioactive (Unstable) Isotopes 174

Summary 176
Questions 176
Key Terms 176
Further Readings 177

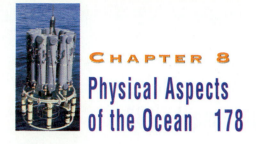

CHAPTER 8
Physical Aspects of the Ocean   178

Introduction   179

Instruments of the Physical Oceanographer   179
   Temperature and Depth Measurements   179

Box 8-1 Eyes in the Sky: Satellites in Oceanography   180

   Continuous Measurements   182

Box 8-2 Observation Stations: Fram, Ice Islands, and Buoys   184

   Acoustical Measurements   186

General Characteristics of the Ocean   186
   Salinity   187
   Temperature   188
   Density   190

Box 8-3 More about Seawater Density   192

Underwater Sound   193
   Sound Velocity   193

Box 8-4 A Shot Heard Around the World   194

Summary   196
Questions   197
Key Terms   197
Further Reading   197

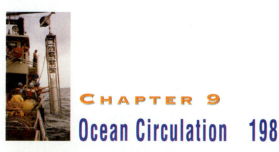

CHAPTER 9
Ocean Circulation   198

Introduction   199

Box 9-1 The World Ocean Circulation Experiment (WOCE)   199

Measurement of Currents   200
   Simple Measurement of Currents   200
   Modern Techniques for Measuring Currents   200
   Indirect Measurements of Currents   204

Solar Energy   204
   Earth's Heat Budget   204

Atmospheric Circulation   206

Box 9-2 The Coriolis Effect   207

Atmospheric Circulation Drives Ocean Circulation   208

Wind-Driven Circulation   208
   Ekman Spiral   210
   Countercurrents and Undercurrents   211
   Eddies and Rings   211
   Upwelling   214

Thermohaline Circulation   216

Box 9-3 Megaplumes   218

Summary   219
Questions   220
Key Terms   220
Further Reading   220

CHAPTER 10
Waves and Tides   221

Introduction   222

Waves: Energy in Motion   222

Wind-Generated Waves   224
   Wind-Generated Waves: Sea, Swell, and Surf   225
   Waves in the Coastal Region   226
   Langmuir Cells   227

Internal Waves   230

Catastrophic Waves   230
   Storm Surges   230
   Landslide Surges   231
   Tsunami   231

Box 10-1 A Historic Tsunami   234

   Stationary Waves   235

Ocean Tides   236
   Causes of the Tides   237
   Tidal Currents   239
   The Importance of the Tides   239

Box 10-2 High Tides   241

X **DETAILED CONTENTS**

Tidal Friction   241

Summary   242
Questions   242
Key Terms   242
Further Reading   243

CHAPTER 11
Life Forms of the Ocean   244

Introduction   245

Benthic Environments   245
   Littoral System   246
   Deep-Sea System   246

Pelagic Environments   246

Box 11-1 Marine Snow   247

   Neritic System   247
   Oceanic System   248

Classifying the Ocean Population   249
   Plankton   249
   Nekton   252
   Benthos   252

Box 11-2 Fish Antifreeze   253

Classifying Organisms by Their Physical
   Characteristics   253
   The Five Kingdoms   254

Bacteria and Blue-Green Algae   256

Plants of the Sea   256
   Kingdom Plantae   257
   Kingdom Protista   258

Animals of the Sea   261
   Kingdom Protista   261
   Kingdom Animalia   261

Box 11-3 Sounds in the Ocean   274

*Box 11-4 Whaling and the International Whaling
   Commission   277*

Summary   277
Questions   278
Key Terms   278
Further Reading   278

CHAPTER 12
**The Marine Biological
Environment   280**

Introduction   281

Biological Oceanography: Sea Food and Ecosystems   281

Instruments of the Biological Oceanographer   282

The Biological Community   284

Box 12-1 A Wonderful Example of Commensalism   285

The Ocean, Seawater, and Their Biological
   Consequences   286

Plants and the Ocean   288
   Photosynthesis   288
   Flotation   288
   Compensation Depth   288
   Nutrient Cycles   289
   Other Factors Affecting Plants   290

Animals and the Ocean   293
   Adaptation to Environment   293
   Origin of Deep-Sea Fauna   294
   Bioluminescence   294
   Color   294
   Deep Scattering Layer   294

Coral Reef Communities   296

Box 12-2 Coral Bleaching   296

   Formation of Coral Atolls   298

Vent Communities   298
   Vent Dwellers and Chemosynthesis   300
   Vent Dwellers: Insight into the Past?   300

Box 12-3 Chemosynthesis   301

Organic Production   302
   Total Organic Production   303
   Measuring Organic Production   303
   Standing Crop   304
   Factors Influencing Organic Production   304
   Grazing   304
   Summarizing Primary Production   305

The Food Cycle   305
   Areas Having Different Food Cycles   305
   Implications of the Food Cycle   307

Summary   308

Questions   309
Key Terms   309
Further Reading   309

Evidence in the Sea   333
Ice Cores and the Great Ocean Conveyer   334

Summary   335
Questions   336
Key Terms   336
Further Reading   336

CHAPTER 13
Climate, the Ocean, and Global Change   311

Introduction   312

Interaction Between the Ocean and the
  Atmosphere   313
  Unequal Heating of Land and Water   313
  Summary of Oceanic-Atmospheric Interaction   315
  Hurricanes   315

Box 13-1 Hurricanes and Global Change   318

Interaction among the Air, Sea, and Climate   320
  Carbon Dioxide and Climate   320
  Ocean Surface Temperature and Climate   320
  Phytoplankton and Climate   321

The Greenhouse Effect and Global Warming   321
  What Is the Greenhouse Effect?   321
  Greenhouse Gases   322
  Carbon Dioxide   323
  Is the Greenhouse Effect Increasing?   323
  The Greenhouse Effect and Our Future   324

Box 13-2 The Mount Pinatubo Eruption   325

Box 13-3 An Experiment with Photoplankton   326

El Niño   326
  What Causes El Niño?   327

Box 13-4 Predicting El Niño with Satellites   328

  Effects of El Niño   329

The Ozone Problem   329
  The Ozone Hole   330
  Ozone near the Ground   330
  Solving the Ozone Problem: A Start   331

Past Climates   331
  The Milankovitch Hypothesis   332

CHAPTER 14
Coastal and Estuarine Environments   337

Introduction   338

Beaches   339
  Longshore Current and Rip Current   339
  Beach Erosion   340
  Seasonal Sand Movement   342
  Glaciers, Sea Level, and Slowly Migrating
    Beaches   342
  Barrier Beaches and Barrier Islands   343
  Movement of Barrier Beaches and Islands   343

Estuaries, Lagoons, Fjords, Marshes, and Mangrove
  Swamps   344

Box 14-1 Cycles of Coastal Erosion   346

  Where Estuaries and Lagoons Occur   348
  Short-Lived Estuaries   348
  Water Mixing in Estuaries   348
  Estuarine Life   350

Deltas   352

A Quick Tour Around the U.S. Coast   353

Coastal Problems   354
  Global Change   354

Box 14-2 Is Sea Level Rising?   355

  Beaches   355
  Estuaries and Marshes   356
  Human Activity in the Coastal Zone   356

Uses of the Coastal Zone   358
  Coastal Zone Facts   358

*Box 14-3 The U.S. Approach to Managing the
  Coastal Zone   359*

Summary   361
Questions   361
Key Terms   361
Further Reading   361

Summary   392
Questions   393
Key Terms   393
Further Reading   393

CHAPTER 15
Resources of the Ocean   363

Introduction   364

Mineral Resources of the Continental Margin   364
    Sediment Deposits   365
    Placers   365
    Sand and Gravel   367
    Phosphates   367
    Other Minerals   368
    Seawater as a Resource   368
    Oil, Gas, and Sulfur Deposits   368

Mineral Resources of the Deep Sea   371
    Seafloor Spreading and Mineral Resources   371
    Metalliferous Muds   372
    Red Sea Metalliferous Muds   372
    Manganese Nodules   374

Box 15-1   Mining of a Deep Sea Vent   375

    Cobalt Crusts   377
    Deep-Sea Muds and Oozes   377
    Mineral Potential of the U.S. Exclusive Economic
       Zone   377

Biological Resources   380
    The Protein Crisis   380
    Phytoplankton and Fish   380
    Fishing Methods   382

Box 15-2   The U.S Fishing Industry   382

    By-Catch   384
    Natural Controls of the Fish Catch   385
    Legal Problems of International Fishing   388
    Other Problems   389

Box 15-3   Sashimi and Sushi   389

Box 15-4   Pollution and Fisheries   390

    Aquaculture: Farming Seafood   390

Physical Resources of the Ocean   392

CHAPTER 16
Marine Pollution   395

Introduction   395
    Marine Pollution   396
    The Great Lakes   396
    Pollution: Defining, Detecting, and Measuring   396

Box 16-1   Monitoring Coastal Environmental Quality   398

    Human Impacts on the Marine Environment   399

Box 16-2   Marine Hitchhikers   400

    Pathways to the Sea   402
    Controlling Marine Pollution: Options   402

Domestic, Industrial, and Agricultural Pollution   403
    Pesticides   404
    Synthetic Organic Compounds   405
    Excess Nutrients   405
    Heavy-Metal Pollution   405

Ocean Dumping   407
    Plastic Problems   408
    Ocean Incineration   408

Oil Pollution   407
    Major Spills   410

Box 16-3   The Persian Gulf Oil Spill and Fires   413

    Acute and Chronic Effects of Oil Spills   415
    Are Platforms Bad Polluters?   415
    How Oil Behaves in the Marine Environment   415

Pollution from Mineral-Resource
   Exploration/Exploitation   416

Radioactive Waste   417

Thermal Pollution   417

Summary   417
Questions   418
Key Terms   418
Further Reading   418

CHAPTER 17
Marine Archeology   420

Introduction   421
   Competing Interests   421
   The Discipline of Marine Archeology   421
   A Capsule History of Marine Archeology   423

Important Discoveries   424
   U.S.S. *Monitor*   424
   Ulu Burun Site   425
   S.S. *Central America*   426
   R.M.S. *Titanic*   427

Box 17-1 Can You Make a Small Fortune from the Ocean?   429

   Bismarck   432

The *Jason* Project   432

The Future of Marine Archeology   435

Box 17-2 Wrecks in U.S. Waters   436

Summary   436
Questions   437
Key Terms   437
Further Reading   437

CHAPTER 18
Future Uses of the Ocean   438

Innovative Biological Uses   439
   Marine Biotechnology and Genetic Engineering   439

Box 18-1 Aquaculture and Genetic Engineering   440

   Live Better—Eat More Fish!   440
   Improving the Harvest   442
   Krill: Protein of the Future?   442

Box 18-2 Fish TV   443

   Military Use of Marine Mammals   443
   Drugs from the Sea   444

Energy from the Ocean   445
   Tidal Power   446
   Wave Power   447
   Power from Major Ocean Currents   448
   Power from Thermal Differences   448
   Artificial Upwelling   451
   Power from Marine Biomass   452
   Power from Gas Hydrates   452
   Other Possible Sources of Energy   453

Freshwater from the Sea   453
   Desalination   454
   Glaciers and Ice Caps   454

Disposal of Nuclear Waste in the Deep Sea   455

Underwater Robotics   457

Satellites for the Sea   457

Box 18-3 Satellite Missions for the '90s   459

Underwater Habitats: Offshore Islands and
  Platforms   459

The Ocean for Tourists   459

Summary   460
Questions   460
Key Terms   461
Further Reading   461

CHAPTER 19
The Law of the Sea   462

Early History of the Law of the Sea   463
   From Early History to Harry S Truman   463

The 1958 and 1960 Geneva Conventions on the Law of
  the Sea   464
   Convention on the Territorial Sea and Contiguous
    Zone   464
   Convention on the Continental Shelf   465
   Convention on the High Seas   466
   Convention on Fishing and Conservation of Living
    Resources of the High Seas   466
   Shortcomings of the 1958/1960 Geneva
    Conventions   466

Third Law of the Sea Conference   467

Basic Issues of the Third Law of the Sea
Conference   468

Box 19-1 Good Marine Boundaries Make Good Neighbors   469

Results of the Third Law of the Sea Conference   470
Exclusive Economic Zones   470
Continental Margins   471
Developed vs. Developing Nations   471
Deep Sea Mining   473

Freedom of Scientific Research   473
Concerns of Developing Nations   474

Summary   475
Questions   475
Key Terms   475
Further Reading   475

APPENDIX
Terms and Statistics   477

GLOSSARY   479

REFERENCES   487

INDEX   489

Preface

Recent human and natural events have focused attention on the 71 percent of the Earth's surface that is the ocean. An important human example is the continuing release of carbon dioxide and other greenhouse gases into the atmosphere. This has led some scientists to speculate that a worldwide temperature increase is in progress, which could lead to a melting of glacial ice and consequently a serious rise in sea level. The data at this time, however, are inconclusive. Because the ocean stores many times more carbon dioxide than the amount that is in the atmosphere, the ocean may play the controlling role in any future warming or cooling of our planet.

Natural phenomena that are directing interest toward the ocean include the occurrence of El Niño conditions every few years in the Pacific Ocean. An El Niño causes major worldwide problems, including floods, drought, and famine. Another phenomenon is the increasing frequency of intense hurricanes, which are driven in part by heat energy from the ocean.

These examples illustrate the diversity and impact of ocean-related phenomena. Studies of such problems and others are giving us powerful insight into the complex interactions among the ocean, land, atmosphere, and biosphere. Early results of these studies are yielding a more global view of the ocean and how it affects land, climate, atmosphere, and the inhabitants of our planet—including us. Such studies probably will dominate the science of oceanography for decades to come.

The oceans are also being considered for various uses. These include harvesting of their biological and mineral resources, disposal of land-derived pollutants in their waters, and further development of the coastal zone. The almost universal acceptance of exclusive economic zones (EEZ), extending 200 nautical miles seaward from a country's coast, has accelerated the trend toward increased ocean use. The EEZs of all coastal countries combined give them jurisdiction over about 36 percent of the ocean, an area that actually exceeds the present land area of our planet. If the countries of the world exercise this new legal control over their EEZs in an intelligent manner, they could improve problems such as overfishing of certain species, marine pollution, and other detrimental uses of the ocean.

The past few years have seen new ocean technologies emerging and exciting discoveries. These include the finding of additional deep sea vents with their uniquely associated fauna, new drugs from marine organisms, and complexities in ocean circulation. The importance of these discoveries is far from being fully appreciated or understood. Innovative underwater-observing devices that operate independent of surface ships offer great promise in further exploration of the ocean. The emerging field of marine archeology has made several dramatic discoveries in recent years using such technology.

In this edition I emphasize the important events and technologies just mentioned, but at the same time maintain a balance with marine science fundamentals. I try to strike a fair position between just concern for the environmental health of the ocean and some of its potential uses. That is, I try to present a realistic view of oceanography today.

This text is a comprehensive revision and provides up-to-date coverage of all aspects of oceanography. I have completely updated and expanded the text from earlier editions and added over 200 new color figures. To interest students from other disciplines and show the breadth and relevance of marine science, I devote entire chapters to climate, the ocean, and global change (Chapter 13); ocean resources (Chapter 15); marine pollution (Chapter 16); marine archeology (Chapter 17); future uses of the ocean (Chapter 18); and the law of the sea (Chapter 19). The book will be understandable and interesting to the student with a modest or limited background in science.

Recognizing that this book may be a student's one and only contact with oceanography, I have enriched it with over 50 boxes. The boxes illuminate either technical topics (for example, light in the ocean, Coriolis Effect, the art of navigation, sounding methods, and so on), or fascinating aspects of marine science (aquaculture and genetic engineering, sounds in the ocean, hurricanes and global change, coral bleaching, and so on). Each chapter also describes recent articles and books to further the learning of interested students. Supplementary materials include an Instructor's Manual/Textbook by Professor William G. Siesser of Vanderbilt College, and a Study Guide by Professor Bob Wallace of Ripon College. Full-color transparencies, slides, and a newsletter describing new marine science findings are also available.

As a practicing oceanographer at the Woods Hole Oceanographic Institution, I have given this book a strong practical streak. I show how oceanographers work and how ocean science relates to daily life. As in earlier editions, I describe oceanography in a manner understandable to nonscience students. I stress the interdisciplinary nature of oceanography and the role of the disciplines outside the physical sciences—law and economics, for example. Both metric and English measurements are used throughout the text, and I define new terms on first use. Examples of marine phenomena are mostly North American because this is the primary audience for this book.

I have been very fortunate to discuss many aspects of this edition with my colleagues in the Woods Hole marine scientific community and elsewhere. Many people have contributed photographs and other materials, and I have acknowledged this throughout the text. I wish to offer special thanks to those who have been especially gracious with their time or information; these include from the Woods Hole area: Don Anderson, David Aubrey, Robert Ballard, Susan Berteaux, Nancy Brink, William Dunkle, Eben Franks, Dave Gallo, Graham Giese, Susumu Honjo, Susan E. Humphris, Shelley Lauzon, Phillip S. Lobel, Laurence P. Madin, Richard Pittenger, Philip L. Richardson, Terry M. Rioux, Ivan Valiela, and Keith von der Heydt. Among those at other areas or institutions are: Mike Champ, of Environmental Systems Development, Cindy Clark of the Scripps Institution of Oceanography, David Duane of NOAA, Mai Edwards of the National Geophysical Data Center, Kim Grasso of NASA, David Hastings of the National Geophysical Data Center at NOAA, Charles E. Herdendorf of the Columbus-American Group, Maria Jacobsen of the Institute of Nautical Archaeology, John Kermond of NOAA, Don Pryor of NOAA, Peter Sloss of the National Geophysical Data Center at NOAA, Susan

van Holk of the Harbor Branch Oceanographic Institution, and Janet Weber of the Lamont Doherty Earth Observatory. Sheri DeRosa, Tracey Crago, and Pam Foster helped me in the preparation of parts of the manuscript.

I would also like to thank the following professors who reviewed the manuscript: Ernest E. Angino, University of Kansas; Marsha Bollinger, Winthrop University (SC); Karl M. Chauff, Saint Louis University; Richard Dame, University of South Carolina, Coastal Carolina; John Ehleiter, West Chester University; William T. Fox, Williams College; Dirk Frankenburg, University of North Carolina, Chapel Hill; Robert T. Galbraith, Crafton Hills College; Robert R. Given, Marymount College (CA); William Hamner, University of California, Los Angeles; William H. Hoyt, University of Northern Colorado; David L. Kan, Massachusetts Maritime Academy; Ernest Knowles, North Carolina State University; Albert M. Kudo, The University of New Mexico; Ronald Johnson, Old Dominion College; Michael Lyle, Tidewater Community College (VA); James E. Mackin, State University of New York, Stony Brook; James M. McWhorter, Miami-Dade Community College; James Mybakken, California State University at Hayward; Harold R. Pestana, Colby College; C. Nicholas Raphael, Eastern Michigan University; Richard A. Roller, Lamar University; William G. Siesser, Vanderbilt University; Stanley Ulanski, James Madison University; Raymond E. Waldner, Palm Beach Atlantic College; and Mel Zucker, Skyline College (CA).

It has been a pleasure working with the people at HarperCollins. I have especially enjoyed the enthusiasm and continued encouragement of Acquisitions Editor Susan McLaughlin, and her talented assistant Juliana Nocker. Others who have been instrumental in putting together various parts of the book or the learning package are: Supplements Editors Kathi Kuntz and Donna Campion, Marketing Managers Andria Ventura and Anita Virgilio, and Project Editors Paula Soloway at HarperCollins and Rachel Youngman of Hockett Editorial Service. My highest thanks go to Fred C. Schroyer of Waynesburg, Pennsylvania, who did a great job of developing the manuscript.

I would like to thank my wife, Edith, who gave me help, encouragement, and privacy when I needed it. I dedicate this book to Paula Reppmann, a gracious woman and wonderful mother to Edith.

DAVID A. ROSS

CHAPTER 1

The Science of the Ocean

From our more restricted view on Earth, it is easy to conclude that land is the predominant environment of our planet. It is not! When astronauts look at Earth from space, they see a more accurate view: a blue planet, on which land is just an interruption in the vast, world-encircling ocean (see opening figure). But even more important than the ocean's extent is how it *affects* our entire planet.

This chapter presents a snapshot of oceanography as a science and its importance to all of us. The story of how marine science developed shows how ocean discoveries were achieved through the persistence and imagination of early explorers. Today, oceanography is a high-tech science involving geology, chemistry, physics, biology, meteorology, energy and resource management, environmental studies, archeology, and law, and has numerous career opportunities.

Whether or not you pursue a career in oceanography or any other science, a knowledge of the ocean is important, for reasons you are about to discover.

Satellite view of Planet Earth. The fact that the sea dominates the planet is quite evident.
(Photograph courtesy National Aeronautics and Space Administration)

Why Study the Ocean?

There are nine planets in our Solar System and probably billions more in the universe. But, to the best of our knowledge, our planet is unique in its biological diversity. Earth's atmosphere, ocean, and thin soil layer operate together to support a wide variety of living creatures. Each part of our planet is closely tied to the others by complex biological, chemical, physical, and geological processes, most of which are driven by energy from the sun. These processes and their applications are all aspects of *oceanography,* the scientific study of the ocean.

The Ocean and the Global Environment

The oceans cover about 71 percent of the surface of this planet and land covers the remaining 29 percent. The ocean affects many aspects of our environment, such as climate, weather, and even sea level. For example, a worldwide increase in temperature could melt glaciers and result in a rise in sea level that would affect all coastal cities to some degree. Even a small rise in sea level would flood extensive portions of low-lying countries such as Bangladesh and the Netherlands and states along our Atlantic, Gulf, and Pacific coasts.

A knowledge of how the ocean interacts with both the land and the atmosphere is thus very important to your understanding of our environment. Scientific research observations, mainly from satellites, show that changes are occurring to the land, atmosphere, and ocean that may affect all life on this planet. These **global changes*** include a possible increase in worldwide temperature, changes in sea level, more frequent hurricanes and drought, and even decreased supplies of fresh water. The point is that *processes within the ocean play a key role in our understanding and control of global change.*

In recent years, powerful interactions among processes on the land, in the atmosphere, and in the ocean have become better understood through major research programs. For example, we have learned that much of the solar energy that reaches Earth remains in the ocean to help drive oceanic and atmospheric circulation. In this manner, the ocean influences worldwide patterns of weather and climate. One mechanism is the evaporation of ocean water into the atmosphere, from whence it falls as rain and snow to feed our streams and rivers. These in turn drain back into the ocean, carrying rock debris from **weathering** and **erosion.**

In this same manner, atmospheric pollution over land may end up in the ocean. Other processes and interactions are less understood, but one thing is clear: the role of the ocean in our lives is enormous, no matter where we live, even in North Dakota, far from any coast. We and future generations are the stewards of our planet and its resources. To be effective stewards, both as students and as citizens, we must better understand the processes and interactions among the sea, land, and air and apply the appropriate information that we learn toward solving various environmental problems.

The Ocean and Its Resources

The oceans are an important source of food, energy, mineral, chemical, recreational, and transportation resources. At present they yield slightly less than 100 million tons of food per year, a large portion of which is used to feed livestock. The ocean directly provides only about 10 percent of the animal protein currently consumed by people, in the form of fish and shellfish. Expansion of such food sources could one day help solve some of the world's food shortages.

Energy already is drawn from hydrocarbon deposits (oil and natural gas) found buried within the sediments of the **continental shelf.** The future promises new methods for exploiting the physical, chemical, and biological processes active within the ocean. These include tides, waves, and temperature differences between surface water and deep water—all part of the ocean's enormous potential.

Certain minerals and rocks that occur on the seafloor are valuable. In some shallow areas, sand, gravel, and gold already are mined, and other potential resources such as nodules rich in manganese exist in deeper parts of the ocean. Biological products of the sea also are useful—some shells, for example, are valuable as building materials.

Seawater itself is a resource. First, it is a source of valuable chemicals, including iodine, bromine, potassium, magnesium, and manganese. Second, salty seawater can be **desalinated** to make fresh water, which is in critically short supply in some areas of the world, such as the Persian Gulf.

The oceans are Earth's widest highways. Most international trade—over 90 percent—is carried on ships. You almost certainly own something—a car, TV, computer, clothing—that was made on another continent and spent time as ship's cargo.

Clearly, the study of **oceanography** (sometimes called **marine science**) is essential so that we may benefit from the ocean's valuable mineral, biological, and energy resources. But the ocean also affects us in less obvious ways. For example, the ocean is important for military reasons; indeed, the seas have been a battlefield for much of recorded history.

Finally, coastal waters are a prime area for recreation as well as for residential or industrial use (Figure 1-1). In the United States, about 75 percent of the population, and a similar percentage of U.S. industry, reside within

---

*Key terms given in bold are listed at the end of each chapter to aid your review. Terms are defined in the text or in the Glossary.

FIGURE 1-1 (a) A coastal marsh in New England. Salt marshes, biologically important for the growth of many fish and shellfish species, are the most biologically productive area of the ocean.

(b) Coastal housing development along Long Island, New York. Note the difference between the undeveloped area in the foreground and the developed area in the background. (Photograph (a) courtesy of John Teal, Woods Hole Oceanographic Institution; photograph (b) courtesy of Dr. Ivan Valiela.)

80 km (50 miles) of the ocean or the Great Lakes. In some other countries, the percentage is even greater. In many nations, the sports of fishing, boating, waterskiing, scuba diving, and swimming attract ever greater numbers. However, uncontrolled pollution could place some of these activities into the past tense. Again, the science of oceanography is and will be critical to the quality of our lives.

The Ocean and Its Secrets

Although the ocean occupies most of the surface of our planet, it is sometimes hostile and inaccessible and does not easily yield its secrets. This very secrecy, the lure and romance of the sea, has drawn many to the oceans. People have always been interested in their environment, and the ocean depths are no exception.

The oceans may hold answers to some very important scientific questions and problems. Within the layers of sediment on the ocean floor, for example, lies a remarkable record of the geologic history of Earth. In fossils preserved within these sediments, Earth's biological history is concealed. By studying ocean sediments, we can also learn how climate has changed, and thus better understand our past, present, and future.

The ocean may carry within it the mystery of life itself. Considerable evidence suggests that life began in the marine environment several billion years ago and gradually evolved to produce the vast variety of organisms now found both in the ocean and on land. Issues like these make oceanography an exciting field of scientific research.

The Early History of Ocean Exploration

Early interest in the sea was usually motivated by practical need. People simply wanted to go from one place to

another in as short a time as possible. Early ocean studies therefore were closely connected to the development of trade. Slowly, this early mapping of Earth and exploring of the ocean and its contents developed into modern oceanography. By 800 B.C. explorers had made voyages around Africa, and by 400 B.C. the rise and fall of tides had been related to the phases of the moon. The Greek astronomer and poet Eratosthenes determined the circumference of Earth with remarkable precision around 250 B.C.

From these early explorations and scientific works, two hypotheses developed among ancient scholars about the distribution of land and water. Eratosthenes and the Greek geographer Strabo believed that the continents of the world formed a single island surrounded by the ocean. The other view, expressed by the Greek astronomer Ptolemy in the middle of the second century A.D., was that the Atlantic and Indian Oceans are enclosed seas, like the Mediterranean (Figure 1-2).

Importantly for world history, Ptolemy also believed that the eastern and western points of the world known at that time were very close together, and that by sailing west one could reach the eastern extremity rather quickly. It was this idea that led Christopher Columbus to sail west to reach Asia. Instead, he ran into a barrier known at that time only to Native Americans and a few early Viking explorers: North America (see Box 1-1).

The Indian Ocean probably was the first open ocean to be used for trade, but the last to be explored in any detail. Because of its winds it is uniquely suited to sailing vessels. During the summer **monsoon** season—a period of heavy winds and local storms—the wind blows from the southwest; during the winter monsoon season, it blows from the northeast. Thus, vessels with the simplest square-rigged sails could travel across the entire ocean in one season and return in the following season. The Pacific Ocean also was explored during

FIGURE 1-2   Claudius Ptolemy, a famous early astronomer, made a map of the known world of his day, first published about A.D. 140. The map shown here is a Renaissance version of Ptolemy's map, published in 1522. The details are very impressive for a chart that was made over 1,800 years ago. The heads around the borders represent the winds.

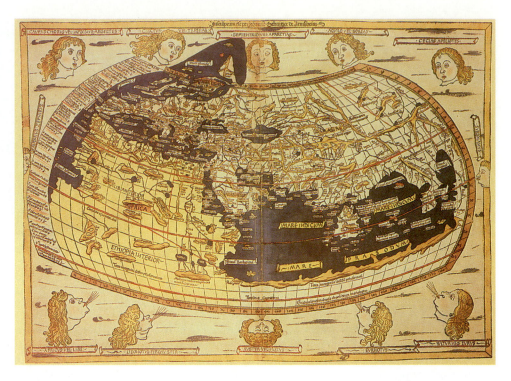

these early times by the Polynesians, who sailed vast areas of it.

After the fall of the Roman Empire in the fifth century A.D., and during the Middle Ages (A.D. 400–1500), trade decreased and European knowledge of the sea declined. However, the Vikings of northern Europe established colonies in Iceland and Greenland prior to A.D. 1000, helped in their travels by a temporary climatic warming that caused the ice cover to retreat, making for easier sailing. By A.D. 982, the Viking Eric the Red had sailed to northeastern Canada, and later his son, Leif Ericsson, reached Newfoundland. It is possible that the Vikings may also have established colonies in the United States; however, by A.D. 1200 the climate had so cooled that all colonies were abandoned due to the advance of glacial ice. Around this time, Arab navigators were sailing across and around the Indian Ocean and had even reached China. There the magnetic properties of the iron mineral called lodestone were discovered, and the compass came into use for navigation.

It was not until the late 1400s that the people of southern Europe again explored the sea. This renewed interest in exploration was driven by the need for new trading routes to the East. The previous routes were effectively closed by the capture of Constantinople (now Istanbul, in Turkey) by Sultan Mohammed II in 1453. The early explorers Bartolomeu Dias, Vasco da Gama, and Christopher Columbus are well known. The Portuguese were among the leaders of European exploration, and Prince Henry (Henry the Navigator) established a center for teaching navigation and other nautical skills to Portuguese sailors, although Henry himself probably never went to sea. This center, overlooking the Atlantic in Sagres, Portugal, may be considered the first oceanographic institution.

The Beginning of Oceanography as a Science

By the early 1500s, it was generally accepted that the Earth is round, not flat. The Ferdinand Magellan voyage around the world (1519–22), which continued *after* the famed explorer's death in 1521, proved this beyond doubt. After the voyages of exploration for new lands, new expeditions began to explore the properties and resources of the ocean. Although most of the examples presented here come from English-speaking countries, important research certainly was done in other areas of the world, but less is known about the activities of these non-English speaking countries.

Charting and Sounding

By the end of the sixteenth century, basic characteristics of the ocean were known. For example, it had been learned that deeper waters are generally cooler than surface waters. It was also known that ocean **salinity** or salt content is similar almost everywhere. One of the most successful of the early explorers was Captain James Cook (1728–1779), who was chosen in 1768 to lead an expedition aboard the *Endeavour* to the South Pacific for an astronomical study. Cook was the first explorer to carry instruments that could accurately measure latitude

BOX 1-1

The Voyage of Columbus

Exactly what *did* Christopher Columbus "discover" in 1492? The story of his journey to the so-called New World is well known; Columbus hoped to find a new passage to Asia by sailing west from Spain. On his initial trip (he eventually made four), his first landfall was somewhere in the Caribbean—but where? The island he first reached was called Guanahaní by its aboriginal inhabitants at the time. But today, no one is certain which island this was. Uncertainties about his navigation and the ambiguity of his description of the island, compounded by the many similar-appearing islands in the region plus the faded trail of an event 500 years old, combine to create this mystery. Columbus named the island San Salvador (Holy Savior) and left for what he believed to be China, but what we today call Cuba.

A 1942 study by the eminent historian Samuel Eliot Morison pointed to Watling's Island, which was renamed San Salvador in 1926. More recent research has tried to resolve the question based on copies of Columbus's carefully kept shipboard logs, along with estimates of ship speed, currents, and winds. In effect, modern researchers are trying to refine Columbus's navigation, 500 years after the fact!

Luis Marden and Joseph Judge, researchers for the National Geographic Society, suggest that the island of Samana Cay, somewhat southeast of San Salvador, was the actual landfall. Steven Mitchell of California State University disagrees, concluding that tiny Conception Island, a little west and south of San Salvador, was the place. Two oceanographers from Woods Hole Oceanographic Institution—Philip Richardson and Roger Goldsmith—have asked whether Columbus might have overestimated his speed, and compute a spot only 15 miles from San Salvador. Figure 1 summarizes the different computed tracks.

Unless some artifacts from Columbus's explorations in the Bahamas are found, the true first landfall may never be known. A difference of only a degree in heading would be sufficient to shift the landfall from San Salvador to Samana Cay or the other possibilities, and a one-degree difference was certainly possible

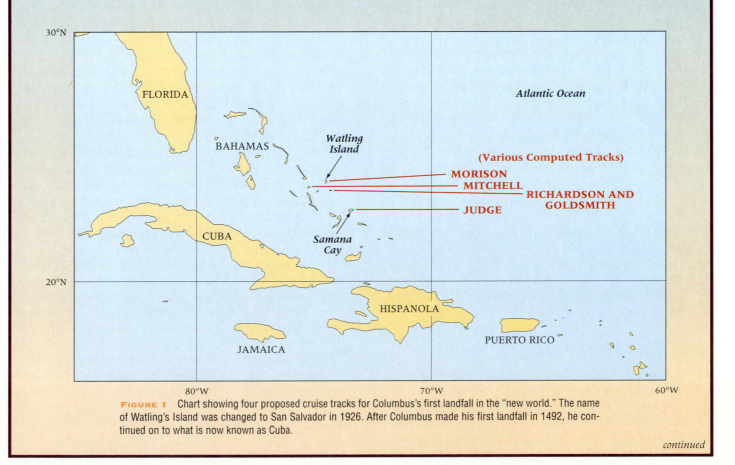

FIGURE 1   Chart showing four proposed cruise tracks for Columbus's first landfall in the "new world." The name of Watling's Island was changed to San Salvador in 1926. After Columbus made his first landfall in 1492, he continued on to what is now known as Cuba.

continued

Box 1-1 continued

with the cruder navigation techniques of Columbus's day. In any case, thanks to the excellent logs of Columbus, we at least have a consensus on *when* he landed—October 12, 1492.

REFERENCES

Cohen, B. I. 1992. "What Columbus 'Saw' in 1492." *Scientific American* 267, no. 6, pp. 100–106. A grim account of what Columbus found in the New World.

Judge, J. 1986. "Where Columbus Found the New World." *National Geographic* 170, no. 5, pp. 567–99.

Marden, L. 1986. "The First Landfall of Columbus." *National Geographic* 170, no. 5, pp. 572–77.

Morison, S. E. 1942. *Admiral of the Ocean Sea: A Life of Christopher Columbus,* Boston: Little, Brown and Co.

Richardson, P. L., and R. A. Goldsmith. 1987. "The Columbus Landfall: Voyage Track Corrected for Winds and Currents." *Oceanus* 30, no. 3, pp. 210.

and longitude. With information from his final voyage (1776–79), the broad general outlines of the world's oceans were known, and only Antarctica remained to be discovered in the early 1800s. However, the depth and character of the seafloor were still generally unknown.

It appears that the first successful **sounding**—or determination of ocean depth using a long line—was made by Sir John Ross in 1818. He obtained a sounding and a mud sample from a depth of 1,050 **fathoms** in Baffin Bay, west of Greenland. Countries that do not use the metric system may use fathoms to indicate depth. One fathom is equal to 6 feet or 1.83 meters. Thus, the depth of Ross's early sounding was 6,300 ft. or 1,919 m. The art of sounding was advanced by Midshipman Edward Brooke of the U.S. Navy in 1854; he fixed a detachable weight to the end of the sounding line that would drop off when the line hit the bottom. The loss of the weight made it easier to judge when the sounding line had reached the bottom.

Nevertheless, sounding results were often inaccurate because even a weighted line can be pulled at an angle by currents rather than going straight down from the ship, thus giving a reading deeper than the true one. It was not until 1925 that accurate soundings, using electronic methods, were first continuously made across the ocean.

Oceanographic research in the United States started in earnest in 1770 when Benjamin Franklin published his famous map of the Gulf Stream (Figure 1-3). Franklin, as Deputy Postmaster General of the American Colonies, observed that mail-carrying ships traveled faster by two weeks to Europe than in the reverse direction. He correctly concluded that a northeastward-flowing current along the east coast of the United States must be responsible.

Darwin, Forbes, and Huxley

An important expedition, although not strictly an oceanographic one, was the voyage of the British ship *Beagle* from 1831 to 1836, with the young naturalist Charles Darwin aboard. Darwin's findings about evolution and other aspects of the natural world stimulated other scientists to explore the ocean further. He also developed a hypothesis about the origin of **atolls** (curious, ring-shaped coral islands) that is still accepted by scientists (see page 299).

Another important contributor to the developing field of oceanography was an English biologist, Edward Forbes (1815–1854), generally considered to be the first *biological oceanographer*. Unfortunately, however, he is best known for one of his errors. In the 1850s he suggested that the ocean is devoid of life below 300 fathoms (548 m, or 1,800 ft.), a depth below which little light can penetrate. Although living organisms had already been dredged up from what he called the azoic zone (literally, "without animal life"), the prestige of Forbes was sufficient to keep his hypothesis alive until the late 1870s.

At about the same time, Thomas Huxley, a close friend of Charles Darwin, was studying deep-sea calcareous (calcium carbonate) **oozes,** which he noted to be very similar to cliffs made of chalk found on land. Indeed, both are composed of the small calcareous shells of animals and plants. Huxley concluded that the chalk cliffs found on land had originally formed as oozes in the deep ocean floor and were eventually compacted and uplifted to their present position. This idea was very nearly correct, as most chalks are originally formed in the marine environment (although not necessarily in the deep sea).

Huxley had another idea that was much more spectacular; it came from the study of samples of deep-sea mud. He noticed in the sample bottle a thin, jellylike layer overlying the mud. After examining this layer under a microscope, he concluded that it contained a basic form of life from which all other life forms had originated. Many years later it was found that this "miraculous" material was nothing but calcium sulfate resulting from the mixing of "spirits" (the alcohol then used in preserving the sediment sample) and seawater. The various hypotheses proposed by Forbes and Huxley, although eventually found to be largely incorrect, created considerable interest in ocean studies and were directly responsible for many later expeditions.

Matthew Fontaine Maury

Another important contributor to oceanography was Matthew Fontaine Maury (1806–1873) of the U.S. Navy.

FIGURE 1-3   A portion of the famous Franklin-Folger chart of the Gulf Stream, 1770. Benjamin Franklin's cousin Timothy Folger, a ship captain from Nantucket, Massachusetts, assisted in making this chart. By traveling in the Gulf Stream rather than outside it, a ship bound for Europe could shorten its travel time from North America by two weeks. (Photograph courtesy of Dr. Philip Richardson, adapted from his 1980 article in *Science*—see references; photocopy of chart provided by Bibliothèque Nationale, Paris.)

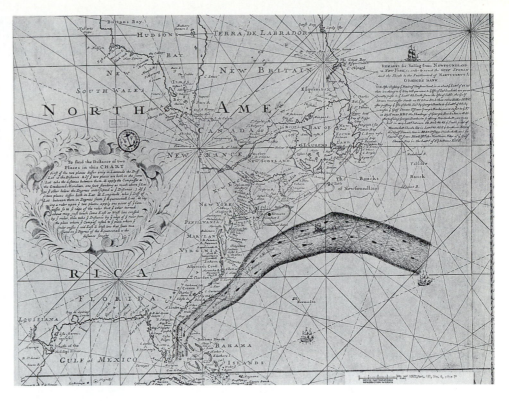

Maury used data from the logbooks of ships that had crossed the Atlantic to establish a relationship between currents and oceanic weather. He published his findings in 1855 in *The Physical Geography of the Sea*, one of the first oceanography books written in English. Maury also accumulated records of deep-sea soundings, and published the first **bathymetric** map (a map that shows **topography** or features of the seafloor) of the North Atlantic Ocean (Figure 1-4). Another American, William Ferrel, intrigued by Maury's book, was the first to

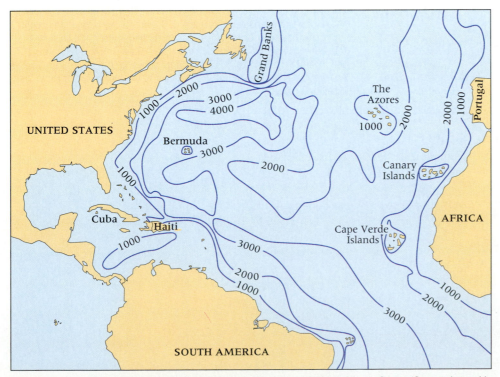

FIGURE 1-4   Matthew Fontaine Maury's 1854 bathymetric map of the North Atlantic Ocean. (Contour interval is in fathoms—6 ft. or 1.83 m.) (From Murray and Hjort, 1912; see also Figure 1-7.)

explain scientifically that the wind causes the motion of the ocean's surface water.

The Challenger Expedition

Although earlier voyages were important in setting the stage for the science of oceanography, deep-sea research is generally thought to have started with the British Challenger expedition (1872–76). Under the direction of Sir Charles Wyville Thomson, this expedition circumnavigated the world (Figure 1-5). *Challenger* was a steam-powered corvette (a fast vessel, somewhat smaller than a destroyer), 68.8 m (226 ft.) long and weighing 2,300 tons. It had a scientific party of 6, covered about 127,000 km (79,000 miles), and made 492 deep soundings and 133 dredgings of the seafloor. The *Challenger* scientists conducted experiments at 362 separate **oceanographic stations.**

Their ambitious program advanced the systematic study of the oceans, for at each station they performed these steps:

1. Measured water depth

a

b

FIGURE 1-5   (a) The research vessel H.M.S. *Challenger.*
(b) The officers and scientists aboard the *Challenger* in 1873: (1) Prof. C. Wyville Thomson, the scientist in charge; (2) R. von Willemoes-Suhm, a naturalist; (3) J. Y. Buchanan, a chemist; (4) J. J. Wild, an artist; (5) H. N. Moseley, a naturalist; and (6) J. Murray, a naturalist. Captain G. S. Nares is the second to the right of Professor Thomson. The remaining individuals are officers of the ship.
(c) Route traversed by *Challenger.* The ship left Great Britain in December 1872 and returned in May 1876. (Source for names in (b): A. L. Rice, *British Oceanographic Vessels 1800–1950,* Institute of Oceanographic Sciences, published by the Ray Society.)

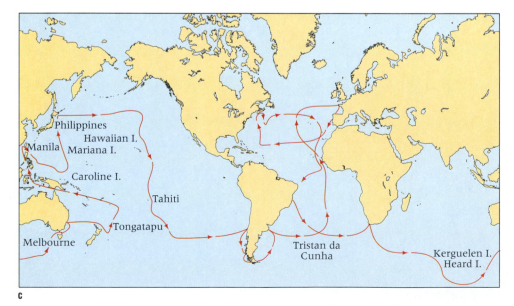

c

2. Obtained a sample of the bottom with a tube in the sounding weight

3. Sampled the animal and plant life in the surface and intermediate waters by towing nets

4. Dredged up a sample of the fauna living on the bottom

5. Measured the water temperature at the surface and at various depths

6. Obtained water samples near the bottom

7. Measured the direction and speed of surface currents and occasionally attempted to measure subsurface currents

8. Made atmospheric and meteorological observations

The volume of data collected was immense. More than 4,700 new species of marine life were discovered (an amazing average of about five new species for each day at sea). The deepest sounding at that time, 8,180 m (26,850 ft.), was made in the Marianas Trench in the South Pacific, in an area now called the Challenger Deep. The published reports of the expedition filled 29,500 pages in 50 volumes and took 23 years to complete. One of the great achievements of scientific exploration, the Challenger expedition also showed how little anyone had previously known about the sea.

After the Challenger expedition, interest in oceanography increased and many countries conducted worldwide expeditions, large and small, which made significant contributions to oceanography. The first U.S. ship especially equipped for oceanographic research was the steamer *Albatross* (Figure 1-6), acquired by the U.S. Commission of Fish and Fisheries. They used the ship to

study the distribution of fish and how their abundance was influenced by temperature, salinity, and water movements such as currents.

Nansen and the Poles

The Scandinavian countries, so dependent on the sea, have long been interested in marine research. One especially exciting Norwegian expedition was under the direction of Fridtjof Nansen during 1893–96. Their vessel, the *Fram,* was a wooden ship constructed so that it could be frozen into the ice of the Arctic Ocean without being crushed. At one time, Nansen and a companion left the *Fram* and tried to reach the North Pole on foot. After 14 months they came within 400 km (245 miles) of the pole, but had to turn back. *Fram* was trapped in the ice for 3 years and during that time drifted 1658 km (1035 miles). This expedition clearly showed that the North Polar area was not a continent, but rather an ocean covered with moving ice.

A recreation of the Fram expedition was attempted in 1979, using a field station built on the ice rather than a vessel frozen into the ice. The results have included new geophysical information from the Arctic Ocean north of Greenland (see also discussion and figures in Chapter 6). Later in his career, Nansen designed various devices, including a water-sampling device called the Nansen bottle (see Figure 7-3, page 157), still used today. Among Nansen's other achievements was aid to refugees from World War I, for which he won the Nobel Peace Prize in 1923.

Modern Oceanography

Columbus, da Gama, and others had no idea that they were sailing over great mountains and valleys. It was not until the early twentieth century that a good general picture of the ocean floor was obtained. The fundamental reason that this waited so long to happen was that people simply could not *see* very deeply into the ocean, and what lay below a few meters remained mostly a mystery. Slowly the general topography of the seafloor was discerned and mapped (Figure 1-7).

A new era of oceanography opened with the German *Meteor* expedition in 1925–27, which made one of the first detailed systematic studies of a specific part of the ocean. *Meteor* made 14 crossings of the South Atlantic during a 25-month interval, collecting data day and night through all weather conditions and seasons. This ship was one of the first research vessels to use an electronic **echo sounder** (see Box 4-1, page 72, for a description of the technique). It measured ocean depths, recording more than 70,000 soundings. These soundings clearly revealed the mountainous ruggedness of the

FIGURE 1-6 The *Albatross*, a 71 m (234 ft.) steamer that was the first American ship designed and built as a research vessel. It was acquired by the U.S. Commission of Fish and Fisheries (precursor of the National Marine Fisheries Service) around 1884. (See Figure 1-12 for illustration of a more modern research vessel.)

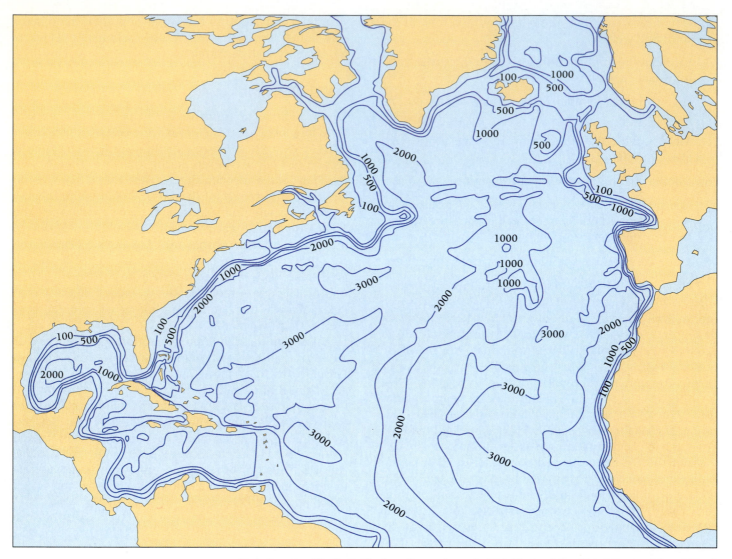

FIGURE 1-7 The bathymetry of the North Atlantic seafloor, as known in 1911. (Contour interval is in fathoms—6 ft. or 1.83 m.) Compare with Figures 1-4 and 2-10. (After J. Murray, and J. Hjort, 1912, *The Depths of the Ocean* [London: Macmillan].)

ocean floor, a fact we take for granted today but which was a surprise then.

In this century, oceanography has become a major science, with many large marine laboratories. In 1912, Scripps Institution of Oceanography in La Jolla, California (Figure 1-8), became part of the University of California and has since grown to be one of the largest oceanographic institutions in the United States.

This author's home base, the Woods Hole Oceanographic Institution in Woods Hole, on Cape Cod, Massachusetts, is among the world's foremost oceanographic research organizations (Figure 1-9). Since 1871 under Spencer Baird (first Commissioner of the U.S. Commission of Fish and Fisheries), the Woods Hole region has developed into a major center for oceanographic research and training. Today it includes three parts: the Woods Hole Oceanographic Institution (founded in 1930), the Marine Biological Laboratory (1888), and a laboratory of the National Marine Fisheries Service of

the U.S. government's National Oceanographic and Atmospheric Administration (NOAA).

Other large oceanographic research institutions are established at several universities, including the University of Rhode Island, Oregon State University, University of Washington, Texas A&M University, University of Hawaii, University of Miami, and the Lamont-Doherty Earth Observatory associated with Columbia University.

Submarine warfare during World War II required far more information about the ocean than anyone possessed, so oceanography expanded rapidly and that expansion has continued to the present. Two remarkable events—the descent of the **bathyscaphe** *Trieste* to nearly 11 km (about 6.7 miles) into a deep-sea trench, and the voyage of the nuclear submarine *Nautilus* beneath the Arctic Ocean's ice pack at the North Pole—showed that *all* parts of the ocean could be explored.

The 1960s and 1970s saw research escalate to an international scale. Large cooperative programs captured

a

b

FIGURE 1-8   (a) The Scripps Institution of Oceanography around 1915. (b) An aerial view looking north of the present Scripps Institution of Oceanography, La Jolla, California. (Both courtesy of Scripps Institution of Oceanography/University of California, San Diego.)

the imagination of the public. One such program, the **Deep Sea Drilling Project (DSDP),** drilled about 1,000 holes in deep regions of the Atlantic, Pacific, and Indian Oceans, using the *Glomar Challenger* (Figure 1-10). The information gained from this project was perhaps as scientifically valuable as that gained by the first Challenger expedition almost 100 years earlier.

A more advanced vessel, the *JOIDES Resolution* (see Figure 5-7a, page 113), can drill the seafloor even in the rigorous environment of the polar regions. Other programs, such as **GEOSECS (GEochemical Ocean SECtionS),** have collected a "water library" with samples from all the oceans. These data are valuable in establishing baseline water chemistry for assessing pollution.

All of these people, ships, and events set the stage for today's sophisticated technological studies of the ocean. Despite our apparent power over the sea, the loss of ships and life, especially during storms, reminds us that the ocean remains a formidable environment which demands our respect and caution.

Oceanography Today

The 1990s are exciting times for marine exploration. The 1985 discovery of the *Titanic,* 73 years after it sank, showed that much older wrecks, dating even to early Roman or Greek times, can be located, studied, and perhaps even recovered. The technologies used in the *Titanic* work have potential for many other types of marine research. Deep-diving vessels called **submersibles,** although not new, have been used to explore the seafloor for valuable minerals, current volcanic activity, and exotic animal life. The more recent use of **remotely**

FIGURE 1-9 (a) A 1937 picture of Woods Hole Oceanographic Institution (right) and research vessel *Atlantis* (right foreground), the Marine Biological Laboratory (central area), and the National Marine Fisheries Service of NOAA (left).

(b) A portion of today's Woods Hole Oceanographic Institution (center) and the Marine Biological Laboratory and the National Marine Fisheries Service of NOAA (upper portion of figure). In the foreground is a pier for ferries to the island of Martha's Vineyard. Two research vessels, *Oceanus* and *Atlantis II*, are tied to the Institution's dock. (Photograph (a) by H. M. Wood; photograph (b) by Terri Corbett; both courtesy of Woods Hole Oceanographic Institution.)

operated vehicles (ROVs) has led to even more discoveries.

The possibility of living and working on the ocean floor has always fascinated scientists and nonscientists alike. With the promise of obtaining resources from the seafloor, the need for direct observation has become critical. Under simulated conditions, divers have been able to work under the pressures of about 610 m (2,000 ft.) of water. However, this is only one-sixth of the average ocean depth of about 3,800 m (about 12,500 ft.). Even today, it is technically easier and safer to walk on the moon than on the deep-ocean floor!

Satellites have become another valuable oceanographic tool. They collect more data in a few hours than a surface vessel might collect in years. Satellites can measure and obtain sea and land surface temperatures, biological data, sea-level information, and wind speeds that can be used to predict the weather (see Figure 1-11). The first satellite launched especially for oceanography, *SEASAT A* in 1978, operated only 3 months but obtained over 8 billion measurements. This number of data points exceeded all those collected by research vessels over the previous century. Recently, satellites have begun to study the increasing carbon dioxide content in the atmosphere and its implications for global changes in climate and sea levels. Global issues such as these promise to promote marine science programs well into the twenty-first century.

FIGURE 1-10 (a) The drilling vessel *Glomar Challenger*, managed by Scripps Institution of Oceanography. The vessel is about 131 m (430 ft.) long and has a drilling derrick with a one-million-pound capacity that stands about 59 m (194 ft.) above the waterline. Forward of the derrick is the automatic "pipe racker," which holds about 7,800 m (25,600 ft.) of 12.7 cm (5 in.) drill pipe.

(b) Part of the core collection taken during the Deep Sea Drilling Project. Cores (cylindrical sections) are split in half longitudinally (lengthwise) and stored in the plastic containers, one of which is being removed by a technician. (Photographs courtesy of the Deep Sea Drilling Project.)

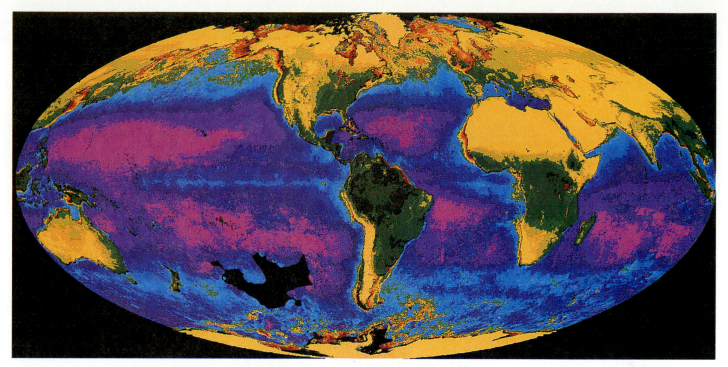

FIGURE 1-11   This remarkable satellite image shows the first truly global view of the Earth's biosphere. Over three years of information using more than 400 billion pieces of data are synthesized in this figure to show levels of chlorophyll (a pigment found in plants). In the ocean, high values are indicated by red and yellow, intermediate values by blue, and low values by purple. On land, greens indicate high levels and yellows are low. (Photograph by C. J. Tucker, National Aeronautics and Space Administration/GSFC.)

In the late 1980s and early 1990s, several large-scale research programs, generally international, focused on global issues such as climate, energy, and mineral resources from the sea; coastal zone use; methods for increasing marine biological productivity; and marine environmental issues. Several of these programs will continue into the twenty-first century. The major ones, detailed later in appropriate sections, are:

- The World Ocean Circulation Experiment (WOCE)
- Tropical Ocean and Global Atmosphere (TOGA)
- Joint Global Ocean Flux Study (JGOFS)
- Ridge Interdisciplinary Global Experiments (RIDGE)

Global change is a major theme in each of these large programs. And, as you can see, oceanographers are big users of acronyms!

The Scope of Oceanography

What, then, is oceanography? One simple definition is the scientific study of the ocean. We might now expand this to read: *Oceanography is the application of all the disciplines and methods of science to understanding ocean phenomena.* The key word in this definition is *all,* because to truly understand the ocean and how it works, you need to know something about all areas of science and their relation to the marine environment. Thus, in many respects, oceanography is not a single pure science but a combination of several.

Most universities and organizations now divide oceanography into five main parts, or subdisciplines:

1. Marine geology and geophysics
2. Chemical oceanography
3. Physical oceanography
4. Biological oceanography
5. Ocean engineering

In recent years, other branches of marine science have evolved. *Marine policy* applies the social sciences and law to ocean issues. *Marine archeology* applies the techniques of archeology to discovering the human history hidden in the seas. These subdivisions are convenient for organizing textbooks, but please remember that ocean study is a coherent entity and not a group of individual subjects.

Marine Geology and Geophysics

Marine geologists are concerned mainly with the **ocean basin**—the sediments and rocks of the seafloor, from the beach out to the deepest depths of the ocean. Later we will see that a large area of research concerns **plate tectonics,** a type of movement within the Earth that explains much about the present configuration of the continents and the ocean basin. **Marine geophysicists** often are most interested in deeper aspects of the Earth's crust, including its structure and physical properties.

The study of seafloor sediment can indicate the geologic history of the ocean and the climate history of the Earth, mainly by studying fossil shells of various species. Sediment study also reveals information about ocean currents that transported the sediments to their present site. Another subject of recent interest has been the interconnected worldwide system of ridges that runs over 65,000 km (40,000 miles) on the ocean floor and how they are formed and evolve. Some parts of these ridges have unique forms of animal life.

Marine geologists and geophysicists employ sophisticated equipment. Among the key tools of the geophysicist are acoustical devices that transmit sound into the sediment and rock layers beneath the seafloor. The characteristics of the returning sound can be used to model the structure and physical properties of the layers. These tools also are used to explore for oil and gas. Marine geologists have benefited greatly from the availability of deep-diving submersibles that can probe and sample the ocean depths.

Chemical Oceanography

Chemical oceanographers study seawater, its composition, and the chemical processes and reactions connected with its animal and plant life. Chemical oceanographers also may be concerned with chemical processes occurring on the seafloor and beneath. An important challenge is to understand how the chemical environment of the ocean has been affected by the by-products of human activities, especially pollution.

Much marine chemistry requires sophisticated techniques for sampling and analyzing ocean water, because it contains many different substances in extremely small quantities, measured in parts per million (ppm) or less. Frequently this means that large volumes of water must be collected to accumulate enough of the rare constituents to be measurable. Other measurements can be made directly in the water using electronic devices.

Physical Oceanography

Physical oceanographers are mainly interested in the physical characteristics of the water in the ocean, such as the motion of the water from the molecular level to the global scale, including ocean currents, eddies,

waves, and tides. The interaction of the ocean with the atmosphere is another important area of inquiry, especially in light of the recent concern about global change.

The main physical properties of seawater are *temperature, salinity,* and *density.* Differences in these properties can cause water movement, both horizontally and vertically. Changes in these properties can strongly affect the plants and animals that live in the ocean. Many of the complex physical processes in the ocean require a balanced program of field work combined with theoretical research to provide better models (conceptual schemes) for scientific study.

Most physical oceanographic measurements are made electronically. In recent years, satellites have provided a worldwide view of the ocean to all oceanographers. This technique is especially valuable to physical oceanographers and allows them a broad view of physical oceanographic processes.

Biological Oceanography

Biological oceanographers study the relations among marine organisms (both plants and animals) and the interaction of these organisms with chemical and physical processes in the ocean. Biological studies include the distribution of organisms, their life cycles, and their responses to the environment and to each other. Biological oceanography extends from coastal marshes to the great depths of ocean trenches, and from the tiniest bacteria to whales. Research may follow **food chains** or **food webs**—the production of organic matter (food), its consumption by organisms, and its eventual return to the water.

Many marine organisms are hard to capture and most cannot survive outside of their natural environment. The creatures of the deep sea are especially challenging to study. For years, at best, they could only be photographed but now deep-diving submersibles and ROVs permit direct observation and experimentation in the creatures' natural habitats.

Ocean Engineering

Ocean engineers develop much of the technology used by marine scientists. Ocean engineering covers technological aspects from deep below the seafloor into the atmosphere above the ocean. Ocean engineers design equipment for sampling, measuring, and observing diverse ocean phenomena. The problems of tremendous water pressure at great depth, seawater's corrosive nature, and the need for highly accurate measurement by delicate instruments pose unique challenges for instrument designers.

For example, the pressure at 10,000 m (32,800 ft.) beneath the ocean surface is almost 1,000 kg/cm^2 (15,000 lb/in^2), an instrument-crushing 1,000 times the atmospheric pressure at sea level. (See Appendix I for common oceanographic units of measure.) Structures placed on the ocean's surface have to withstand high winds, strong currents, corrosion, and vandals. Satellites provide new ways to study the ocean, as have sophisticated buoys that acoustically transmit data to facilities on shore.

Marine Policy

Marine policy researchers come from disciplines such as law, economics, anthropology, and political science. They study the human factor in ocean use, dealing with concerns such as the legal or economic constraints on mining sand and gravel deposits in the nearshore region. Another example is determining the economic and social impact of altering the regulations on fishing in a region (number or size of fish permitted to be caught, season when fishing is permitted, size of boat or net, and so on).

Marine Archeology

Marine archeology increasingly depends on ocean science. Research submersibles and advanced navigational and surveying techniques are used to detect and explore marine wrecks and ancient sites of habitation. There have been several recent discoveries of famous marine wrecks, including the *Titanic*, the World War I vessel *Bismarck*, and the civil war battleship *Monitor*. Details of these exciting finds are presented in Chapter 17.

Oceanography: The Unified View

Although the divisions just discussed seem to fragment oceanography into neat little niches, in practice this is a highly integrated science. For example, geologic deposits on the seafloor are intimately influenced by the chemistry, physics, and biology of the water above. A marine geologist sampling the sediment from under the equatorial Pacific would obtain material composed mainly of the shells of dead microscopic organisms that once lived in the surface water. To find and interpret these sediments, this geologist must be fluent not only in geology but in chemistry, physics, biology, the history of life, and procedures for sampling, not to mention knowing a good bit about geography and navigation just to get to the sampling site!

As this example shows, the divisions of oceanography are artificial. The ocean must be visualized as a coherent entity, and an oceanographer must be versed in all these fields. Oceanography is advancing so rapidly, however, that it is impossible to be an expert in all its aspects, so an oceanographer generally specializes in one of its subdivisions.

But keep in mind that the different fields are closely related and that the answers to important scientific and policy questions often require the skills of oceanographers from several fields. To meet future environmental challenges, the expertise of land scientists and atmospheric scientists will also be needed.

How to Use This Book

This book treats separately the broad aspects of ocean science—biology, chemistry, physics, geology, and geophysics. These subjects overlap so much, however, that you will often need to apply points made earlier in the course to material you study later. Regarding ocean engineering, detailed coverage is beyond the scope of this book, but sections on important oceanographic instrumentation are placed throughout the text where they apply. Later chapters cover marine policy, especially regarding the coastal zone, mineral resources, pollution, archeology, and ocean law.

Oceanography as a Career

Whether doing deep-sea research or local offshore work, most marine scientists spend part of their time at sea acquiring data, sometimes under adverse conditions. Generally, oceanographic cruises last from a few days to several months. Seagoing time has decreased in recent years, in part because satellites now perform some oceanwide measurements and other devices can measure or observe ocean phenomena while left unattended for months or longer periods. However, ships will always be needed in oceanographic work to collect biological, chemical, and geological samples, as well as to make many types of measurements.

In the United States, approximately 3,500 people are trained as marine scientists. This is roughly one marine scientist for every 70,000 people in our population of over 250 million. Oceanographers are employed by universities, research laboratories, state agencies, environmental groups, industry, and the federal government.

In the United States, the largest federal agency concerned with the marine environment is the **National Oceanic and Atmospheric Administration (NOAA).** This organization has over 10,000 employees, numerous ships (Figure 1-12), an annual

FIGURE 1-12 The *Malcolm Baldridge* (previously known as the *Researcher*), one of the ocean research vessels maintained by the National Oceanic and Atmospheric Administration (NOAA). The 84.7 m (278 ft.) *Malcolm Baldridge* accommodates 70 officers, scientists, and crew, is highly automated for all types of marine research, and can stay at sea for more than a month. (Photograph courtesy of National Oceanic and Atmospheric Administration.)

phenomena, an aspiring *research* oceanographer would do best to obtain sound undergraduate training in a basic science (chemistry, biology, geology, physics, and math) and then specialize in the marine application of that science in graduate school.

About 80 U.S. colleges and universities offer an undergraduate degree in some form of marine science; many are in coastal studies or coastal biology. Some schools offer innovative "land-and-sea" programs in which you can get a wonderful exposure to many aspects of the ocean, plus learning how to sail (Figure 1-13). About 10 institutions offer undergraduate degrees in ocean engineering. And over 150 universities, colleges, and junior colleges offer at least a few courses in marine science or related fields. Anyone interested in a career in oceanography should see Box 1-2, "Is a Career in Science a Good Opportunity?"

budget exceeding $1.9 billion, and a broad charter of responsibilities. Other U.S. government agencies involved in marine science are the Office of Naval Research, the National Science Foundation (which mostly provides funds for research), the U.S. Geological Survey, the U.S. Department of Energy, the U.S. Environmental Protection Agency, and the U.S. Coast Guard.

Marine science employment opportunities also exist at the state and local levels. For example, most coastal states have a coastal-zone management program and marine environmental and monitoring agencies. Many coastal cities and towns have marine-related concerns and turn to consulting companies for advice. These consulting and marine technology firms thus offer another type of marine-science employment. Many are small and focus on marine instrumentation or environmental issues.

Canada has marine science opportunities similar to those of the United States. Major government departments exist for fisheries, the ocean, and the environment. Canada also has several outstanding marine-science research laboratories and university programs.

College graduates having bachelor's degrees in a science related to oceanography who want employment in marine science sometimes work as laboratory assistants or research assistants, often while continuing advanced training. Training or working opportunities may also exist with consulting companies, especially those working on problems of environment or coastal use. Other possibilities are working for a government agency—federal, state, or local—that regulates or manages marine issues.

Because oceanography is the application of all the disciplines and methods of science to understanding ocean

FIGURE 1-13 The *Corwith Cramer*, an educational sailing vessel, part of a marine education program run by the Sea Education Association (SEA) in Massachusetts. The 40.8 m (134 ft.) vessel can take 25 students to sea for up to 6 weeks. (Photograph courtesy of Susan E. Humphris, SEA.)

Box 1-2

Is a Career in Science a Good Opportunity?

Are you considering your career plans? This is a difficult decision, driven by factors such as family pressure, your anticipated life-style, economic benefit, contributions you might hope to make, where you'd like to live, and so forth. Another major difficulty is determining where future opportunities might be. Will there be a shortage of doctors? Will there be an excess of lawyers? Will scientists and engineers be in short supply? The answer to the last question appears to be "yes!" for several reasons.

A career in science can be extremely satisfying. I can speak best about marine science. I hope that in this text and through the media you will discover that the many challenges facing our environment may have solutions that lie in knowing more about the ocean. Economically, you probably will not become wealthy as a scientist, but you can live a comfortable life. Today, the fields of science and engineering are especially open to women, minorities, and disabled individuals—indeed, to anyone.

Perhaps the key question is: Will jobs be available when you are ready? ("Ready" could mean when you graduate from college or upon finishing graduate school—two additional years for an M.S. degree or about five more years for both an M.S. and a Ph.D.) Several studies of the demand for scientific personnel in the United States in the late 1980s concluded that in the 1990s companies and universities may have problems recruiting young scientists and engineers to replace employees who are approaching retirement. (And I suspect the situation will be similar in many other countries.) More recently, however, reduction in the defense industry, continuing recession, and the influx of foreign students and scientists have confused future job scenarios.

Nevertheless, several reasons remain for an impending shortage of scientists. One is the low birth rate during the 1960s and 1970s, which has reduced both the present and recent college-age population. A minimum of 24 million people of age 18 to 24 is expected in the U.S. population by the mid-1990s, compared to the 25 percent larger peak of over 30 million that existed in 1980. Another reason for the shortfall is that only a small portion of our population actually earns a B.S. degree—about 5 percent in 1990. If that percentage remains low, a shortfall of scientists and engineers will occur.

Another way of looking at this problem is that, based on a 1987 study, only 1,800 out of every 10,000 high school sophomores (18 percent) have indicated a career interest in natural science or engineering. Only 850 will actually earn a bachelor's degree in these fields, and only 20 will earn a Ph.D.

One solution is to attract more women and minorities into science and engineering careers. The number of women considering science careers seems to have peaked in recent years, but women still compose only about 15 percent of the present workforce. Historically, the reasons for fewer women than men in science were lower salaries and fewer job opportunities, but recent years have brought considerable improvement. U.S. women now receive about one-third of all science doctorates; however, many tend to be in the social sciences and psychology.

For blacks and Hispanics, the percentage is even less: blacks constitute 12 percent of the population, but hold only about 2 percent of the scientific and engineering positions. Hispanics constitute about 9 percent of the population and are similar in percentage to blacks in the scientific and engineering workforce.

These percentages are a shame. Excellent scientific opportunities exist for women and minorities in science, and they should be considered. A positive way to view this personnel shortage is as an *opportunity*—career positions exist, and you can train to fill one!

In the hope of motivating you, I recommend the following publication, which describes most of the marine programs in U.S. colleges and universities and is updated every three or four years (it should be available in your school's library):

University Curricula in Oceanography and Related Fields: A Guide to U.S. Academic and Technical Programs, 1988–91. Compiled by the Marine Technology Society, 1828 L Street, N.W., 9th Floor, Washington, D.C. 20006.

The University of Delaware's Sea Grant Program in 1993 produced a video and publication, "Marine Careers." Copies are available from the College of Marine Studies, University of Delaware, 700 Pilottown Road, Lewes, DE 19958, or call (302) 831-8083.

See the reading list at the end of this chapter (page 20) for recent brochures and articles about careers in marine science.

Oceanography for the Nonoceanographer

Knowledge of the ocean is too important to be left only to the scientific community. The use and maintenance of our marine environment affects everyone. Almost daily, decisions are made nationally and locally concerning the marine environment. These decisions range from national policies on waste disposal (after all, waste dumped in Minnesota travels the Mississippi River and eventually enters the Gulf of Mexico) to local issues such as coastal development or protection (see Figure 1-1). The ocean also offers tremendous potential for food and mineral production and as a new source of medicines. Informed citizen-voters can ensure that present and future decisions concerning ocean use are made openly and carefully with a proper balance between ocean use and ocean protection. You will also understand how harvesting ocean resources can be beneficial if proper management and conservation are used.

The coming years promise increasing opportunity for volunteers to work alongside scientists and public officials in many marine activities. Not only will this help to preserve our fragile coastal environment; it also will create great personal satisfaction. In the author's hometown

FIGURE 1-14 Local citizens collecting water samples as part of a program to monitor environmental conditions in their waters. (Photograph courtesy of Terri Corbett, Woods Hole Oceanographic Institution.)

of Falmouth, Massachusetts, on Cape Cod, there is a citizens' group that monitors coastal ponds (Figure 1-14). More than 60 citizens measure water and air quality throughout the year. One result of this program has been stronger local environmental rules concerning development near the ponds. Many towns in other states have similar programs.

SUMMARY

Oceanography is the application of all the disciplines and methods of science to understanding ocean phenomena. Oceanography generally is divided into five main areas: chemical oceanography, biological oceanography, physical oceanography, geological and geophysical oceanography, and ocean engineering. Marine policy and marine archeology are growing new fields. A knowledge of all fields is necessary for much oceanographic research.

The ocean is one of our last unexplored frontiers and is an extremely important one. In the coming years, competing nations will increase their use of the ocean for its biological and mineral resources. Scientists everywhere will study the ocean for better understanding of climate and weather. Global change issues will influence many research programs in the 1990s and beyond.

Early ocean study was motivated by practical reasons: people wanted to sail from place to place as quickly as possible. Early U.S. contributions included Benjamin Franklin's map of the Gulf Stream and Matthew Fontaine Maury's charts of ocean currents. European

contributions included the research of Charles Darwin on the *Beagle,* Edward Forbes, and Thomas Huxley. The voyage of the *Challenger* (1872–76) gave a major thrust to the start of oceanography as a science by making systematic measurements around the world. The period of modern technological oceanography began in 1925, when the *Meteor* performed a comprehensive study of the southern Atlantic using electronic echo-sounding. Also during this time, many oceanographic institutions were founded.

Oceanography expanded rapidly following World War II, partly as a result of advances in technology, and of military uses of the ocean. By the 1960s and onward, oceanographic research became international, involving many countries. The future will see such large international efforts continuing, with more emphasis on applied problems such as climate, energy from the sea, and especially issues of global change. A knowledge of the ocean and the processes incurring within it are important to everyone.

QUESTIONS

1. What are the different categories of marine science? What type of study is done in each field?

2. What are some important societal reasons for studying the ocean?

3. How does one become an oceanographer? Where are jobs available?

4. What role can an average citizen play in ocean-related matters?

5. What were some motivations and difficulties of early ocean explorers?

6. What were the key contributions of Germany, Great Britain, and Scandinavian countries to early oceanography?

7. Describe the importance of the Challenger expedi-

tion of 1872–76 to the development of oceanography as a science.

8. Describe differences between early marine expeditions and modern ocean studies.

9. How will future oceanography differ from oceanography of 10, 50, or 100 years ago?

KEY TERMS

atoll
bathymetric
bathyscaphe
Benjamin Franklin
biological oceanography
Challenger expedition
Charles Darwin
chemical oceanography
coastal waters
continental shelf
desalinated
DSDP
echo sounder

Edward Forbes
erosion
fathom
food chain
food web
Fram
Fridtjof Nansen
GEOSECS (*GE*ochemical *Ocean SEC*tion*S*)
global change
Glomar Challenger
JOIDES Resolution
marine archeology

marine geology
marine geophysics
marine policy
marine science
Matthew Fontaine Maury
Meteor expedition
monsoon
NOAA
ocean basin
ocean engineering
oceanographic station
oceanography

ooze
physical oceanography
plate tectonics
remotely operated vehicle (ROV)
salinity
satellite
sounding
submersible
Thomas Huxley
topography
weathering

FURTHER READING

General

Charton, Barbara. 1989. *Seas and Oceans. A Collins Reference Dictionary.* Glasgow, Scotland: Collins. Defines many common marine terms.

Graves, Don. 1989. *The Oceans: A Book of Questions and Answers.* John Wiley and Sons, New York. A series of general-interest questions and answers.

Hsu, K.J.H. 1992. *Challenger at Sea: A Ship that Revolutionized Earth Science,* Princeton, N.J., Princeton University Press. One scientist's highly personal view of the Deep Sea Drilling Project.

Humphris, Susan. 1990. "The Ocean as a Classroom." *Oceanus* 33, no. 3, pp. 46–51. A compelling description of the practical learning experience of working and studying at sea.

1990. "Marine Education." *Oceanus* 33, no. 3. An entire issue devoted to marine education.

National Geographic Society. 1992. *Atlas of the World.* 7th ed. Washington, D.C. Besides some beautiful charts of the ocean, this atlas also contains many interesting facts about our planet.

1990–91. "Naval Oceanography." *Oceanus* 33, no. 4, An entire issue devoted to describing the U.S. Navy's interest in the ocean.

National Research Council. 1992. *Oceanography in the Next Decade.* Washington, D.C. A group of noted marine scientists give their collective views on new directions in oceanography over the next decade.

Oceanus, a magazine produced by the Woods Hole Oceanographic Institution, contains articles that are nontechnical and easily understood. Here is an interesting series of issues on the various disciplines of oceanography:

"Ocean Engineering and Technology." *Oceanus* 34, no. 1 (Spring 1991).

"Marine Chemistry." *Oceanus* 35, no. 1 (Spring 1992).

"Physical Oceanography." *Oceanus* 35, no. 2 (Summer 1992).

"Biological Oceanography." *Oceanus* 35, no. 3 (Fall 1992).

"Marine Geology and Geophysics." *Oceanus* 35, no. 4 (Winter 1992–93).

University of New Hampshire. "Marine, Oceanographic, Coastal and Estuarine Laboratories, Centers and Research Stations/Facilities in and near the United States." Available from the University of New Hampshire, Sea Grant Program, Durham, NH, 03824. Describes numerous large and small marine laboratories and research facilities.

History of Oceanography

Darwin, C. 1959. *The Voyage of the Beagle.* Edited by Millicent E. Selsam. New York: Harper and Row. The story of Darwin's famous and important expedition.

Deacon, D.E.R. 1962. *Seas, Maps, and Men: An Atlas-History of Man's Exploration of the Oceans.* Garden City, N.Y.: Doubleday. A very readable history of early oceanography written by the daughter of a famous British oceanographer.

Deacon, Margaret. 1971. *Scientists and the Sea, 1650–1900: A Study of Marine Science*. London: Academic Press. A good account of the early history of oceanography.

Linklater, Eric. 1972. *The Voyage of the "Challenger."* London: John Murray (Publishers), Ltd. Story of the famous Challenger expedition.

Mills, Eric, L. 1989. *Biological Oceanography: An Early History, 1870–1960*. Ithaca, N.Y.: Cornell University Press. Includes the stories of Huxley, Forbes, and other early pioneers in biological oceanography.

Murray, J., and J. Hjort. 1912. *The Depths of the Ocean*. London: Macmillan. Story of early attempts to measure the depths of the seafloor.

Richardson, P. L. 1980. "Benjamin Franklin and Timothy Folger's First Printed Chart of the Gulf Stream." *Science* 207, pp. 643–45. Describes finding of the Gulf Stream chart shown in Figure 1-3.

Sears, M., and D. Merriman, eds. 1980. *Oceanography: The Past*. New York: Springer-Verlag. A series of papers, some very detailed, given at a meeting celebrating the 50th anniversary of the Woods Hole Oceanographic Institution.

Schlee, Susan. 1973. *The Edge of an Unfamiliar World: A History of Oceanography*. New York: Dutton. Very well-written book on the development and history of oceanography.

Viola, H. J., and C. Margoles, eds., 1985. *Magnificent Voyagers: The U.S. Exploring Expedition*. Washington, D.C: Smithsonian Institution Press. Story of one of the first U.S. oceanographic expeditions.

Careers in Marine Science

American Geophysical Union. 1986. Careers in Geophysics, Solid Earth, Hydrologic, Oceanic, Atmospheric, and Space Sciences. Write for a copy: 2000 Florida Avenue, N.W., Washington, D.C. 20009.

———. 1986. *Careers in Oceanography.* Write for a copy: 2000 Florida Avenue, N.W., Washington, D.C. 20009.

Griscom, Sarah B. 1984. *Marine Geology: Research Beneath the Sea*. U.S. Department of Interior/Geological Survey. To obtain copies or further information, write either: (a) Eastern Distribution Branch, U.S. Geological Survey, 604 South Pickett Street, Alexandria, VA 22304, or (b) Western Distribution Branch, U.S. Geological Survey, Box 25286, Federal Center, Denver, CO 80225.

Knauss, John A. 1988. "Academic Oceanography: How We Got from There to Here." *Marine Technology Society Journal* 22, no. 1, pp. 5–11. Describes how academic marine science has grown since World War II.

Norwell, A. R. M., and C. D. Hollister. 1988. "Graduate Students in Oceanography: Recruitment, Success, and Career Prospects." *Oceanography Report,* September 6, 1988. Article features statistics about students in various areas of oceanography.

Wunsch, Carl. 1993. "Marine Sciences in the Coming Decades." Science 259, p. 296–97. Describes some changes the author feels are necessary for marine science success in future years.

CHAPTER 2

The Origin and Evolution of Earth and Its Oceans

THE ORIGIN AND EVOLUTION OF THE UNIVERSE, OUR SOLAR SYSTEM, AND PLANET EARTH HAVE MYSTIFIED AND CHALLENGED US FROM EARLY TIMES. NO CONCLUSIVE THEORY HAS YET BEEN FORMULATED, BUT A VIABLE HYPOTHESIS FOR THE ORIGIN OF OUR EARTH, ATMOSPHERE, AND OCEAN IS PRESENTED IN THIS CHAPTER.

SINCE THE FORMATION OF EARTH ABOUT 4.7 BILLION YEARS AGO, OUR PLANET HAS GONE THROUGH MOMENTOUS CHANGES. THESE INCLUDE CHANGING POSITIONS OF THE CONTINENTS AND OCEAN BASINS AND ENVIRONMENTAL CHANGES SUCH AS MAJOR CYCLES OF CLIMATE AND THE RISE AND FALL OF SEA LEVEL. THESE CHANGES CONTINUE TODAY.

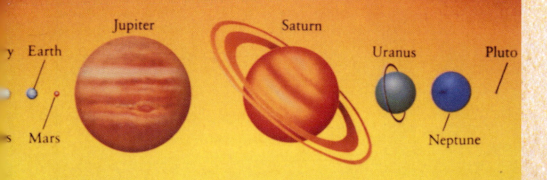

The Solar System, consisting of the sun and its planets, drawn to scale. The four inner planets, closest to the sun—Mercury, Venus, Earth, and Mars—are relatively small and are composed of rock. The next four—Jupiter, Saturn, Uranus, and Neptune—are larger and are composed mainly of gas.

(Courtesy of Prof. William J. Kaufmann III, 1993, Discovering the Universe, 3d Edition [New York: W. H. Freeman and Co.].)

The Setting

Although planet Earth is an insignificant speck in the cosmos, its origin and evolution obviously are of considerable interest to us. Earth is part of a system that travels through space and consists of nine planets and one sun—our **Solar System.** The nine planets revolve around the sun, which is a star (see opening figure). Our sun is just one of over 100 billion billion stars that we believe populate the **Universe.** Some of these stars may have associated planets. The rest of the Universe is a vast area of essentially empty space. Our Solar System is also part of a smaller collection of stars called the **Milky Way Galaxy,** which itself is composed of about 100 billion stars.

The dimensions of the Universe, our Milky Way Galaxy, and even our Solar System are mind-boggling. Expressing the distances in kilometers requires enormous numbers that are clumsy to work with. To simplify these numbers, the **light-year** was developed. A light-year is simply the distance that light can travel in a year. Light travels at 300,000 km (186,000 miles) *per second,* often called the "speed limit" of the universe. Thus, in just one second, light travels a distance equal to circumnavigating the Earth 7.5 times. There are more than 31 million seconds in a year, so in one year, light travels nearly 9.5 trillion km (about 6 trillion miles).

In the shorthand method used in science, a trillion is 1×10^{12} (1 followed by 12 zeros: 1,000,000,000,000). If you are one of the many students needlessly intimidated by these numbers, see Box 2-1.

Box 2-1

The Power of 10

Oceanographers frequently deal with numbers that are both very large and very small, as do scientists in other fields. Rather than use a cumbersome sequence of zeros, a standard system of notation has been developed. Called the "power of ten," it is expressed as the number 10 followed by a number that tells how many zeros to add if you wanted to write out the full number. For example:

Power of Ten		Actual Number
10^0	=	1
10^1	=	10
10^2	=	100
10^3	=	1,000
10^4	=	10,000
10^6	=	1,000,000 (one million)
10^9	=	1,000,000,000 (one billion)
10^{12}	=	1,000,000,000,000 (one trillion)

and so on.

Putting this to work, we can express "4 million kilometers" by placing a 4 in front of the 10, and the unit of measure after it: 4×10^6 kilometers. This is spoken as "four times ten to the sixth kilometers."

The same concept can be applied to numbers that are less than 1 by using a minus sign in front of the exponent. The minus sign indicates the number of times you must *divide* by 10. For example:

Power of Ten		Actual Number
10^{-1}	=	0.1
10^{-2}	=	0.01
10^{-3}	=	0.001
10^{-6}	=	0.000001 (one-millionth)
10^{-9}	=	0.000000001 (one-billionth)

and so on.

So, 4×10^{-6} cm is 0.000004 cm, or four millionths of a centimeter, spoken as "four times ten to the minus six centimeters."

By using the power of 10, it is also easier to multiply or divide numbers, because you just add or subtract the exponents. For example:

$$10^4 \times 10^3 = 10^7$$

$10^3 \div 10^6 = 10^3/10^6$ or $10^3 \times 10^{-6}$, which equals 10^{-3}.

The metric system is built on the power of ten:

$$10 \text{ millimeters} = 10^1 \text{ mm} = 1 \text{ cm}$$
$$100 \text{ centimeters} = 10^2 \text{ cm} = 1 \text{ meter}$$
$$1,000 \text{ meters} = 10^3 \text{ m} = 1 \text{ kilometer}$$
$$1,000 \text{ grams} = 10^3 \text{ g} = 1 \text{ kilogram}$$

or

$$1 \text{ millimeter} = 10^{-1} \text{ centimeter}$$
$$1 \text{ centimeter} = 10^{-2} \text{ meters}$$
$$1 \text{ millimeter} = 10^{-3} \text{ meters}$$
$$1 \text{ meter} = 10^{-3} \text{ kilometers}$$

The distance from our star, the sun, to the next closest star is over 4 light-years (about 43 trillion km or 27 trillion miles). The diameter of our Solar System, from the sun to the outermost planet, Pluto, is about 13×10^9 (billion) km or 8×10^9 (billion) miles. It takes close to 12 hours for light from the sun to reach Pluto!

Origin and Evolution of Our Solar System

For centuries, scientists, philosophers, and theologians have asked questions concerning the origin of the Universe, our Solar System, and our planet. No one knows the origin for certain, but scientists have proposed various hypotheses, which have led to a few theories.

First, we must distinguish between a hypothesis and a theory, because the terms often are confused and interchanged. A **hypothesis** begins with observations and facts, and is an assumption or general proposition offered to explain the facts. A hypothesis could be right or wrong; it is a starting point for study. If the hypothesis is confirmed through experiments or more data, it may then be considered a theory. A **theory** is a generally accepted idea or principle. Thus, a hypothesis is the initial idea that requires further confirmation to become a theory. If a theory turns out to be true in every case, it becomes regarded as a **scientific law.**

There is presently no completely agreed-upon theory for the origin of the Universe, Solar System, or Earth. But we do have some good hypotheses.

In considering our Solar System, several key points must be accounted for in any viable hypothesis. These include the relative abundances of elements in the planets and in the sun, the orbital patterns of the planets and their distances from the sun, and the variations in mass and size among the planets.

The four "inner" planets, closest to the sun—Mercury, Venus, Earth, and Mars—are small, rocky, and relatively dense (4 to 5.5 times the density of water). They are called **terrestrial planets** because they are somewhat similar to Earth (*terra* in Latin). The next four "outer" planets—Jupiter, Saturn, Uranus, and Neptune—are very large and gaseous, and have relatively low densities (0.7 to 1.7 times the density of water). They are called **gas giants** because they are huge balls of gas. These planets are similar to the sun, but are not big enough to sustain the nuclear reaction that makes the sun so hot. Pluto, the ninth planet, is relatively small compared to its neighbors. Little is known about it, and it is an "oddball" planet in many ways.

Other key points are that the sun and giant planets have high abundances of the light gases hydrogen and helium, whereas the terrestrial planets have as their most common elements iron, oxygen, silicon, and manganese. The planets together have only about 0.1 percent of the total mass of the Solar System; the sun has the remaining 99.9 percent. The planets all revolve around the sun in the same direction. Except for Venus and Uranus, they rotate (spin on their axes) in a similar direction as they go around the sun.

In their orbits around the sun, the planets trace an elliptical pattern, and all the orbits essentially lie in a similar plane. There also is a pattern to the distance between the planets; each planet is about twice as far from the sun as its immediate inner neighbor. This remarkable collection of facts allows scientists to develop hypotheses on the origin of the Solar System.

The Condensation Hypothesis

The hypothesis that best accounts for all the facts just presented is this: our Solar System began about 4.7 billion years ago as a vast rotating cloud of cosmic dust and gas condensed (Figure 2-1a). The condensation of the cloud slowly increased three things: its rotational speed, temperature, and gravity. Rotational speed increased for the same reason that ice skaters spin faster when they move their extended arms inward against their bodies while spinning. Temperature increased due to the compression of the accumulating material.

Gravity increased in the cloud because each particle attracts other particles. The principle at work here is that particles attract each other with a force that is directly proportional to the product of their masses, and inversely proportional to the distance squared between the particles. Therefore, as the distance between the particles decreases, the force of gravity increases. This in turn improves the chances of attracting other particles.

As the cloud of cosmic dust and gas continued to spin, it underwent further contraction and formed a disk shape (Figure 2-1b). With this further contraction, gravitational forces increased even more, and material moved toward the center of the disk, forming the early protosun. Continued increases in gravitational forces eventually compressed and heated the protosun until it reached temperatures greater than 1 million degrees **Celsius** (more than 1.8 million degrees Fahrenheit). These temperatures were sufficient to start thermonuclear reactions, releasing large amounts of energy and forming our present sun.

As the sun was developing, other parts of the disk also were changing. Gases and dust particles collided and formed various solid minerals and particles. These in turn gradually clumped with other particles, eventually forming larger bodies called **planetesimals** (Figure 2-1c). As the planetesimals grew, their gravity increased, so they attracted and coalesced with still other planetesimals.

Those planetesimals forming closer to the sun were exposed to its intense heat, making them lose elements with low boiling points, such as the gases helium and hydrogen. These mostly boiled away, leaving behind in

FIGURE 2-1 How the Solar System may have formed: the condensation hypothesis. This is a very schematic illustration, not drawn to scale, showing the general stages of the hypothesis.
(a) A slowly rotating cloud of cosmic dust and gas.
(b) As the cloud condenses, its rotational speed increases, forming a disk-like shape. Most of the material slowly migrates toward the center of the disk, forming the early sun, or protosun.
(c) Later the remaining dust and gas particles condense to form planetesimals.
(d) Eventually, the planets evolve, revolving in orbit around the sun (arrows show direction).

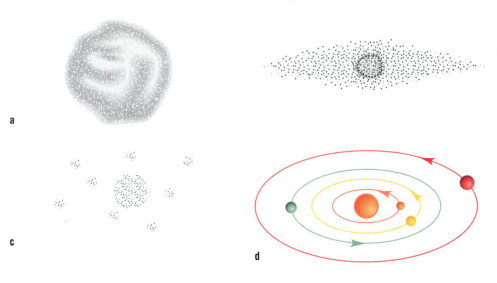

the planetesimals those elements with high boiling points, like iron. This point is demonstrated by Mercury, the planet closest to the sun, which, as expected, has the highest iron content of all planets.

Planetesimals farther from the sun were sufficiently cool so that the gaseous components from the original cloud did not boil away. Because the original dust and gas cloud was composed mainly of hydrogen and helium to begin with, these elements eventually concentrated in the outer gas giant planets. As the process continued, the present planets of our Solar System were eventually formed (Figure 2-1d).

This hypothesis is just a possible explanation and requires considerably more evidence before it can become an acceptable theory. The correctness of this **condensation hypothesis** is being tested by observing other stars, much like observing the life cycle of animals by studying young and old specimens. The millions of stars viewable from Earth are in various stages of their lives, so we can observe "young" stars and "elderly" stars to test the condensation hypothesis. In addition, since our sun is thought to be similar to other stars, it seems logical to expect that some of the hundreds of billions of other stars might also have planets (see Box 2-2).

Box 2-2

Do Other Stars Have Planets?

The question of whether planets exist around stars outside of our Solar System is a fascinating one. Until recently, however, detecting planets orbiting around other stars was nearly impossible because of the vast distance of these stars from Earth (the closest is about 4 light-years away), combined with the probable smallness of any associated planets. Nevertheless, in 1983, a remarkable observation was made from *IRAS,* the *Infrared Astronomical Satellite.* (Infrared means "below red," or light that is not visible to humans because it has a slightly longer wavelength than visible red light.)

This orbiting satellite, which is essentially a robot laboratory, detected a large source of infrared energy being emitted around the star Vega, about 26 light-years from Earth. Vega is thought to be a relatively young star (a few hundred million years old) compared with our sun's much greater age of about 4.7 billion years.

The infrared emissions appear to be coming from a large group of objects, some perhaps as big as planets in

our Solar System, and having temperatures considerably lower than that of Vega. Could these be planets, or perhaps planetesimals in the process of condensing into planets? We don't know yet. Perhaps the most interesting thing about this discovery is that Vega is a relatively ordinary star, similar to millions of others, like our sun. Does this mean that planets are fairly common in the Universe?

Another discovery was evidence of a Solar System around a star called Beta Pictoris. Astronomers actually photographed a disk-shaped cloud of particles around this star (see Figure 1). This computer-enhanced figure is the first direct evidence of a cloud of particles around a star and may represent the early stage of formation of a Solar System.

In 1992, astronomers using the orbiting *Hubble Space Telescope* discovered 15 so-called infant stars, young stars surrounded by flattened disks of dense dust. It is not known whether planets will actually result from

Box 2-2 continued

these disks, but the observation fits the type of phenomenon that is presently believed to form new planets. Later, in 1994, astronomers noted surprising variations in the sound received from a star that is more than 11,200 trillion kilometers (7,000 trillion miles) from the Earth. The variations are interpreted as being due to the presence of two or three planets orbiting the star, but no visible observations have yet been made. So, again, some indirect evidence exists for planets outside of our Solar System.

These recent findings continue to suggest that planets may be fairly common. The next question, the really big one, is: Are any of these planets inhabited, and if so, by whom?

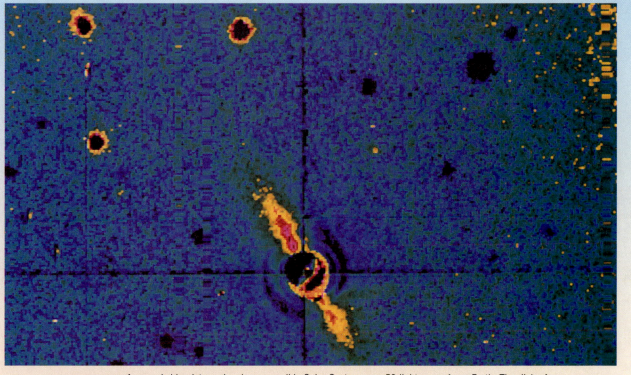

FIGURE 1   A remarkable picture showing a possible Solar System over 50 light-years from Earth. The disk of material extends 64×10^9 km (40×10^9 miles) around the star Beta Pictoris (center). The disk, seen nearly edge on, is thought to be relatively young, perhaps a few million years old. It is interesting to compare the shape of this disk with that proposed in the condensation hypothesis for the origin of our Solar System (see Figure 2-1b). The disk is probably composed of ice, carbonaceous organic substances, and silicates, similar to the material believed to have formed our Solar System. (Courtesy of the University of Arizona and the Jet Propulsion Laboratory.)

The Early Earth

The early Earth was mainly a mixture of iron, manganese, and silicon compounds, with lesser amounts of the other elements. Our planet was essentially homogeneous at first, having a similar composition from its center to its surface. The early Earth also was cool, but slowly was heated by three phenomena: heat generated from the impact of particles upon Earth, heat from compression of the new accumulating material, and heat from the decay of radioactive material contained within the Earth. (When a radioactive element decays, the element disintegrates, losing parts of its atomic structure, changing into a different element, and emitting a small amount of heat.)

The temperature of the early Earth was greatest in the interior of the planet, and eventually exceeded the melting point of iron. It may have taken about 1 billion years before such a temperature was reached (about 2,000°C or 3,600°F). When the iron melted, it migrated toward the center of the planet (the **core**) because of its greater density. Less dense materials migrated outward toward the surface of the planet (Figure 2-2a). Eventually the material became segregated in concentric layers around the core, according to the density of each material.

When the hot, outward-moving material reached the surface, it produced volcanic activity. The resulting

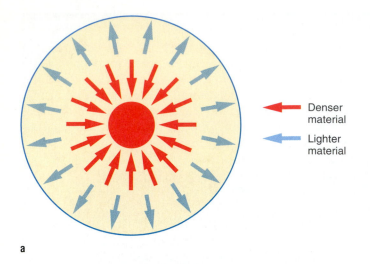

a

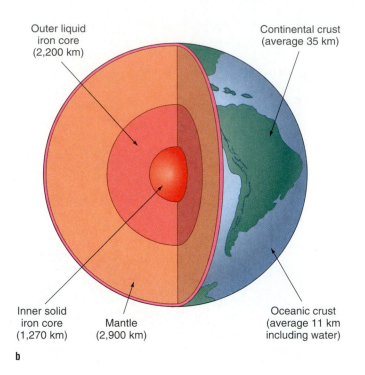

Outer liquid
iron core
(2,200 km)

Continental crust
(average 35 km)

Inner solid
iron core
(1,270 km)

Mantle
(2,900 km)

Oceanic crust
(average 11 km
including water)

b

FIGURE 2-2   Early Earth was essentially homogeneous and not layered internally, as it is today. However, over many years, and with Earth's increasing temperature and gravity, a differentiation process began.
(a) Denser materials slowly migrated (sank) toward the center of the Earth, while lighter materials migrated (floated) toward its surface.
(b) The eventual result was an Earth with concentric layers: the dense core, lighter mantle (from a German word meaning "coat"), and even lighter crust. The average thickness of the different layers is indicated. (See also Figure 3-1, page 41.)

magma eventually cooled and solidified into parts of Earth's crust. During the early history of Earth, this process happened repeatedly, eventually differentiating the planet into layers. Today, the result of this **differentiation** is a dense core, rich in iron and nickel; a less-dense zone called the **mantle,** rich in iron and magnesium; and a thin **crustal layer,** rich in aluminum, magnesium, and silicon (Figure 2-2b). The general characteristics of these layers are summarized in Table 2-1. For oceanographers, the crustal layer is generally of most interest, because this is the layer that directly underlies the ocean and contains the continents.

The process of differentiation was extremely important in Earth's history, for it led to the initial formation of the continents. Likewise, the origin of Earth's atmosphere and ocean are believed to be closely related to this continuing sequence of melting, differentiation of elements, and volcanic activity, as you now shall see.

Origin of Earth's Atmosphere

Our present atmosphere is mainly nitrogen (about 78 percent) and oxygen (about 21 percent). It is very different from the atmosphere of the early Earth. Early Earth had insufficient gravity to retain the original atmosphere of hydrogen and helium, so they diffused into space. Volcanic activity continued as Earth

TABLE 2-1

General Characteristics of Earth's Main Layers

Layer	Average Density (g/cm^3)	General Composition	Average Thickness km (miles)	Percentage of Total Mass of Earth
Crust	2.8 (continents) 3.0 (oceans)	Magnesium and aluminum silicates	35 (21.7) (continents) 11 (6.8) (ocean, including the water)	0.4
Mantle	4.5	Iron and magnesium silicates	2,900 (1,800)	68.1
Core: outer	11.8	Liquid iron and nickel	2,200 (1,390)	31.5
inner	17.0	Solid iron and nickel	1,270 (790)	

evolved, releasing various gases, mainly water vapor, carbon dioxide (CO_2), chlorine, nitrogen, and hydrogen (Figure 2-3). These gases formed the next phase of our atmosphere.

Eventually, as temperatures decreased, much of the water vapor in this phase of the atmosphere condensed as rain, which formed the early oceans. This phase is believed to have started about 4 billion years ago. With the loss of all this water vapor from the atmosphere to the early ocean, the remaining atmosphere was mainly CO_2 and lesser amounts of nitrogen, similar to the present atmospheres of our neighbor planets, Mars and Venus.

Interestingly, the early atmosphere and ocean contained little free oxygen (oxygen uncombined with other elements). It's easy to understand why: oxygen is very active chemically, combining with many other elements, and the hotter the environment, the greater its activity. Free oxygen, which is necessary for animals to breath, did not appear in significant quantities until the evolution of plants such as algae. These plants used solar energy to combine water and CO_2 to make their food (a sugar) and to produce free oxygen as a byproduct in the process called **photosynthesis.**

The photosynthetic process (described in more detail in Chapter 7) effectively removed most of the CO_2 from the atmosphere, ultimately incorporating it into hydrocarbon deposits (such as oil and natural gas) or carbonate rocks (such as limestone). The photosynthetic process helped to produce our present atmosphere of nitrogen and oxygen, with minor amounts of water vapor, carbon dioxide, other gases, and dust. Photosynthesis began about 3 billion years ago and has profoundly changed the characteristics of the early atmosphere and the ocean.

Origin of the Ocean

The question of the origin of the ocean really has two parts. Where did all the water come from? And how did it get its unique concentration of elements?

Our best hypothesis is that the origin of the water is directly related to the warming and differentiation of elements within Earth during its early history. In these processes, water was released from various chemical compounds contained within Earth and moved to the surface along with volcanic material. Once at the surface, much of the water entered the atmosphere as water vapor (steam), eventually condensing as rain, filling Earth's lower areas to form the ocean (see Figure 2-3).

The concentration of elements resulted from weathering and erosion of the early continents by the rainfall. The runoff carried eroded debris into the developing ocean, where it was deposited on the seafloor. Thus, many of the elements in seawater came from the weathering of the volcanic rocks composing the continents. These elements include **cations** (atoms or molecules having a positive charge) such as sodium, magnesium, calcium, and zinc. They also include **anions** (atoms or molecules with a negative charge) such as chloride and sulfide that were released through volcanic activity and eventually entered the ocean. These processes continue today as the atmosphere receives additional gases from more than 600 active surface volcanoes. Seawater itself directly receives other elements and gases from several hundred more underwater volcanoes along ocean ridges.

Recently, a new hypothesis for the origin of ocean water has been proposed by Louis Frank, a physicist at the University of Iowa. He suggests that ocean water has accumulated from the collision of small, icy comets with

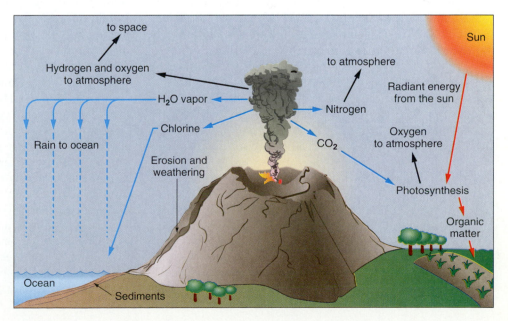

FIGURE 2-3 How volcanoes probably formed the early atmosphere and ocean. The various components in volcanic gas behave in different ways after leaving the volcano. Lightweight hydrogen gas eventually escapes Earth's gravity into space, while heavier nitrogen remains in the atmosphere. Water vapor condenses to form the ocean, and some water molecules are split into hydrogen and oxygen by solar energy. Through photosynthesis, carbon dioxide is converted by plants into organic material and free oxygen gas. Elements such as chlorine accumulate in the ocean. All of these processes continue today.

Earth. Frank feels that as many as 10 million small comets each year reach our atmosphere, and that such a frequency could, over several billion years, have supplied sufficient water to form the oceans. This hypothesis is controversial, however, due to the large numbers of comets required. More data and observations of comets are required before it can be considered correct.

This discussion concerning the origin of ocean water and how it attained its present composition obviously is a simplification of an extremely complex subject. In summary, it is evident that the oceans, in some form, have been around for several billion years. Good evidence also exists that volcanic activity is the principal source of the water and its anions, while the cations come mainly from the decomposition and weathering of volcanic rocks.

As various elements reach the ocean, many different chemical and physical processes determine their ultimate fate, and thus the relative amount of each element remaining in seawater. These processes take place mainly at the major interfaces, or boundaries, of the ocean:

• The water-atmosphere interface

• The water-biosphere (plant or animal) interface

• The water-sediment interface (seafloor)

The reactions occurring at each interface are described in later chapters on the physics, biology, and chemistry of the ocean.

Further Development of Earth and Its Ocean

Little is known about the early history of the ocean, but some facts are emerging. Recently, for example, Australian scientists found rocks about 3.5 billion years old that are of similar composition and structure to marine rocks that formed much more recently. (Techniques used to date ancient rocks are described in Chapter 7.) This indicates that some sort of marine environment existed at that time (see Box 2-3). Marine fossils similar to present-day living forms but about 600 million years old suggest that some ancient marine environments may have been similar to present conditions.

Through their study of rocks and fossils, Earth scientists have divided Earth's history into the **geologic time scale** (Table 2-2). The long interval of time from Earth's origin to about 590 million years ago is called **Precambrian** time, where little evidence has been preserved of the primitive life that existed. The most recent 590 million years of Earth's history is divided into three eras, based on the life forms that appear in the fossil record. Each era is further divided into periods.

Important geological and biological events in each period are shown in Table 2-2.

Over geologic time, Earth's geography has undergone remarkable changes, including the creation, movement, and sometimes destruction of continents and ocean basins, due to *seafloor spreading.* Seafloor spreading is an exciting scientific concept described in detail in the next chapter.

Simply said, magma from Earth's interior slowly rises in the deep ocean between the continents. On reaching the surface, it slowly spreads the seafloor apart (Figure 2-4). In the spreading process along the ocean ridge, new seafloor is being formed by the chilled magma. In other places, the seafloor either is being recycled back into magma by being dragged into Earth's interior under a continent, or is being appended to a continent. The process of seafloor spreading is responsible for the present shape and position of the oceans and continents.

Over 225 million years ago, a single supercontinent existed, called **Pangaea** (named for *Gaia,* the Greek goddess of the Earth). This supercontinent, surrounded by ocean, combined all the land area of today's continents and extended essentially from the North Pole to the South Pole (see Figure 3-4a, page 46). Studies of rocks deposited at that time indicate that large ice sheets extended over much of what today is India, Australia, South America, and Africa. Arid desert conditions dominated the mid-latitude and low-latitude regions. About 200 million years ago, seafloor spreading began to break up Pangaea, and this continuing movement has led to the present configuration of continents and oceans.

Climatic Changes, Glaciers, and Sea Level Changes

Widespread climatic cooling occurred during Precambrian time, the Carboniferous-Permian Periods, and in our present time period, the Quaternary. This cooling resulted in extensive freezing and widespread formation of glaciers. The Quaternary is divided into two epochs: the **Pleistocene epoch,** which began about 1.6 million years ago, and the **Holocene epoch** (or Recent), which began about 18,000 years ago, near the end of the most recent major ice age.

Studies of recent geological deposits left by the glaciers, of soil types, and of seafloor sediment reveal that numerous intervals of glaciation occurred, followed by milder interglacial intervals where the climate was similar to present-day conditions. (In other words, we are presently living in a mild interglacial period.) The reasons for these major changes in climate have been the source of much speculation by oceanographers, geologists, climatologists, biologists, and others. Indeed, about a decade ago, some felt we might be poised to enter a frigid glacial period. But now, because of an increase in

Early Earth and Favorable Conditions for Life

Most scientists accept that Earth formed about 4.7 billion years ago. What were the first billion years like? Images of the pockmarked moon and other planets suggest that both volcanism and collisions have been common over geologic time. The collisions occurred with meteorites, asteroids (planetlike bodies, mostly smaller than 1 km), and comets. If Earth and other planets are all of the same age, it is reasonable to assume that similar events have occurred on Earth. But on Earth, the remains and scars of these impacts have been obliterated by erosion or covered with sediment. Both of these processes involve water, which is lacking on the moon and on other planets. Plate tectonics (the slow but destructive movements of Earth's crust) can also remove evidence of previous impacts.

Looking at the moon, with its thousands of impact craters on top of other impact craters, gives some feeling for the intensity of activity that must have occurred in Earth's early history (see Figure 1). Earth is bigger and has six times the gravity of the moon, so it had an even greater chance of being hit harder by more objects from space, more often. It is thought that in the first billion years or so of our Solar System, there were many times more possibilities for collision among planets, asteroids, and meteors than at present, simply because more debris existed in interplanetary space than after the planets formed.

Two of the large craters on the moon suggest impacts by objects about 100 km (about 60 miles) in diameter. Such a collision of a similar-sized object with early Earth certainly would have had a major impact on its environment. Several recent studies suggest that such an impact on Earth would have been catastrophic, vaporizing any ocean that might have existed. If so, the increased water vapor in the atmosphere would have retained heat (greenhouse effect), raising atmospheric temperatures dramatically.

Geologist James Kasting has suggested that the water vapor in the atmosphere from such an impact would have elevated temperatures to well over 1000°C

FIGURE 1 This picture was taken by the astronauts aboard *Apollo* about 16,000 km (10,000 miles) from the moon. Note the many large dark areas, called *mare*, and numerous craters. Over 30,000 craters are visible with a telescope, ranging in diameter from 1 km (0.6 mile) to over 100 km (60 miles). The moon itself has a diameter of 3,476 km (2,129 miles), a little less than the distance from New York to San Francisco. (Photograph courtesy of National Aeronautics and Space Administration.)

continued

Box 2-3 continued

(1800°F) and essentially "sterilized the whole planetary surface." If any life existed at that time, it would have been destroyed. Eventually, the temperature would cool and the water vapor would condense to refill the oceans by rainfall. This entire process could have happened many times in the early history of our planet.

Clearly, such an event would have been catastrophic for any early life forms. Smaller asteroid impacts might not have converted all of the ocean into water vapor, but only the water near its surface (see also Box 2-4, "Why Mass Extinctions?"). This suggests that life perhaps started in the deep ocean, where living forms were protected from obliteration from such major impacts. The

most likely places in the deep ocean would be at "vents" on mid-ocean ridges, the areas where hot water is being discharged and where some unique forms of life recently have been discovered. (These vents are described in Chapter 7, but see pages 167–170 at this time.)

This concept of the deep ocean as the site of the origin of life competes with the more generally accepted idea of life beginning in a tranquil warm ocean or pond containing various organic compounds (the famous "primordial broth"). The true answer demands much more study and some very sophisticated experiments. But most scientists concur: life began somewhere in the ocean.

---

the *greenhouse effect* (described later in this chapter), the prognosis is that Earth is warming instead.

During the most recent glaciation, which reached its maximum about 18,000 years ago, much of Canada and large parts of the northern and central United States were covered with ice, snow, and glaciers. In North America, at least four major glacial advances and withdrawals (interglacial stages) occurred over parts of the continent during the Pleistocene epoch. These periods of glaciation had two distinct effects on the ocean. Because glaciers tied up a significant volume of Earth's water in ice, they lowered sea level when they formed and raised

sea level when they melted. Melting glaciers also lowered the temperature of the surface water by several degrees.

The amount that sea level was lowered during the most recent ice advance has been studied by determining the ages of deposits that formed at or near sea level. These deposits include peat (decaying plants in the earliest stage of becoming coal) and shells of shallow-water animals. (Ages were measured using radiometric techniques, described in Chapter 7.) Because sea level has subsequently risen due to the melting of the ice, these deposits are now underwater. By dating this material

---

TABLE 2-2
Geologic Time Scale and Some Major Events

Era	Period	Events	Began Millions of Years Ago
CENOZOIC ("Recent Life")	Quaternary	Our present geologic period. Evolution of humans. Numerous glacial advances.	1.6
	Tertiary	Increase in mammals. Appearance of primates. Mountain building in many areas.	65
MESOZOIC ("Middle Life")	Cretaceous	Extinction of dinosaurs. Increase in reptiles and flowering plants.	144
	Jurassic	Birds. Mammals. Dominance of dinosaurs. Mountain building in western North America.	213
	Triassic	Beginning of dinosaurs and primitive mammals.	248
PALEOZOIC ("Ancient Life")	Permian	Reptiles spread and develop. Evaporite deposits. Glaciation in Southern Hemisphere.	286
	Carboniferous	Abundant amphibians. Reptiles appear.	360
	Devonian	Age of fishes. First amphibians. First abundant forests on land.	408
	Silurian	First land plants. Mountain building in Europe.	438
	Ordovician	First fishes and vertebrates.	505
	Cambrian	Age of marine invertebrates.	590
PRECAMBRIAN TIME		At least six times longer than all of the following geologic times above. Primitive one-celled life began during the Precambrian era.	4,700

FIGURE 2-4   How the seafloor-spreading hypothesis works. Hot magma from the mantle rises and reaches the seafloor. It pushes through the seafloor as lava and cools and solidifies to form the mid-ocean ridge. The newly formed seafloor and underlying crust slowly spread away on both sides of the ridge area. In some instances, a continent may be carried along with the moving seafloor (right side of figure). Alternatively, the seafloor may eventually be thrust beneath the continent (subducted) and become part of the continent, also forming a trench in the process (left side of figure). The rate at which all this happens is typically a few centimeters a year. Chapter 3 presents a more detailed discussion of seafloor spreading.

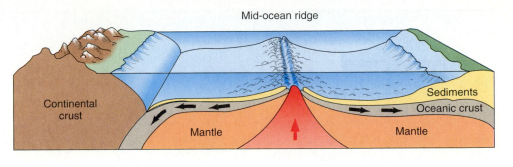

and noting the depth at which it was collected, it is possible to construct a curve of sea-level changes over the past 35,000 years (Figure 2-5).

These data show that about 18,000 years before present (B.P.), sea level was more than 130 m (426 ft.) lower than today. From 18,000 B.P., sea level rose relatively rapidly until about 7,000 B.P., when it reached a level roughly 10 m (33 ft.) below present sea level. Since 7,000 B.P., sea level has risen slowly and irregularly.

During times of lowered sea level, the continents of Asia and North America were connected by a narrow stretch of land that today is covered by the shallow Bering and Chukchi Seas off Alaska. When sea level was this low, it has been suggested that people and animals migrated via this land bridge from Asia to North America, and onward to South America. Indeed, genetic studies show that Native Americans are of Asian decent, providing evidence for extensive migration from Asia into North America that such a land bridge would have helped.

Extinctions and the Greenhouse Effect

Studies of fossils and evolution have led to the discovery of massive worldwide extinctions of animal and plant life. An **extinction** refers to the disappearance of an entire species, or group of species, of living things. In modern times, humans have caused extinctions of individual species by destroying their environment or by overhunting. But before humankind came on the scene, mass extinctions may have been caused by major changes in sea level, disruptions in the food chain, or other environmental perturbations.

One remarkable event appears to have been a widespread extinction late in the Permian Period, about 240 million years ago, due to coalescing of the continents into the supercontinent Pangaea. This caused the world's shallow sea habitat to be reduced nearly 70 percent in area, and appears to have led to a mass extinction of as much as 96 percent of the species of marine life.

A more recent mass extinction occurred between the end of the Cretaceous Period and the beginning of the Tertiary Period, about 65 million years ago. At this time, as much as 52 percent of existing marine species may have become extinct. Dinosaurs, which had dominated their environment for about 130 million years, abruptly disappeared (see Box 2-4 for more on this).

Some changes still are occurring. For example, in some parts of the world, sea level continues to slowly rise. If all the present ice on Earth were to melt, sea level would rise by 60 m (about 200 ft.). Such a rise would be catastrophic, since a large portion of the world's population lives along coasts, at or near sea level. Alternatively, but much less likely, we could enter another period of glaciation, with decreasing temperatures that would cause extensive freezing and lower the sea level.

These possibilities, especially a rise in sea level, now are receiving serious scientific consideration. This interest is mainly due to new awareness of increasing carbon dioxide (CO_2) and other gases in our atmosphere that retain heat. These increase the **greenhouse effect**, possibly causing a global warming.

The greenhouse effect might also be called the "car-in-the-sun effect." If you leave a car parked in direct sunshine with its windows up, the interior turns into an oven. The reason is that the window glass is trans-

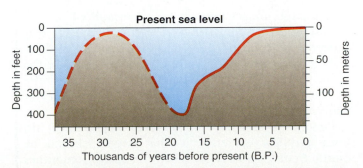

FIGURE 2-5   Changes in sea level over the last 35,000 years. Where did all the water go when the sea was lowered so dramatically 18,000 years ago? Note the slowing of the sea level's rate of rise, about 7,000 years ago. The dashed lines indicate uncertainty about the exact positions of sea level.

BOX 2-4

Why Mass Extinctions?

What caused the Cretaceous mass extinction about 65 million years ago, including the demise of the dinosaurs? The magnitude and apparent quickness of this mass extinction has challenged scientists for many years, and many hypotheses have been put forth. One fascinating suggestion was made in 1980 by Luis Alvarez, his son Walter, and their colleagues. They detected unusually high concentrations of the rare element *iridium* and other rare elements in a layer of clay that was deposited worldwide at about the end of the Cretaceous Period. Asteroids generally have a concentration of these rare elements. Could a large asteroid have collided with Earth, scattering this iridium evidence of itself worldwide, and somehow caused the mass extinction?

The impact, they suggested, could have caused large clouds of dust to be injected into the atmosphere, leading to three months or more of global darkness. Because photosynthesis requires light, the plants that were the food supply for many dinosaurs would have been greatly reduced. Thus, these creatures and others could have starved to death (see Figure 1).

Although the disappearance of the dinosaurs has received the most publicity, the organisms living in the

FIGURE 2 Location of the impact crater believed to have been caused by collision with an asteroid about 65 million years ago. The asteroid was about the size of the island of Manhattan in New York. The crater, buried beneath sediments, is centered near the town of Chicxulub in Mexico's Yucatán Peninsula. Evidence of a monster wave that was generated by this impact has been found in Mimbral and Haiti, to the east.

ocean at that time were also seriously affected. Examination of deep-sea sediments deposited at the end of the Cretaceous Period show essentially a complete absence of plant fossils, indicating that plant growth in the surface waters had almost completely stopped. The fossil record also indicates that it took about 500,000 years before the surface conditions returned to normal.

This collision hypothesis, although exciting, has some problems, as do other explanations of the mass extinction. For example, such a worldwide period of darkness also would have had a catastrophic effect on tropical plants, but the fossil evidence does not reflect this. Likewise, no unequivocal physical evidence of such a large impact initially was detected. Of course, the asteroid could have landed in the ocean, so evidence of its impact could have been obliterated by seafloor spreading or burial by sediments. Because the ocean covers almost three-quarters of the planet, the probability of an asteroid hitting the ocean is greater than for land.

Regardless of where an asteroid might have hit, some physical evidence of this impact should be found. Geologists have searched the globe, sometimes using satellite imagery, looking for a 65 million-year-

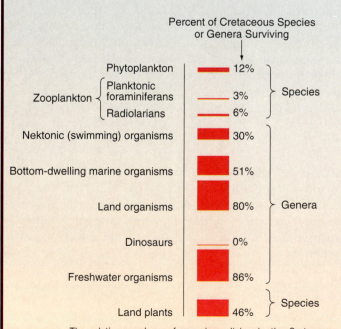

FIGURE 1 The relative numbers of organisms living in the Cretaceous Period that survived into the beginning of the Tertiary Period (about 65 million years before present). Note that none of the dinosaurs survived. (Courtesy of Prof. Michael A. Arthur, now at Pennsylvania State University.)

Box 2-4 continued

old crater. At one time, Cuba's Isle of Pines region was thought to be a possibility, but certain distinctive minerals that form during such an impact were not found.

After about 15 years of searching, scientists now seem fairly sure that the remains of an impact crater 180 km (112 miles) wide exists off the Yucatán Peninsula along the east coast of Mexico (see Figure 2). Drilling in the region has obtained samples of those distinctive minerals that form during an asteroid impact. The size and character of the crater indicates that it resulted from an asteroid 10 km (6 miles) wide that hit the area around 65 million years ago. This discovery, and its time of formation, are very supportive of the Alvarez hypothesis.

Nevertheless, other marine scientists, such as Charles Officer and Charles Drake from Dartmouth College, see things a little differently. They suggest that the species that became extinct did not all die at exactly the same time. They suggest that the extinctions were not the result of an instantaneous impact, but rather of gradual processes occurring over 10,000 to 100,000 years. These scientists propose a series of relatively intense volcanic events, leading to environmental changes that caused the extinction. A large asteroid impact itself could even have stimulated increased volcanic activity. Thus, both sides in this decade-long debate might be essentially correct.

parent to incoming sunlight, but the sunlight is converted to heat when it strikes the car's interior, and the window glass is relatively opaque to the heat, trapping it inside. This is how a greenhouse is heated by solar energy; hence the name greenhouse effect. It also is how the Earth's surface is heated by solar energy, with the atmosphere acting like window glass (see Figure 13-13, page 322).

The opacity of the atmosphere is made greater by increasing amounts of certain gases in the atmosphere, especially CO_2, methane, and water vapor. The greater the opacity of the atmosphere, the more the atmosphere will absorb heat and in turn increase the temperature of Earth's surface. Thus, the increase in these gases may have a dramatic and perhaps devastating effect on the world's climate in coming decades.

The General Character of Earth's Present Ocean

The next chapter describes the continuing pattern of growth and recycling of both seafloor and the continents by the process of seafloor spreading. But before turning to that important subject, let us first consider the general features of the ocean.

The ocean is the dominant surface feature of our planet, as is demonstrated by some numbers (see also Box 2-5):

- The ocean covers 70.7 percent of the planet (362×10^6 km², or 139×10^6 miles²), whereas land covers only 29.3 percent (150×10^6 km², or 57.9×10^6 miles²).

- Ocean volume is approximately $1,350 \times 10^6$ km³, or 318×10^6 miles³.

- Average ocean depth, including adjacent seas, is about 3,800 m (2,077 fathoms, 12,464 ft., or 2.4 miles).

- If Earth's surface were smooth, with no mountains or valleys, it would be covered by about 2,400 m (8,000 ft.) of water.

Perhaps the name "Earth" is a misnomer—because most of our planet is covered by it, it should have been named Water!

The distribution of elevation over the world is shown in Figure 2-6. This type of representation is called a **hypsographic curve.** The curve shows the amount of Earth's surface above or below any selected elevation or depth. Two important facts are evident from Figure 2-6:

- Two regions dominate Earth's surface: one just above sea level, at about 100 m (328 ft.), and the other below sea level and broader, from about 4,000 to 5,000 m depth (about 13,100 to 16,400 ft.).

- There is a sharp intermediate zone between these elevations.

The two elevations clearly indicate two different parts of Earth's crust: the portion of land at or just above sea level, and the deep seafloor. (Note that if all the water were removed from the ocean, these differences would still exist.) The intermediate zone is the transition area between the continental and the oceanic regions.

We divide the seafloor into two principal provinces: the continental margin and the ocean basin or deep sea. The **continental margin** (Figure 2-7a) is the area adjacent to the continent and includes the marine portion of the coastal region and the continental shelf, continental slope, and continental rise. The continental slope is the sharp intermediate zone noted from Figure 2-6. The **ocean basin** (Figure 2-7b) makes up about 79 percent of the total seafloor. It is dominated by three

---

BOX 2-5

Some Units of Measure

Oceanographers use a wide variety of units of measure. In general, however, metric terms are used most frequently.

Depth usually is measured in meters (m) or kilometers (km), although sometimes feet (ft.), miles, or even fathoms (6 feet or about 1.8 m) are used. In the United States, depths on coastal charts used for navigation are frequently shown as fathoms or feet.

One meter equals about 3.28 feet. Meters can be divided into **microns** (μm, or a millionth of a meter), millimeters (mm, or 0.001 meter), centimeters (cm, or 0.01 meter), and so on. Meters can be expressed in convenient groups, such as kilometers (km, or 1,000 meters).

The main conversion factors from the English system to metric units are:

Multiply	By	To obtain
inches	2.54	centimeters
feet	0.3048	meters
fathoms	1.828	meters
miles	1.609	kilometers

Some commonly used measures of distance and height are:

$$100 \text{ meters} = 328 \text{ feet}$$

$$1 \text{ kilometer} = 3,280 \text{ feet or } 0.6214 \text{ mile}$$

Weight or mass is mainly expressed as grams (g); 1,000 grams is equivalent to 1 kilogram (kg), which is equivalent to 2.20 pounds (lb.). One pound is equivalent to 453.6 grams. The expression *grams/kg* is equivalent to *parts per thousand*. The salinity of seawater, discussed in later chapters, is also expressed in parts per thousand (o/oo).

Speed is generally expressed as kilometers per second (km/s), centimeters per second (cm/s), or meters per second (m/s). One term somewhat unique to oceanography is knots, which is nautical miles per hour. A nautical mile is 6,080 feet. Therefore, a speed of one knot equals 6,080 feet per hour, or 50 cm/s. Some common conversions are:

$$1 \text{ cm/s} = 1.19 \text{ ft./minute}$$

$$1 \text{ mile/hour} = 0.447 \text{ m/s}$$

Temperature in oceanography is generally expressed in Celsius (°C) (formerly called centigrade), rather than Fahrenheit (°F). Conversion from Fahrenheit to Celsius (or vice versa) is done with these equations:

$$°F = (1.8 \times °C) + 32$$

$$°C = (°F - 32) \div 1.8$$

If you are not familiar with the metric system, this may seem a little confusing, so equivalents are given throughout the text. (Also see Appendix I for more information on units and terms.)

---

components: the abyssal plains, trenches, and ocean ridges. The ocean basin contains the deepest parts of the ocean. These provinces are presented in detail in Chapter 4.

Earth's Ocean and Earth's Oceans

The continents and ocean basins are unevenly distributed on Earth's surface. Examine a globe and you can see that a continental area usually has an oceanic area on the opposite side. Likewise, most of Earth's land surface is north of the equator (Figures 2-8 and 2-9). Less than 35 percent of the land occurs south of the equator, and between the southern latitudes of 50° and 65° there is essentially no land at all. Figure 2-9 makes it very clear that the individual "oceans" we think of are actually fully interconnected to form the global ocean. But functionally and historically we divide the *world*

ocean into four main oceans: the Pacific, Atlantic, Indian, and Arctic, plus numerous smaller seas and gulfs (Figure 2-9).

The Pacific Ocean is the largest and deepest, having both a surface area and a volume greater than the other three oceans combined (Table 2-3). The Pacific is essentially oval in shape, with numerous deep trenches and associated islands around its perimeter. This perimeter is also the site of frequent earthquakes and active volcanoes, Mount Saint Helens and Mount Pinatubo being well-known examples.

Several large **marginal seas** flank the western Pacific. These are so named because they are "on the margin" of the main ocean. They are usually shallow and somewhat isolated from the main ocean by land or by shallow submerged areas. Marginal seas include the Sea of Okhotsk, Sea of Japan, Yellow Sea, and East and South China Seas (Figure 2-9).

FIGURE 2-6 Hypsographic curve shows the percentage of Earth's surface above and below sea level. (*Hypso-* is a Greek word meaning "height"; oceanographers use either the term *hypsographic* or *hypsometric* to describe the curve.)

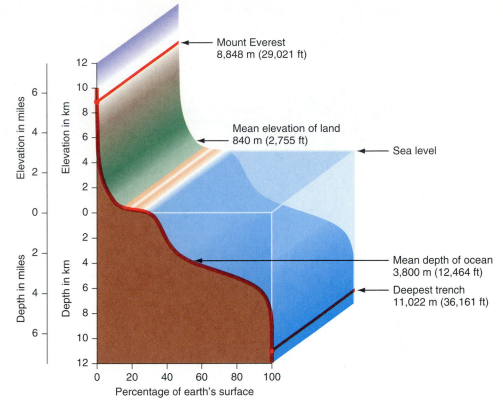

Mount Everest
8,848 m (29,021 ft)

Mean elevation of land
840 m (2,755 ft)

Sea level

Mean depth of ocean
3,800 m (12,464 ft)

Deepest trench
11,022 m (36,161 ft)

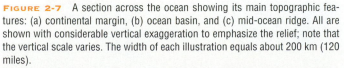

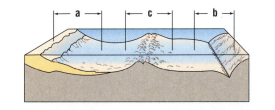

Continental shelf

Continental slope

Continental rise

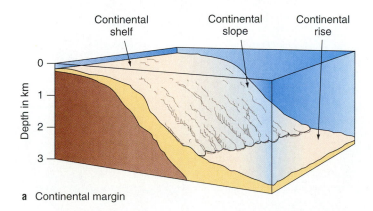

a   Continental margin

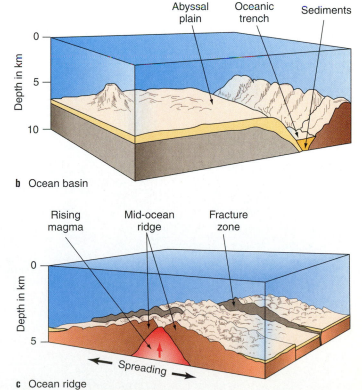

Abyssal plain

Oceanic trench

Sediments

b   Ocean basin

Rising magma

Mid-ocean ridge

Fracture zone

Spreading

c   Ocean ridge

FIGURE 2-7 A section across the ocean showing its main topographic features: (a) continental margin, (b) ocean basin, and (c) mid-ocean ridge. All are shown with considerable vertical exaggeration to emphasize the relief; note that the vertical scale varies. The width of each illustration equals about 200 km (120 miles).

(a) Continental margin: note the continental shelf, which fringes every continent; the shelf break, where the slope begins; the relatively steep continental slope; and the rise, which slopes gradually outward into the ocean basin.

(b) Ocean basin: basically a horizontal plain of sediments, interrupted by various volcanic features and sometimes a trench.
(c) Mid-ocean ridge: formed by cooling magma.

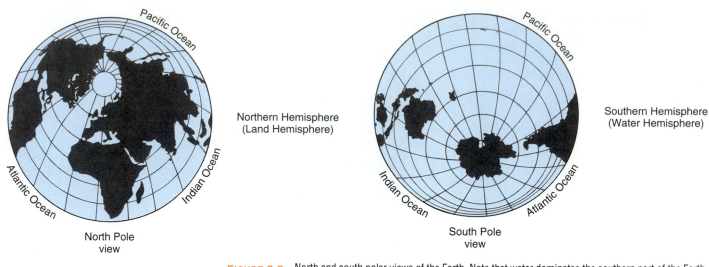

FIGURE 2-8 North and south polar views of the Earth. Note that water dominates the southern part of the Earth, whereas land prevails in the northern part. Note also that the major oceans are connected around the Antarctic continent.

The Atlantic Ocean is relatively narrow. Its surrounding continents—Africa, South America, Europe, and North America—appear as if they would fit together like pieces of a jigsaw puzzle if the ocean were removed and the continents were pushed together. This symmetry was one of the key points that led to the development of the concept of seafloor spreading described in the following chapter.

Several large but shallow seas are adjacent to and are actually part of the Atlantic Ocean. These include the Mediterranean, Baltic, North, Norwegian, and Caribbean Seas, and the Gulf of Mexico (Figure 2-9). Several of the largest rivers of the world, such as the Amazon, Congo, and Mississippi, discharge their sediment and freshwater into the Atlantic.

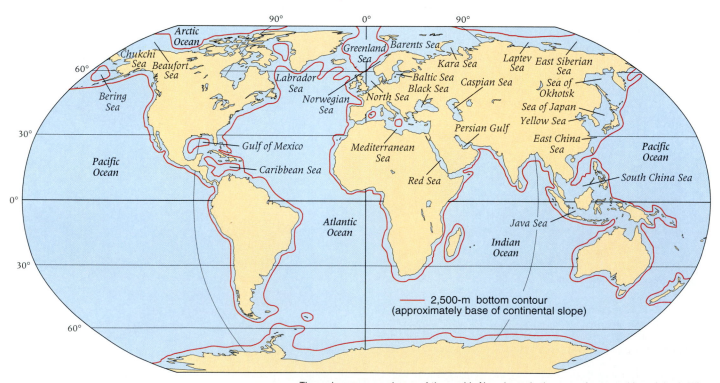

FIGURE 2-9 The major oceans and seas of the world. Also shown is the approximate position of the 2,500 m (8,200 ft.) bottom contour, which is the approximate base of the continental slope.

TABLE 2-3
Area, Volume, and Mean Depth of the Ocean

Ocean and Adjacent Seas	Area (10^6 km²)	Volume (10^6 km³)	Mean Depth (m)	Maximum Depth (m)
Pacific	181.34	714.41	3,940	11,022
Atlantic	94.31	337.21	3,575	8,605
Indian	74.12	284.61	3,840	7,450
Arctic	12.26	13.70	1,117	4,600
Totals and mean depth	362.03	1,349.93	3,800	

Source: Data from various sources including Menard and Smith, 1966.

The Indian Ocean is somewhat triangular and situated mainly south of the equator (Figure 2-9). Although it was one of the first oceans to be explored and sailed, it remains one of the least understood. Three especially large rivers—the Ganges, Indus, and Brahmaputra—discharge into the Indian Ocean. The Persian Gulf and the Red Sea are well-known marginal seas in the northwestern Indian Ocean.

The Arctic Ocean is somewhat circular in shape and is considerably shallower than the other oceans. Some question whether it should be classified as a separate ocean, or rather as an almost landlocked arm of the Atlantic or possibly the Pacific (see Figure 2-8).

A distinguishing feature of all these oceans is an **ocean ridge** that nearly encircles the globe (Figure 2-10). This feature often is called the mid-ocean ridge because in many locations it is situated in the central part of the ocean. It is by far the longest mountain chain on Earth—65,000 km (40,000 miles) long. This essentially continuous ridge goes by several different names, depending on which ocean it is in: the Mid-Atlantic Ridge, the East Pacific Rise, or the Mid-Indian Ocean Ridge. This ridge is the most distinctive topographic feature of Earth, yet its existence and extent were confirmed only a few decades ago. The origin of the ridge is explained in the next chapter.

FIGURE 2-10 Digital shaded-relief map of Earth's surface. The elevations are from topographic maps; ocean floor data are from data sets produced by the Navy and other organizations. The combined data were colored to emphasize large features such as oceanic trenches, abyssal plains, continental shelves, plateaus, and mountains. Shading emphasizes smaller features. The projection used for this map exaggerates the size of features near the north or south pole. For example Australia is considerably larger than Greenland (which is closer to a pole than Australia is). (Figure courtesy of Dr. David A. Hastings, NOAA National Geophysical Data Center, Boulder, Colorado.)

SUMMARY

Much has been learned recently about the origin of our own Solar System, but many unknowns still exist. There is general agreement among scientists that Earth started to form about 4.7 billion years ago from a rotating cloud of cosmic dust and gas. As the Earth formed it grew hotter and, by the process of differentiation, developed a dense core surrounded by a less-dense mantle and a thin crust. The differentiation processes, including volcanism, also contributed to the development of our atmosphere and provided the water for the early ocean. The early atmosphere and ocean lacked sufficient amounts of free oxygen until the evolution of photosynthesis.

The cations in the ocean come mainly from the weathering of rocks on land, whereas the anions result primarily from volcanic activity. The relative amount of each element in seawater is determined chiefly by chemical and physical processes at the three major interfaces of the ocean: water-atmosphere, water-biosphere, and water-sediment.

Little is known about the early history of the ocean, but the similarity of fossil forms to living organisms suggests that environments similar to present-day conditions may have existed about 600 million years ago. Over Earth's long history, considerable variation in the position of the continents and oceans has occurred. Caused by seafloor spreading, these variations also affected climate and oceanic circulation.

Among the more dramatic recent changes in the ocean was a series of glacial advances and withdrawals. In the last 18,000 years, sea level has risen about 130 m (426 ft.) and has inundated the previously exposed continental shelves. If all the remaining ice were to melt (perhaps due to global warming from an increasing greenhouse effect), sea level could rise an additional 60 m (about 200 ft.).

The ocean has a volume of 1,350 million km³ (318 million miles³) and an average depth of about 3,800 m (12,464 ft.). Oceans dominate the Southern Hemisphere, while land dominates the Northern Hemisphere. About 71 percent of Earth is covered by water.

QUESTIONS

1. What important evidence must be considered in a hypothesis for the origin of the Solar System?

2. What important evidence must be considered in a hypothesis for the origin of the Earth?

3. What role did volcanic activity play in the formation of continents, of the oceans, and of the atmosphere? Are any of these processes still active?

4. What are the main provinces and subcomponents of the seafloor?

5. What makes Earth unique within our Solar System?

6. Where did the water in the ocean come from, and where did the cations and anions in the oceans come from?

7. Describe the different stages in the evolution of Earth's atmosphere.

8. Discuss how environmental conditions on planet Earth have varied over geological time. What have been some of the major changes?

9. What was unique about the Pleistocene epoch? Could such conditions occur again?

KEY TERMS

anion	gas giant	mantle	Pleistocene epoch
cation	geologic time scale	marginal sea	photosynthesis
Celsius	gravity	micron	ocean basin
condensation hypothesis	greenhouse effect	mid-ocean ridge	scientific law
continental margin	Holocene epoch	Milky Way Galaxy	Solar System
core	hypothesis	Pangaea	terrestrial planet
crustal layer	hypsographic curve	Precambrian	theory
differentiation	light-year	planetesimal	Universe
extinction	magma		

FURTHER READING

General

Broecker, W. S., and G. H. Denton. 1990. "What Drives Glacial Cycles." *Scientific American* 262, no. 1, pp. 49–56. A general article on a new hypothesis concerning the advance and retreat of ice sheets.

Kaufmann, W. J., III. 1993. *Discovering the Universe.* 3d ed. New York: W. H. Freeman and Co. Solid textbook that will tell you everything you want to know about astronomy.

National Geographic Society. 1988. *Exploring Our Living Planet.* Washington, D.C. A general, well-illustrated book on the history and development of the Earth.

Peterson, Ivers. 1991. "State of the Universe: If Not with a Big Bang then What?" *Science News* 139, no. 15, pp. 232–35. Nice, readable summary of the various hypotheses concerning the origin of the Universe.

Stewart, Keith. 1991. *Living Fossil: The Story of the Coelacanth,* New York, W. W. Norton and Co. The story of the coelacanth, a fish thought to be extinct for hundreds of millions of years, that was discovered 50 years ago off South Africa.

On the Extinction Debate

Alvarez, L. W. July 1987. "Mass Extinctions Caused by Large Bolide Impacts." *Physics Today,* pp. 24–33. The straight story of the impact hypothesis from the scientist who conceived it.

French, B. M. 1990. "25 Years of the Impact-Volcanic Controversy," *EOS* 71, no. 17, pp. 411–14. Discusses the pros and cons of the impact hypothesis versus volcanic hypothesis.

Gore, Rick. 1989. "Extinctions." *National Geographic* 175, no. 6, pp. 662–99. A very reader-friendly discussion of the various extinction hypotheses.

Murphy, J. B., and R. D. Nance. 1992. "Mountain Belts and the Supercontinent Cycle." *Scientific American* 266, no. 4, pp. 84–91. Authors propose that supercontinents, like Pangaea, form every 500 million years or so, and then break apart.

Officer, C. B., and C. Drake. March 1985. "Terminal Cretaceous Environmental Events." *Science* 227, pp. 1161–66. An alternative view that this mass extinction resulted over considerable time from volcanic activity.

Patrusky, B. 1987. "Mass Extinctions: Volcanic, or Extraterrestrial Causes, or Both?" *Oceanus* 30, no. 3, pp. 40–48. A good general discussion of the various extinction hypotheses.

Sammon, Rick. 1992. *The Seven Underwater Wonders of the World.* Charlottesville, Va.: Thomasson-Grant. A well-written and beautifully illustrated work on seven underwater wonders, including the Great Barrier Reef, deep-ocean vents, and the northern Red Sea.

York, D. 1993. "The Earliest History of the Earth." *Scientific American* 268, no. 1, pp. 90–96. Describes how radioactive techniques are used to learn more about Earth's first 1.5 billion years.

CHAPTER 3

Reshaping the Ocean: Plate Tectonics

IT HAS BEEN YEARS SINCE THE CONCEPTS OF SEAFLOOR SPREADING AND PLATE TECTONICS WERE INTRODUCED IN THE MID-1960s, YET THESE STILL ARE SOME OF THE MOST EXCITING IDEAS IN MARINE SCIENCE. SIMPLE TO UNDERSTAND, THESE CONCEPTS EXPLAIN HOW MOST OF THE MAJOR GEOLOGIC AND GEOPHYSICAL FEATURES OF EARTH'S CRUST FORMED AND MOVED OVER GEOLOGIC TIME. MORE RECENTLY, EMPHASIS HAS SHIFTED FROM SIMPLY DESCRIBING THE MOTION OF EARTH'S CRUST TOWARD UNDERSTANDING THE COMPLEX PROCESSES THAT CAUSE THE UPPER LAYERS OF EARTH TO EVOLVE. INTERESTINGLY, ONCE IT WAS DISCOVERED, THE PROCESS OF SEAFLOOR SPREADING WAS RELATIVELY OBVIOUS, LIKE SUDDENLY FINDING A HIDDEN OBJECT IN A PICTURE THAT YOU HAVE BEEN STUDYING FOR A LONG TIME.

Lava lake within a volcanic crater, photographed during the May 1989 eruption of Kilauea volcano on Hawaii. Lava is about 100 m (330 ft.) from the top of the crater and is flowing from the far side and exiting the crater through tubes beneath photographer. Crater diameter is about 500 m (1,640 ft.).

(Photograph courtesy of Wilfred B. Bryan, Woods Hole Oceanographic Institution.)

Early Understanding of Earth's Structure

Accurate information about the structure of Earth was known early in this century. For example, scientists understood that Earth could be divided into three distinct layers: a thin **crust,** a thick **mantle,** and a dense **core** (Figure 3-1). The core itself was understood to have two parts, an outer liquid core and a solid inner core. These layers were formed by the differentiation processes active in the formation of early Earth (see Chapter 2 and Figure 2-2, page 26).

Much of the knowledge about Earth's internal structure came from analysis of **earthquakes,** including patterns of where they occur and the energy they produce (see Box 3-1). For example, analysis of the vibrations produced by earthquakes showed that a distinct boundary or difference exists between the crust and mantle. Named after its discoverer, Croatian geophysicist Andrija Mohorovičić, this boundary is called the Mohorovičić discontinuity, or more simply, the **Moho.**

It also was fairly well understood early in this century that Earth's crust had formed from mantle material by sequences of volcanic eruptions followed by the processes of weathering and erosion. But several very important questions remained. Why was the crust thicker under continents, averaging about 35 km (22 miles), compared to crust

beneath the oceans, which is 5–10 km thick (3–6 miles)? If the oceans are simply sunken portions of the continent, or if the continents are simply elevated portions of the seafloor, why does their crustal structure differ so much?

Another question came from the observation that fossils of marine origin and ancient beach deposits are found high above sea level within the rocks that form mountain ranges. How did they get there? This surely implied that parts of Earth have undergone some very major movement.

Still other important observations concerned the linearity and location of mountain ranges, and the nearly matching shapes, like puzzle pieces, of opposite parts of the Atlantic coastline. These observations and questions eventually led to an early hypothesis, called continental drift, that attempted to address these points.

Continental Drift: A Hypothesis to Explain Earth's Crust

Some eighteenth-century geologists, noting the long, linear trends of many mountain ranges and the thick, folded sequence of sediments that they contained, suggested that the Earth was contracting. They proposed

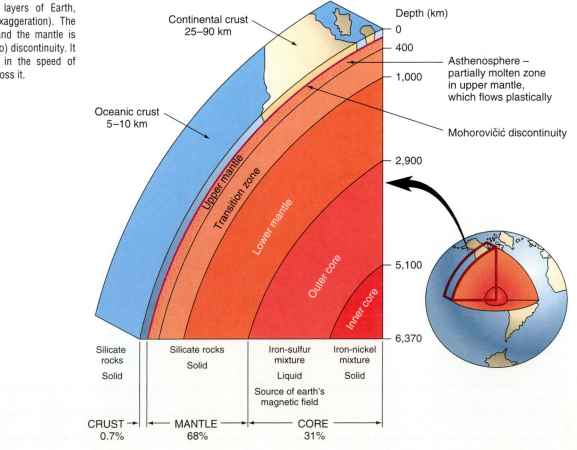

FIGURE 3-1 The different layers of Earth, drawn to scale (no vertical exaggeration). The boundary between the crust and the mantle is called the Mohorovičić (or Moho) discontinuity. It is distinguished by a change in the speed of earthquake vibrations going across it.

Continental crust 25–90 km

Oceanic crust 5–10 km

Depth (km)
0
400
1,000

Asthenosphere – partially molten zone in upper mantle, which flows plastically

Mohorovičić discontinuity

2,900

5,100

6,370

Upper mantle

Transition zone

Lower mantle

Outer core

Inner core

Silicate rocks	Silicate rocks	Iron-sulfur mixture	Iron-nickel mixture
Solid	Solid	Liquid	Solid

Source of earth's magnetic field

CRUST 0.7% — MANTLE 68% — CORE 31%

Box 3-1

Earthquakes and the Internal Structure of Earth

The interior of Earth is essentially inaccessible to direct observation. The deepest wells rarely exceed 10 km (6 miles) and thus cannot pass through the crust to deeper zones. Volcanoes are our only "window" into Earth's interior. It is only by the analysis of volcanic activity and study of the resulting rock can we actually see direct evidence of Earth's internal composition.

Fortunately, there are indirect ways to learn about Earth's interior. One especially valuable method is by studying the energy released by earthquakes, which result from the sudden release of built-up stresses within Earth. This release produces a vibration that can last from a few seconds to a minute or more. The vibration motion can cause considerable damage and loss of life. The actual subsurface location where an earthquake originates is called the **focus,** and the position on the surface of Earth directly above the focus is called the **epicenter.**

Earthquake vibrations, called *seismic waves,* are detected using a measuring device called a *seismograph.* Three distinct types of waves are produced by an earthquake:

- An **L wave** travels only on the surface and can be compared to waves on a lake or ocean surface. L waves travel out in all directions from an earthquake epicenter, just like waves travel out in rings from a splash in a pond.
- A **P (primary) wave** has a motion similar to that of a coil spring being alternately compressed and stretched out. Thus, the P wave oscillates (compresses and stretches) in the same direction in which the wave is traveling. This is the type of wave that does most of the damage in an earthquake.
- An **S (secondary) wave** has a motion similar to that of a rope being shaken side-to-side. The S wave oscillates (moves side-to-side) in a direction perpendicular to that of its travel direction (Figure 1a).

Both P and S waves travel through Earth (Figure 1b), but the P wave travels twice as fast as the S wave. Because of this difference in speed, P waves reach seismographs before the S waves. By measuring the time difference between the arrival of these two waves, it is possible to determine the distance to the epicenter. By measuring this time difference at several different seismograph stations, and comparing the distance of each from the earthquake, it is possible to locate the epicenter of the earthquake.

When seismic waves were first being measured, scientists observed that stations on the other side of Earth from an earthquake epicenter received only P waves and no S waves. Because the side-to-side oscillation of

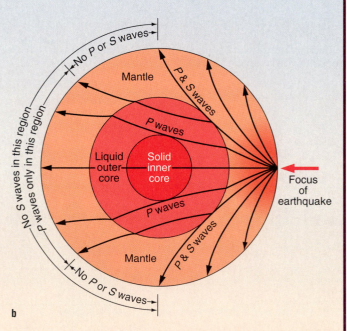

(b) The paths that P and S waves follow away from an earthquake. Note the *shadow zone,* where neither wave type penetrates. Also note that S waves cannot penetrate the outer liquid core. By measuring the time of arrival and pattern of earthquake waves at various locations around Earth, the location of an earthquake can be determined. Likewise, the study of these patterns has been used to deduce the internal structure of the Earth.

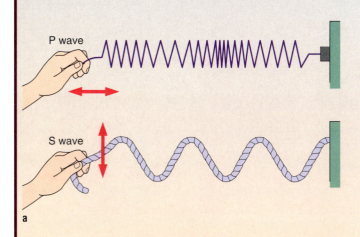

FIGURE 1   (a) Earthquakes produce two types of waves that travel through the Earth. P, or primary, waves are compressional and are similar to pushing and pulling on a spring. S, or secondary, waves have a side-to-side motion similar to shaking a rope.

that, as Earth contracted, its outer skin or crust became wrinkled like an old apple, and in this manner the existing sediments became folded into mountain ranges.

Later, geologists discovered rift valleys, both on the ocean floor (especially running along ridges like the Mid-Atlantic Ridge) and on land (in East Africa, for example). They argued that only an *expanding* Earth, creating tension on the crust and thus pulling it apart, could form these features.

But there was a larger question than who was right or wrong: How could *both* types of features—compressional and tensional—be present?

Another early observation of scientists was the similar shape of the coastlines on either side of the Atlantic. This is especially notable for western Africa and eastern South America. This apparent fit led some to suggest that these two continents once were connected and that they had split apart some time in the geologic past. The movement would have left a depression that filled with water to form the Atlantic Ocean, and correspondingly reduced the size of the Pacific Ocean.

The Continental Drift Hypothesis

The **continental drift** hypothesis probably was first advanced by German naturalist Alexander von Humboldt in the early 1800s, with subsequent modification by others. But its main advocate was German geophysicist Alfred Wegener in the 1920s. The hypothesis argued that the continents essentially float and move around on the deeper and heavier subcrustal or mantle material (Figure 3-2). The notion that continents and oceans are "floating" on a denser mantle comes from a larger idea called **isostasy** (see Box 3-2).

Later refinements to the continental drift hypothesis were that *convection currents* deep within Earth's mantle slowly stirred the hot, plastic rock (Figure 3-3). **Convection** refers to motion in a fluid due to differences in its density or temperature, like the motion observed when heating a pot of soup (Figure 3-3a). In Earth, convection currents are driven by the uneven heating of Earth's mantle, which is hotter near the core and cooler near the thin crust. Thus, convection circulation cells developed (Figure 3-3b).

These slow movements are the driving force of continental drift, dragging the lighter crust (continents) along with the moving mantle and concentrating them over the downward-moving part of the convection cells. Ocean ridges would then form at the upward-moving part of the convection cell. The pulling-apart or tensional force split the surface, and magma welled up, building the ridge. On a smaller scale, similar

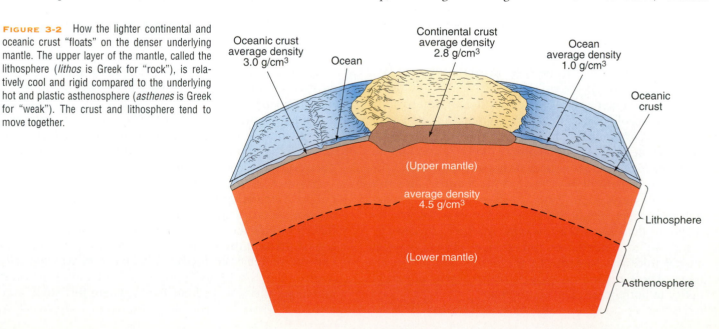

FIGURE 3-2 How the lighter continental and oceanic crust "floats" on the denser underlying mantle. The upper layer of the mantle, called the lithosphere (*lithos* is Greek for "rock"), is relatively cool and rigid compared to the underlying hot and plastic asthenosphere (*asthenes* is Greek for "weak"). The crust and lithosphere tend to move together.

Oceanic crust average density 3.0 g/cm^3

Continental crust average density 2.8 g/cm^3

Ocean

Ocean average density 1.0 g/cm^3

Oceanic crust

(Upper mantle)

average density 4.5 g/cm^3

(Lower mantle)

Lithosphere

Asthenosphere

BOX 3-2

Isostasy

Why does an inflated beach ball float higher in the water than a person? Density and gravity. Gravity attracts denser objects more, and less-dense objects less. Exactly the same phenomenon occurs when continental crust and oceanic crust float on Earth's denser mantle.

Lighter continental crust floats higher than the denser oceanic crust on the slightly fluid mantle. The concept is similar to that of icebergs floating in seawater. The icebergs are less dense than the seawater and individual blocks of ice will sink until they displace an amount of water equal to their weight. The remaining part of the iceberg floats above the seawater. Bigger icebergs are both higher above the water surface and deeper below it than smaller icebergs (see the figure to understand this surprising aspect). This is similar to

mountain ranges on land: the higher the mountain, the deeper its root. This concept of oceanic and continental blocks floating on the denser mantle is called **isostasy** (Figure 1). Isostasy comes from the Greek words *isos,* meaning "equal," and *statis,* meaning "standing."

If a large weight is added to land, an isostatic adjustment occurs, depressing the land in the region. If the weight is later removed, the land slowly rises back to its original height. Widespread glaciation during the Pleistocene epoch provides an excellent example of an added weight. As glaciers spread and thickened, the rock beneath them was depressed many meters by the weight. As the glaciers melted and retreated, removing the weight, the land slowly began rising (called *glacial rebound*). This is happening today in northern parts of Canada and Europe.

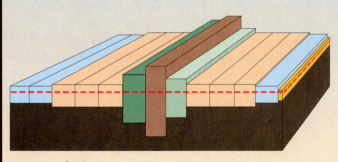

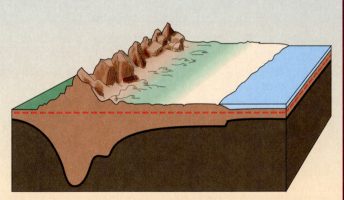

FIGURE 1   Isostasy.
(a) With blocks of similar density, the taller ones float higher and extend deeper. This also applies to continental blocks and icebergs, as discussed in the text.

(b) The concept of isostasy also applies to mountain ranges and explains why high mountains have deep roots.

---

convection currents can be seen in some types of lava flows (Figure 3-3c).

When Wegener proposed the continental drift hypothesis, some geological observations supported it, and some did not. Many scientists argued that the force needed to move whole continents was simply too large and that the evidence for drifting was poor or nonexistent. For example, if a continent plowed around the mantle like a ship moving through ocean water, fresh seafloor without any sediment cover should be left behind in its wake, and the seafloor obviously is not this way. Another objection focused on the differences, mentioned earlier, between crustal thickness underneath the oceans and the continents. These differences suggest some degree of permanency to both oceans and continents.

The hypothesis that Earth is expanding suggested that at one time Earth was covered with a thick continental crust. Expansion split it along weak areas that now are occupied by the ocean basins. This sounded good, but one problem is that the ocean basins should then contain considerably more sediment than they do. (This is discussed further on page 116.)

Another objection to continental drift was that it required dramatic movement of the continents (thousands of kilometers), which seemed unreasonable considering that no evidence of such displacement had been observed. In fact, the evidence was there, but no one had yet interpreted it correctly. This objection was partially overcome in the 1950s by studies off the California coast and along the San Andreas Fault, where just such large

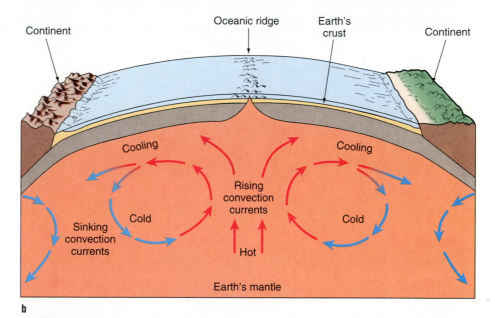

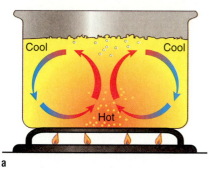

FIGURE 3-3 Convection current cells.
(a) The basic concept of convection. Heat from below, in this instance applied to a pot of soup, causes warming and expansion of the soup, which thus becomes less dense. The warmer, less-dense soup rises by convection and starts to cool, increasing density once again. With the increase in density, the soup starts to sink. The entire process creates a convection cell similar to what happens in Earth's mantle or in a lava lake; see b and c.
(b) General model of convection cells operating within Earth's mantle. Notice that the continents are over the downward, or sinking, part of the convection cell and the oceanic ridges are over the upward-moving part of the cell.
(c) Convection-induced subduction (sinking) and rifting (spreading) occurring in the surface crust of a lava lake formed during a 1989 Kilauea eruption (see also opening photo to this chapter). The dark line to the left of center is the sinking area and the red area (lava) to the right of center is the upwelling or rifting region. (Photograph courtesy of Wilfred B. Bryan, Woods Hole Oceanographic Institution.)

displacements have occurred.

Proponents of the continental drift hypothesis often differed over details, but they usually concurred that the continents were once joined in the single supercontinent **Pangaea,** mentioned in Chapter 2. They explained that Pangaea started to split apart about 200 million years ago. Indeed, when the position of the continents at that time is reconstructed (Figure 3-4), many large features of Earth's surface seem to fit together, and some features even extend across several continents (see Box 3-3).

The main difficulty with the continental drift hypothesis is that the structure and character of the continents and oceans are so different. Unless one assumes an expanding Earth, for which there is no evidence, it is very difficult to see how the two types of crust could evolve or change into one another. The central portions of most continents are quite old (rocks nearly 4 billion years old have been found). Further, they are composed of sediment and rock that have been folded, deformed, melted, and remelted. These rocks are a very complex mixture of types resulting from intricate processes, and this makes understanding them very difficult. In contrast, the oceans have much younger rocks and sediments (rarely older than 200 million years). They have undergone comparatively little deformation.

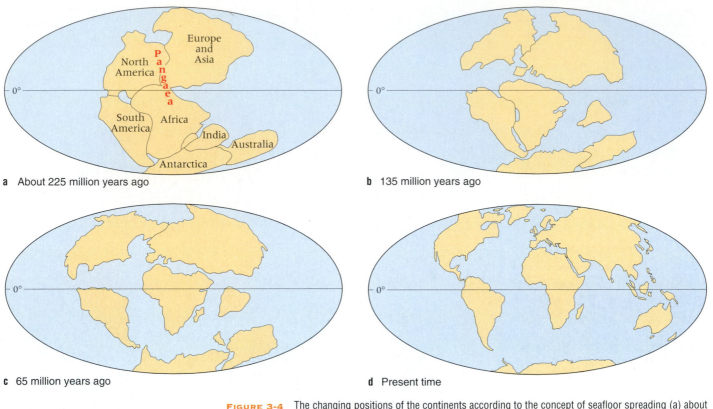

a About 225 million years ago

b 135 million years ago

c 65 million years ago

d Present time

FIGURE 3-4 The changing positions of the continents according to the concept of seafloor spreading (a) about 225 million years ago, (b) about 135 million years ago, (c) about 65 million years ago, and (d) present time. Note the long trip that India has taken.

The most critical question was: Why does one portion of the Earth—the continents—have a thick crust (average thickness 35 km, or 22 miles), while such a thin crust exists over the remainder (the crust under the ocean basins is generally 5–10 km thick)? The answer was to be found in the ocean. Here, the erosional and depositional processes that are often so dramatic and variable on land are relatively constant and continuous. Thus, it is not surprising that the rocks and sediments in the ocean would yield the important evidence needed to understand the origin and evolution of major features on Earth: seafloor spreading and plate tectonics.

Seafloor Spreading and Plate Tectonics: A Theory to Explain Earth's Crust

The new hypothesis, called **seafloor spreading,** was first proposed by Harry Hess and Robert Dietz in separate articles published in the early 1960s. The hypothesis incorporated several older ideas and was able to overcome most objections to continental drift. Seafloor spreading has considerable supporting data and can be tested—something that was difficult to do with earlier hypotheses. In fact, seafloor spreading is now generally accepted by marine scientists as a workable *theory*.

The fundamental concept is that the seafloor *itself* is moving, with the motivating force being convection currents. The **plate tectonics** concept comes from a later refinement that showed how the outer portion of Earth is actually composed of a series of rigid plates that all move relative to one another. **Tectonics** is a general term for the study of the construction and deformation of Earth's structure. It comes from the Greek word *tektonikos,* which means "to construct."

In seafloor spreading, the spreading movement starts at the **ocean ridges** and continues away from them in opposite directions. The resulting gap at the ridge is filled with new seafloor, formed from lava that wells up from deep within the Earth. (Revisit Figure 2-4, page 31, to see this process; Figure 3-5 shows some of the types of rock formed by volcanic activity.)

This process is analogous to two conveyor belts moving away from the ocean ridge, toward the continents that flank the ocean. Eventually they either move under the continents or carry the continents along with them (Figure 3-6). The oceanic ridge therefore is an area of crustal upwelling, or **divergence,** where new volcanic

How We Know the Continents Were Once Joined Together

It is a fascinating idea that the continents were joined together over two hundred million years ago as one supercontinent called Pangaea, and then split apart and moved to their present positions. But what data support the idea? Certainly one piece of evidence is the fit of coastlines when the continents are rejoined. But there should also be some more specific geologic evidence. For example, the fossil record of ancient plants and animals should show strong similarities in areas that once were connected. Likewise, the patterns of rock types and structure should be similar.

In fact, substantial geological evidence exists to support the idea that the continents were once attached. Numerous fossils of land species have been found across the continents in rocks that formed while the continents were attached (Figure 1). Likewise, few or no such similarities are found in the rocks that formed after the continents separated. (Marine species do not work well in this comparison because they can migrate via the ocean from one continent to another after the continents separated.)

Similarities in rock type also occur. The best example is the distribution of glacial deposits that occurred prior to separation. These glacial deposits align around what would have been the South Pole at the time the continents were joined. If the continents were not joined, one is left with the hard-to-explain fact that India and parts of Africa were glaciated when they were situated in tropical climates. Another piece of evidence is the coal deposits that extend across the boundaries when the continents are refitted.

The evidence for Pangaea and its separation about 200 million years ago is therefore very solid.

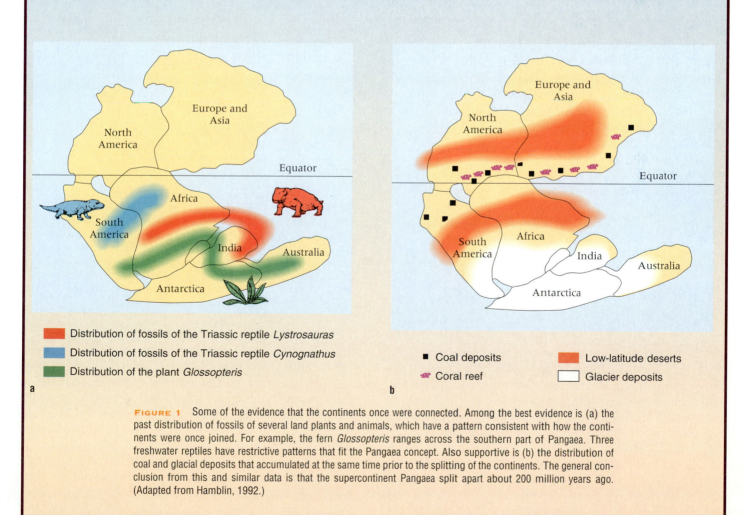

Distribution of fossils of the Triassic reptile *Lystrosauras*
Distribution of fossils of the Triassic reptile *Cynognathus*
Distribution of the plant *Glossopteris*

■ Coal deposits   Low-latitude deserts
Coral reef   Glacier deposits

a   b

FIGURE 1   Some of the evidence that the continents once were connected. Among the best evidence is (a) the past distribution of fossils of several land plants and animals, which have a pattern consistent with how the continents were once joined. For example, the fern *Glossopteris* ranges across the southern part of Pangaea. Three freshwater reptiles have restrictive patterns that fit the Pangaea concept. Also supportive is (b) the distribution of coal and glacial deposits that accumulated at the same time prior to the splitting of the continents. The general conclusion from this and similar data is that the supercontinent Pangaea split apart about 200 million years ago. (Adapted from Hamblin, 1992.)

FIGURE 3-5 Pillow lava is a rock produced by recent underwater volcanic activity. As the lava cools from the outside, the molten lava inside pushes outward on the cooling crust, giving it a pillowlike shape. The equipment in the foreground is part of the research submersible, *Alvin*. Area shown is about 5 m (16 ft.) wide. (Photograph courtesy of Wilfred B. Bryan, Woods Hole Oceanographic Institution.)

material is continually added from the deeper parts of Earth's crust and mantle. The relatively frequent seismic activity at the oceanic ridges (see the Plate Tectonics and Earthquakes section, page 54) is a consequence of this process.

Plenty of Evidence

The hypothesis of seafloor spreading is strongly supported by magnetic observations made across the seafloor. These show that long, linear **magnetic patterns** tend to parallel the ocean ridges. What is more, these patterns, when followed outward from the ridge at right angles, show *matching* **magnetic reversals** of polarity on both sides of the ridge, like mirror images (Figure 3-6).

Earth's magnetic field is known to have reversed polarity many times in our planet's history. The *geographic* North and South Poles have of course remained in place, but the *magnetic* north and south poles have changed polarity continually. Marine geophysicists interpreted the magnetic seafloor patterns as indicating that the rock either solidified during a time when Earth's magnetic field was like it is today (a "positive" value) or solidified at another time, when the field was reversed (a "negative" value). In a way, the formation of these magnetic patterns in rocks is similar to a tape recording: the changes in Earth's magnetic field are the signal, which is being recorded on two very slowly moving "tapes" that spread in opposite directions from the ridge.

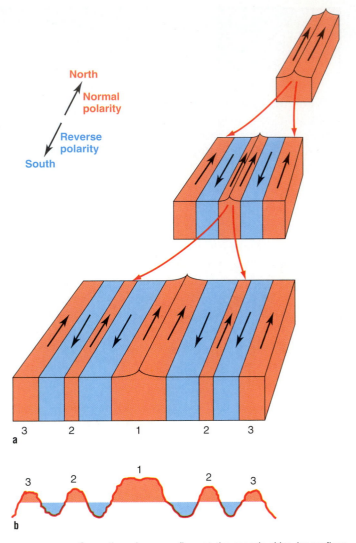

FIGURE 3-6 Generation of new seafloor at the oceanic ridge by seafloor spreading. At the ocean ridge the rising magma solidifies into new seafloor and becomes magnetized with the polarity of Earth's current magnetic field (north-south or south-north, just like a bar magnet). This locks in a "permanent record" of how Earth's magnetic field was oriented at the time. Because the magnetic field reverses from time to time, the resulting magnetic patterns can be used to estimate the rate of seafloor spreading. Red areas indicate magnetization that has the same polarity as the present ("positive"), and blue areas indicate the opposite magnetization ("negative" or reversed).
(b) Hypothetical magnetic field that would be generated by the scenario shown in (a). The numbers refer to specific positive periods of magnetization. See also Figure 3-7.

Thus, according to the seafloor-spreading hypothesis, slow convection currents within Earth's mantle gradually transport molten rock (magma), which contains magnetic minerals, to the surface, where it erupts from undersea volcanoes to form the ocean ridges (see Box 3-4). As this material cools in the deep ocean water, it starts to solidify, at which time the magnetic minerals behave like a compass and align themselves parallel to Earth's magnetic field. Material cooling and solidifying at the present time is oriented to align with the present magnetic field and we

Box 3-4

A Little Geology

Geologists classify rocks into three main categories:

1. **Igneous rocks** (*ignis* is the Latin word for "fire") are the result of magma cooling and solidifying, either above ground or beneath the surface. When the magma solidifies at Earth's surface the rocks formed are called *volcanic rocks,* or more technically, *extrusive igneous* *rocks.* Basalt is an example of an extrusive igneous rock. However, if the magma cools and solidifies below the surface, the rocks are called *intrusive igneous rocks;* granite is an example. When the magma flows on the surface it is called *lava.* Intrusive and extrusive igneous rocks constitute about 95 percent of all the rocks in the upper 10 km of Earth.

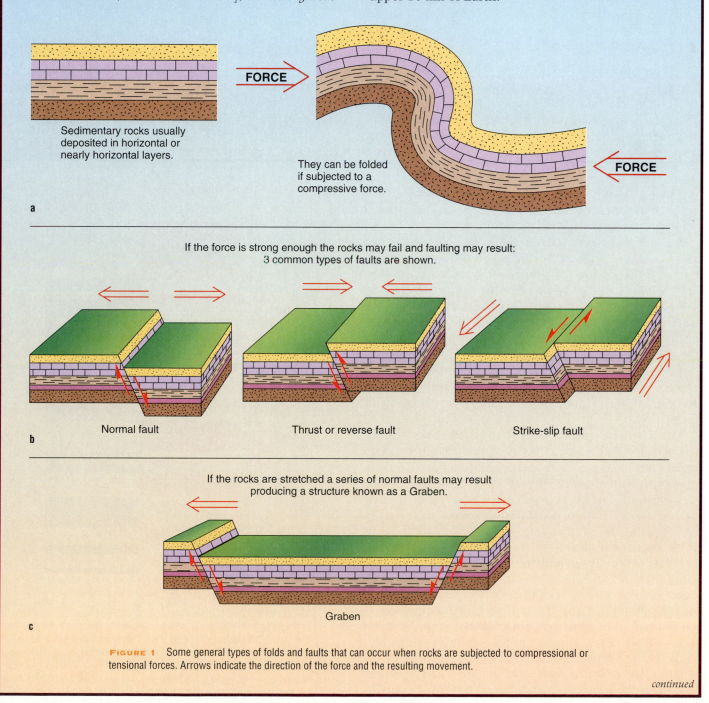

FIGURE 1   Some general types of folds and faults that can occur when rocks are subjected to compressional or tensional forces. Arrows indicate the direction of the force and the resulting movement.

continued

Box 3-4 continued

2. **Sedimentary rocks** (*sedimentum* is the Latin word for "settling") are composed of particles of pre-existing rocks. These particles are carried by water, wind, or ice to localities where the particles (sediments) settle out and form a sedimentary deposit. If such a deposit becomes cemented together, it becomes a sedimentary rock. Sandstones and shales are examples of such rocks. Sedimentary rocks can also form by chemical precipitation of particles; limestones are an example. Sedimentary rocks generally have a distinctive layering pattern, frequently horizontal or gently sloping layers that parallel the slope of the surface on which they were deposited.

3. **Metamorphic rocks** (in Greek, *meta* means "change" and *morphe* means "form") are existing rocks that have been changed in form—altered by physical or chemical changes caused by high pressure, heat, or contact with fluids such as water. Examples are slate

(metamorphosed shale) and marble (metamorphosed limestone).

When rocks are subjected to compressional or tensional (pulling apart) forces, several things can happen. With compression (but not sufficient compression to cause fracturing), rocks frequently undergo bending or **folding** (Figure 1). Folds can range from microscopic to hundreds of kilometers long.

If the force is sufficient, the rocks may actually break, with one side of the break moving relative to the other side. The resulting fracture is called a **fault.** Depending on the direction of the force, three types of faults can occur, illustrated in Figure 1b: normal, reverse (thrust), or strike-slip faults. (Another type of fault, common near ocean ridges and called a transform fault, is described later.) If tensional forces cause the rocks to be pulled apart, a rift or *graben* (German for "ditch") may develop.

---

say it has a "positive" or "normal" pattern. Rocks solidified in the geologic past when the poles were reversed have a "negative" or "reversed" pattern (Figure 3-7).

Because the rising basaltic material is slowly carried away from the ridge in conveyor-belt fashion, measurements of the magnetic pattern will show an alternating pattern of normal and reversed values. The patterns observed also show a high degree of symmetry around the ridge area: the pattern on one side is the reverse image of that on the other side. This is to be expected, because the material is moving out and away from either side of the ridge (see Figure 3-6). Note here that we are talking about the *volcanic rock,* not the overlying sediments that are continuously deposited atop the now-solid volcanic rock.

The magnetic patterns around the ocean ridges can be correlated with magnetic reversals observed in rocks on land. These rocks have been dated by radioactive techniques, so we know their actual ages (Figure 3-7). When a correlation is possible, a specific age can be ascribed to various parts of the seafloor pattern. From this, we can estimate the rate of spreading.

This procedure has been used in many areas, revealing that a typical **spreading rate** is about 1 to 5 cm (about 0.4 to 2 in.) per year. (This is actually an average rate, for the motion probably is not continuous.) A spreading rate of 1 cm (0.4 in.) per year is equivalent to the formation of 10 km (about 6 miles) of new seafloor in a million years. Such a rate of motion is compatible with the idea that the continents on either side of the Atlantic were joined over 200 million years ago, and since that time the continents have been slowly moving apart, forming the Atlantic Ocean in the process (see

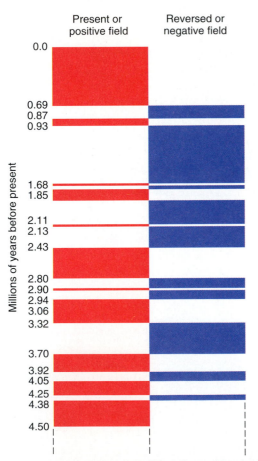

FIGURE 3-7 The geomagnetic time scale. This scale was determined by measuring the age of reversed and positive (or present) anomaly patterns in igneous rocks on land, and then extrapolating to patterns observed in ocean rocks along the seafloor (see Figure 3-6). The red areas indicate times when Earth's poles were in their "normal" (present-day) positive position; the blue areas indicate when the poles were reversed.

Figure 3-4). In other words, if South America and Africa spread apart at a steady rate of 2.5 cm/yr., in 200 million years this would spread them 5,000 km apart (about 3,100 miles), the actual distance today.

The rising volcanic rock at the oceanic ridge is hot and causes an upward expansion of the ridge area. However, as the rocks move away from the ridge axis, they gradually cool and contract, causing the seafloor to sink. This phenomenon explains why the ridge areas are so much higher compared to the adjacent regions.

There are numerous places along the oceanic ridge where there is a change or offset in the location of the center of spreading. The offset creates a **fracture zone** that has a **transform fault** situated between the active spreading areas of the ridge (Figure 3-8). Fracture zones can form long cliffs or escarpments that extend over 1,000 km (over 600 miles), some even reaching into the continents (see Figures 4-17, 4-19, and 4-20, pages 93, 95, and 97).

Plate Tectonics

Recent ideas concerning seafloor spreading and the origin and evolution of the ocean basins are incorporated into an even more encompassing concept called **plate tectonics.** In plate tectonics theory, Earth's entire crust is composed of about a dozen large plates, up to 160 km (100 miles) thick. These plates make up the **lithosphere.** Each moves essentially as a rigid block over Earth's surface. The lithosphere rests upon the less rigid part of Earth called the **asthenosphere** (see Figures 3-1 and 3-2).

The plates can be visualized as curved caps covering a large ball, or perhaps as sections of peel on an orange. Boundaries of the individual plates are usually areas of high seismic (earthquake) activity. Deformation, such as faulting, folding, or shearing (see Box 3-4) mainly occurs at the boundaries between the plates. Because the surface area of Earth is essentially constant, spreading in one place must be balanced by sinking or **subduction** somewhere else.

The appearance of the edge of a continent is often influenced by its relation to areas of seafloor spreading or subduction. If a continent's edge is at the boundary of a subducting oceanic plate, a trench will form, and sometimes an **island arc** or coastal mountain range (Figure 3-9a). A **trench** is a deep, long, narrow depression in the seafloor, with very steep slopes. An island arc is a curved pattern of mainly volcanic islands and a trench. It is generally curved in a convex direction (bows out) toward the ocean, with the trench on the seaward side. The Aleutian Islands and Aleutian Trench in the north Pacific is a good example of a trench–island arc system (see Figure 4-19, page 95).

The subducting plate forms the trench. In the process of submerging beneath the other plate, much of its sediment is scraped off onto the landward side of the trench. This material and the subducting plate may be reheated, perhaps even melted, eventually becoming part of the continental plate and forming a coastal mountain range. These are events that occur when oceanic crust is subducted beneath lighter continental crust. A similar sequence occurs during the collision of two ocean plates.

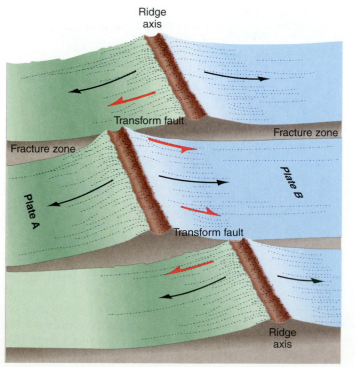

a

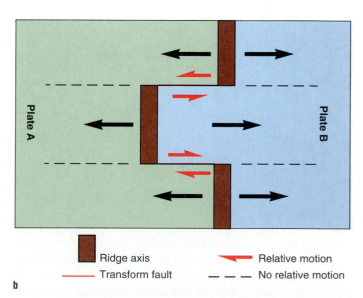

b

FIGURE 3-8 Transform faults occur in the region where ridge axes are offset and where relative motion occurs between plates. Figure (a) is a perspective view, Figure (b) is a plan (overhead) view. Transform faults are generally sites of considerable earthquake activity because of the opposing motion between the two plates. Fracture zones extend beyond the active parts of the transform fault. However, the part of the fracture zone beyond the transform fault lacks any relative or opposed motion; thus, no earthquake activity is generated.

FIGURE 3-9 Possible types of collision of Earth's crustal plates, caused by seafloor spreading.

(a) An *oceanic/continental* plate collision, resulting in a subduction or consumption zone. Denser oceanic crust is subducted under less-dense continental crust. Marine sediments are frequently scraped from the seafloor and acreted onto the continent as the oceanic plate is subducted. This type of collision is typical along the west coast of South America.

(b) An *oceanic/oceanic* plate collision, typical along western parts of the Pacific Ocean. Usually one of the plates is the oceanic edge to a continental crust. When this type of collision occurs, the result is a typical island-arc system with a marginal or inland sea.

(c) A *continental/continental* plate collision. In this case, because the plates are of the same density, one cannot readily slide beneath the other. Instead, like two trucks colliding in slow motion, rocks will crumple and fold, intense faulting will occur, and a mountain range may result. The Appalachian, Himalayan, Alps, and Zagros mountains (in Iran) were formed by this type of collision (see also Figure 3-20b). In all types of collision, earthquake activity is greater where the plates are in contact (see Figure 3-12).

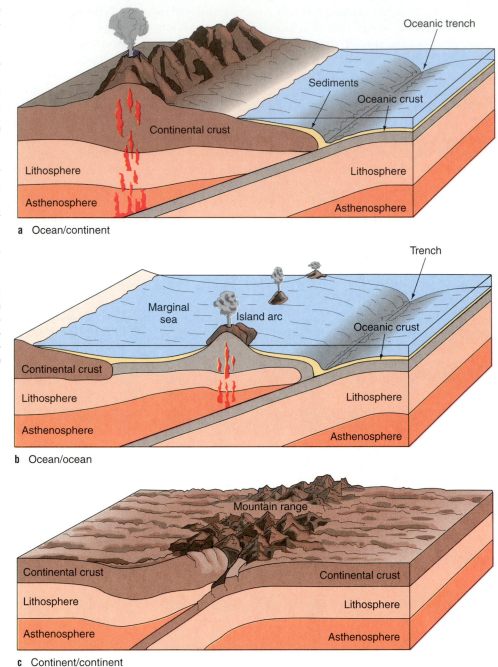

a   Ocean/continent

b   Ocean/ocean

c   Continent/continent

The main difference is that an inland or marginal sea may form behind an island arc (Figure 3-9b).

Results can be quite different when two plates of continental composition collide. In this case, both plates are of similar density, so neither easily overrides the other. The result is unusually intense folding and faulting of the sediments and rocks, often leading to the formation of a major mountain range (Figure 3-9c). The Himalayan Mountains are still being formed in this manner by the collision of India (continental crust) with Asia (continental crust).

An important intellectual advantage of the hypothesis of seafloor spreading is that marine scientists are no longer faced with the embarrassing problem of making a continental area become an oceanic one. Instead, new seafloor is constantly being produced at the ocean ridges, and old seafloor is disappearing at areas of subduction (see Figure 3-11b). Continents can grow at their edges by the accumulation of material (as with the formation of coastal mountain ranges). Continents are not "destroyed" and can "survive" with their central parts being their oldest areas.

Dating the Seafloor

As the "conveyor belt" carries older marine sediments beneath the continents, or accretes it onto the continents, the sediment remaining in the ocean is only what has been deposited within the last 150 million to 200 million years. As the process of seafloor spreading continues, a slow, continuous rain of shells of microscopic organisms and clay particles falls from surface waters to the depths. These materials accumulate slowly, at only a few centimeters per 1,000 years. Where the seafloor is new, the overlying sediment is relatively thin; it increases in thickness toward older areas (they have existed longer to receive more sediment). In this manner, sediment thickens away from the spreading region, away from the oceanic ridge.

Sometimes the initial sediment that accumulates on new seafloor just after it forms can be dated, using radioactive-isotopic techniques or by knowing the ages of the fossil shells in the sediment. From this the age of that seafloor's formation can be approximated. This age should closely match that indicated by the magnetic method of seafloor dating (Figures 3-6 and 3-7).

Because the youngest part of the ocean basin is the ridge area, this is also the area where we expect the thinnest sediment. Studies of sediment thickness have indeed found relatively small quantities of sediment in the ridge area and have determined that this thickness and age increases as we move away from the ridge. Likewise, the age of the underlying rock becomes older as we move outward away from the ridge axis (Figure 3-10). The discovery that the age of the underlying rock increases away

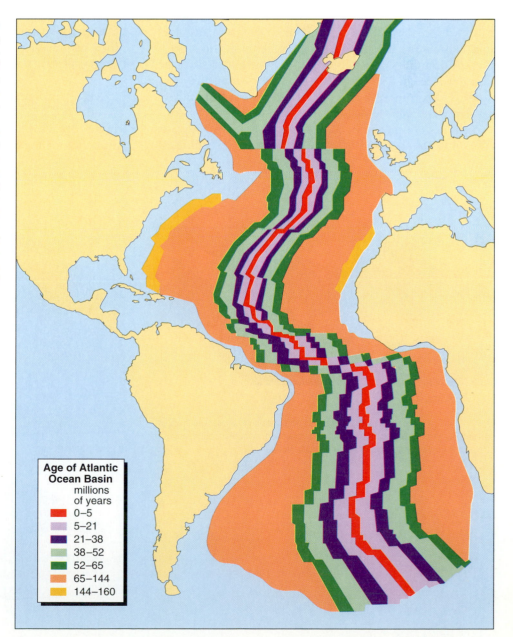

FIGURE 3-10 The age of volcanic seafloor rocks in the Atlantic Ocean. The youngest rocks are along the ocean ridge, growing older with increasing distance from the ridge. This pattern of increasing age as one moves away from the ocean ridge is consistent with that expected from seafloor spreading (see Figure 3-6). A similar pattern is found in the Pacific and Indian Oceans. The differences in width of seafloor of similar age in various areas simply reflect differences or changes in the spreading rates. The faster the spreading rate, the more new seafloor is created in a given period of time. The small east-west trending offsets in the patterns are due to transform faults and fracture zones that have displaced the seafloor (see Figure 3-8). Data for this illustration come mainly from analyses of samples collected by drilling into the seafloor, and by determining magnetic patterns of seafloor rocks and relating them to dated magnetic patterns of rocks found on land (see Figure 3-7).

Age of Atlantic Ocean Basin
millions of years

■ (red)	0–5
■ (light purple)	5–21
■ (dark purple)	21–38
■ (light green)	38–52
■ (dark green)	52–65
■ (orange)	65–144
■ (yellow)	144–160

from the ridge axis was made by drilling into the seafloor (see Chapter 5, page 108), and is very strong evidence to support the seafloor-spreading concept.

Plate Tectonics: The Motions

To summarize briefly, the basic mechanism whereby "old" seafloor is consumed or subducted (recycled is also an appropriate term) is well documented. When regions of seafloor collide with continents, the lighter continents override the heavier ocean floor (see Figure 3-9a), forming a trench. Sediment on the seafloor may be "scraped" off and accreted onto the continental plate.

However, if the seafloor and continents become joined and move together, which is the case on the eastern and western sides of the Atlantic Ocean, no subduction occurs. In this instance, the necessary subduction occurs in the Pacific, along the west coast of North and South America. In the Atlantic, thick sedimentary sequences (derived mainly from land) usually pile up at the base of the continental slope, forming the continental rise (see Chapter 4, page 88). If subduction occurs later, the sediment that formed the rise will be carried into the newly formed trench and possibly accreted onto the continent.

Subduction, although not as spectacular a process as spreading, is an essential part of seafloor spreading. If subduction did not occur, Earth would have to expand continually to accommodate the new seafloor being formed along the ridge axis. With seafloor spreading, however, the amount of new crust formed by spreading is essentially balanced by the material consumed, either through subduction or through mountain building.

A different type of motion, called **strike-slip,** can occur when plates move parallel to and past each other. This creates a shear or **transform fault** zone. In this instance, neither plate overrides nor is thrust beneath another. The three basic types of plate motion—spreading, subduction, and strike-slip—are shown in Figure 3-11.

The region where mantle material rises and forms new seafloor is usually referred to as the *constructive* part of a plate boundary (Figure 3-11a). The boundary where one plate is thrust or subducted under another is generally referred to as the *destructive* plate margin (ocean trenches, for example). A third category, *conservative* plate margins, occurs when plates just slide by each other (strike-slip motion), and there is neither construction nor destruction.

Plate Tectonics and Earthquakes

Studies of seismic activity on a worldwide scale clearly show that most earthquakes are restricted to narrow but

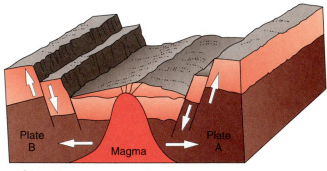

a Spreading

FIGURE 3-11   The three principal types of plate motion that occur in seafloor spreading: (a) spreading, or extension; (b) subduction, or consumption; and (c) strike-slip. Arrows indicate direction of relative motion.

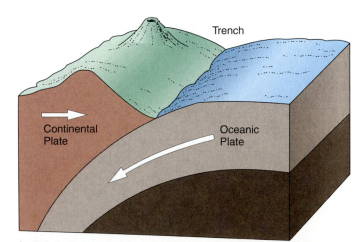

b Subduction

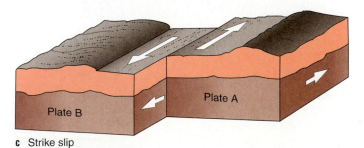

c Strike slip

continuous zones that surround large, relatively stable zones (Figure 3-12). Most earthquakes occur along or very near plate boundaries. In areas of spreading motion and strike-slip motion, seismic activity tends to be relatively shallow and moderate. In subduction zones, however, earthquakes may occur 100 km (62 miles) beneath the surface or even deeper. This is consistent with the deep and down-plunging aspect of this type of plate boundary (see Figure 3-9a).

The seismic activity is caused by seafloor spreading and the addition of new crustal material along the ocean ridges, by the sliding motion along plates (strike-slip), and by subduction in trench regions. The plates are thus outlined by seismic activity (compare Figures 3-12 and 3-13).

Earthquake activity and the type of motion indicated usually confirms the relative motion of the plates, as predicted by the plate tectonic concept. **Compression** movements are indicated by earthquake studies from areas where plates collide, whereas **tension** or a pulling apart motion is indicated where plates move apart. This relation between seismic activity and plate boundaries is valuable in earthquake prediction, as most earthquakes occur along plate boundaries. The interior portions of plates are among the seismically quietest areas of Earth.

One plate boundary, well known to those living along the West Coast of the United States, is the San Andreas Fault, a zone of active faulting (of the strike-slip variety) and earthquakes.

Hot Spots

Perhaps 60 or more long-lasting areas of volcanic activity exist within a plate, distant from its boundaries. These areas, called **hot spots,** have an almost continuously rising plume of magma that reaches the surface, forming volcanoes and eventually islands. A hot spot remains essentially stationary as a plate slowly moves overhead. The area of the plate directly over the hot spot experiences volcanic activity, which decreases and eventually stops as that area of the plate moves on. The result is a chain of volcanoes or islands (Figure 3-14).

Several such chains exist in the Pacific Ocean—the Hawaiian Islands Chain, the Emperor Seamount Chain, and the Line Islands. A quirk of the Hawaiian Islands–Emperor Seamount chain reveals how plate motion can change direction. You can see in Figure 3-14 that the island chain is straight but bends around Midway Island. We believe that this happened because the Pacific Plate changed its direction of motion

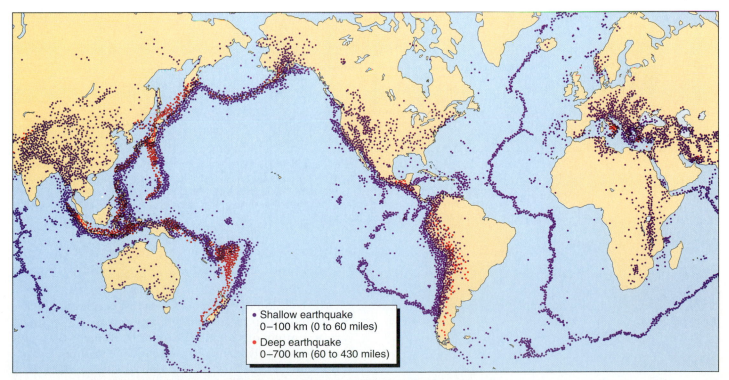

- Shallow earthquake
  0–100 km (0 to 60 miles)
- Deep earthquake
  0–700 km (60 to 430 miles)

FIGURE 3-12 Seismically active areas of Earth, showing earthquakes over a 9-year period. The earthquakes are divided into two groups:

Shallow earthquakes, originating 0 to 100 km (0 to 60 miles) below the surface.

Deep earthquakes, originating 100 to 700 km (60 to 430 miles) below the surface.

(Data from National Earthquake Information Center.)

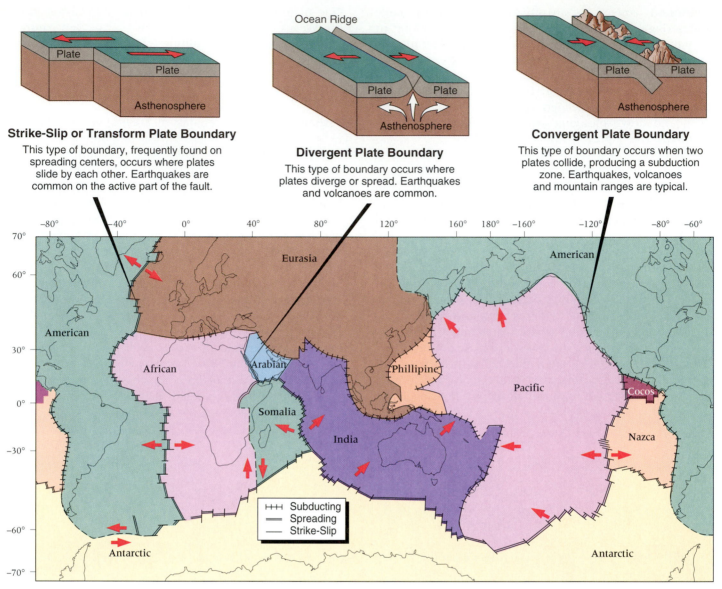

Strike-Slip or Transform Plate Boundary
This type of boundary, frequently found on spreading centers, occurs where plates slide by each other. Earthquakes are common on the active part of the fault.

Divergent Plate Boundary
This type of boundary occurs where plates diverge or spread. Earthquakes and volcanoes are common.

Convergent Plate Boundary
This type of boundary occurs when two plates collide, producing a subduction zone. Earthquakes, volcanoes and mountain ranges are typical.

FIGURE 3-13   Earth's major crustal plates. Note how their boundaries coincide with the seismically active areas shown in Figure 3-12. The three types of plate boundaries are indicated (spreading or divergent, where two plates spread apart; subducting or convergent, where two plates collide; and strike-slip, where two plates slide by each other). Arrows show relative motion of the plates. Numerous smaller divisions of these plates can be made; some classifications show as many as 26 plates.

millions of years ago as it crept ponderously over the hot spot.

Ocean Basins and Mountains

The seafloor spreading and plate tectonics hypotheses indicate that the ocean basins are not fixed in their position or size. Rather, they are continually opening in one area and possibly closing in another. At present, the Atlantic Ocean is opening, whereas the Pacific Ocean is closing.

The collision of ocean plates against continental plates forms mountain ranges along many coasts. These ranges are accumulations of ancient marine sediment that have been folded and thrust above sea level. A similar situation could result if a subduction zone started forming along the continental rise of the East Coast of the United States. The sediments there could be folded and thrust up to form a long mountain range that would parallel the East Coast, similar to the old Appalachian Mountains. Indeed, that is how the Appalachians are thought

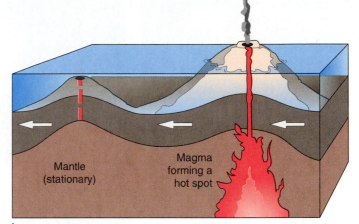

a

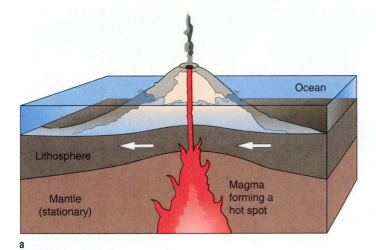

b

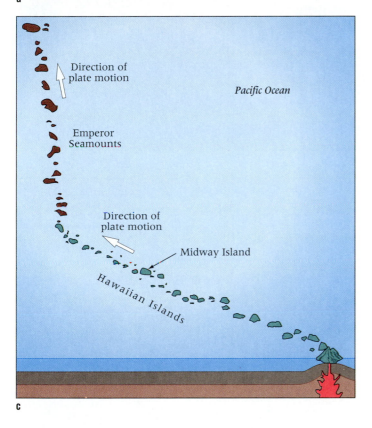

c

d

FIGURE 3-14 The process whereby a hot spot forms a volcano (a) and ultimately a volcanic island chain (b and c). A stationary hot spot in the Earth's mantle generates slowly rising magma that invades the slowly moving overhead crustal plate. The part of the plate currently over the hot spot experiences volcanism, forming volcanic islands such as Hawaii (d). As the plate creeps along, moving perhaps a kilometer every 100,000 years, the volcanic area at the surface shifts to a new volcano (b), eventually forming a string of seamounts and volcanoes like the Hawaiian Islands–Emperor Seamount chain that we see today in the Pacific Ocean (c). Note that a directional change of the plate motion occurred in the geologic past, causing a change in the lineation of the island chain. (d) A series of volcanic cones at Kilauea, Hawaii. This area on the flanks of giant Mauna Loa volcano became active in 1790 and continues today. (Photograph courtesy of U.S. Geological Survey, Hawaii Volcano Observatory, distributed by the National Geophysical Data Center, NOAA.)

to have formed, during a prior period of opening and closing of the Atlantic Ocean.

A Tour Around One Plate

Perhaps the best way to visualize plate tectonics and the motions involved is to look at a single plate. A good example is the Arabian Plate, a relatively small one situated between the larger African and Eurasian Plates (Figures 3-13 and 3-15).

Seafloor spreading is obvious in the Red Sea region, so it is often used to exemplify what an ocean should look like early in its evolution. The opening of the Red Sea has resulted from the movement of the Arabian Peninsula away from Africa (Figure 3-16). *Seismic reflection profiling* is a geophysical technique used to determine subsurface geological structure (see page 77 for details of this technique). Seismic reflection profiles from the Red

a

b

FIGURE 3-15   The Arabian Plate Region.
(a) A computer-generated global view of the Arabian Plate region, showing the topography of both land and seafloor. The image was generated from a land-and-sea database, using a 5-minute latitude/longitude grid, except for ocean areas that have not been surveyed in such detail.
(b) The real thing, an *Apollo 17* picture of the northeastern portion of the African continent. Key parts of the Arabian Plate are obvious in this unique picture, including the entire Red Sea, Persian Gulf, Gulfs of Oman and Aden, and even parts of the Mediterranean Sea (at the top of the figure).
(c) The Arabian Plate, showing location of the following figures. Toothed areas in the Gulf of Oman and along the Zagros thrust zone are regions of subduction. Spreading is indicated by full arrows, as in the Gulf of Aden and the Red Sea. Strike-slip motion is shown by half-arrows, as along the Levant fracture zone and the Owen fracture zone. (Image (a) courtesy of Dr. Peter W. Sloss, National Oceanic and Atmospheric Administration/National Geophysical Data Center; image (b) courtesy of National Aeronautics and Space Administration.)

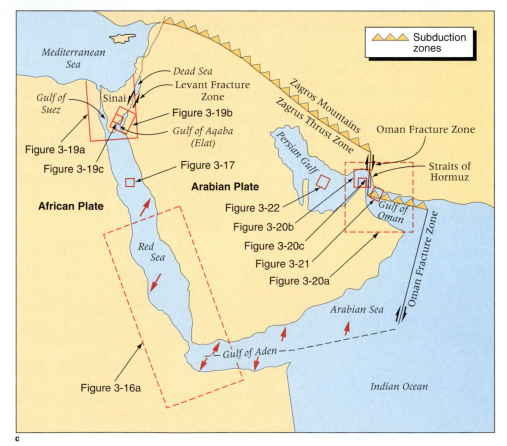

c

a

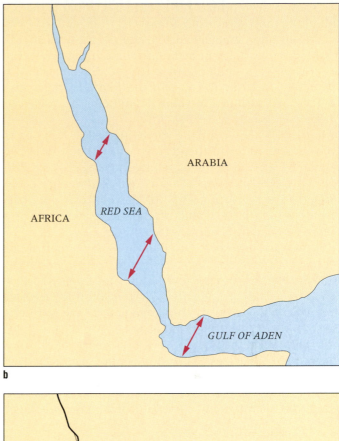

b

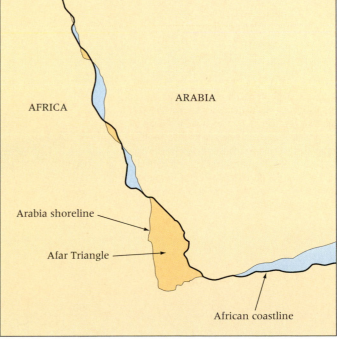

c

FIGURE 3-16 The Red Sea and Gulf of Aden.
(a) *Gemini 11* satellite image of the Red Sea and Gulf of Aden from an altitude of about 620 km (390 miles). Both the Red Sea and the Gulf of Aden are forming by seafloor spreading in the directions indicated in b. North is to the top of the figure. Satellite equipment can be seen in the upper left.
(b) Line drawing of the present-day configuration of Red Sea and Gulf of Aden. Arrows indicate direction of present seafloor spreading.
(c) Fit of Arabian and African shorelines. The area of overlap, the Afar Triangle, has many characteristics of the ocean floor and may simply be an uplifted part of the ocean. If these interpretations are correct, the fit of the two shorelines is almost perfect. (Image (a) courtesy of National Aeronautics and Space Administration.)

Sea show a deep central rift (Figure 3-17) that is the result of the most recent spreading.

Bottom photographs taken in the rift show clear evidence of recent volcanic action. In some regions, hot water discharges in similar fashion to spreading areas in the Pacific and Atlantic Oceans. In the Red Sea, these discharges are forming a valuable mineral deposit (see Chapter 15, pages 372–374).

Further evidence for the spreading of the Red Sea is seen in similarities in the geology of the land on either side of the sea and in the matching shape of the coastline (see Figure 3-16). It appears that the Red Sea probably formed from a "doming up" of the entire region, followed by subsidence and rifting, then eventually by seafloor spreading (Figure 3-18).

Spreading has also occurred in the Gulf of Aden at the southern end of the Red Sea (Figure 3-16). Here, the Arabian Plate is moving toward the northeast, the Red Sea and the Gulf of Aden are slowly opening at 1 to 2 cm (0.4 to 0.8 in.) per year. The movement at the northern end of the Red Sea in the Gulf of Aqaba (or Elat) is of the strike-slip variety and the region is marked by very rugged relief (Figures 3-19b and 3-19c, page 62). This type of motion continues, forming the Dead Sea and extending into Syria. A similar type of motion is occurring along the Owen Fracture Zone and Oman Fracture Zone.

As the Arabian Plate moves toward the northeast, its eastern margin is meeting a relatively immovable object, the Eurasian Plate. The result is a continent-to-continent collision. This is best seen in the southern end of the Persian Gulf and in the Zagros Mountains of Iran, which have been formed by this collision (Figure 3-20, page 63). In this case, because both colliding crusts are continents, neither is thrust under the other. Rather, the edges of both are folded and thrust upward to form a mountain range.

In the Gulf of Oman, the collision is between two oceanic parts, but the folding has not yet reached the mountain-forming stage (Figure 3-21, page 64). The Persian Gulf is far from the present subduction activity, and therefore sedimentary layers are relatively undeformed (Figure 3-22, page 64).

Plate Tectonics: Economics, Life-Styles, and Environment

Seafloor spreading and plate tectonics concepts help us find oil and natural gas. For example, we know that the thick sedimentary sequences necessary for petroleum

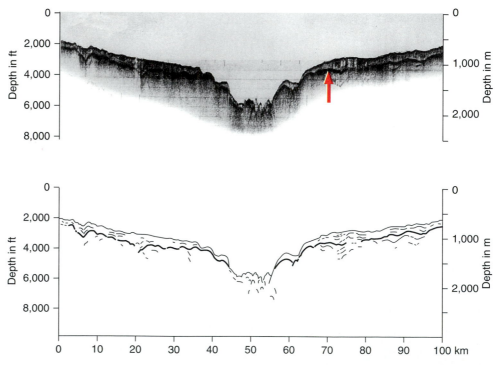

FIGURE 3-17 A seismic reflection profile (top) and interpretation (bottom) across the Red Sea. The central rift area (from about 40 to 65 km on the distance scale) is believed to have resulted from recent spreading that occurred over the past 3 million years. The strong reflector indicated by the arrow in the upper figure comes from rocks that were deposited when the Red Sea evaporated about 5 million years ago.

FIGURE 3-18 Simple model for the origin of the Red Sea. Compare phase c with Figure 3-17. In general, this model works for any new ocean.

20–30 million years ago

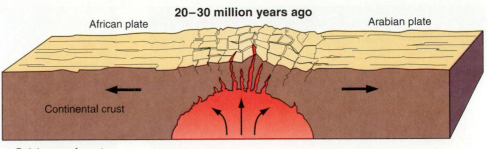

a Bulging up of crust

2–3 million years ago

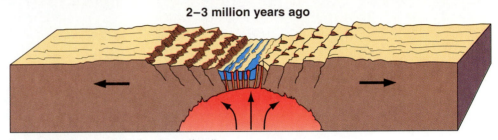

b Collapse of central area forming a rift valley

Present conditions

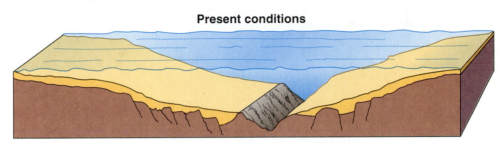

c Spreading, formation of central rift, filling with water, and sedimentation

generation are rarely found in subduction zones, so that is not the place to seek them. However, they are common in continental margins in marine areas where subduction does *not* occur. We'll look further at resources of the ocean in Chapter 15, but see Figure 15-7, page 000, at this time.

Knowledge of plate tectonics can help find other types of mineral deposits. An early reason for thinking that the continents once were joined is the similarity of rocks on either side of the South Atlantic. This observation can be applied in searching for resources. For example, mineral or oil deposits occurring on coastal or nearshore regions on one side of the Atlantic, and formed before the continents spread apart, may correspondingly be found in similar deposits on the other side. In other words, the spreading may have simply divided the resources.

In addition, the actual process of spreading or subduction can form deposits enriched in metals such as zinc, silver, or copper, such as the previously mentioned Red

Sea example. Recent discoveries at other spreading centers also indicate mineral deposits in early stages of development, as discussed later in this chapter.

Seafloor spreading also affects ocean chemistry. Magma chambers under the spreading areas can establish small convection systems in which cold ocean water flows into the chamber region, becomes heated, and returns to the ocean through crustal rock. During this passage, the water is involved in chemical reactions. An example is the removal of magnesium from seawater and its precipitation in the crust rock. This process is described in detail in Chapter 7 (see especially Figure 7-11, page 170).

Seafloor spreading even affects life-styles. Living near a plate boundary means living with a high potential for earthquakes and volcanoes. Parts of California are undergoing a series of movements similar to those occurring along the Gulf of Aqaba. The lower portion of California is splitting or moving away from the rest of California along

a

c

b

FIGURE 3-19 Sinai Peninsula.
(a) Satellite image of the Sinai Peninsula (center of figure), with the Gulf of Suez to the left and the Gulf of Aqaba (Elat) to the right. The Mediterranean Sea is in the background and the Red Sea is in the foreground. Note also the Nile River and its delta to the left.
(b) Satellite image of the narrow, steep Gulf of Aqaba (see also c). Jordan and Saudi Arabia are to the right, Egypt and Israel to the left. The Gulf of Aqaba may be the newest part of the ocean to have been formed.
(c) Rugged topography along the western shore of the Gulf of Aqaba. (Images (a) and (b) courtesy of National Aeronautics and Space Administration.)

the San Andreas Fault. In California's Imperial Valley, the motion is causing many California earthquakes and concern about a major destructive quake in the near future.

The entire Pacific coast of North America is highly active, with numerous earthquakes, frequent faulting, and enough volcanic activity to put Pacific coast states in the headlines. The most famous example of the latter is Mount Saint Helens, which erupted in May 1980 (Figure 3-23). This volcanic activity results from the North American Plate overriding the adjacent Pacific seafloor (Figure 3-24). The plate boundaries off the Oregon and Washington coasts are very irregular, with several fracture zones cutting the main spreading center. The Juan de Fuca Ridge off Washington and Oregon has been intensely studied and may have some economic potential (discussed later in this chapter). Alaska also frequently experiences volcanic eruptions.

Clearly, seafloor spreading and plate tectonics have far-reaching implications. They affect resources. In the past they have affected ocean circulation, climate, and sea level, and we have every reason to believe that these effects will continue. The flow of seawater has been changed and diverted by new ridges, the opening of new basins, and the establishment or closing of land bridges by spreading and tectonic activities. The ocean ridges themselves can prevent east-west movement of deep water from one side of the ridge to the other, as you shall see in Chapter 9. The past position of continents and their subsequent movement certainly affected global climatic patterns.

We know that changes in sea level have occurred in the past. In some instances, however, these were not related to changes in the volume of water, as was caused by melting or forming of glaciers. Rather, sea level shifted in response to changes in the shape of the ocean basin itself. Reducing the dimensions of the ocean basin (especially depth) by increasing the rates of subduction or seafloor spreading could cause a rise in sea level worldwide. An increase in the size of the ocean basin could cause a drop in sea level.

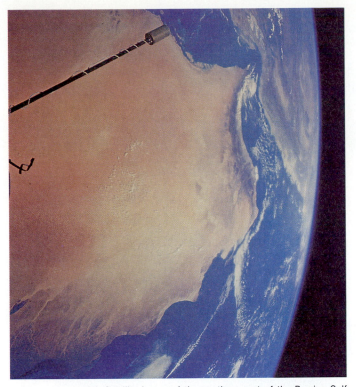

FIGURE 3-20   (a) Satellite image of the southern part of the Persian Gulf (top of image), Gulf of Oman (right center), and Arabian Sea. The land area is mainly desert. Note the narrow opening (the Straits of Hormuz) to the Persian Gulf, caused by faulting along the Oman Fracture Zone (see Figure 3-15c).

(b) Satellite image of the Straits of Hormuz (upper right of figure; the peninsula is Oman), Persian Gulf (middle and lower part of figure), and the Zagros Mountains of Iran. North is toward the top of the image, which is about 320 km (200 miles) wide. Some equipment from the satellite is visible at the lower part of the figure.

(c) Uplifted blocks in the Straits of Hormuz area on the Oman Fracture Zone.

(d) An aerial view of the Zagros Mountains of Iran.

(Images (a) and (b) courtesy of National Aeronautics and Space Administration.)

The Ridge Program and Some Unanswered Questions

Questions concerning seafloor spreading and plate tectonics still exist. But again, an important advantage of these concepts is that they can be tested. One test is to drill into the seafloor. One of the first results from scientific drilling in the Atlantic seafloor showed that the deepest sediments, immediately atop the volcanic rocks, are systematically younger as one approaches the crest of the Mid-Atlantic Ridge (see Figure 3-10). This systematic change in sediment age indicates a spreading rate similar to that determined from the magnetic pattern. These data strongly support the seafloor spreading concept.

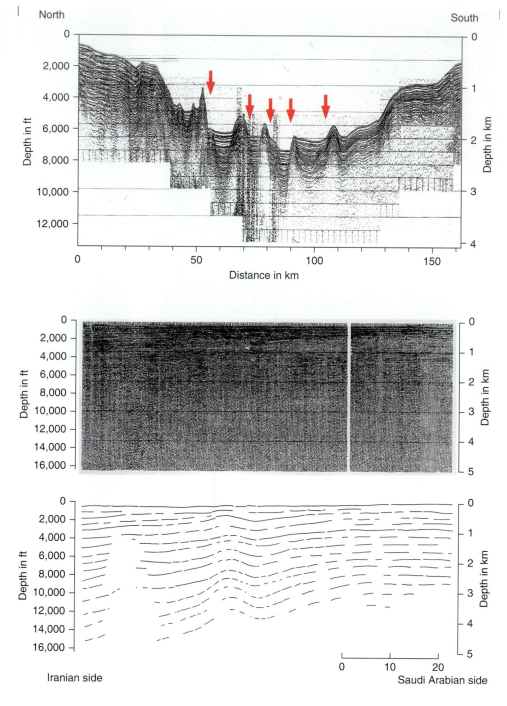

FIGURE 3-21 Seismic profile across the Gulf of Oman, an area of subduction. Note the extensive faulting (some faults are indicated by small arrows), especially when compared with essentially horizontal layers typical of the Persian Gulf (see Figure 3-22).

FIGURE 3-22 A seismic profile (top) and interpretation (bottom) across the Persian Gulf. Note the general flatness to the layering and the lack of folding.

The exact mechanism for seafloor spreading is hard to determine unequivocally. However, it seems clearly related to convective movements within Earth's crust and mantle (see Figure 3-3). Hot material rises along the ridge, moves away from the ridge, cools, and ultimately sinks. Some feel that the subduction does not result from the conveyor-belt effect but rather is the result of sinking, with the leading edge of the dense oceanic plate simply pulling along the remainder of the plate.

Another question centers on whether the plates move continuously, or whether relatively large movements occur irregularly. Seafloor spreading can be directly observed in Iceland—a large island of recent volcanic material produced by the Mid-Atlantic Ridge—that sits directly over the ridge. In Iceland, volcanic activity indeed comes in spurts. A typical pattern seems to be about 100 to 150 years of quiet, followed by a 5-to-20-year period of spreading and volcanic activity.

FIGURE 3-23   Two views of Mount Saint Helens.
(a) For most of this century, Mount Saint Helens was a beautiful snow-covered mountain. All this changed when it erupted in May 1980. The initial eruption and subsequent others caused over $1.2 billion in damage and killed 62 people. Over 400 km² (about 155 mi.²) of forest was leveled by the explosion. This figure shows the 1980 eruption as viewed from the east. (Photograph courtesy of Department of Natural Resources, distributed by the National Geophysical Data Center, NOAA.)
(b) Mount Saint Helens continued to be active after the initial eruption in 1980. This view shows the volcano three years later, in May 1983. (Photograph courtesy of University of Colorado, distributed by the National Geophysical Data Center, NOAA.)

FIGURE 3-24   How the North American Plate is overriding the Pacific Plate. This process has weakened the North American Plate, allowing magma to reach the surface. In this case, it reached the surface at Mount Saint Helens (see Figure 3-23). Recent volcanic activity is also occurring along the Juan de Fuca spreading center (see Box 9-3, page 218).

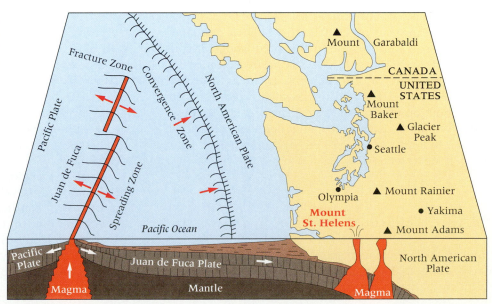

The forces necessary to drive the plates are still widely debated, as is the actual strength or rigidity of the plates. Many important questions remain about mineralization and economic mineral potential associated with spreading and subduction.

These questions, and others in marine biology and chemistry, led the marine scientific community to develop an international program called RIDGE (Ridge Inter-Disciplinary Global Experiments). The goal of RIDGE, stated in typically scientific terms, is to understand the physical, chemical, and biological causes and results of energy transfer within the global ocean ridge system, through time and through space.

The RIDGE program has six primary objectives:

1. To understand the flow of mantle material and the generation and transport of magma beneath mid-ocean ridges

2. To understand the processes that transform magma into ocean crust

3. To understand the processes that control the shape and timing of ridge formation

4. To understand the physical, chemical, and biological processes between circulating seawater and the lithosphere.

5. To determine the interactions of organisms with physical and chemical environments at the mid-ocean ridges.

6. To determine the distribution and intensity of mid-ocean hydrothermal vents, and the interaction of venting with the ocean environment

The RIDGE program will probably require a decade to complete and will certainly lead to new discoveries—and new questions.

A long-term goal of RIDGE is to build a permanent seafloor observatory near an active ocean ridge. It could be similar to that used to study land volcanoes; in Hawaii, eruptions of the volcano Kilauea have been monitored since 1912 by the Hawaiian Volcano Observatory, built on the volcano's rim. The ocean ridge observatory would be automated and operate without observers, but would be visited by submersible or be remotely operated.

Recent Research

One key goal of scientists who are studying plate tectonics is actually to document the small movements that are occurring between plates. We are probably a few years away from making such absolute measurements, but some scientists believe that they already have found direct evidence of seafloor spreading. The measurements are very preliminary, but measurements with lasers, radio telescopes, and other technology suggest that the Atlantic Ocean is widening by about 1.5 cm (0.6 in.) per year and that Hawaii and North America are coming closer by about 5 cm (2 in.) per year.

Another focus of recent research is the question of whether plate movements occurred prior to the formation of Pangaea about 225 million years ago. An especially interesting question is whether seafloor spreading was an important process during the early periods of Earth's history. It appears that heating from radioactive decay, which could be the source of heat that drives convection currents, could have been as much as three times greater in this early period. What effect this greater internal heat might have had is unknown. Perhaps it caused more rapid spreading, or formed different types of plates. Recent geological studies of North America indicate that plate tectonics may have been active for at least the past 2 billion years.

Important discoveries have recently been made off the coast of Oregon in the general region of the Juan de Fuca Ridge (see Figure 3-24):

- In 1986, researchers found a plume of hot water above a vent field. The plume was nicknamed "megaplume" because it was 20 km (12.4 miles) in diameter and 700 m (2,300 ft.) thick (see Box 9-3, page 218).

- In early 1990, a series of underwater volcanoes was discovered 480 km (300 miles) off the Oregon coast. Some are more than 30 m (98 ft.) high. They extend along a line about 16 km (about 10 miles) long and apparently formed within the last 10 years. They may be related to the volcanic activity that formed the "megaplume."

Detection of the new volcanoes came by comparing results of a 1990 bathymetric survey with an earlier 1981 survey, which showed that the seafloor had become shallower by up to 30 m (98 ft.) in some places. Dives in the research submersible *Alvin* discovered fresh lava flows covering a fractured area in older seafloor, which occurred in an area of active seafloor spreading and was the first clear evidence of an underwater eruption on a ocean ridge.

A similar situation may be occurring along the flanks of the island of Hawaii. Loihi Seamount, located about 35 km (22 miles) southeast of Hawaii, is a young but growing volcano that some feel could become an island (that is, attain sea level) in about 50,000 years. The top of the seamount is currently at about 1,000 m (3,280 ft.) depth. The volcano is forming as the crust moves over the hot spot that is currently making Hawaii volcanically active.

For the past two decades, marine scientists have discovered much about plate tectonics through extensive studies of ocean ridges. Some of these studies have used submersibles such as *Alvin* and sophisticated photographic systems. Some results have been startling and exciting.

A series of dives in 1977 along the Galápagos Rift (between the Cocos and Nazca Plates of the Pacific Ocean, see Figure 3-13) discovered many new species, including long tubeworms and clams over 30 cm (12 in.) in length (see Figure 12-19, page 300). These unique creatures were concentrated around vents along the oceanic ridge that discharge hot water. The discharged water is clearly associated with cracks or rifts on the seafloor. Later, scientists were to learn that these organisms have a unique mechanism for surviving on the seafloor in such an extreme environment.

These discoveries were followed in 1979 by even more amazing ones from the East Pacific Rise, at about 21°N. Similar animals, although somewhat larger, were found, but the highlight was the discovery of vents that dramatically discharge black clouds of minerals (see Box 2-3 and Figure 7-10, page 168). These vents, up to 10 m (33 ft.) high, discharge water as hot as 350°C or about 662°F (beause of the high pressure, the water remains a liquid even at these high temperatures rather than becoming steam). The vents were depositing minerals enriched in copper, zinc, cobalt, lead, silver, and other metals. When these hot waters came in contact with the surrounding cold seawater (temperature about 2°C or 36°F) at a depth of about 2,800 m (9,300 ft.), the minerals precipitated, forming a chimneylike structure. Scientists appropriately nicknamed these **black smokers.**

Initially, scientists believed these vents to be restricted to the faster spreading centers in the Pacific, as none had been found in the Atlantic Ocean, where spreading is relatively slow. In the Atlantic, the annual spreading rate is approximately 2.5–3.8 cm (1–1.5 in.), compared to 7.6–12.7 cm (3–5 in.) a year in the Pacific regions where the vents were found. However, this opinion changed with the finding in 1985 of vents and similar fauna along the Mid-Atlantic Ridge at 28°N (Figure 3-25). Some fauna in the Atlantic were different, but the phenomena were clearly similar. It is evident that many new and exciting discoveries are yet to come (see also Box 7-4, page 169).

The discovery of these vent areas received media attention, often with elaborate claims as to their potential value. Though it is premature to speculate on the economic implications of these phenomena, it is certainly an interesting series of oceanographic discoveries. It must be emphasized that *considerably less than 1 percent of the 65,000-km-long (40,000 mile) oceanic ridge has been studied in detail.* Certainly, other areas of the worldwide ocean-ridge system will be explored intensively in the coming years.

FIGURE 3-25 The hot vent area recently discovered along the Mid-Atlantic Ridge.
(a) Vent surrounded by white shrimp (a new genus of shrimp), which are about 10 cm (about 4 in.) long.

(b) Isolated shrimp and a few crabs around an active vent. Note the shimmering aspect to the escaping hot water near the center of the figure. (Photograph (a) by Dr. Geoffrey Thompson; photograph (b) by Dr. Susan E. Humphris; both of Woods Hole Oceanographic Institution. Images courtesy of Woods Hole Oceanographic Institution.)

SUMMARY

The combined concepts of seafloor spreading and plate tectonics are one of the most exciting scientific ideas of this century. The simplicity of the hypotheses and confirming geological and geophysical data argue very strongly for its acceptance as a theory. Aspects of the seafloor spreading and plate tectonics concepts may also be used to ascertain areas favorable for marine resources.

The outer part of Earth's crust is a series of plates that are in motion relative to each other. Along oceanic ridges, the seafloor is slowly spreading apart along the boundary of two plates, and crust of new volcanic material is slowly being added to the ocean floor.

In other areas, plates are colliding. Where a continental plate meets an oceanic plate, the lighter continental plate overrides the denser oceanic plate. The result is usu-ally subduction, causing a deep-sea trench and sometimes an island-arc system. Where plates of similar composition (oceanic-oceanic or continental-continental) collide, a mountain range may result. A third type of motion, strike-slip, occurs when plates slide parallel to each other.

The concepts of seafloor spreading and plate tectonics seem to answer almost all questions about the origin and evolution of the structure of ocean floor. But some points, such as the type and size of the moving forces, remain unexplained. Recent discoveries of vents discharging hot water along the spreading centers in the Pacific and Atlantic Oceans show how dynamic and spectacular the seafloor-spreading processes actually can be and emphasize how much still remains to be learned about the ocean.

QUESTIONS

1. Draw a cross section across an ocean showing the key parts of the seafloor spreading processes. Show where earthquakes are common.

2. What are the main components to the seafloor spreading–plate tectonics concept?

3. How can seafloor spreading influence the characteristics of a continental margin?

4. Why are there more trenches in the Pacific Ocean than in the Atlantic Ocean?

5. Why is the sediment thickness in the ocean considerably less than anticipated?

6. What happens when plates collide?

7. What happens when plates slide past each other?

8. How can data from earthquakes be used to learn about the internal structure of Earth?

9. What can mantle plumes tell us about processes within Earth?

10. Why are the Red Sea and the Persian Gulf so different from each other?

11. How do we know that seafloor spreading is a dynamic and active process, and not just a historical curiosity?

KEY TERMS

asthenosphere	epicenter	L-wave	S-wave
basalt	fault	magnetic pattern	seafloor spreading
black smoker	focus	magnetic reversal	sedimentary rock
compression	fold	mantle	spreading rate
continental crust	fracture zone	mantle plume	strike-slip
continental drift	granite	metamorphic rock	subduction
convection	hot spot	Moho	tectonics
core	igneous rock	ocean ridge	tension
crust	island arc	P-wave	transform fault
divergence	isostasy	Pangaea	trench
earthquake	lithosphere	plate tectonics	

FURTHER READING

Anderson, R. N. 1986. *Marine Geology: A Planet Earth Perspective.* New York: John Wiley. A good introduction to marine geology and geophysics.

Ballard, R. D. 1983. *Exploring our Living Planet.* Washington, D.C: National Geographic. A well-illustrated book ranging from the origin of Earth to dramatic examples of hot spots and seafloor spreading.

Canby, T. Y. 1990. "Earthquake Prelude to the Big One." *National Geographic* 177, no. 5, pp. 76–91. The Loma Prieta earthquake through the eyes of *National Geographic.*

Dvorak, J. J., C. Johnson, and R. I. Tilling. 1992. "Dynamics of Kilauea." *Scientific American* 267, no. 2, pp. 46–53. A discussion of one of the most thoroughly studied volcanoes, and how scientists are learning to understand and predict where eruptions might occur.

Frohlich, C. 1989. "Deep Earthquakes." *Scientific American* 260, no. 1, pp. 48–55. A good general description of earthquakes and what causes them.

Hamblin, K. W. 1992. *Earth's Dynamic Systems.* 6th ed. New York: Macmillan Publishing Co. A superb introductory Earth-science textbook with excellent illustrations.

Kennett, J. 1982. *Marine Geology.* Englewood Cliffs, N.J: Prentice-Hall. A good but advanced introduction to the fields of marine geology and geophysics.

Macdonald, K. C., and P. J. Fox. 1990. "The Mid-Ocean Ridge." *Scientific American* 262, no. 6, pp. 72–79. The definitive article on ocean ridges.

Menard, H. W. 1986. *The Ocean of Truth.* Princeton, N.J.: Princeton University Press. Development of the concepts of seafloor spreading and plate tectonics, described by one of the scientists involved in the formation of these ideas.

Murphy, J. B., and R. D. Nance. 1992. "Mountain Belts and the Supercontinent Cycle." *Scientific American* 266, no. 4, pp. 84–91. Authors propose that supercontinents, like Pangaea, form every 500 million years or so, and then break apart.

Ross, D. A. 1979. "The Red Sea: A New Ocean." *Oceanus* 22, no. 3, pp. 33–39. A general description of how the Red Sea formed and developed.

Moores, Eldridge, ed. 1990. *Shaping Earth: Tectonics of Continents and Oceans.* New York: W. H. Freeman and Co. A collection of fourteen articles from *Scientific American* concerning Earth and its geological development.

Van Andel, T. 1985. *New Views on an Old Planet: Continental Drift and the History of Earth.* New York: Cambridge University Press. A well-written book detailing the evolution of the recent hypotheses concerning the development of our planet.

White, R. S., and D. P. McKenzie. 1989. "Volcanism at Rifts." *Scientific American* 261, no. 1, pp. 62–71. Describes the processes that lead to volcanic activity at ocean rifts.

CHAPTER 4

Characteristics of the Seafloor

THE SEAFLOOR OWES ITS SHAPE, ORIGIN, SEDIMENT COVER, AND SEDIMENT TYPE TO A VARIETY OF PROCESSES. THESE RANGE FROM LOCAL ENVIRONMENTAL CONDITIONS TO THE BROAD, OCEANWIDE EFFECTS CAUSED BY SEAFLOOR SPREADING AND PLATE TECTONICS. SEAFLOOR SPREADING AND PLATE TECTONICS HAVE GIVEN EARTH SCIENTISTS NEW INSIGHT INTO HOW THE OCEAN BASINS AND CONTINENTS FORMED, AND HOW THEY CONTINUE TO EVOLVE. FROM THIS, AND FROM A BETTER UNDERSTANDING OF THE MARINE ENVIRONMENT, MANY OF THE GEOLOGIC IDEAS DEVELOPED OVER THE PAST TWO CENTURIES ARE BEING REVISED.

GEOLOGISTS TEND TO BE INTERESTED IN EARTH'S HISTORY AND PROCESSES. GEOPHYSICISTS MORE OFTEN FOCUS ON EARTH'S INTERNAL STRUCTURE. DIFFERENCES BETWEEN THE TWO FIELDS OFTEN ARE REALLY MINOR. THE MARINE GEOLOGIST (OR GEOLOGIC OCEANOGRAPHER) AND THE MARINE GEOPHYSICIST BOTH ARE INTERESTED PRIMARILY IN THAT PORTION OF EARTH NOW COVERED BY WATER. THIS EXTENDS TO BEACHES, MARSHES, AND TIDAL AREAS THAT ONLY SOMETIMES ARE SUBMERGED, AS WELL AS TO THE CONTINENTAL MARGIN AND DEEPER PORTIONS OF THE OCEAN.

This computer-generated image of the seafloor is based on bathymetric data acquired by a Sea Beam mapping system on *R.V. Conrad*, a vessel of the Lamont-Doherty Earth Observatory. The image shows part of a trench (dark blue, on the left) off the continental margin of Chile (green and yellow, on the right).

(Image created by Joyce Miller and Scott Fergusson, University of Rhode Island; courtesy of Lamont-Doherty Earth Observatory.)

Introduction

Major objectives of *marine geology* and *marine geophysics* are (a) to describe the ocean floor and to determine its origin and underlying structure, and (b) to ascertain the thickness, composition, and origin of the sediments and underlying rocks in the ocean. This chapter is concerned with the character and structure of the seafloor. Chapter 5 focuses on the sediments and underlying rocks. Chapter 14 examines the coastal region, an extremely important portion of the ocean that is especially influenced by physical oceanographic processes.

Depending on where you live, many of the land features you see every day may originally have formed in a marine environment, or were later modified in a marine environment, or both. Thus, by studying the geology of the ocean, you can better understand your own habitat. To support our study of the ocean's geology, we have extremely accurate seafloor charts based on new mapping and surveying technologies, such as Sea Beam (see opening figure of this chapter, and Box 4-1).

Early attempts to map the seafloor were limited by technologies that allowed only one depth reading to be taken at a time. A major advance occurred with the German *Meteor* expedition in 1925–27, one of the first expeditions to use electronic echo-sounding equipment rather than determining depths by the tedious and time-consuming method of lowering a line with heavy weights to the bottom. Working in the South Atlantic, *Meteor* made over 70,000 depth soundings that clearly showed the ocean bottom was not a single featureless plain but consisted of mountains, valleys, and flat areas. In later years, oceanographers discovered that the ocean floor has a **topography** (the shape of its surface) every bit as diverse as that of land.

After World War II, large quantities of explosives were available, so oceanographers put them to use for seismic refraction studies at sea. An explosion is generated underwater and the shock waves from it are recorded as they return from the sea bottom. These techniques were first used on land in the 1920s, and twenty years later were tried in shallow parts of the ocean. In the following years, improved technologies made these and other techniques possible in the deep sea, permitting its previously unknown structure to be determined.

Tools for Studying the Structure of the Seafloor

Scientists have developed a very sophisticated array of instruments to study the internal structure of Earth. Some instruments measure Earth's gravity or magnetic field; others detect subsurface layers and structure beneath the ocean. In some instances, instruments designed for use on land can be adapted for use at sea with some modification. Usually, however, the ocean environment necessitates a new instrument system.

Measuring Earth's Gravity

An example of an adaptable instrument is that used for measuring **gravity.** Although we think of gravity as a constant force, it is not. Gravity varies from place to place, due to variations in density and thickness of the underlying rock and sediment. We measure gravity by noting either the period of a pendulum's swing or the pull of gravity against a delicately calibrated spring. On land, these measurements can be made easily and quickly. At sea, however, waves and wind cause up-and-down accelerations. These can produce errors in the gravity measurements that are thousands of times greater than the anticipated variations in gravity.

A Dutch scientist, F. A. Vening Meinesz, partially solved this problem by devising a pendulum system that could be used on a submarine, below the disturbing movement of waves. Gravity-measuring instruments, or gravimeters, have been successfully used at sea when mounted in devices that keep them effectively motionless, regardless of the movements of the ship.

Gravity differences are usually interpreted as being caused by some difference in the underlying geologic structure. The reasons for such geological variations are many, so gravity data alone cannot enable a precise interpretation of the subsurface structure. It must be used in combination with other geophysical data, such as magnetic and seismic refraction data. Worldwide measurements of gravity have recently been achieved using satellite systems (Box 4-2, page 76).

Measuring Earth's Magnetic Field

Measurement of Earth's **magnetic field** at sea is relatively simple. A **magnetometer** can be towed through the ocean by ship, or over the ocean by airplane. Magnetic measurements can reveal information about the composition of the rocks that form the upper parts of Earth's crust. For example, igneous rock bodies such as volcanoes can be identified by their distinctive magnetic properties, which depend on the presence and amount of magnetic minerals in the rock (such as magnetite), the thickness of the rock body, and its depth below the surface.

Magnetic measurements also can indicate the direction and strength of Earth's magnetic field at the time the rock cooled. A rock's magnetization is similar in direction and intensity to Earth's magnetic field at the time that the rock solidified or was deposited. This fact is

BOX 4-1

Sounding Methods, from Rope to Sea Beam

Before the development of electronic techniques, ocean depth was determined by *sounding,* which simply involved lowering a heavy line of known length and noting how much line had to be played out until it reached the ocean bottom. This tedious technique, which requires hours for a single depth measurement in the deep sea, frequently did not result in correct depth values. When a line is lowered to the seafloor, it does not necessarily go straight down but can be deflected by currents or by movements of the surface vessel, erroneously adding to the depth measurement. Another disadvantage is that this technique gives only one depth with each sounding, rather than a continuous picture of the bottom.

Echo sounding, a technique using electronically controlled sound impulses, solved these problems (although it also introduced other, smaller difficulties). This technique is also known as **sonar,** which stands for *so*und *na*vigation *a*nd *r*anging. Echo sounding works by sending an outgoing signal or sound pulse from a ship. It travels through the water to the ocean bottom, is reflected off the bottom, and travels back to the ship (Figure 1). The time required for the signal to make this trip is accurately measured. Knowing the speed of

sound in water, one can determine the depth. The water depth equals half the travel time (because the total travel time covers the two-way trip to the bottom and back), multiplied by the speed of sound in the water. The speed of sound in water is about 1,450 m/sec. (4,756 ft./sec.), which is over four times faster than the speed of sound in air.

A conventional echo-sounding device produces a permanent graphic record of the returning sound (Figure 2a), which gives the oceanographer a visual image of the character of the ocean floor. Echo-sounding records from the ocean reveal that the topography of the seafloor is at least as irregular as that of land.

However, there are some problems in interpreting the echo-sounding record. First is the exaggeration in scale. The surface vessel is traveling at about 10 or 12 knots (18.5 to 22.2 km per hour), which represents the horizontal scale of the record. (If the ship were motionless, the echo-sounder would simply keep reading the distance to the same point below.) The vertical scale of the record shows the depth, commonly in hundreds of fathoms or meters. The vertical scale usually is smaller and is exaggerated (amplified to show detail better) relative to the horizontal scale. In other words, the vertical scale (water depth) is in hundreds of meters, whereas the horizontal scale (distance across the seafloor) is in thousands of meters. If, however, both scales were made with the same dimensions, it would be very difficult to observe details of bottom features (Figure 2b).

A second echo-sounding problem is determining the actual speed of sound in water. The figure quoted above is truly "ballpark." Sound velocity increases with increasing temperature, salinity, and depth (all discussed in more detail in Chapter 8). These parameters must be known and adjusted for if a very accurate determination of water depth is desired.

A third problem is the shape of the outgoing sound beam, a wide cone that covers a relatively large circular area when it hits the ocean bottom. The first returning echo therefore comes from the point closest to the ship. However, in the case of underwater mountains, the first returning echo can be from a mountain top that is closer than the deeper floor directly beneath the ship. The following echoes may be masked or obscured by the first one. This effect makes it difficult to define small features on the seafloor accurately and to be sure that the feature is directly below the ship.

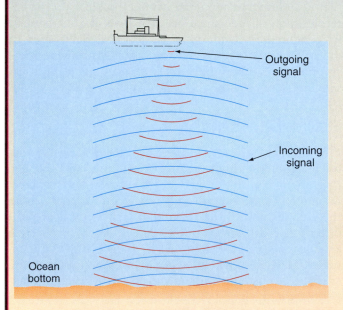

Outgoing signal

Incoming signal

Ocean bottom

FIGURE 1   Echo-sounding technique. Sound from the ship travels to the ocean bottom and is reflected back to the ship. The time the sound takes to make the trip is measured, and these data are used to calculate the water depth.

Box 4-1 continued

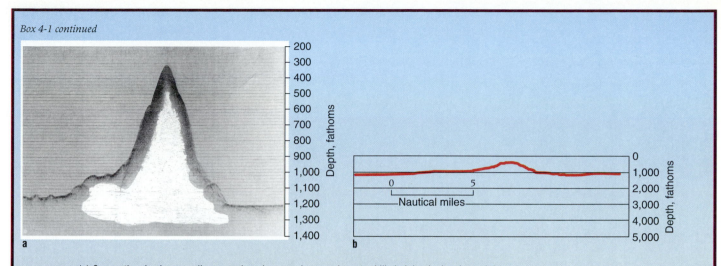

FIGURE 2   (a) Conventional echo-sounding record made over a large underwater hill. Judging by its shape, it is probably a submerged volcano. The vertical exaggeration of this record is 12 times. (b) The same data shown with no vertical exaggeration.

In recent years, the desire to learn the morphology (shape) of the seafloor has led to new technologies for echo sounding. These systems use numerous narrow sound beams sent in a fan pattern to reduce the effect of the third problem mentioned. The narrowness of the beams, combined with their pattern, can be used to produce a strip contour chart of the area beneath the vessel. One such system, Sea Beam, uses 16 beams (8 per side) and can map an area over a mile wide with each pass (Figure 3). These systems are very sophisticated and can produce a colored chart in "real time" as the ship passes over the seafloor. Such charts, which have considerable detail, have made a major impact on our understanding of the processes that shape the ocean floor (Figure 4).

Oceanographic vessels routinely use their echo-sounding equipment at sea, although most do not have sophisticated systems such as Sea Beam. Many crossings over an area can be used to produce a model or bathymetric chart of the ocean floor. Echo-sounding has become so common today that it is even used on pleasure boats, both to avoid hitting the bottom and to detect fish.

FIGURE 3   Schematic of the Sea Beam system in use by a surface vessel, showing the area of seafloor being sounded by the 16 separate but narrow beams. (Drawing by Stefan Masse; courtesy of Woods Hole Oceanographic Institution.)

continued

Box 4-1 continued

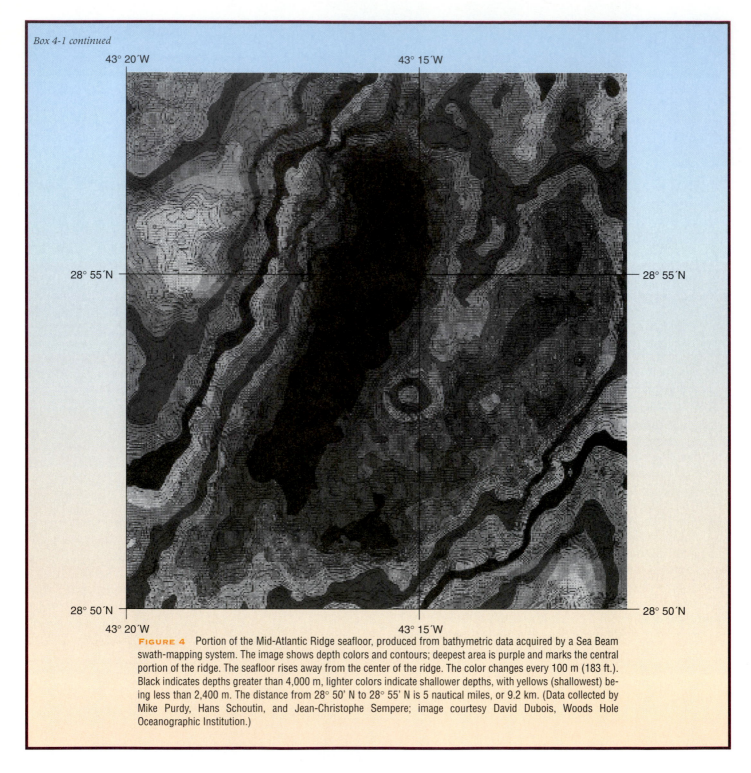

FIGURE 4   Portion of the Mid-Atlantic Ridge seafloor, produced from bathymetric data acquired by a Sea Beam swath-mapping system. The image shows depth colors and contours; deepest area is purple and marks the central portion of the ridge. The seafloor rises away from the center of the ridge. The color changes every 100 m (183 ft.). Black indicates depths greater than 4,000 m, lighter colors indicate shallower depths, with yellows (shallowest) being less than 2,400 m. The distance from 28° 50' N to 28° 55' N is 5 nautical miles, or 9.2 km. (Data collected by Mike Purdy, Hans Schoutin, and Jean-Christophe Sempere; image courtesy David Dubois, Woods Hole Oceanographic Institution.)

Box 4-2

Satellites, Bathymetry, and Gravity

Measurement of seafloor depth was one of the first challenges for early oceanographers. Box 4-1 describes simple sounding by early researchers, who lowered a line with a heavy weight and "felt" when it hit bottom. A later improvement was to trigger an explosion and measure the time it took for the sound to make the trip to the seafloor and return. By estimating the speed of the sound, the depth could be calculated. However, only one depth measurement was possible per explosion. Later, electronic techniques created the sound and more accurately measured the time it took for the round trip. Today, even more sophisticated and accurate devices have been developed, such as Sea Beam. Marine gravity measurements have undergone a simi-

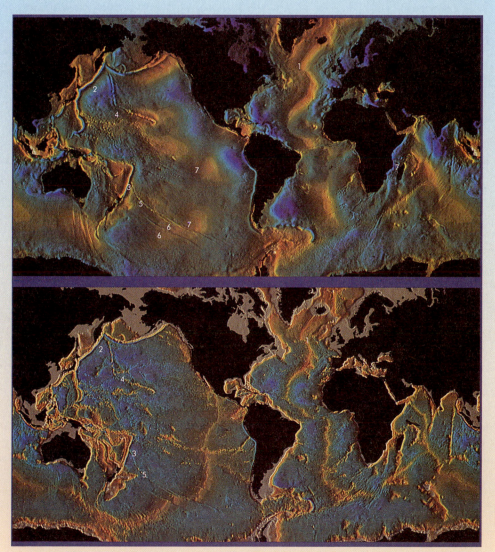

FIGURE 1   Mean sea level (upper) and general bathymetry of the ocean (lower). The mean sea level chart was prepared using data from an altimeter on a satellite. The bathymetric chart was produced with conventional mapping techniques. See the text for more on the techniques and what the numbers on the chart represent. (Both by Dr. William F. Haxby, Lamont-Doherty Earth Observatory, Columbia University; courtesy of National Aeronautics and Space Administration.)

continued

Box 4-2 continued

lar evolution. In both instances, however, a ship is required to traverse the area where the measurement is to be made.

Earth-orbiting satellites, however, permit worldwide measurement of both bathymetry and gravity. The system requires an **altimeter,** a device that very accurately measures the height of the satellite above the sea surface. The position of the satellite also can be accurately determined. Thus, it is possible to measure the height of the sea surface. An altimeter on the *SEASAT* satellite was able to measure sea surface elevations to a precision of about 5 cm (about 2 in.), which is akin to your estimating the distance to an object on the horizon to the nearest millimeter!

From such measurements, the general ocean bathymetry can be inferred. It works this way: We think of the sea as being level, and if no forces acted upon it other than a uniform gravity, it would be perfectly level. However, sea level actually varies with waves, tides, and currents. It also varies over large hills and valleys, due to variations in gravity. For example,

where there is a large concentration of mass, such as the mid-ocean ridges, the pull of gravity is stronger and the water surface will "pile up" above it, perhaps as much as 5 m (16 ft.) higher. Likewise, over trenches, where the gravity is less, the sea surface can be lower by as much as 60 m (200 ft.).

Dr. William F. Haxby, a scientist at the Lamont-Doherty Earth Observatory, has produced a fascinating study showing bathymetry produced by conventional techniques (see Figure 1 lower part) and a map of sea level (upper part). In both figures, yellows indicate shallow areas and blues indicate deep areas. Several features are indicated by numbers: on the lower portion (bathymetry), 1 is the Mid-Atlantic Ridge, 2 and 3 are deep trenches, and 4 and 5 are ridge systems. On the sea-level chart, these same features are seen, plus some fracture zones (6) and some interesting low-level bumps (7). It has been suggested that these bumps may indicate plumes of molten rock welling up from Earth's interior and may be an important driving force in seafloor spreading.

---

very important, because Earth's magnetic field often has changed direction (reversed magnetic poles) over geologic time. Thus, rocks on the seafloor have magnetic patterns that are characteristic of the conditions at the time they were deposited. This point is especially relevant in the seafloor-spreading concept (see Chapter 3 and Figure 3-6, page 48).

Measuring Ocean Bottom Layers and Structure

Studies of the layering and structure of the ocean bottom are made mainly by **seismic reflection** and **seismic refraction** techniques. Both techniques are based on the same principle. Sound waves generated by an explosion travel to the ocean bottom and through it to subsurface layers. Some energy is directly reflected back as an echo to the surface ship (*reflection,* Figure 4-1a). Some energy also travels along the subsurface layers in the crust and is refracted back to the surface, where it can be received by a second ship or an automated recording device (*refraction,* Figure 4-1b). The receiving devices are called **hydrophones.**

A plot of the distance between the energy-source point and the receiving point and the time of first arrival of refracted sound energy can be used to determine the depth and velocity of sound in the different layers. By

knowing how fast sound travels in a specific layer, it is possible to speculate about its composition (for example, sound travels faster in volcanic rocks than in sedimentary rocks).

Another technique, called *continuous seismic profiling,* uses the reflection technique with up to several hundred hydrophones towed in an array (generally in one cable) behind the ship. The ship can travel at 8 knots (about 15 km per hour) or more, collecting data as it proceeds. The sound source, generally either an electrical discharge or the release of air under high pressure from an air gun, is usually fired every 10 or 12 seconds (Figure 4-2). The returning signals, after being amplified and filtered, are printed by a recorder (Figure 4-3).

Seismic profiling techniques have given considerable insight into the origin of continental margins and ocean basins. Recent technological developments have yielded even deeper penetration and better resolution, although cost has increased. Much seismic profiling work now is done by seismic exploration companies, usually in the employ of the petroleum industry.

Another effort of marine geophysicists has been to study the motion of small natural seismic waves as they pass through Earth's ocean crust. They are studied by placing sensitive listening packages, called ocean-bottom hydrophones, on the seafloor. Such studies also produce

FIGURE 4-1 Seismic sounding.
(a) Procedure for seismic *reflection.* This common geophysical technique was used to produce the records shown in Figures 3-17, 3-21, 3-22, and 4-3.
(b) Procedure for seismic *refraction.* The receiving ship can be replaced by buoys that detect the refracted signals.

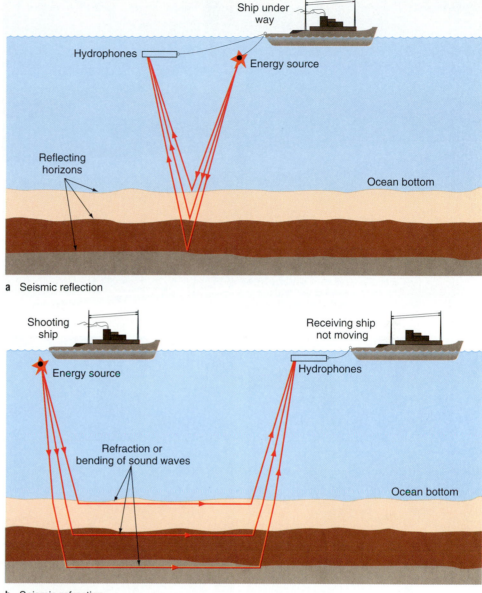

Ship under way

Hydrophones

Energy source

Reflecting horizons

Ocean bottom

a Seismic reflection

Shooting ship

Energy source

Receiving ship not moving

Hydrophones

Refraction or bending of sound waves

Ocean bottom

b Seismic refraction

information about Earth's structure and about the physical properties of the crust. The hydrophones can be placed in various patterns on the seafloor to detect regional variations in crust composition and structure.

Measuring Movement

As oceanography and other sciences grow more sophisticated, it becomes possible to measure Earth movements that are ever-slighter and ever-slower. A key to such measurements is the ability to determine with great precision where something is, called **positioning.** A recent development is an improved positioning technique called the **Global Positioning System,** which uses military satellites and permits an amazing degree of precision in measuring the distance between two points. Accuracies of 1 part in ten million (for example, 1 cm over a distance of 100 km) or even better can be achieved.

The Global Positioning System is an important geophysical tool. Its precision may make it possible to detect the strain developing in Earth's crust prior to an earthquake, or to detect the tiny, centimeters-a-year movements between the tectonic plates that form Earth's surface.

a

b

c

FIGURE 4-2 Parts of a continuous seismic profiling operation.
(a) Surface air bubbles resulting from discharge of an air gun (the source of energy) towed behind a ship and about 10 m (about 30 ft.) below the surface.
(b) Seismic array being readied for towing.
(c) Air gun being prepared for launch. (All photographs courtesy of Lamont-Doherty Earth Observatory.)

FIGURE 4-3 A continuous seismic reflection profile made across the outer continental shelf and slope off the west coast of Mexico. The time scale in seconds shows the time required for sound to travel through the water to a reflecting horizon and back to the ship (see Figure 4-1b). Note the abrupt break in slope between the continental shelf and the continental slope. The contorted reflections on the lower slope may be material that slumped down from above.

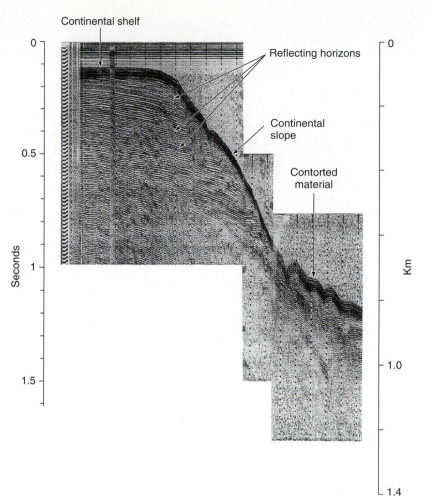

Continental shelf

Reflecting horizons

Continental slope

Contorted material

Seconds

Km

Crustal Structure

Since early in this century, scientists knew that the Earth could be divided into three basic layers: crust, mantle, and core. (The general characteristics of these layers were discussed in Chapter 2 and summarized in Table 2-1, page 26.) From seismic studies made by the 1950s, oceanographers determined that the ocean's crust could also be divided into three layers, based on their sound velocity (Figure 4-4):

1. Sound velocity in the **unconsolidated sediment** is slow, about 2 to 3 km/second (1.2 to 1.8 mile/second). Direct sampling tells us that this layer is sediment. A relatively thin layer overall, it is approximately 300 m (about 1,000 ft.) thick in the Pacific and 600 m (2,000 ft.) in the Atlantic.

2. Sound velocity in the *second layer* is about 3.5 to 6.0 km/second (about 2.2 to 3.7 mile/second), averaging about 5.0 km/second (about 3.1 mile/second). The thickness of this layer is about 1.7 km (about 1.05 miles). Both sound velocity analysis and subse-

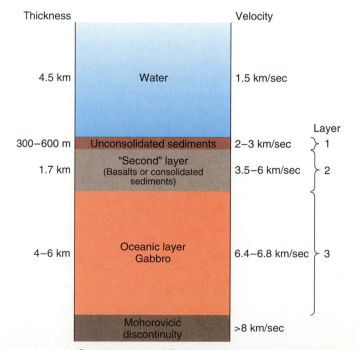

Thickness		Velocity	
4.5 km	Water	1.5 km/sec	
300–600 m	Unconsolidated sediments	2–3 km/sec	Layer 1
1.7 km	"Second" layer (Basalts or consolidated sediments)	3.5–6 km/sec	2
4–6 km	Oceanic layer Gabbro	6.4–6.8 km/sec	3
	Mohorovičić discontinuity	>8 km/sec	

FIGURE 4-4 General structure of Earth's crust under the ocean, as determined by seismic refraction and reflection studies. These are just the major layers; numerous sublayers are possible within these larger ones. The scales are approximate.

quent seafloor drilling have shown this layer to be mainly basaltic rock. **Basalt,** a common rock in oceanic area, is formed from volcanic processes and is relatively rich in iron and magnesium.

3. Sound velocity in the **oceanic layer** is between 6.4 and 6.8 km/second (3.9 and 4.2 mile/second). Thickness is between 4 and 6 km (2.5 and 3.7 miles). The layer's uniformity suggests that it is a major feature of the oceanic crust. This layer sometimes occurs near the surface, at places like the Mid-Atlantic Ridge.

These three layers constitute the oceanic crust of Earth. Oceanic crust and continental crust are fundamentally different, both in thickness and composition (Figure 4-5). Oceanic crust is between 5 and 10 km (3–6 miles) thick, whereas continental crust averages about 35 km (21.7 miles) thick. The main rock type in the continents is **granite,** whereas that of the oceanic crust is **gabbro.** Granites and gabbros both form from cooling magma. Generally, granites are lighter in color, are less dense, and have slower sound velocity than gabbros. Gabbros also contain more manganese and less silica compared with granites.

Mohorovičić Discontinuity

At the bottom of the third or oceanic layer, and at the base of the continental crust, is the *Mohorovičić (Moho) discontinuity,* which marks the base of the crust and the beginning of Earth's mantle. At this discontinuity, the speed of sound increases to about 8.0 km/second (about 4.9 mile/second). There is considerable controversy over whether this layer represents a *chemical change* (that is, from rocks composed of one set of elements to rocks composed of another set) or a *physical change* (from the solid phase to the liquid phase, like ice to water) due to increased pressure and temperature.

The Moho generally occurs about 6.5 km (about 4 miles) below most of the ocean. There is, however, some variation. In most marginal seas (such as the Red Sea or the Caribbean Sea), the Moho occurs at a depth intermediate to that of the ocean and the continents. The Moho also occurs at intermediate depths along the margins of the continents.

Density and Isostasy

Let us compare the densities of continental crust, oceanic crust, and Earth's mantle (in grams per cubic centimeter, or g/cm³):

2.8 g/cm³, continental crust rocks (average)

3.0 g/cm³, oceanic crust rocks (average)

4.5 g/cm³, mantle density (approximate)

The basic principle here is that low-density things float, and high-density things sink. The fact that the continents are higher than the ocean basins is explained by *isostasy,* which was described in Box 3-2 (page 44). The concept is that various parts of the crust "float" on the denser mantle at a height dependent on their mass. This is just like the height at which objects float in water, which depends on mass (compare the floating heights of dense wood and a beach ball). Ocean basins float lower than the continents because they are composed of denser material. This explains the basic difference in elevation between the higher continents and lower ocean basins (see Box 3-2, page 44).

Trenches, however, are one portion of Earth's crust that does not completely follow the isostasy concept, going deeper than density alone would account for. This is because subduction depresses the trench further.

The ocean floor can be divided into two main parts, based on either its depth (see Figure 2-7, page 35) or on its crustal structure (Figure 4-5). These two main parts are the continental margin and the ocean basin (Figure 4-6). We will examine the continental margin first.

The Continental Margin

The **continental margin** comprises the coastal region (including beaches), continental shelf, continental slope, and continental rise or borderland—in other words, that portion of the ocean immediately adjacent to the continents. In area, continental margins make up only about 21 percent of the total ocean (Table 4-1). However, to people, they are the most valuable part.

Continental margins may be classified by their role in seafloor spreading and plate tectonics as either active or passive. A *passive margin* shows no obvious motion, although it is moving along with the plate. It also has limited seismic activity. The North American Plate spans the area from the Mid-Atlantic Ridge to the U.S. West Coast. The entire East Coast of the U.S. and the Gulf of Mexico (the western edge of the Atlantic) is a passive margin. Spreading along the Mid-Atlantic Ridge must cause subduction somewhere, and indeed it occurs on the active Pacific side of the plate, but not on the passive East Coast (see Figure 3-13, page 56). An *active margin*, on the other hand, has some evident plate motion. The motion can be divergent, convergent, or transform (see Figure 3-11, page 54). California, Oregon, and Washington together are an active continental margin where the motion is mainly of the transform variety.

The following eight sections examine portions of the continental margin: the coastal region; the economically and politically important continental shelves; an under-

FIGURE 4-5 Crustal structure under the land and the ocean basin, as determined by geophysical techniques. Note that the abrupt change between continent and ocean structure occurs in the continental slope area.

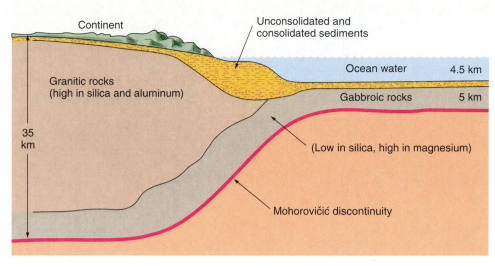

water tour of continental margins of the three U.S. coasts (East, West, and Gulf); continental slopes; continental rises; majestic submarine canyons; mysterious turbidity currents; and marginal seas.

The Coastal Region

The **coastal region** is that part of the continent immediately adjacent to the ocean, and therefore much influenced by it. The coastal region includes the coast and shoreline, beaches, estuaries, lagoons, marshes, and deltas. We will examine this in detail in Chapter 14, but suffice it to say here that, although the coastal zone occupies only a small portion of the continental margin, it is one of the most important parts of the marine environment.

For example, over 50 percent of the U.S. population lives within 80 km (50 miles) of the ocean or of the Great Lakes. In other parts of the world, this percentage is often higher. Most large cities of the world are near the ocean; many are on an **estuary,** where a river meets the ocean. Estuaries, besides being good areas for ports, are often breeding areas for many marine species. Beaches are valuable both as recreational areas and as easily mined areas of sand and gravel. Deltas often have large accumulations of oil and gas.

Continental Shelves

Continental shelves are the shallow part of the seafloor immediately adjacent to and surrounding the land (Figure 4-7). They can be considered as the flooded part of the continent, for they are continental crust that sea level covers. Continental shelves are relatively smooth platforms, gently sloping or almost flat, that terminate seaward with an abrupt change in slope (the

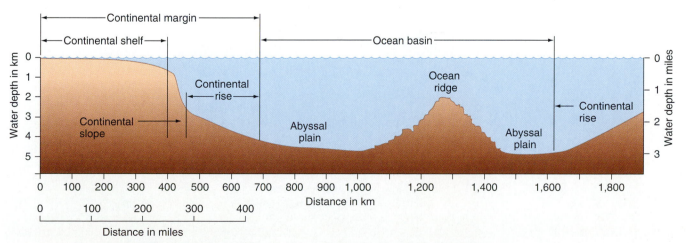

FIGURE 4-6 Diagrammatic profile showing the main features of the continental margin and the ocean basin. Note that the vertical exaggeration is about 50 times.

TABLE 4-1
Proportion of Oceans Occupied by Continental Margin and Ocean Basin

Feature	Pacific Ocean and Adjacent Seas	Atlantic Ocean and Adjacent Seas	Indian Ocean and Adjacent Seas	Arctic Ocean and Adjacent Seas	Total World Ocean
CONTINENTAL MARGIN[1]	**15.8%**	**25.1%**	**14.8%**	**89%**	**20.6%**
Continental shelves and slopes	13.1	17.1	9.1	68.2	15.3
Continental rise and partially filled sedimentary basins	2.7	8.0	5.7	20.8	5.3
OCEAN BASIN	**77.9%**	**72.2%**	**79.4%**	**4.2%**	**74.5%**
Abyssal plains	42.0	39.9	49.2	0	41.8
Mid-ocean ridges	35.9	32.3	30.2	4.2	32.7
OTHER AREAS	**6.3%**	**2.7%**	**5.8%**	**6.8%**	**4.9%**
Total Area[2]	**100%**	**100%**	**100%**	**100%**	**100%**
Percentage of Total World Ocean	**(50%)**	**(26%)**	**(20.5%)**	**(3.4%)**	

[1]The continental margin has a total area of about 74.5 million km^2 (28.8 million mi^2).

[2]Total area percentages do not add to exactly 100% due to rounding.

Source: Menard and Smith, 1966.

shelf break, or shelf edge), which leads to the continental slope (Figure 4-7).

Defining the continental shelves has long been important. In the past, certain rights of the sea, such as free passage of ships, mineral rights, fishing rights, and military jurisdiction, were determined by the position of the continental shelf. These rights have been changed in the recently concluded Law of the Sea Treaty (see Chapter 19).

In the past, continental shelves were legally defined by distance from land (the "three-mile limit" you may have heard about in the news) or by water depth. However, neither criterion was geologically acceptable, because some shelves are exceptionally wide or narrow, and others are unusually shallow or deep. Under the new treaty, each coastal country now can claim an Exclusive Economic Zone that is 200 nautical miles (370 km) wide (see Figure 19-6, page 472). Within this zone, each country can control mineral and fish exploration and exploitation, as well as marine scientific research.

The basic topography of the continental shelf is fairly well known in most areas, and some generalizations can be made. Shelf topography is usually smooth, interrupted by many small hills, valleys, and depressions. Francis P. Shepard, who some consider to be the father of marine geology, noted in 1973 these statistics of continental shelves:

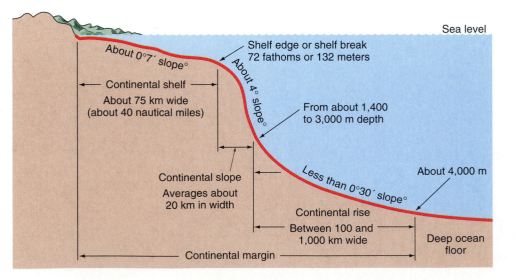

FIGURE 4-7 General characteristics of the continental margin (not drawn to scale, and the vertical exaggeration is great). See Figures 14-1 and 14-2 for some coastal zone features.

1. The average shelf width is 75 km (40.5 nautical miles).

2. The average shelf inclination is 0°07' (7 minutes, or only 7/60 of a degree, a slope so gradual as to be undetectable by the unaided eye).

3. The average depth of the flattest portion of the shelf is about 60 m (197 ft.).

4. The average depth where the greatest change of slope occurs is 130 m (426 ft.).

Although the continental shelves of the world have many variations, two very general types prevail: (1) glaciated shelves and (2) unglaciated shelves (Box 4-3). Unglaciated shelves are the more common due to the limited extent of glaciation, and there are many

Box 4-3
Glaciers on the Continental Shelf

During the Pleistocene epoch, numerous advances and withdrawals of glaciers occurred over the northern parts of North America, Europe, and Asia. During the most recent Pleistocene glaciation, which attained maximum extent about 18,000 years ago, northern areas that today are Chicago, Montreal, Detroit, and even Cape Cod were covered by glacial ice that frequently exceeded 1 km (0.6 miles) in thickness.

A glacier is essentially a large mass of flowing ice. A glacier can grow when steady snowfall adds to the mass at its upper end, and it can shrink when its leading edge melts faster than the rate of snow accumulation. It can also remain fairly constant in size if the snow addition and the melting occur at balanced rates. Regardless, a glacier moves continuously due to the force of gravity.

Glaciers have considerable erosive power. As they move, they scour, abrade, and erode the underlying bedrock and sediment, incorporating this eroded material into the moving ice (Figure 1). At the leading edge of the glacier, the ice melts and a drainage pattern develops, with a system of small streams and channels to carry away the meltwater. Included within the meltwater is finer sediment that eventually settles out, form-

ing a relatively smooth feature called an *outwash fan*.

During the Pleistocene, when glaciers reached the exposed parts of the continental shelf, they caused considerable erosion and change to the character and topography of the underlying rock and sediment. After the glacier melted back, the previously smooth and easily erodible continental shelf had a wide variety of distinctive, irregular topographic forms typical of glaciated areas on land.

Much of the sediment and rock carried by a glacier is deposited as a **moraine** at the glacier's edge. Moraines, which include a broad mixture of sediment rock types and sizes, can also change the character of the exposed parts of the coastal region. For example, Cape Cod and Long Island are composed of moraines formed of rocks and sediments carried by the Pleistocene glaciers originally from as far away as Canada.

On the shallower parts of the shelf, waves and currents move and redistribute the seafloor sediment and slowly smooth and obscure many of the glacial features. Areas deeper and less sensitive to this smoothing retain their glacial characteristics, as does much of the shelf off the northeastern United States and most of Canada (see Figure 4-8).

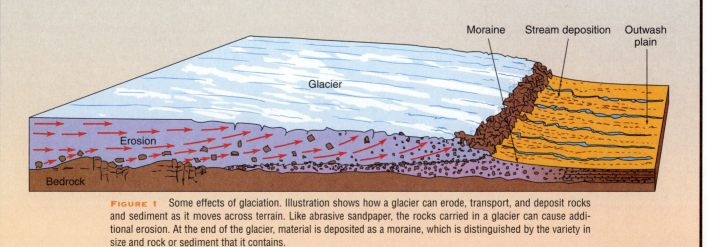

FIGURE 1   Some effects of glaciation. Illustration shows how a glacier can erode, transport, and deposit rocks and sediment as it moves across terrain. Like abrasive sandpaper, the rocks carried in a glacier can cause additional erosion. At the end of the glacier, material is deposited as a moraine, which is distinguished by the variety in size and rock or sediment that it contains.

varieties. Recall that most shelves were recently *above* sea level because of the lowered sea level in the Pleistocene epoch, when glaciation was extensive (see Figure 2-5, page 31). When the glaciers started melting and sea level started rising about 18,000 years ago, the shoreline migrated from what is now the submerged outer edge of most continental shelves to its present position.

Some shelf areas have been geologically unstable in that they have moved up or down, independent of sea-level changes. For example, parts of the California coast where beaches formed a few thousand years ago are now several meters above present sea level, indicating that this area has been rising faster than sea level.

A Model: The U.S. Continental Margin

Following is a brief description of the continental margin of the United States, presented as a model of margin characteristics worldwide. Exploring all the fascinating variations of continental margins is beyond our scope here, so if you are interested in learning more, see Shepard's book, *Submarine Geology*, cited in the "Further Reading" section.

EAST COAST. Both glaciated and unglaciated shelves have been affected by the rise of sea level. Figure 4-8 is a three-dimensional view from northeastern Canada to the Gulf of Mexico. In the northern part of this area, glaciers covered and eroded the shelf (Box 4-3). This erosion is shown by the numerous depressions, topographic highs, basins, and valleys on the shelf.

The irregular shelf topography off Nova Scotia and in the Gulf of Maine indicates that the glaciers may have extended to Georges Bank. Georges Bank, however, is very shallow and exposed to the smoothing effect of

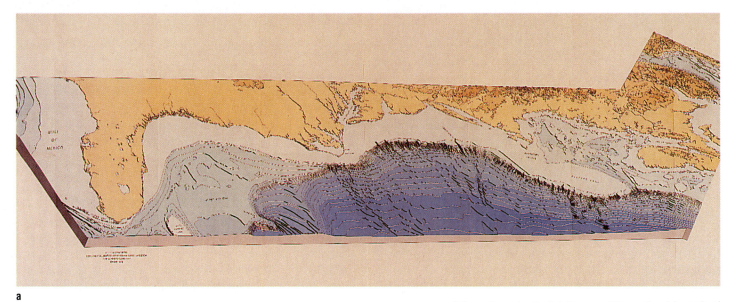

a

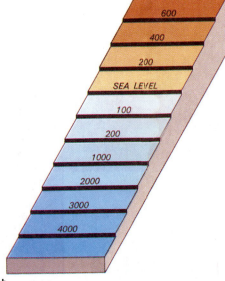

b

FIGURE 4-8 A three-dimensional relief diagram of the continental margin of eastern North America, with the Gulf of Mexico to the left and the Bay of Fundy to the right. The three-dimensional appearance results from the contour intervals being shown as steps or layers. The contour interval is 200 meters, except that the 100-meter contour below sea level is shown. (Figure courtesy of Dr. David Monahan from his 1971 paper.)

waves and tides, so little topographic evidence of the glaciation remains. The channel between Georges Bank and the Nova Scotian shelf was cut by a glacier.

South of Cape Cod, particularly in the Long Island area, there is an abrupt change in the character of the shelf. The deep basin, channels, and topographic highs seen to the north are absent, and the topography becomes relatively smooth. This is an **unglaciated shelf,** which is generally smoother than **glaciated shelves,** although some local relief may appear because of strong currents, submarine canyons, or folding and faulting.

Off coasts where very strong currents flow, the continental shelf can be narrow or almost absent. An example is the east coast of southern Florida, where the Gulf Stream, flowing at speeds up to about 11 km per hour (6 knots), comes close to the mainland. The current is strong enough to prevent deposition of most sediment and apparently has not permitted normal development of a shelf in this area. The Gulf Stream is so strong here that, even at depths of several hundred meters, it has scoured and removed much of the bottom sediment.

GULF COAST. Another variety of continental shelf occurring near large rivers is a **delta.** For example, the Mississippi River channel has extended across most of the continental shelf and is presently supplying sediment to the deeper parts of the Gulf of Mexico (Figure 4-9; see also Figure 14-20c, page 000). A similar example is the Nile River Delta building out into the Mediterranean Sea.

East of the Mississippi Delta is a broad continental shelf with a steep outer cliff, or *scarp*. The scarp is part of an extinct **reef** that formed in the Jurassic Period, over 140 million years ago. Offshore in the central part of the Gulf of Mexico is a relatively flat, deep basin that contains numerous areas where salt has risen up through the overlying sediment, forming **salt domes.** These are sites where oil or gas deposits often occur (see Figure 15-4, p. 369).

WEST COAST. The West Coast of the United States differs considerably from the Gulf of Mexico and the East Coast. The differences reflect the nature of the continental margins: passive (eastern) versus active (western). Much of the character of the East Coast and Gulf Coast results from sedimentary processes. Along the West Coast, how-

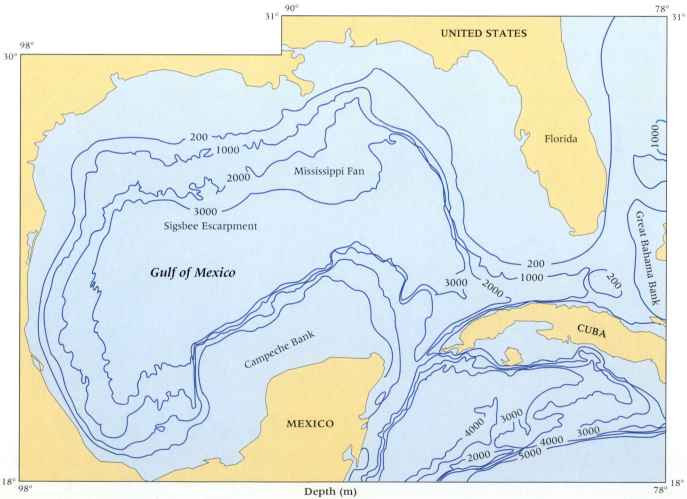

FIGURE 4-9   General bathymetric chart of the Gulf of Mexico. Contours are in meters.

ever, tectonic activities predominate—faulting, earthquakes, and seafloor spreading. Volcanic activity occurs along much of western North and South America, from Alaska to Chile. Evidence of this is recent eruptions of Mount Saint Helens and volcanoes in Alaska and Chile (see Figures 3-23 and 3-24, page 65).

Off California, Oregon, and Washington, a series of small northwestward-trending narrow basins exists (Cascadia Basin is an example), forming a **continental borderland** (Figure 4-10). Two fracture zones (Blanco and Mendocino) and numerous submarine canyons also cross the region (Figure 4-11). Still farther north, off

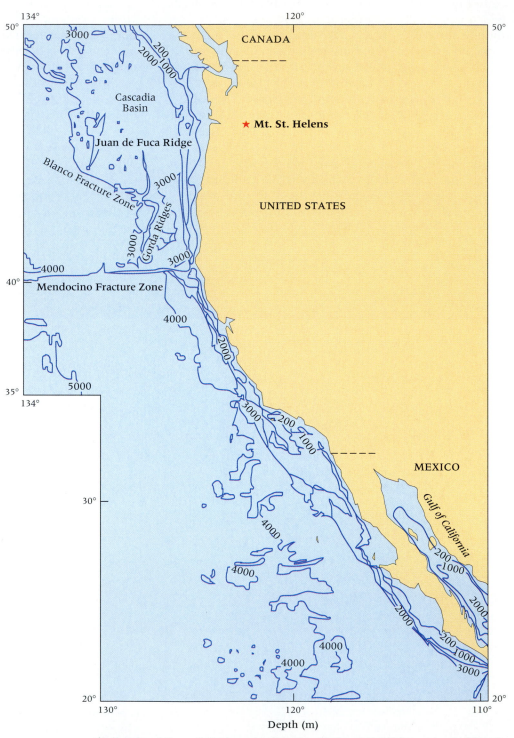

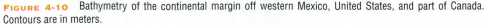

FIGURE 4-10 Bathymetry of the continental margin off western Mexico, United States, and part of Canada. Contours are in meters.

Canada and eventually Alaska, the more typical shelf topography continues, eventually ending in the deep Aleutian Trench.

In general, the shelves off the West Coast of the United States are unglaciated and relatively narrow. They often are covered with only a thin veneer of sediment, overlying rocks. Some glaciation is evident off the coast of Washington and British Columbia.

The islands of the Pacific generally have narrow continental shelves, or none at all. These islands often occur in a chainlike pattern (the Hawaiian Islands are an excellent example) and are volcanic in origin, as explained in Chapter 3. Their surface and flanks are sometimes covered with thick sequences of coral deposits. Indeed, some are **coral atolls,** circular series of coral reefs surrounding a central lagoon (see Figures 12-15 to 12-18, pages 397–399).

Continental Slopes

Continental slopes occur between the two major topographical features of our planet: the land (including the continental shelf) and the ocean basin (see Figures 2-6 and 4-7). The continental slope extends from the outer part of the continental shelf and usually down to the deep seafloor. In some instances, the junction with the deep seafloor may be hard to define, either because of a gradual decrease in slope or because a deep sea trench borders the continental shelf. The region where the slope flattens out, near the deep seafloor, is called the **continental rise** (see Figure 4-7). In some regions, such as off California and on the Blake Plateau off southeastern Florida, an intermediate continental borderland exists between the deep sea and the continental shelf.

The average inclination of the continental slope is about 4°. Figure 4-7 shows this with considerable exaggeration; a 4° slope actually is like walking down a modest hill, descending about 7 meters (23 ft.) for every 100 meters (328 ft.) walked. However, in some areas the inclination can be 20° or more, equivalent to walking down a steep hill. The inclination of the continental slope off large rivers and deltas, by contrast, can be as gentle as 1°. Shepard noted that slopes off coasts that were formed by faults average about 5.6°; slopes off mountainous coasts average about 4.6°; slopes off stable coasts, about 3°; and slopes off major deltas, about 1.3°. The Pacific coast, which follows a major earthquake zone, tends to have somewhat steeper slopes than the Atlantic or Indian Oceans.

The structure of some continental slopes has been examined by continuous seismic profiling. Results show considerable variation, but most slopes exhibit some **slumping** or sliding of the overlying sediments, sometimes exposing the underlying rock. The general explana-

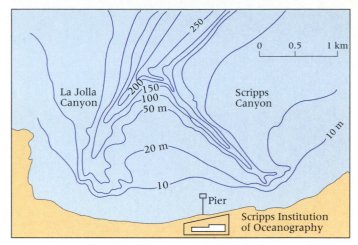

FIGURE 4-11 Two submarine canyons, the La Jolla Canyon and Scripps Canyon, near the Scripps Institution of Oceanography in La Jolla, California. Contours are in meters. (After Shepard and others, 1964.)

tion for the origin of most continental slopes is that they are due to a combination of faulting and/or slumping.

Continental Rises

Many continental slopes end in a gently inclining, broad topographic feature called the **continental rise,** which usually has an inclination of less than 0.5° (for example, a 1 m drop over a distance of 180 m as you move away from the coast). Compare this to the average inclination of the continental slope of about 4°. The rise, where present, typically extends seaward between 100 and 1,000 km (60 and 600 miles). The continental slope is far less extensive, averaging about 20 km (12 miles) in width (see Figure 4-7). The relief of most rises is very smooth, generally interrupted only by **submarine canyons** or **seamounts.**

Continuous seismic profiles from continental rises generally show a wedge-shaped sequence (layers) of sediments that thickens toward the continental slope. These sediments can be up to 10 km (6 miles) thick and frequently contain material that has slumped down from the continental slope.

The presence of continental rises is strongly controlled by seafloor spreading. Where the ocean floor is subducting under the continents, rises generally do not form (see Figure 3-9, page 52). But where such motion is absent, thick continental rises may occur. For example, a continental rise occurs along 85 percent of the Atlantic and Indian Oceans but along less than 30 percent of the Pacific Ocean (Figure 4-12). Petroleum potential exists in areas where these thick sediment sequences occur (for more on this, see Chapter 15, page 368).

The thick sediment sequence in continental rises is comparable to thick sedimentary rock sequences found in many mountain ranges. Indeed, given enough time

FIGURE 4-12 Location of continental rises (dark brown areas) around the world.

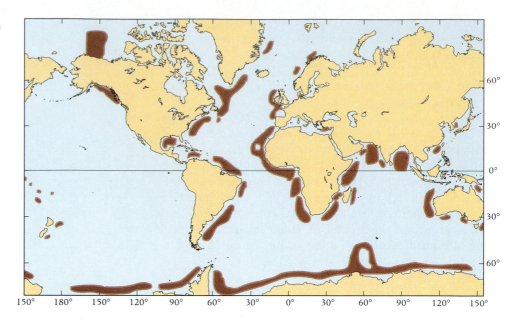

(millions of years) and the right tectonic situation, continental rises and continental shelves may be uplifted above sea level and folded to form a mountain range of sedimentary rocks. The long, linear trend of both shelves and rises, and their parallel position to the edges of continents, are similar to the position and form of many existing mountain ranges before they were uplifted, such as the Appalachian Mountains in the eastern United States and the Sierra Nevada in California. This idea, which is consistent with the seafloor-spreading concept, indicates that continents may grow at their borders by the accretion of the sediments from the continental shelf and continental rise.

Submarine Canyons

Some regions of the continental shelf and slope are cut by submarine canyons. These features sometimes even cross the continental rise and extend into the deep-ocean basins. Some canyons or valleys are of *glacial* origin or *fault* origin. However, the term **submarine canyon** is generally used for those canyons having winding, rock-walled, V-shaped profiles in cross section, often with tributary canyons that extend down the continental slope. The dimensions of some submarine canyons are impressive: Monterey Canyon off central California has a relief (top-to-bottom depth) of about 1,500 m (4,900 ft.), similar to that of the Grand Canyon. Hudson Canyon off New York's Hudson River extends seaward 240 km (150 miles) into the Atlantic and then continues another 240 km as a channel across the continental rise.

Among the best-studied canyons are those off the coast of southern California, especially two that are within a mile of the Scripps Institution of Oceanography:

La Jolla Canyon and its tributary, Scripps Canyon (Figure 4-11). Scripps Canyon lies about 200 m (650 ft.) offshore of a canyon on land. Sediment often accumulates in the "head" (nearshore) parts of this submarine canyon.

During certain times of the year, especially during storms, the sediment moves abruptly from the canyon into deep water because of slumping and strong currents. This sediment movement has enlarged and scoured Scripps Canyon. La Jolla and Scripps Canyons continue as steep-walled features, increasing in depth to about 305 m (1,000 ft.). Both extend about 1.6 km (1 mile) from land, where they join. A little farther seaward, the relief of the precipitous walls is less, and the canyon becomes more of a small channel cut into the underlying sediments. The channel continues seaward until it ends in a thick sedimentary deposit.

Submarine canyons along the northeast coast of the United States generally do not have tributaries. They usually extend straight down the continental slope (Figure 4-13), unlike most California canyons, which cross it at an angle. Atlantic-coast canyons also tend to have a more V-shaped profile and contain rock outcrops.

What causes submarine canyons? Their origin is not easily resolved, and several factors may contribute. Numerous indications of erosion by strong currents have been observed by divers and from submersibles (Figure 4-14). These strong currents may be *turbidity currents*, discussed in the next section. The large accumulation of terrestrial sediment (sediment originated on land) at the seaward end of most canyons also argues for the existence of turbidity currents.

Some scientists have noted the similarity of submarine canyons to land canyons cut by streams. They suggest that the submarine canyons were cut in normal sur-

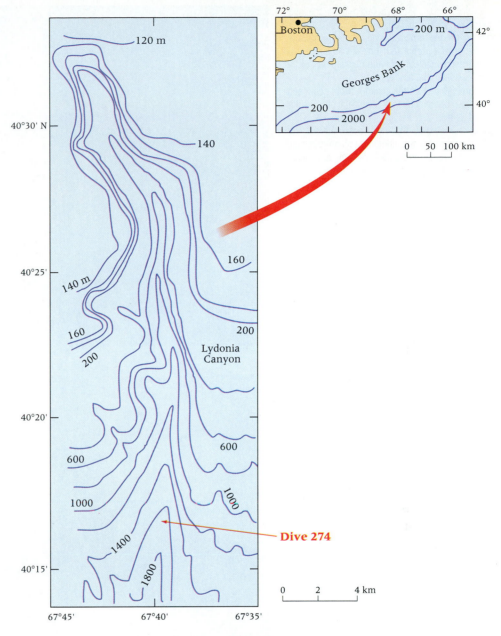

FIGURE 4-13 Lydonia Canyon, on the seaward side of Georges Bank southeast of Boston. Note the generally straight trend of the canyon. The arrow points to the location of a dive (number 274) the author made with *Alvin* into the axis of the canyon (see also Figure 4-14). Contours are in meters.

face-erosion fashion during times when lowered sea level exposed these areas. At such times, rivers would have crossed and cut into most continental shelves and parts of some continental slopes. However, proponents of this hypothesis have difficulty reconciling that many canyons exist below the depth of even the most extreme estimates of lowered sea level.

On the other hand, it is hard to imagine how turbidity currents could cut a winding canyon, or be strong enough to erode the hard rock into which some canyons are carved. It is possible that both hypotheses are partially correct: the upper parts of some canyons could have been cut during times of lowered sea level, and then subsequently enlarged and deepened by turbidity

currents. Slumping or faulting may also be a contributing factor (Figure 4-15).

Turbidity Currents

Marine sediments can be agitated and become resuspended in the water. When this happens, they produce a dense mixture that flows downslope along the ocean bottom, something like a mudflow. When such a flow results, it is called a **turbidity current** (because the water is turbid, which means dense, cloudy, and disturbed). If the velocity and turbulence of the current are great enough, or if the slope is steep enough to prevent settling of the sediment particles,

FIGURE 4-14 Pictures taken from the submersible *Alvin,* in Corsair Canyon, another canyon on the seaward side of Georges Bank, northeast of Lydonia Canyon (see Figure 4-13). Note the mechanical arm and instruments of the submersible in the figures.
(a) Ripple marks on the ocean-bottom sediments in the axis of the canyon at a depth of about 1,604 m (5,262 ft.).
(b) A rock outcrop that has been undercut. The ripple marks and the undercutting are probably due to strong currents. The shadows are caused by the light behind the mechanical arm.

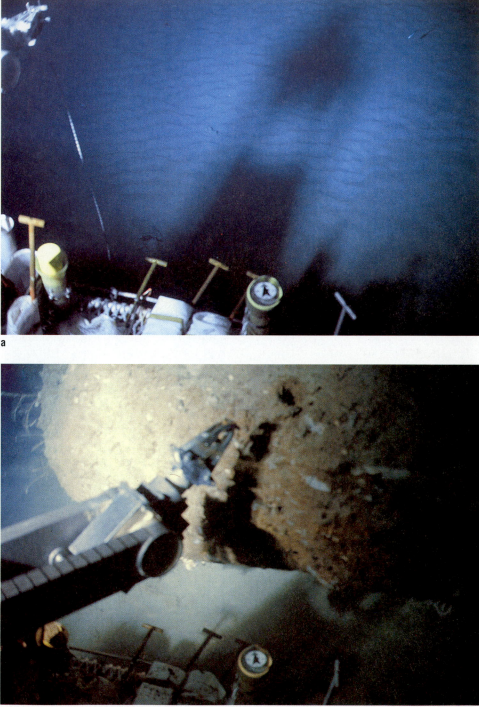

the current can flow for distances exceeding 100 km. Eventually the current stops, losing energy so that its sediment load is deposited. The resulting deposit is called a **turbidite.**

Turbidity currents, their origin, and their effects are controversial topics in marine geology, especially because few direct observations have been made. There is, however, considerable indirect evidence that turbidity currents exist: the loss of instruments placed where these currents are thought to flow, and movement of large bodies of sediment.

One of the first indications of turbidity currents was noted in 1929 after an earthquake on the Grand Banks of Newfoundland. The earthquake somehow broke submarine telephone cables that provided communication between Europe and North America at that time. Geolo-

FIGURE 4-15 Diagrammatic representation of the structure in the axis of Lydonia Canyon (see also Figure 4-13) at a depth of about 1,600 m (5,250 ft.). This interpretation is based on a dive made into the area with *Alvin*. It appears that the structure here is mainly formed by slumping.

gists B. C. Heezen and M. Ewing called attention to a definite sequence in the breaking of the cables: the closer the cables were to the earthquake's epicenter, the sooner they broke. The relatively slow time between the breaks eliminated the earthquake itself as the direct cause, for earthquake waves travel much faster than the rate at which the cables were broken.

The cable breaks may instead have been caused by a turbidity current moving down a slope, snapping communication cables sequentially as it went. Subsequent observations in other areas also have indicated that earthquakes or a series of strong waves can induce movement of large amounts of sediment that possibly result in turbidity currents. In the Grand Banks case, 13 transatlantic telephone and telegraph cables broke in a

sequence along a path 720 km long, continuing after the earthquake. Scientists deduced that the quake had triggered turbidity currents that rolled along, breaking the cables.

The speed of turbidity currents is often far greater than that in river floods. Scientists have computed turbidity currents to have maximum speeds to 120 km/hour, based upon the elapsed time between breaks of submarine cables. Average speed, however, is closer to 50 km/hour.

The best evidence for turbidity currents is provided by numerous deep-sea cores collected from the abyssal plains of the ocean. These cores often contain turbidites (which are generally layers of coarse-grained, sand-sized sediment) interlayered between the normal fine-grained clay deposits of the deep ocean. The sand layers, because of their grain size and mineral composition and their intermixture with numerous shells of shallow-water organisms, are probably derived from a nearshore source.

The only known mechanism that could transport the coarse-grained sediments so far and deposit them as layers is a turbidity current. The sand layers generally have finer-sized material toward the top of the layer and coarser-sized material toward the bottom of the layer, which is what one would expect if the sediment were settling out of a current.

Many areas of the ocean are protected by trenches or are distant from land, and thus have no turbidity current deposits. The topography of these areas generally is hilly and rough (Figure 4-16a). Deep-sea regions where turbidity current deposits have smoothed the topography by filling in the low areas are called **abyssal plains** (Figure 4-16b). These sediment-laden currents, originating from the shallower continental shelf and slope, cover and smooth out most previously existing topographic irregularities as they deposit their sediments. Continuous seismic profiles taken across abyssal plains clearly show that a deeper irregular topography has been buried by sediment.

FIGURE 4-16 Turbidity current deposition in the ocean.
(a) In a trench area, the turbidity current is intercepted, slowed, loses its sediment load (turbidites), and thus cannot deposit sediment in more seaward areas.
(b) In an abyssal plain area, turbidity currents travel much farther, depositing turbidites to some distance beyond the base of the continental slope.

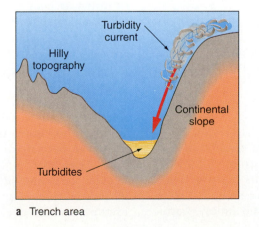

a Trench area

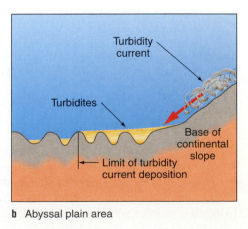

b Abyssal plain area

Marginal Seas

Marginal seas are a common feature of the ocean—for example, the Mediterranean, Red, Black, and Caribbean Seas (see Figure 2-9, page 36). Interestingly, they are not easily classified as either part of the continental margin or as ocean basin. These seas often have a crustal structure intermediate in thickness between continental margin and ocean basin. Their water depths are usually more similar to ocean basins. Some marginal seas, such as the Bering Sea and the Sea of Okhotsk (along the eastern coast of Russia), are associated with volcanic islands. Others, such as the Mediterranean, Red, and Black Seas and the Gulf of Mexico, are almost surrounded by land. Most lie near present or past structurally active areas.

Marginal seas generally have thick accumulations of sediment for three reasons: their age (many have been around a long time to accumulate sediment); their closeness to continents (sources of sediment); and the many large rivers flowing into them (vehicles for transporting the sediment into them).

Many marginal seas have a restricted connection with the ocean and thus, have a restricted exchange of water with the open ocean. This is important because it favors accumulation of organic material. (The lack of water exchange means that no oxygen replenishment is provided by bottom waters carried in from the ocean; such oxygen-rich bottom waters would oxidize the organic material.)

This accumulation of organic material, combined with common coarse-grained sediment occurring in marginal seas, creates an environment favorable for production and accumulation of oil and gas. The actual formation of these resources can take millions of years, of course. Many of these present seas may hold large undiscovered quantities of hydrocarbons. Significant concentrations have already been found and exploited from the Persian Gulf and in the Gulf of Mexico.

The Ocean Basin

The following four sections explore the four largest divisions of Earth's great ocean: the Atlantic, Pacific, Indian, and Arctic Oceans.

The Atlantic Ocean

The Atlantic Ocean is the second-largest division of the world ocean, but is probably the most explored. Viewed from space, the Atlantic's most obvious feature is its shape—an elongated, sinuous basin that extends north to south over 11,000 km (close to 7,000 miles). Early bathymetric studies showed that its topography is dominated by the Mid-Atlantic Ridge, which runs continuously down the central part of the ocean.

Scientists have long been fascinated by the symmetrical shape of the ocean on either side of this ridge (Figure 4-17), and by the jigsaw-puzzle fit of the adjacent continents if they were pushed together (see Figure 3-4, page 46). These observations support the hypothesis of continental drift and, more recently, the hypothesis of seafloor spreading.

Further studies of the Mid-Atlantic Ridge have shown it to be part of an essentially continuous worldwide oceanic feature that can be traced more than 65,000 km (over 40,000 miles) through the Atlantic, Indian, Arctic, and Pacific Oceans (Figure 4-18). The ridge is actually the most significant topographic feature on Earth, although it is hidden by water. In the Atlantic Ocean, the ridge is especially wide and occupies the middle third of the ocean.

The ridge is highly fractured, consisting of numerous mountains and hills. The most rugged topography occurs in the central part of the ridge, where a large central rift or crack appears on most cross sections. The central rift can be as much as 2,000 m (6,500 ft.) below the adjacent mountain peaks, which often rise to within about 300 m (1,000 ft.) of sea level. In some instances, the mountains actually protrude through the surface of the ocean, forming islands; Iceland is the largest example.

Studies of the central rift in the Mid-Atlantic Ridge, using submersibles, show considerable volcanic activity and faulting. Recently some hot vent areas have been discovered along the ridge (see Figure 3-25, page 67). This type of activity is expected in the central rift region because of its seafloor-spreading origin. Further evidence of seafloor spreading is seismic activity in the ridge area and the observation that most of these earthquakes have their focus beneath the central rift area. In many places, the ridge is offset in an east-west direction by transform faults (see Figure 3-8, page 51). These faults are east-west trending and cause a horizontal offset of the ridge crest (see Figure 4-17).

Rocks dredged or drilled from the ridge are the product of volcanic activity. They are basaltic or gabbroic igneous rocks, distinct from typical igneous rocks on the continents, which are granite. Sediments on the ridge are generally young and thin compared to locations away from the ridge.

On either side of the ridge are large, flat basin areas, the abyssal plains (see Figures 4-6 and 4-17). They are common to many parts of the deep sea, comprising about 42 percent of all seafloor worldwide (see Table 4-1). These plains, however, are not necessarily connected; they may be separated by ridges or other topographic features. The flatness of abyssal plains has been ascribed to the effects of turbidity currents, as noted

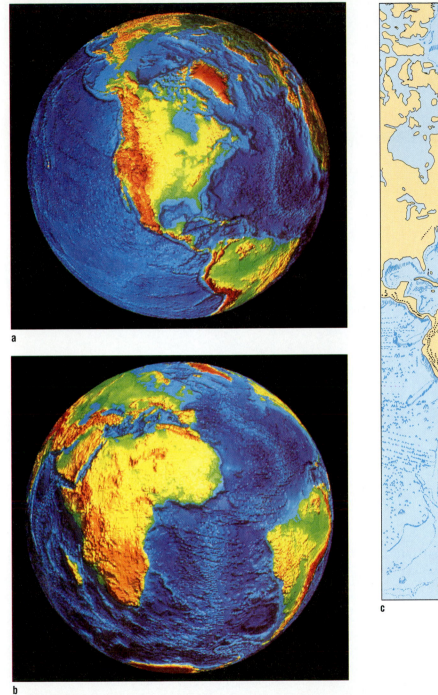

a

b

c

FIGURE 4-17 Computer-generated global view of the Atlantic Ocean, showing land and undersea topography.
(a) View from a position over Wisconsin, at 90°W longitude, centered on 45°N latitude.
(b) View from a position over the South Atlantic Ocean off the African coast, at 0°E longitude, centered on 0° latitude (the Equator).
(c) Drawing of general topographic features of the Atlantic Ocean. Note how the Mid-Atlantic Ridge dominates the topography. Slopes are highly exaggerated. (Images (a) and (b) generated from a land and sea database using a 5-minute (0.05) latitude/longitude grid, except for areas of the ocean that have not been surveyed in such detail; courtesy of Dr. Peter W. Sloss, National Oceanic and Atmospheric Administration/National Geophysical Data Center.)

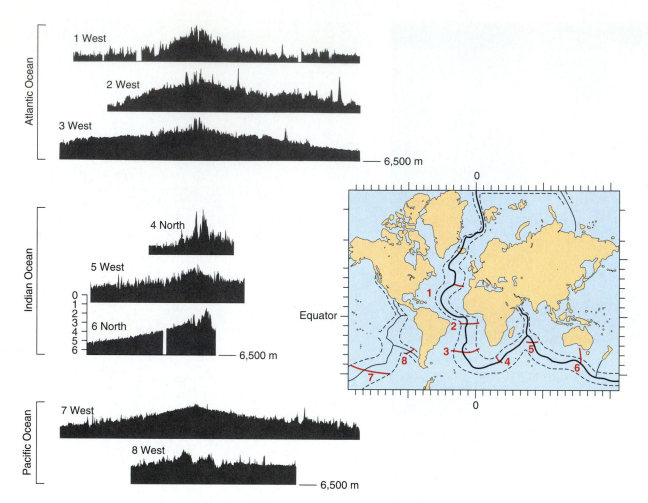

FIGURE 4-18   Profiles across the Mid-Atlantic Ridge, Indian Ocean Ridge, and East Pacific Rise. For all profiles, baseline is 6,500 m (21,326 ft.). The depth scale is in kilometers. (From Heezen and Ewing, 1963.)

earlier. Typical slopes are 1:1000 or less; for example, 1 meter in 1,000 meters or 10 meters in 10,000 meters.

The western continental margin of the Atlantic is generally passive, but a couple of exceptions are creating deep-sea trenches: the Puerto Rico Trench in the Caribbean region and the Sandwich Trench off southern South America. The larger Puerto Rico Trench is about 1,550 km (970 miles) long and 8,605 m (28,224 ft.) at its deepest. Impressive as this depth may seem—nearly as deep beneath the sea as Mount Everest is high above the sea—the trenches of the Atlantic are small compared to their greater counterparts in the Pacific Ocean (Table 4-2).

The Pacific Ocean

The Pacific, the largest division of the world ocean, has a nearly circular shape. The topography and structure of the Pacific are different from those of the other oceans. Its most striking topographic feature is the almost continuous series of deep trenches along its outer edge (Figure 4-19). On the continent side of these trenches,

there usually exist folded mountain ranges or island arcs (see Figure 3-9a, page 52). The trenches are seismically active, and many earthquakes occur in their axes, or landward of them.

TABLE 4-2
Dimensions of Some Oceanic Trenches

Trench	Ocean	Depth (m)	Length (km)
Marianas	Pacific	11,022	2,550
Tonga	Pacific	10,882	1,400
Kuril-Kamchatka	Pacific	10,542	2,200
Philippine	Pacific	10,497	1,400
Puerto Rico	Atlantic	8,605	1,550
Peru-Chile	Pacific	8,055	5,900
Java	Indian	7,450	4,500
Middle America	Pacific	6,662	2,800

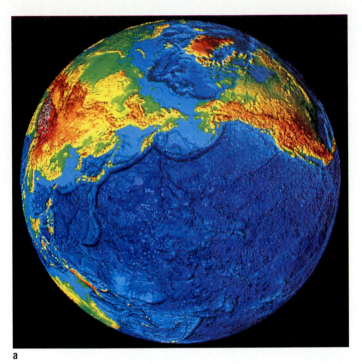

a

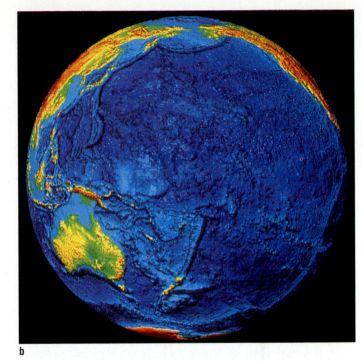

b

FIGURE 4-19 Computer-generated global view of the Pacific Ocean, showing land and undersea topography.

(a) View from a position over the Bering Sea off Alaska at 180°E longitude, centered on 45°N latitude.

(b) View from a position over the Pacific northeast of Australia at 180° E longitude, centered on 0° latitude (the Equator).

(c) Drawing of general topographic features of the Pacific Ocean. Note the trenches along the western and eastern parts of the ocean and along the coast of South America. Note also that the East Pacific Rise trends into the Gulf of California. Slopes are highly exaggerated. (Images (a) and (b) generated from a land and sea database using a 5-minute (0.05) latitude/longitude grid, except for areas of the ocean that have not been surveyed in such detail; courtesy of Dr. Peter W. Sloss, National Oceanic and Atmospheric Administration/National Geophysical Data Center.)

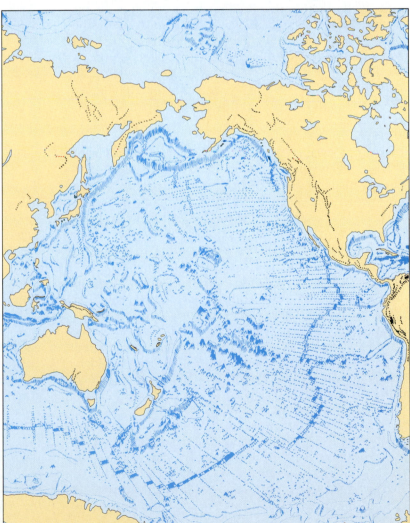

c

The circular area around the Pacific has been called the *ring of fire* because of the many volcanoes and earthquakes that occur. There is a gap in the ring along the West Coast of the United States and Canada where no trench exists. Here, earthquake incidence seems to be less, but volcanic activity still occurs, as evidenced by Mount Saint Helens (see Figure 3-23, page 65).

The deepest trench, and thus the deepest area of the ocean, is the Marianas Trench (Table 4-2). Here the bathyscaphe *Trieste* made its famous descent in 1960 to 10,910 m (35,795 ft.) near the Challenger Deep, the lowest point on Earth at 11,022 m (36,161 ft.). This is the bottom of the hypsographic curve in Figure 2-6 (page 35). It is interesting to compare the lowest and highest points on Earth: the Challenger Deep at −11,022 (−36,161 ft.) and Mount Everest at +8,848 m (+29,028 ft.), a total relief (difference) of 19,870 m, or nearly 20 km (65,190 ft., or over 12 miles).

Another major feature of the Pacific is the **East Pacific Rise,** a continuation of the oceanic ridge system in the Atlantic (the Mid-Atlantic Ridge). The ridge extends northeasterly as a broad rise from off New Zealand to the Gulf of California. It differs from the Mid-Atlantic Ridge in three significant ways: it extends across the edge of the ocean basin rather than down its center, its topography is relatively smooth, and it generally has a less-distinct central rift valley (see Figure 4-18). The crest rises in places as much as 3,000 m (10,000 ft.) above the general floor of the Pacific. Like the Mid-Atlantic Ridge, however, its crest is offset in many places by transform faults.

It was along the East Pacific Rise, near the Galápagos Islands to the west of Ecuador, that a remarkable discovery was made in 1977. Here a group of scientists using the submersible *Alvin* came upon a series of vents. Discharging from these vents was hot water, up to 400°C (750°F). The water was enriched with dissolved copper, zinc, and iron, as were the sediments in the adjacent area. Perhaps even more exciting was the unique biological life associated with the vents (see Figures 12-19 to 12-21, pages 168–170) and the associated chemical processes. These have changed many previous ideas about ocean chemistry (see pages 167–170). Later expeditions found similar vents along other parts of the East Pacific Rise and eventually along the Mid-Atlantic Ridge. The biology and chemistry of these vents are described further in later chapters.

Abyssal plains occur in some areas of the Pacific, but they are not so common as in the Atlantic, for two reasons. First, abyssal plains are formed when turbidity current deposits cover the previous topography (see Figure 4-16). The presence of an almost continuous belt of deep-sea trenches around the Pacific Ocean prevents many turbidity currents from reaching the more sea-

ward ocean basins, which are less deep. In other words, the current would have to run uphill to get out of the deep trench axis.

The second reason is that less sediment is carried by rivers into the Pacific than into the Atlantic (more of Earth's large rivers flow into the Atlantic). It therefore follows that the incidence of turbidity currents (containing land-derived sediments) should be considerably less in the Pacific than in the Atlantic. Turbidity currents can, however, also occur far from land when marine sediments slump. The resulting deposits tend to cover existing topography and can produce abyssal plains locally. This mechanism is less important, however, than the effects of land-derived turbidity currents.

Because most of the Pacific has not been covered by thick sequences of turbidity-current sediment, its topography has not been smoothed as much as has occurred in the Atlantic, and much of the original topography remains exposed. Consequently, a prominent feature is **abyssal hills,** rising 30 to 1,000 m (100 to 3,300 ft.) above the seafloor and up to several kilometers wide. In some areas, abyssal hills occur at the seaward end of abyssal plains, where they have not been completely covered by turbidites (see Figure 4-16).

A notable geological difference between the Atlantic Ocean and the Pacific Ocean is the relative absence of **seamounts** and coral atolls in the Atlantic. Seamounts are isolated submarine hills with steep sides, often standing more than 1,000 m (3,000 ft.) above the surrounding seafloor. Seamounts can be very large. One of the largest is the Great Meteor Seamount in the northeastern Atlantic, about 110 km (68 miles) in diameter at its base and attaining over 4,000 m (13,100 ft.) elevation above the seafloor. The area of its submerged summit is about the size of Rhode Island.

Rocks dredged from seamounts are usually basalt, suggesting a volcanic origin. Seamounts also have typical volcano shapes. Some seamounts have flattened tops; such a feature is called a **guyot** (pronounced *ghee-yo*). The flattening apparently resulted when the seamount was near sea level and subjected to erosion by waves. This could have happened when sea level was lower in the past, or when the seamount was initially formed at or near sea level and later subsided to a greater depth. If the seamount is located in the equatorial area, is shallow enough, and the right oceanographic conditions prevail, a coral atoll may develop on it (see Figure 12-18, page 299; this process is described further in Chapter 12).

The Indian Ocean

Somewhat less is known about the Indian Ocean than the Atlantic or Pacific. The continental margins (shelf, slope, and rise) of the Indian Ocean are similar to other oceans (Figure 4-20). Submarine canyons, although apparently fewer in number, do occur. The fact that fewer canyons are known to exist may not reflect reality but instead that fewer surveys have been made of the Indian Ocean. Some very large submarine canyons exist off the Indus and Ganges Rivers. They are cut into extensive areas of thick sediment that have been deposited by these rivers.

The Java Trench, with a maximum depth of about 7,450 m (24,436 ft.), extends northwest from western Australia. Shallower than the Atlantic trenches, it is the only major trench in the Indian Ocean (Table 4-2). However, the Java Trench spawns the active island-arc volcanoes in Indonesia that occasionally make headlines. It also created the famous Krakatoa volcano, which exploded in 1883, killing 33,000, causing one of the loud-

est noises in recorded history, and creating climatic disturbances for months afterward.

The Indian Ocean basin floor is dominated by abyssal plains having gradients from 1:1,000 to 1:7,000 and a relief usually of only a few meters. In some areas, small channels, possibly caused by turbidity currents, cut across the abyssal plains.

The most striking feature of the Indian Ocean is its mid-oceanic ridge, which appears to split in the center of the ocean, forming an inverted Y. One arm of the Y runs south of Australia into the Pacific, where it continues as the East Pacific Rise. The other arm of the Y runs south of Africa and continues as the Mid-Atlantic Ridge. The base of the inverted Y extends into the Gulf of Aden and the Red Sea.

The ridge is similar to the Mid-Atlantic Ridge. It is rugged, seismically active, and may have a continuous

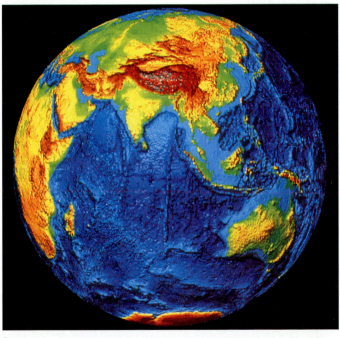

a

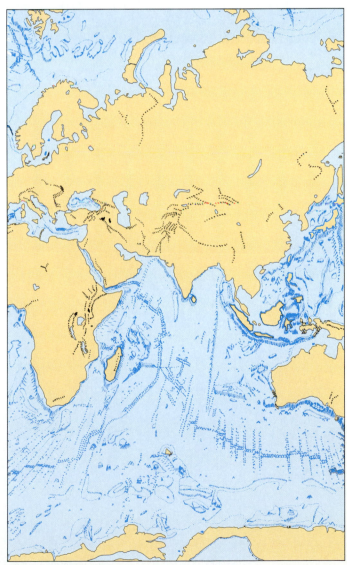

b

FIGURE 4-20 Computer-generated global view of the Indian Ocean, showing land and undersea topography.
(a) View from a position over the Indian Ocean south of India at 90°E longitude, centered on 0° latitude (the Equator).
(b) Drawing of general topographic features of the Indian Ocean. Note the large number of abyssal plains (blank area of figure) and that the mid-ocean ridge splits, with one limb going into the Red Sea. Slopes are highly exaggerated. (Image (a) generated from a land and sea database using a 5-minute (0.05) latitude/longitude grid, except for areas of the ocean that have not been surveyed in such detail; courtesy of Dr. Peter W. Sloss, National Oceanic and Atmospheric Administration/National Geophysical Data Center.)

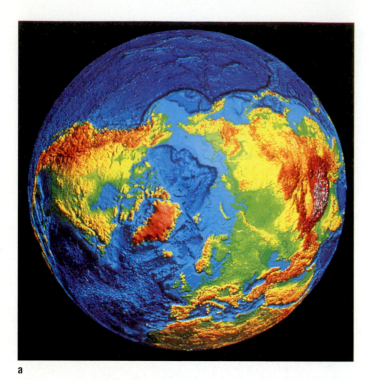

a

FIGURE 4-21   Computer-generated global view of the North Pole, showing land and undersea topography of the Arctic Ocean.
(a) View from a position directly over the North Pole at 90°N latitude.
(b) Drawing of general topographic features of the Arctic Ocean. (Image (a) generated from a land and sea database using a 5-minute (0.05) latitude/longitude grid, except for areas of the ocean that have not been surveyed in such detail; courtesy of Dr. Peter W. Sloss, National Oceanic and Atmospheric Administration/National Geophysical Data Center.)

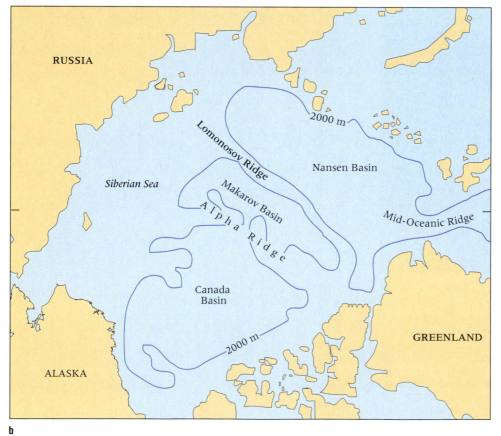

b

median rift (see Figure 4-18). The Indian Ocean Ridge also is situated essentially in the central part of the ocean.

Several long, linear ridges exist in the Indian Ocean (Figure 4-20). They differ from the mid-oceanic ridge in their lack of seismic activity and in their general shape, being somewhat higher and more blocklike. The origin of these features has not been determined. Fracture zones are evident in the Indian Ocean, in some instances containing small trenches.

The Arctic Ocean

The Arctic Ocean is the smallest division of the world ocean. Most of it is covered by ice, so even today it has not been fully explored. Some observations have been made by Russian and American nuclear submarines traveling under the sea ice, but some of this information is not available, for military reasons. The Arctic Ocean is roughly circular and is divided by three submarine ridges: the Alpha, Lomonosov (named after a Russian scientist), and a possible extension of the Mid-Atlantic Ridge (Figure 4-21). These ridges divide the ocean into a series of basins or deeps.

Between one-third and one-half of the Arctic floor consists of continental shelves, some of which are about 300 m (1,000 ft.) deep, as off the Greenland coast. This unusual depth may result from depression of the area from the tremendous weight of the thick ice sheet that covers Greenland.

The basins or deeps of the Arctic Ocean are separated from the Pacific and Atlantic Oceans by shallow submarine ridges. The basin depths and general topography are similar to those of the other oceans. The extension of the Mid-Atlantic Ridge into the Arctic Ocean seems apparent but needs more study to detail its exact position.

The "Southern Ocean"

We have just examined the four largest divisions of Earth's great ocean, the Atlantic, Pacific, Indian, and Arctic. Although the ocean waters surrounding Antarctica generally are not recognized as a fifth "ocean,"

they nevertheless have some common characteristics and are very important. It is in this area where the waters of the Indian, Atlantic, and Pacific meet in an area sometimes called the **southern ocean** (Figure 4-22). These waters are mixed by the Antarctic Circumpolar Current, one of the strongest currents in the world ocean. (See Chapter 9 for further discussion.)

Geologically, the area is surrounded by several sections of the worldwide mid-ocean ridge, and Antarctica. The continental shelf around Antarctica is even deeper than those in the Arctic. The reason for this depth is the same: depression of the area due to the weight of the massive continental ice sheet that covers the continent.

FIGURE 4-22 Computer-generated global view of the "southern ocean," showing land and undersea topography around Antarctica.
View from a position directly over the South Pole, at 90°S latitude. (Image generated from a land and sea database, using a 5-minute (0.05) latitude/longitude grid, except for areas of the ocean that have not been surveyed in such detail; courtesy of Dr. Peter W. Sloss, National Oceanic and Atmospheric Administration/National Geophysical Data Center.)

SUMMARY

The disciplines of marine geology and marine geophysics are principally concerned with the history, processes, and structure of the seafloor and the material beneath it.

The major objectives of both sciences are (a) to describe the ocean floor and to determine its origin and underlying structure, and (b) to ascertain the thickness, composition, and origin of the sediments and underlying rocks in the ocean. Many early ideas have been dramatically changed by the concepts of seafloor spreading and plate tectonics and how they explain the evolution of the ocean basins. Recent technological advances have made possible more direct observation and mapping of the ocean floor and its underlying layers.

The seafloor is divided into two principal components, based on either depth or structure: the continental margin and the ocean basin. The continental margin, which is strongly influenced by the adjacent land, includes the coastal region and the continental shelf, slope, and rise. It constitutes about 21 percent of the total ocean. The ocean basin comprises the deeper portions of the ocean, including abyssal plains, abyssal hill regions, trenches, and oceanic ridges.

Continental shelves and slopes generally mark the boundary between oceanic and continental structure. They are the site of many mineral deposits and most fishing activity in the ocean. Most parts of the con-tinental shelf recently were above sea level when the ocean was lowered as a result of glacial advances. With the recent rise of sea level, starting about 18,000 years ago, many shelves still are adjusting to the new environmental conditions. Studies of continental shelves and slopes show that there is not a unique method of formation; more often, various processes are responsible.

Marginal seas, small areas isolated from the main oceans, are common, although they are not easily classi-fied geologically. In structure, they are generally inter-mediate between ocean basin and continental margin. Some marginal seas, such as the Gulf of Mexico and the Persian Gulf, are active sites for exploration and ex-ploitation of marine resources.

The most dominant feature of the ocean basin is the continuous oceanic ridge system that winds through the Atlantic, Pacific, Indian, and Arctic Oceans. This ridge re-sults from seafloor spreading, is earthquake prone, and can be traced for over 65,000 km (over 40,000 miles) worldwide.

QUESTIONS

1. What are some of the difficulties in determining the correct depth of the seafloor?

2. Describe the features of Earth's crust and of the lay-ers below the crust.

3. What are the main features and general characteris-tics of the ocean floor?

4. Draw a cross section from coast to coast of either the Pacific Ocean or Atlantic Ocean. Label the key fea-tures and approximate water depths.

5. Describe the general characteristics of the various components of the continental margins.

6. What are the main similarities and differences among the Atlantic, Pacific, Indian, and Arctic Oceans?

KEY TERMS

abyssal hill	delta	magnetic field	shelf break
abyssal plain	East Pacific Rise	magnetometer	slumping
altimeter	echo sounding	marginal sea	sonar
basalt	estuary	moraine	southern ocean
coastal region	gabbro	oceanic layer	submarine canyon
continental borderland	glaciated shelf	positioning	topography
continental margin	Global Positioning System (GPS)	reef	trench
continental rise	granite	salt dome	turbidite
continental shelf	gravity	seamount	turbidity current
continental slope	guyot	seismic reflection	unconsolidated sediment
coral atoll	hydrophone	seismic refraction	unglaciated sediment

FURTHER READING

Anderson, R. A. 1986. *Marine Geology: A Planet Earth Perspective.* New York: John Wiley. A somewhat advanced marine geology text.

Hamblin, K. W. 1992. *Earth's Dynamic Systems.* 6th ed. New York: Macmillan Publishing Co. A superb introductory Earth science textbook with excellent illustrations.

Kennett, James. 1982. *Marine Geology.* Englewood Cliffs, N.J: Prentice-Hall. An advanced marine geology text containing nearly everything you might want to know about the subject.

Macdonald, K. C., and P. J. Fox. 1990. "The Mid-Ocean Ridge." *Scientific American* 262, no. 6, pp. 72–79. The definitive basic article on ocean ridges.

"Marine Geology and Geophysics." *Oceanus* 35, no. 4 (Winter 1992–93). A special issue on recent research in marine geology and geophysics.

"Mid-Ocean Ridges." *Oceanus* 34, no. 7 (Winter 1991–92). A special issue devoted to the geology, geophysics, and biology of ocean ridges.

Monohan, D. 1971. "Three-Dimensional Representation of Submarine Relief: Continental Margin of Eastern North America." Marine Science Paper 9, Marine Science Branch, Dept. of the Environment, Ottawa, Canada. Explains how the interesting three-dimensional map shown in Figure 4-9 was prepared.

National Geographic Society. 1988. *Exploring Our Living Planet.* Washington, D.C. A general, well-illustrated look at Earth's history and development.

Shepard, F. P. 1973. *Submarine Geology.* 3d ed. New York: Harper & Row. Although a bit dated, this work contains the best description of the world's continental shelves.

Tarbuck, E. J., and F. K. Lutgens. 1994. *Earth Science.* 7th ed. New York: Macmillan. A good elementary Earth science text.

CHAPTER 5

Marine Sediments

The sediment and rock of the seafloor can reveal many facts about the history and evolution of our planet. The sediments come from various sources, including erosion and weathering processes on land; biological, geological, or chemical processes occurring in the water or on the seafloor; or even from outer space. Various physical aspects of the ocean, such as waves and currents, can redistribute the sediment. Nevertheless, sedimentation in the marine environment is generally more continuous (uninterrupted) than on land. This makes the marine record more complete and usually much easier to understand.

The RV *Atlantis II* and the deep diving submersible *Alvin*.
(Courtesy Woods Hole Oceanographic Institution.)

Introduction

Chapter 4 discussed the seafloor's topography, underlying structure, and origin. This chapter's focus is the material that lies *on* the seafloor—the **marine sediments** and rocks—and techniques for sampling and studying them. These sediments and rocks contain a record of the last 200 million years of Earth's history, so their study can lead to an understanding of past oceanic and climatic conditions.

Techniques for dredging up rocks and sediment from the seafloor were developed as early as 1750. By 1773, mud had been obtained from a depth of 1,246 m (4,095 ft.) in the Arctic region, and by 1870 more than 9,000 samples of the ocean bottom had been collected.

Photography of the seafloor became possible in the early 1940s, due in large part to the interest in detecting and photographing enemy submarines in shallow water during World War II (1939–45). Today's sophisticated cameras take hundreds of color pictures of the deepest ocean during a single lowering (see Box 5-1).

The fields of marine geology and geophysics have benefitted from several major technological developments, some already mentioned in Chapter 4. For the study of marine sediments, the most noteworthy advance is the use of the deep-ocean drilling vessel *Glomar Challenger* and its successor the JOIDES *Resolution*. These vessels have recovered sediment and rock samples from essentially all parts of the world ocean. Submersibles, another exciting tool for marine geologists and geophysicists, have been used to explore ocean ridges and have enabled some remarkable discoveries.

Underwater Cameras

The old axiom that one picture is worth a thousand words is especially true in oceanography. Underwater photography has become very important in oceanographic research. Ocean-bottom photographs are used to study sediment and rock on the seafloor, to determine biological activity, and to observe indirect evidence of currents, such as ripple marks and scour marks.

Camera systems, protected against the great pressure from overlying water, have been lowered by cable into the deepest parts of the ocean (Figure 1). Because most daylight is absorbed within the upper 100 to 200 m (330 to 650 ft.) of the ocean, these camera systems must carry their own powerful light sources. They also must have a "pinger" so that the camera can be positioned accurately above the seafloor.

Pingers are devices used to position instruments precisely that are suspended on cables. A pinger is attached to the cable, usually near the instrument, and emits sound pulses that sound like "pings," usually exactly 1 second apart. The ship's echo-sounder receives two ping signals: the direct signal from the pinger, and a reflected signal from the seafloor. The difference in arrival times between the direct and reflected signals indicates the height of the pinger above the bottom (Figure 2). Note that in spite of all precautions, things sometimes do not work out right at sea (Figure 3).

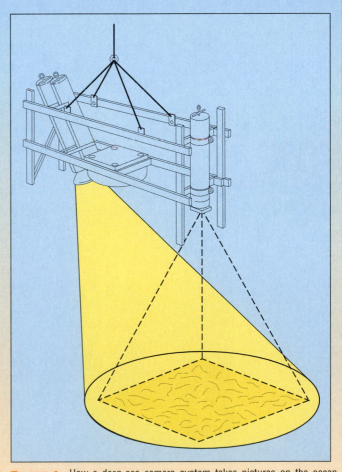

FIGURE 1 How a deep-sea camera system takes pictures on the ocean floor.

continued

Box 5-1 continued

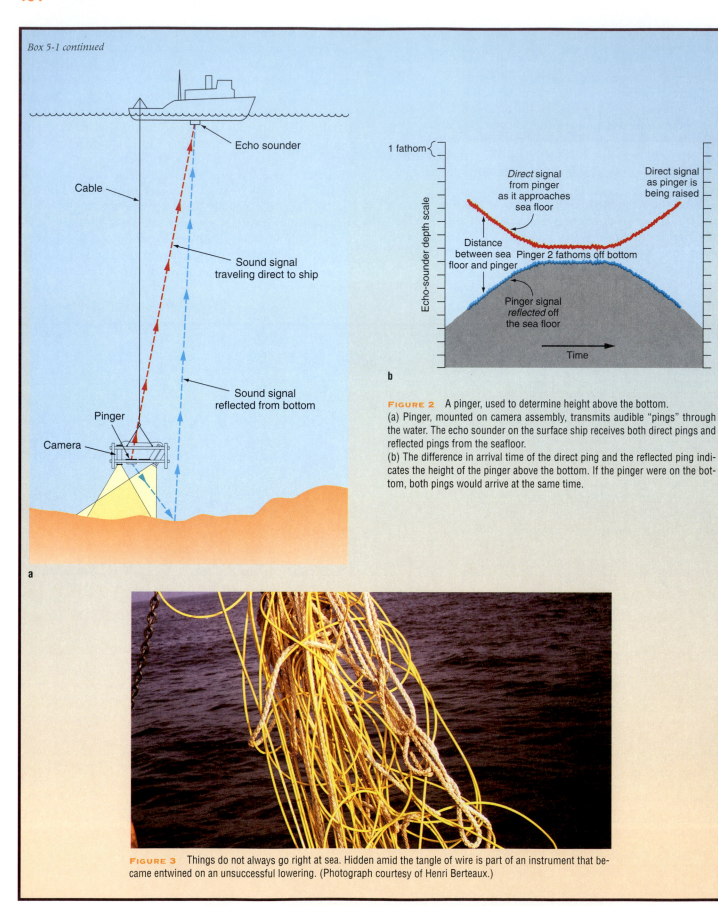

Echo sounder

Cable

Sound signal traveling direct to ship

Sound signal reflected from bottom

Pinger

Camera

a

1 fathom

Echo-sounder depth scale

Direct signal from pinger as it approaches sea floor

Direct signal as pinger is being raised

Distance between sea floor and pinger

Pinger 2 fathoms off bottom

Pinger signal *reflected* off the sea floor

Time

b

FIGURE 2 A pinger, used to determine height above the bottom.
(a) Pinger, mounted on camera assembly, transmits audible "pings" through the water. The echo sounder on the surface ship receives both direct pings and reflected pings from the seafloor.
(b) The difference in arrival time of the direct ping and the reflected ping indicates the height of the pinger above the bottom. If the pinger were on the bottom, both pings would arrive at the same time.

FIGURE 3 Things do not always go right at sea. Hidden amid the tangle of wire is part of an instrument that became entwined on an unsuccessful lowering. (Photograph courtesy of Henri Berteaux.)

Box 5-1 continued

a

FIGURE 4 ANGUS, an underwater survey camera.
(a) A small version (Mini-ANGUS) being tested. The device contains cameras, light sources, and supporting electronic components. Note the rugged steel enclosure to protect the device when it hits the bottom.
(b) ANGUS near the seafloor, photographing a hydrothermal vent. The position of the camera is determined by reference to the three acoustic transponders on the seafloor. (Photographs courtesy of Woods Hole Oceanographic Institution.)

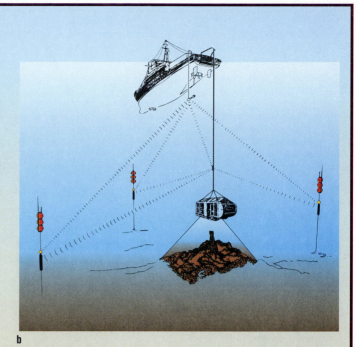

b

To afford close views, pictures are usually taken only a few feet above the bottom. Most camera systems can take several hundred pictures with each lowering. When the instrument is near the bottom, the winch operator tries to maintain camera position a few feet above the ocean floor, guided by pinger returns displayed on the ship's echo-sounder. This is important to keep the ocean bottom in focus. In some instances, two cameras are used to obtain stereoscopic pictures. These can be used to measure the dimensions of small features on the seafloor.

Newer, more sophisticated cameras are used as surveying systems. These devices have more intense lighting, and thus can take pictures over larger areas. One, called ANGUS (Acoustically Navigated Geological Underwater Survey), is designed to work in very rugged terrain (Figure 4). ANGUS contains three 35mm cameras with large film capacity and can photograph a portion of the seafloor 60 m (200 ft.) wide. In some instances, cameras like this are used with bottom acoustical transponders, devices used to locate instruments or submersibles relative to the seafloor. Using such devices allows very accurate location of the camera, which in turn permits the photographs to be correlated precisely with the seafloor.

Side-Scan Sonar

Side-scan sonar devices are similar to conventional sonar or echo-sounding equipment (see also Box 4–1, page 72). However, side-scan sonars transmit sound at an angle from the ship rather than just straight down, as in echo-sounding (Figure 5a). The difference in intensity of the returning sound signals from the bottom can be used to distinguish types of sediment (for example, sand from gravel) or to observe other features on the seafloor. Side-scan sonar is very useful in submarine archeological investigations and in the search for vessels lost at sea (Figures 5b and 5c).

Echo-Sounding

Simple echo-sounding can provide information about the sedimentary layers beneath the ocean floor. When sound energy reaches certain layers, it is reflected back toward the surface, received, and recorded. The resulting record can show a cross section of the layering of the sediments on the ocean floor (see Figure 6). Similar marine geophysical techniques, using considerably greater sound energy, can obtain reflections from layers hundreds of meters (a mile or more) below the ocean floor (see Figure 4–3, page 79).

continued

Box 5-1 continued

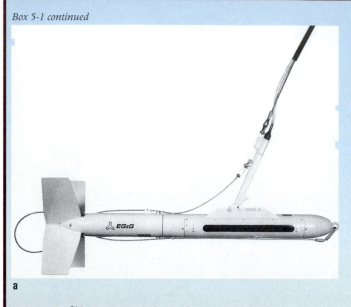

a

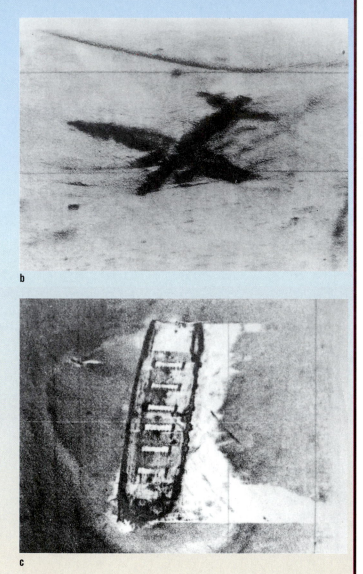

b

c

FIGURE 5 Side-scan sonar device and images.
(a) A towed side-scan sonar device.
(b) Side-scan sonar record showing an aircraft on the bottom of Loch Ness in Scotland. The device unfortunately did not spot the famous but elusive Loch Ness monster.
(c) Side-scan sonar record showing an old wooden sailing barge in the Great Lakes. (Photograph (a) courtesy of EG&G Company; photographs (b) and (c) courtesy of Klein Associates, Inc.)

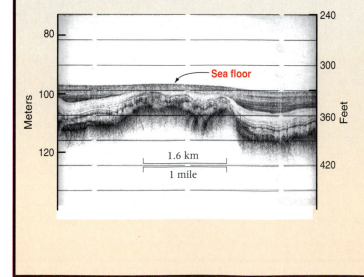

FIGURE 6 Echo-sounding record from the Baltic Sea showing layering below the surface sediments. These layers probably result from differences in sediment type—for example, a change from a clayish sediment to a sandy sediment. Vertical exaggeration is about 50 times.

Techniques for Sampling and Observing the Seafloor

One way to obtain sediment and rock samples from a surface vessel is by drilling into the ocean floor. Three other methods are (1) snappers or grab-type samplers, (2) coring devices, or (3) dredges.

FIGURE 5-1 Campbell grab sampler.
(a) Sampler lying on its side. The camera is triggered when a compass (in foreground) hits the bottom, but before the sampler disturbs the sediment.
(b) Picture taken by a sampler (note compass) in the Gulf of Maine. The coarse-grained sediments depicted are typical of glacial deposits (see Box 4-3). The compass in the lower right-hand corner is about 10 cm (about 4 in.) in diameter. Its purpose is to determine the orientation of features on the seafloor, such as ripple marks.

Snappers or **grab samplers** (Figure 5-1) will obtain only a surface sample, which is generally disturbed during the process. These devices are mostly used in shallow water. If a rock or some other object gets wedged in the jaws of the sampler, it may not close properly and the sediment sample can be lost.

Coring devices obtain a long vertical section (or core) of the sediment by forcing a long pipe vertically into the sediment. The simplest coring device is the **gravity corer,** a weighted pipe that penetrates the bottom by its own momentum (Figure 5-2). This instrument will usually obtain only short cores, about 1.8 to 3 m (6 to 10 ft.) long.

To obtain longer cores, a **piston corer** (Figures 5-3 and 5-4) is used. A piston inside the core tube increases the penetration of the device during the coring operation. Thus, it is sometimes possible to collect cores 18 m (60 ft.) or longer. The piston corer is usually triggered by a gravity corer. These sampling devices can be positioned above the bottom with the use of a **pinger** (see Box 5-1).

Some types of research require relatively undisturbed samples. One way of obtaining them is with a **box corer** (Figure 5-5), a thin-walled, box-shaped device that is

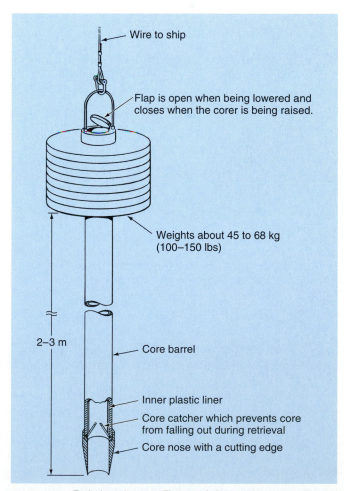

FIGURE 5-2 Typical gravity corer. The corer is lowered from the surface ship at a high speed. Its weight and momentum help the core barrel to penetrate sediments on the seafloor.

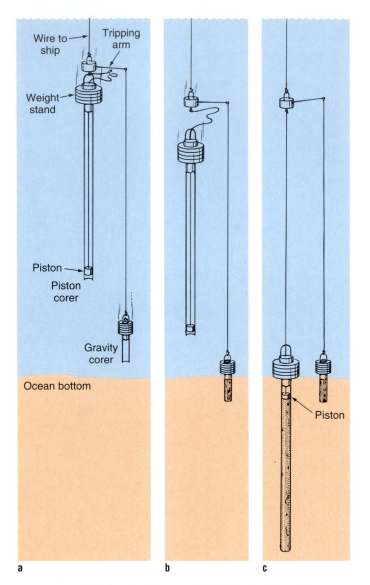

FIGURE 5-3 Operation of a piston corer that is tripped with a gravity corer.
(a) Lowering position.
(b) When the gravity core hits the bottom, the piston corer is released and free-falls to the bottom. At the moment of impact, the line to the piston tightens and the core barrel slides past it; this reduces friction within the core barrel.
(c) Completion of the coring operation prior to retrieval of core from inside the barrel.

pushed into the sediment by a heavy weight. This technique generally does not seriously disturb the sediment layers.

A new instrument that combines the techniques of piston coring with drilling is the *hydraulic piston corer,* which can be used to take a continuous series of cores 10 m (about 33 ft.) long to an eventual depth of several dozen meters below the bottom. The device stops when the sediment becomes so consolidated that the corer cannot be pushed deeper by water pressure. The cores obtained by this process, besides being nearly continu-

ous, are well preserved. In contrast, conventional drilling usually disturbs the relatively soft sediments.

The **dredge** is a rock-sampling device dragged at slow speed along the ocean floor. Dredges generally consist of a large-diameter pipe partially closed at one end, or a large metal frame with a chain "bag" at its end (Figure 5-6). If the open end of the dredge encounters a rock, the pulling power of the ship on the rock generally will break it off, so part of the rock is caught in the dredge.

Obtaining a satisfactory sample of the seafloor can be very difficult. Often it does not suffice to just lower a sampler and obtain whatever it happens to hit. Many sophisticated studies require precise location of the sample. To do so requires careful positioning of the ship over the appropriate area (see Box 5-2), plus considerable care in placing the sampler over the desired part of the seafloor. This is not easy, because the sampler is frequently attached to a long cable that can be moved by currents—which, of course, can also move the surface ship. Among the new techniques used to reduce some of these problems are drilling ships and research submersibles.

Drilling Ships

Drilling ships can obtain samples of sediment and rock from well below the seafloor. These ships have been used for scientific investigation in the ocean since the early 1960s, starting with *Project Mohole,* which was an attempt to drill completely through Earth's crust, through the Mohorovičić discontinuity, to the mantle. Although ultimately abandoned, this work showed that the technology for drilling in the deep sea was available and useful.

From 1968 to 1983, a scientific program called the **Deep Sea Drilling Project (DSDP)** drilled hundreds of holes into major geological features in all the oceans. This international program included the U.S., West Germany, Japan, France, the former Soviet Union, and the United Kingdom. DSDP used the drilling vessel *Glomar Challenger* (see Figure 1-10a, page 12), named after the original *Challenger.* The principal objective of DSDP was to obtain long cores of the sediment and rock of the seafloor. By the end of the program, holes had been drilled at 624 sites, some to as deep as 1,525 m (5,000 ft.) below the seafloor. The sediment and rock collected have been studied by hundreds of scientists from dozens of countries.

The successor to the *Glomar Challenger* is the JOIDES *Resolution* (Figure 5-7). The new program, called the Ocean Drilling Program (ODP), is run by Texas A&M University on behalf of a consortium of 10 U.S. oceanographic institutions. The consortium is called JOIDES, an acronym for Joint Oceanographic Institutions Deep Earth Sampling. Several other nations also are or were

FIGURE 5-4 Piston coring operation.
(a) Lowering of a piston corer; note that rigged tripping arm gravity corer is already below the water.
(b) Piston corer after recovery; the weight stand is secured and the gravity corer is already on board.
(c) Scientists preparing to remove the plastic liner within the piston corer, which contains the core. (Photographs (a) and (b) courtesy of Jim Broda and (c) courtesy of Eben Franks, both of Woods Hole Oceanographic Institution.)

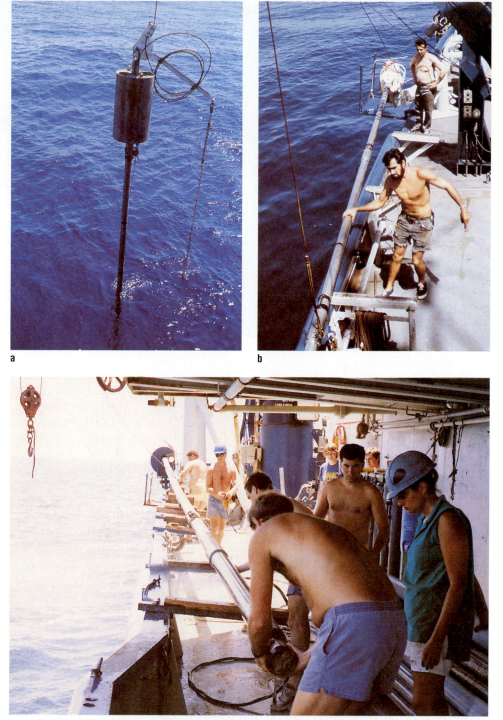

a

b

c

contributing $2.5 million per year to be a part of this group, including Australia, Canada, France, Germany, Japan, United Kingdom, a consortium of 12 other European countries, and former nations of the Soviet Union. The first core sample was raised by the *Resolution* in January 1985, and since then it has continued to drill in the four major oceans and around Antarctica.

Research Submersibles

Submersibles are extremely valuable research tools. They come in varied shapes, sizes, and capabilities, and perform different and sometimes very specialized tasks. Their key value is the opportunity to observe and sample the seafloor. Initial interest in submersibles started in the

FIGURE 5-5 A box corer. (a) A long box corer being launched in the Black Sea; the Caucasus Mountains are in the background. (b) Scientists back on land, sampling the sediments obtained by the box corer. Note the well-preserved sediment layering.

a

b

FIGURE 5-6 A chain bag dredge being prepared for lowering. The dredge will be dragged along the seafloor in an attempt to have it catch and break off a piece of any rock it may encounter. (Photograph courtesy of Woods Hole Oceanographic Institution.)

late 1960s and early 1970s, following the accidental sinking of the U.S. submarines *Thresher* and *Scorpion*. The U.S. Navy was interested in deep-submergence programs to rescue possible survivors. Nowadays, submersible use is driven more by scientific curiosity, especially to explore the vents that emit hot water along the mid-ocean ridges of the Pacific and Atlantic Oceans.

One of the first uses of submersibles for science was by William Beebe in the early 1930s. He and colleagues built a **bathysphere** (a strong sphere with viewing ports) that was lowered by cable from a surface ship. His record dive to 923 m (3,028 ft.) in 1934 did much to excite interest in the ocean. A different type of submersible was developed by Swiss physicist Auguste Piccard. Piccard's device was not attached to a surface ship, nor was it powered, but it was capable of adjusting its buoyancy to make it sink or float. After developing several versions, he eventually took one to 4,175 m (13,700 ft.) in the Mediterranean in 1954.

Following this effort, Piccard and his son Jacques built the *Trieste*. The U.S. Navy purchased it and used it in 1960 to make a record dive of 10,910 m (35,795 ft.) in the Challenger Deep near Guam in the south Pacific. This exploit clearly showed that scientific submersibles could essentially be used anywhere. However, the *Trieste* did not have the ability to move horizontally; it could move only up or down.

In the 1960s and 1970s, many new submersibles were built, and by 1970 about 60 were operational. Most were used for economic ends—inspecting underwater oil well and pipeline construction—rather than for scientific stud-

Box 5-2
The Art of Navigation

Navigation is knowing your position on Earth, or finding your way to a specific position. Navigation is fundamental in oceanographic work. However, most navigational techniques are not very accurate, except modern methods using electronic or satellite devices.

One of the simplest methods of navigation at sea is called *dead reckoning*. The initial position of the vessel is determined by reference to an object on land or by some other method. Then a course is plotted that will lead the ship to its intended objective. By knowing the speed of the vessel, you can estimate its position during any part of the voyage. The accuracy of this technique is poor because, due to currents, winds, and waves, it is difficult to move in a steady straight line on the ocean (Figure 1a). Departures from the predicted position and arrival time (called the *set*), however, can provide information concerning currents and surface winds.

Within visual or radar sight of land (usually less than 80 km, or 50 miles) several navigation techniques can be used. Using a sextant (a device that measures angles), the horizontal angle between three well-located objects on land can be used to locate the observer's position (Figure 1b). Radar reflections from known objects on land can provide a range (distance) and bearing (angle). These can be used, like sextant sights, to obtain a position (Figure 1c). Positions having an accuracy of a hundred meters or so are possible with a good radar system, although in practice the error is generally larger.

If the bottom topography of an area is well known and charted, the chart can be used as a navigational aid. For example, a ship's position can be determined from the chart when the ship passes over a known feature on the seafloor. When the ship is out of sight of land or range of electronic equipment such as radar, celestial navigation may be one method of obtaining a position. Determining a ship's position by star sighting has been used for many centuries. The accuracy of this method depends on the observer's skill and at best is usually within a mile or so. Shortcomings of celestial navigation are that the sky must be clear enough to see the stars, and sightings must be done at dawn or dusk.

A widely used electronic navigation system is **Loran** (*Long Range Navigation*). Loran can sometimes be used out to 1,600 km (about 1,000 miles) from land. Like most electronic positioning systems, Loran depends on an accurate measurement of the time required for a radio signal to travel from a transmitter at a known place on land to a receiver on a ship. The arrival time is a measure of the distance the ship is from the station. Using two or more Loran land stations, a very accurate estimate of the ship's position can be made, within about 50 m (165 ft.) or better (Figure 2). The main shortcoming of this and other electronic techniques is that land-based stations are not available in some parts

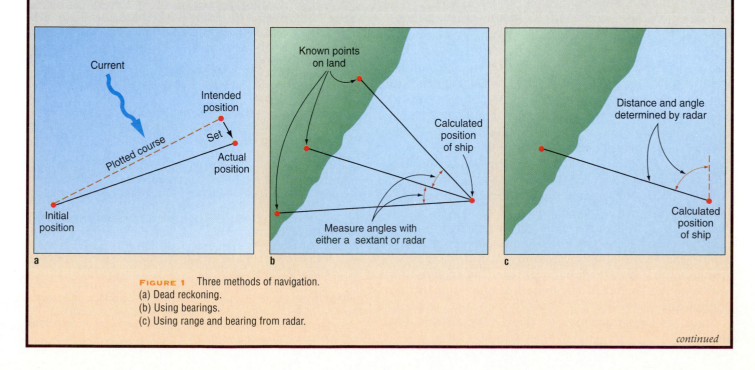

a

b

c

FIGURE 1 Three methods of navigation.
(a) Dead reckoning.
(b) Using bearings.
(c) Using range and bearing from radar.

continued

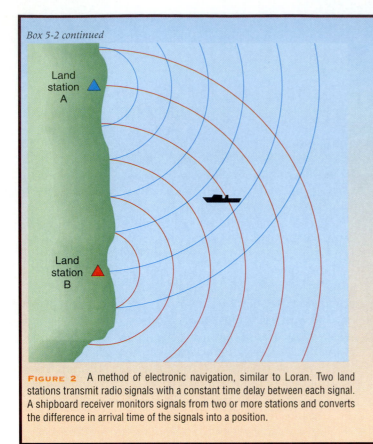

Box 5-2 continued

FIGURE 2 A method of electronic navigation, similar to Loran. Two land stations transmit radio signals with a constant time delay between each signal. A shipboard receiver monitors signals from two or more stations and converts the difference in arrival time of the signals into a position.

of the world. An advantage is that they can be used in most kinds of weather.

Nowadays, satellite navigation systems are often used by oceanographic research vessels. This technique requires sophisticated electronic equipment that measures the change in radio frequency of the satellite as it moves away from or toward the ship, just like the sound of an automobile horn changes pitch with motion toward or away from you. These changes can be used to calculate the ship's position fairly accurately, to a tenth of a kilometer or less.

Recently, an even more sophisticated system has become available for general navigation. As with many technological advances, its first application was restricted to the military. The satellite-based **Global Positioning System (GPS)** can have an accuracy of 1 part in 10 million, equivalent to an error of 1 cm in measuring the distance between objects 100 km apart. This system depends on a series of satellites, some of which will always be within line of sight. The amazing accuracy of this system, besides allowing pinpoint navigation, also could be used to measure tiny movements associated with earthquakes, faulting, and even seafloor spreading.

ies. In the last decade, however, research submersibles have made some remarkable discoveries, especially along the mid-ocean ridges. Submersibles used include the *Alvin* (Figure 5-8) and the *Johnson Sea Link II* (Figure 5-9).

A Japanese submersible named *Shinkai 6500* recently completed a dive to 6,527 m (21,212 ft. or 4.06 miles), which may be the deepest dive by a vessel that has the ability to move while on the seafloor. Japanese scientists are also considering building a fully mobile submersible that can attain depths of 11,000 m (36,000 ft.). Russia has two deep-diving submersibles, *Mir I* and *Mir II*, built in 1987 by a Finnish Company (Figure 5-10). The Russian submersibles cost about $25 million each. They can dive to 6,000 m (19,680 ft.) and can remain on the bottom for up to 20 hours. Only the U.S. Navy's *Sea Cliff* (a sister ship to *Alvin*), a French submersible, and the Japanese submersible can go as deep.

In a recent cooperative effort with scientists from the U.S. and Canada, the two *Mirs* worked together in water depths exceeding 5,000 m (16,400 ft.), a first for such an operation. The *Mirs* have been used to explore the hull of a Russian nuclear submarine that sank in the Barents Sea.

Some of the adventures of these submersibles are amazing:

FIGURE 5-7 JOIDES *Resolution* (opposite page).

(a) The vessel is 143 m long and 21 m wide (470 by 70 ft.). The ship's derrick towers 62 m (202 ft.) above the waterline. A computer-controlled dynamic positioning system, using twelve powerful thrusters and two main shafts, maintains the ship over a specific location while drilling in water depths to 8,230 m (27,000 ft.). The ship has a crew of 65 and a scientific party of up to 50. It can lower up to 9,150 m (30,000 ft.) of drill pipe to obtain core samples. A seven-story laboratory and other scientific facilities located fore and aft occupy 12,000 square feet.

(b) Re-entering a hole previously drilled on the deep seafloor. Sound waves are sent from a sonar beacon on the seafloor to the ship's hydrophones to locate the drilling site on the ocean floor, as powerful thrusters hold the ship in position. A remote TV camera aids in positioning the drill pipe into a metal cone at the top of the hole from which cores will be taken.

(c) When the JOIDES *Resolution* is on site, drilling operations continue 24 hours a day. Most drilling operations take place on the rig floor, a platform about one meter above the bridge deck in the center of the ship.

(d) The core is split lengthwise. Half is designated as the *working section*, and the other half is the *archive section*, which is described and photographed for reference. Macroscopic and microscopic descriptions of the recovered core material are made in the core laboratory. Scientists take samples from the working half for analysis on ship and on shore.

(e) Currently, more than 160 km (100 miles) of core are stored at three Ocean Drilling Program repositories: Lamont-Doherty Earth Observatory in Palisades, NY, Texas A&M University at College Station, TX, and Scripps Institution of Oceanography, La Jolla, CA. (All photographs courtesy of Ocean Drilling Program, Texas A&M University.)

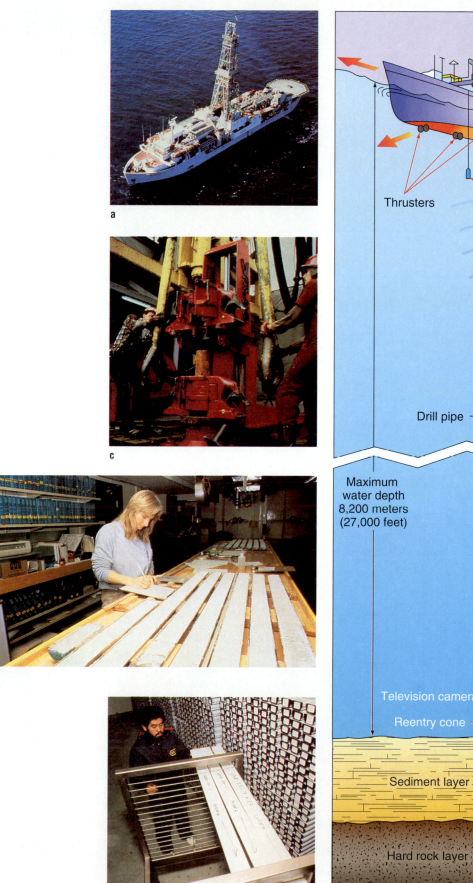

a

c

d

e

b

Thrusters

Thrusters

Hydrophones

Drill pipe

Maximum
water depth
8,200 meters
(27,000 feet)

Television camera

Reentry cone

Sonar
beacon

Sediment layer

Hard rock layer

a

b

FIGURE 5-8 *Alvin*, a research submersible.
(a) *Alvin* is supported by the National Science Foundation, the Office of Naval Research, and the National Oceanic and Atmospheric Administration. It is operated by the Woods Hole Oceanographic Institution (see also opening figure of this chapter, which shows *Alvin* with its support ship, *Atlantis II*). *Alvin* can dive to 4,500 m (14,760 ft.) and normally can stay submerged for 6–10 hours.
(b) Front view of *Alvin*, showing some of the equipment it may carry.
(c) Cutaway view of *Alvin*. Total length is 7.6 m (25 ft.). The submersible uses a ballast system consisting of interconnected pressure-proof aluminum spheres and collapsible rubber bags partially filled with oil. As oil is pumped from the spheres into the bags, the amount of seawater displaced by the vehicle is increased, thus increasing the buoyancy and making the submersible lighter, although the weight of the vehicle remains the same. (Photographs (a) and (b) by Rodney Catanach; all images courtesy of Woods Hole Oceanographic Institution.)

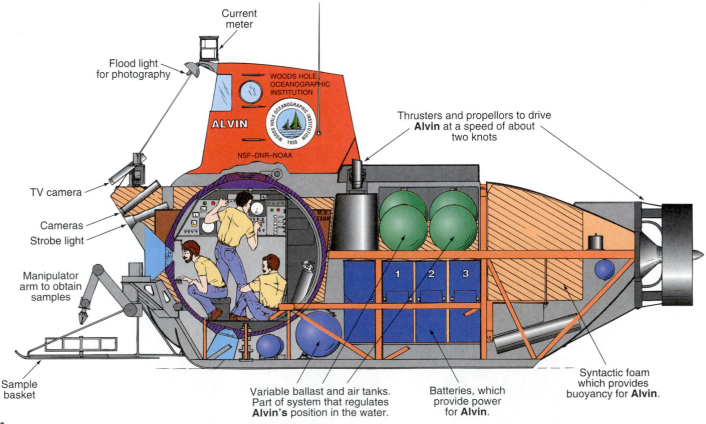

Current meter

Flood light for photography

WOODS HOLE OCEANOGRAPHIC INSTITUTION

ALVIN

NSF–DNR–NOAA

Thrusters and propellors to drive **Alvin** at a speed of about two knots

TV camera

Cameras
Strobe light

Manipulator arm to obtain samples

Sample basket

Variable ballast and air tanks. Part of system that regulates **Alvin's** position in the water.

Batteries, which provide power for **Alvin**.

Syntactic foam which provides buoyancy for **Alvin**.

c

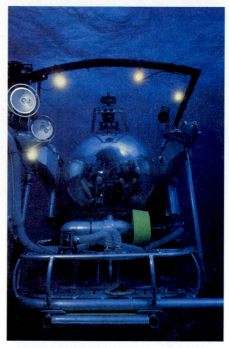

FIGURE 5-9 Two views (above and below the water) of the *Johnson Sea Link II* submersible. This submersible, which is 7 m (22 ft.) long, can be operated to a depth of 914 m (3,000 ft.). A pilot and one passenger can be in the forward section and two people in the aft section. The latter two can leave using SCUBA gear and return to the vessel while it is submerged. (Photographs by Tom Smoyer; courtesy of Harbor Branch Oceanographic Foundation.)

FIGURE 5-10 The Russian deep sea research vehicle *Mir*. It is 7.8 m (25.5 ft.) long, weighs 18.7 tons out of the water, has a maximum operating depth of 6,000 m (19,680 ft.), and can carry 3 people. The Russian word *mir* means "peace." (Photograph courtesy of Rauma-Repola.)

- In 1966, *Aluminaut* (an aluminum submersible no longer in use) and *Alvin* participated in the successful recovery of a hydrogen bomb (H-bomb) lost off the Spanish coast.

- *Alvin*, while on a dive on the Blake Plateau off Florida, was attacked by an apparently nearsighted swordfish (Figure 5-11).

- *Alvin* also once sank at sea, fortunately without the loss of any life, and about a year later was recovered from the ocean bottom. The sinking was due to a break in a cable that supported *Alvin* during its launch. The recovery was accomplished using *Aluminaut* and *Mizar* (a surface vessel); ironically, these three vessels were previously teamed in the H-bomb recovery.

PROS AND CONS OF SUBMERSIBLES. A submersible possesses numerous advantages over a surface vessel. The most important is direct observation, permitting a scientist to see and photograph directly what is being measured or sampled. Another advantage is the submersible's ability to operate independently of the sea surface and to explore small features that may not even be detectable from a surface vessel. In a submersible, a scientist can make measurements in the same place for several hours, whereas an unanchored surface ship may drift away from a given area.

Another advantage is that a submersible can return to the same place on the bottom (if marked), whereas a surface vessel is limited by navigational constraints in finding any given area (see Box 5-2). This, coupled with some of the technology about to be described, has been

FIGURE 5-11   A swordfish attacked *Alvin* and got stuck in the external hull of the submersible. The fish, about 2.5 m (8 ft.) long and weighing about 91 kg (200 lbs.), ended up making a good meal for the crew of the *Alvin* and its support ship! (Photograph courtesy of Woods Hole Oceanographic Institution.)

FIGURE 5-12   ROPOS (Remotely Operated Platform for Ocean Science). An ROV (Remotely Operated Vehicle) being lowered into the ocean from the NOAA vessel *Discoverer*. Equipped to map and photograph the seafloor and obtain samples of sediment, rock, animals, or water, ROPOS can operate to a depth of 3,500 m (11,480 ft.). Once the launch package is on the seafloor, the ROV leaves its protective cage (like a car leaving a garage) and maneuvers along the bottom, attached by a tether 200 m (656 ft.) long. Aboard the surface ship, observers control its movements and watch via TV. (Photograph courtesy of Institute of Ocean Sciences, Canadian Dept of Fisheries and Sciences, Sidney, British Columbia, who operate the ROV.)

used to advance the field of underwater archeology (discussed in Chapter 17).

On the other hand, submersibles depend on surface ships to tow or carry them to dive sites. Difficulties in launching and retrieving submersibles often limit their use to periods of mild sea conditions. Because submersibles often are very small, observers inside them must remain in uncomfortable positions for long periods. The cost of each dive usually is very high, and submersibles generally can explore only a small area at a time. Despite these disadvantages, submersibles are becoming one of the most important research tools of the oceanographer.

REMOTE-OPERATED VEHICLES (ROVs).   To reduce cost and solve some of these problems for certain types of submersible operations, a new underwater device was developed. It does not carry people and can remain submerged for weeks. These *remote-operated vehicles (ROVs)* have varied uses, such as underwater photography or inspecting underwater facilities (Figure 5-12).

One of the more advanced remote systems is ARGO/JASON (Figure 5-13), developed and used by Dr. Robert Ballard and his colleagues at the Woods Hole Oceanographic Institution. Ballard is a marine geologist who used this remote system to discover and explore the *Titanic* and the German battleship *Bismarck* (see pages 432–434). ARGO (Figure 5-13b) is an unmanned device with various sensors to measure temperature, Earth's magnetic field, and other parameters. It can be towed deep in the ocean and thus is suitable for deep-ocean surveying.

ARGO is attached by cable to the surface ship and can operate to 6,100 m (20,000 ft.). From cameras on ARGO, TV pictures of the bottom are transmitted via the cable to the surface ship, as are other data. JASON (Figure 5-13c) is an ROV operated from ARGO. It has its own television camera and is capable of close-up inspection and precision sampling. The samples can be stored in ARGO while both devices continue to work on the seafloor for weeks or more.

SCUBA.   For some types of research, a *person* can function almost as a submersible with the use of **SCUBA** gear (Self-Contained Underwater Breathing Apparatus). Exploration of the sea bottom to 60 m (about 200 ft.) has been performed routinely by experienced divers and scientists using SCUBA gear. Newer techniques, using a mixture of gases and portable decompression chambers or special suits, have allowed people to work for an extended time at even greater depths (see Box 5-3).

The Missing Sediments

After World War II ended about 50 years ago, seismic studies of the ocean floor increased greatly. One of the first questions explored was the amount of sediment on the ocean floor. There had been considerable speculation on this question. Most scientists expected the sediment to be between 2 and 3 km (1.2 to 1.8 miles) thick, based on an average sedimentation rate measured in the ocean

FIGURE 5-13   The ARGO/JASON system being used from a surface vessel. (a) ARGO (see b) is attached to the surface vessel, and the smaller JASON operates from and is attached to ARGO. The three "poles" on the bottom are acoustic transponders used for navigation.
(b) ARGO on land, being readied for sea duty.
(c) JASON photographed while on the dock. (Schematic (a) by Stefan Masse; photograph (c) by Terri C. Corbett; images courtesy Woods Hole Oceanographic Institution.)

and extrapolated back through geologic time. Another basis was estimates of the quantity of material removed from the continents.

But seismic studies made in the early 1950s revealed a surprise. They showed that the ocean crust could be divided into three main layers on the basis of the sound velocity within these layers (see Figure 4-4, page 81). Sampling and drilling have shown the upper layer to be unconsolidated sediment, commonly about 300 m (1,000 ft.) thick in the Pacific and 600 m (2,000 ft.) thick in the Atlantic. Most scientists had expected a considerably thicker accumulation, so they had a new mystery to solve: where were the missing sediments?

From the new information, it would have taken no more than 400 million years for 300–600 m of sediment to accumulate. This is only 10 percent of the 4,700 million years that the ocean is thought to have existed. If the assumptions are correct, and the evidence is good, only two answers to this dilemma exist: either some sediments have been consolidated and occur as a deeper layer within the earth's crust, or some sediments have been removed from the ocean.

When *Glomar Challenger* started drilling in the different oceans of the world, scientists found that the second main layer of the crust was not consolidated sediment but was mainly basaltic rock, formed from volcanic processes. Some consolidated sediment was found, but it generally constituted only a minor portion of the second layer. Thus, marine geologists were left with the alternative that large amounts of sediment somehow had been removed from the oceans.

Initially, this was a discomforting idea. But it was later found to fit very well into the then-developing concept of seafloor spreading. Basically, older marine sediments had been removed by subduction as the crust on which they rested was thrust under a continental plate. The sediment presently in the ocean is that which has been deposited during the last 200 million years or so, and has not yet reached a subduction area.

BOX 5-3
Walking on the Seafloor

It is commonly said, "We know more about the surface of the moon than we know about the deep-ocean floor." It is true that six *Apollo* crew members have walked on the moon's surface (see Figure 1), and that no similar event has yet occurred on the deep seafloor. It also is true that scientists have directly observed the moon with telescopes and satellites, and that until very recently we had better topographic maps of the moon than of the world ocean floor. However, the statement that we know more about the moon than about the seafloor certainly is debatable. At least a million pictures have been made of the ocean floor, and scientists have observed it from submersibles and with SCUBA and special diving suits. But indeed, there are no human footprints yet on the seafloor below about 600 m (2,000 ft.).

Humans can safely dive to about 90 m (about 300 ft.) using SCUBA equipment. Greater depths, of course, are possible in submersibles. Some submersibles allow a diver to exit for short periods (see Figure 5-9). Nevertheless, many scientists and engineers desire to walk and work directly on the seafloor. There are practical reasons: for example, salvage work, maintaining seafloor equipment, and marine archeology might better be done by scientists who are free to stroll the seafloor. These interests have led to the development of

FIGURE 1 *Apollo* astronaut exploring a large rock on the moon. (Photograph courtesy of National Space Science Data Center.)

Box 5-3 continued

one-person devices tethered to a surface ship. One of the first, developed in the 1930s, was called *Jim*, for Jim Jarrett, the first person willing to try it. It was an articulated and armored diving suit.

A more advanced device is a one-person diving suit capable of working to 600 m (1968 ft.). The device is attached to a surface ship and is called **WASP** because it resembles one (Figure 2). WASP has been used mainly by the oil industry to inspect and maintain oil rigs. However, some scientific work has been done with these devices, especially by well-known diving scientists such as Bruce Robison of the Monterey Bay Aquarium Research Institute in California and Sylvia Earle, recent Chief Scientist of NOAA. By using a WASP suit, researchers can study animals and plants directly in their habitats in the water column.

The resulting observations have changed many views of ecology and of the behavior of organisms as well as giving a more realistic three-dimensional view of the ocean realm. In oceanography, as in all human endeavor, it is often true that "you have to be there."

REFERENCE
Hanson, Lynne Carter and Sylvia A. Earle. 1987. "Submersibles for Scientists." *Oceanus* 30, no. 3, pp. 31–39.

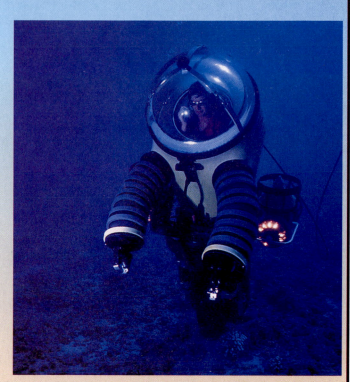

FIGURE 2 A diver in a WASP suit maneuvering along the seafloor. In this instance the diver is Graham Hawkes, a designer of the WASP piloting system. (Photograph courtesy Dr Sylvia Earle, Deep Ocean Engineering.)

Marine Sediments

Sediment can be described in various ways—its source, how it was transported, its chemical makeup, and its particle size. The size of particles that comprise the sediment is particularly useful and is generally called the **grain size.** One widely used grain-size scale is shown in Table 5-1. The grain-size ranges of many sediment types mentioned in this section are shown in Figure 5-14.

Generally, the coarser (larger) grain sizes occur where the energy level is great enough to carry them—for example, where strong currents or waves exist. Finer particles, such as clay, settle from the water where the energy level is too low to keep them moving, frequently far from shore. This phenomenon is called **sorting.** *Unsorted* sediments are a mixture of particles of various size. If similar-sized grains predominate, the sediment is said to be *well-sorted.* For example, beach sands are well sorted, whereas glacial deposits are poorly sorted because they generally contain particles ranging from boulders to colloids (see also Box 4-3). It should be emphasized that terms like *sand, silt,* or *boulders* relate only to the *size* of grains, not to their composition or origin.

After deposition, a sediment can be *reworked* by bottom currents or waves, resulting in better sorting. Generally, this happens by removing the finer-sized particles. This is why beach deposits are well sorted; they are constantly reworked by waves and currents in the nearshore area. Intensive reworking can result in an entire deposit being moved and redeposited elsewhere. The finer the grain size, the farther it can be carried by water or wind.

A more basic classification of marine sediment is based on the *origin* of the particles and, to a lesser degree, on their method of deposition. In classification by origin, marine sediments are divided into four broad groups: terrigenous, biogenous, hydrogenous, and cosmogenous sediments. Let us now examine each type and its distribution.

Terrigenous Sediments

Terrigenous sediments are derived from existing rock (*terra* refers to Earth). The rock is weathered and eroded, and the resulting sediment is transported to the ocean. During the weathering process, some parts of the rock may become dissolved and enter the ocean in solution,

TABLE 5-1

The Wentworth Scale of Sediment Grain Size*

Particle Name	Minimum Diameter (mm)
Boulder	256
Cobble	64
Pebble	4
Granule	2
Sand	
Very coarse sand	1
Coarse sand	1/2
Medium sand	1/4
Fine sand	1/8
Very fine sand	1/16
Silt	
Coarse silt	1/32
Medium silt	1/64
Fine silt	1/128
Very fine silt	1/256
Clay	
Coarse clay	1/640
Medium clay	1/1024
Fine clay	1/2360
Very fine clay	1/4096
Colloid	<1/4096

*Source: Adapted from Wentworth, 1922.

via rivers. Other components are carried to the ocean as various-sized sedimentary particles by wind, ice, or flowing water (Figure 5-15).

Rivers are the main suppliers of terrigenous material to their estuaries, their deltas, or adjacent marginal seas, and these are the places where most of their transported sediment is initially deposited. Thus, terrigenous sediments tend to be most abundant near their source, the continent. Logically, they prevail on the shelf, slope, and rise.

In some areas, terrigenous sediments may be removed and redeposited. An example of the mechanism might be movement by slumping and moving downslope, or by movement as turbidity currents (see page 89). In both cases, they can be carried into the deep sea, far from land and their place of origin. Wind can also carry large amounts of sediment into the ocean, especially from desert areas where little or no protective cover of vegetation exists to prevent erosion.

Mud is a general term applied to sediment of clay size (see Table 5-1). Terrigenous muds are generally variable in color because of differences in their source or because of conditions in the area of their deposition. For example, black muds are common in areas where there is a large supply of organic matter but insufficient oxygen to oxidize it; red or brown muds are typical of well-oxygenated areas.

Terrigenous muds generally contain minerals common in land rocks. They are transported to the ocean mainly by rivers and, to a lesser extent, by wind. The rate of deposition in the marine environment varies, depending on nearness to the source and the amount being moved. Deposition rates are, however, several times higher than those deposited far from land in the deep

FIGURE 5-14 The grain-size range of common components in marine sediments. Grainsize is the diameter of an individual grain in millimeters (1/1,000 of a meter or 1/10 of a centimeter, which is about 0.39 inches).

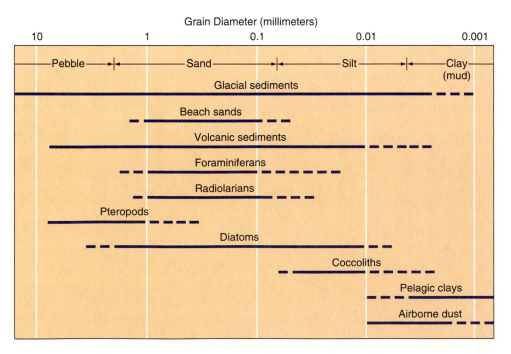

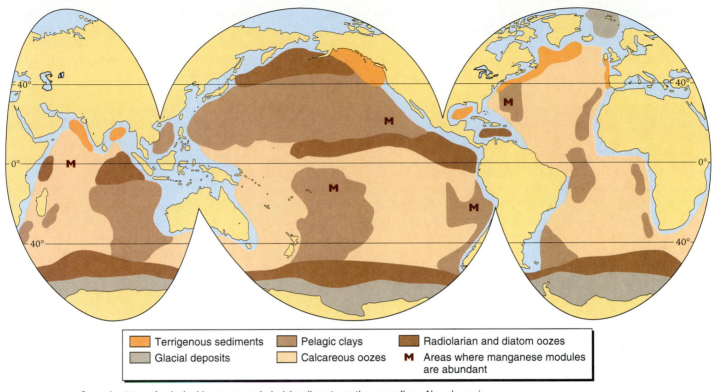

Terrigenous sediments   Pelagic clays   Radiolarian and diatom oozes
Glacial deposits   Calcareous oozes   M Areas where manganese modules are abundant

FIGURE 5-15 General pattern of pelagic, biogenous, and glacial sediments on the ocean floor. Also shown is the distribution of terrigenous sediments and areas where manganese nodules are abundant. Data from various sources.

ocean. Near large rivers, a rate of 100 cm (40 in.) or more is common per 1,000 years. Deep-sea rates are typically a few centimeters per 1,000 years.

SLUMP DEPOSITS. Slump deposits have moved or slumped down from a topographically higher area. Although difficult to distinguish by conventional sampling techniques, they can sometimes be identified as slumps from seismic reflection profiles. Slump deposits are common on the continental slope and rise where large blocks of deformed sediment occur that have slumped down from shallower areas (see Figure 4-3, page 79). Under some conditions, slumped material may form a turbidity current, and the sediment will move more as individual particles rather than as a large mass. Sediments in slump deposits or turbidites can be of any origin, but they are generally terrigenous.

TURBIDITES. Layers of coarse-grained sand and silt found interbedded with the typical fine-grained muds of the deep sea are generally attributed to turbidity currents. These deposits are called **turbidites** and occur in many areas of the deep sea. Turbidite beds range in thickness from a few centimeters to 3 m (10 ft.) or more. They generally have a variable grain size, with coarser material at the bottom of the layer and finer material above. This grain-size variation is caused by natural sorting, in which the coarser-sized grains settle out first.

Turbidites may contain wood fragments, shells of organisms that live in shallow water, and other materials that indicate a shallow-water origin for the sediment. In many instances, especially off large submarine canyons, large parts of the deep sea are blanketed with these sediments, which form flat abyssal plains (see Figure 4-16, page 91).

GLACIAL SEDIMENTS. Glacially derived sediments (see Figure 5-1b) are common to many parts of the continental shelf in regions of middle latitude to high latitude. Glacial sediments also are found, but less commonly, in the deep sea, where they can be recognized by their considerable content of sand, silt, and occasional gravel compared to the finer clay content typical of the deep ocean. How did these coarser-grained sediments get into the deep ocean? The mechanism is really quite simple: the sediments were initially carried by glaciers. As glaciers enter the ocean, parts break off and form icebergs. Icebergs are moved by currents, and when they eventually melt, any sediment or debris in the iceberg falls to the seafloor.

VOLCANIC DEPOSITS. Sediments of volcanic origin, such as volcanic dust, are common in some parts of the ocean. The source volcanoes may have been on land, but some are submarine. In either case, the particles are carried by the wind and water currents and eventually settle out, falling to the ocean floor. In some areas, sufficient volcanic dust settled out so that a thin but distinct ash layer occurs on the seafloor. If a layer results from a historical volcanic eruption, it can serve as a convenient time marker for dating anything on the seafloor that is in the layer, just above it, or just below it.

Modern volcanic explosions, such as Krakatoa in Indonesia in 1883, Mount St. Helens in Washington State in 1980 (see Figure 3-23, page 65), and Mount Pinatubo in the Philippines in 1991 (see Box 13-2, page 325), formed large dust clouds. In the case of Krakatoa, the dust clouds eventually covered much of the world. Following the rules of sorting, the heavier material settled and accumulated near the eruption area. The ash layers near the eruption site are generally composed of shards of volcanic material and smaller fragments of volcanic rock. Farther away, finer-grained volcanic minerals are found.

Sediments of the Deep Sea: The Pelagic Environment

The sediments of the deep ocean basin are generally called **pelagic sediments.** Pelagic comes from the Greek word meaning *open sea*. Pelagic sediments are so named because they originate more from the ocean water itself than from faraway sources of land sediment. Some types of pelagic sediment settle down through the water in the absence of strong currents; other types form directly on the seafloor.

Pelagic sediments are generally composed of microscopic skeletal material of plants and animals, fine-grained clays, or a mixture of the two. The clays can come from land and be carried to their deposition site by wind, water, currents, or some combination of these agents. Other sources of fine-grained clays are volcanoes and fragments from meteorites, and possibly even from comets (recall the hypothesis for the origin of Earth's ocean on page 27).

Biogenous Sediments

Biogenous sediments are generated by biological processes. In simplified terms, they are the shells and skeletons of dead creatures. Shells and external coverings of organisms are generally composed of the chemical compounds *calcium carbonate* ($CaCO_3$) or *silica* (SiO_2), and thus these are the components that commonly accumulate as fine-grained biogenous sediment (see Figure 5-15).

So, biologically generated sediments in the open ocean are called *biogenous pelagic deposits*. If they contain more than 30 percent skeletal material (fragments or shells of various plants or animals), they are called *oozes*. We may find the word *ooze* humorous, but it is an Old English word meaning "mud," so it is appropriately applied by geologists and oceanographers to this material. Oozes may also contain some material of nonbiological origin,

such as clay particles (Figure 5-16). An ooze is named after the organism that is most prevalent within it.

Common types of oozes are composed of tiny organisms such as coccoliths, diatoms, foraminiferans, pteropods, or radiolarians (Figures 5-17 and 5-18; see also Figures 11–4 (page 250) and 11–12 (page 260). Coccoliths and diatoms are plants (phytoplankton) and the others are animals (zooplankton). Diatoms and radiolarians have shells made of silica; the others have calcium carbonate shells. The terms **siliceous shell** and **calcareous shell** are frequently used for these two types of shells.

Most organisms that form biogenous sediments live in the surface waters of the ocean. After death, their shells settle to the ocean floor, forming the sedimentary deposit. The distribution of these sediments on the ocean bottom (see Figure 5-15) generally corresponds to the condition of surface waters overhead. In other words, the skeletons of cold-water creatures lie on the ocean bottom beneath surface cold-water areas. This means that the rate at which the shells settle to the deep seafloor is rapid enough to prevent them from being moved sideways very far by water currents.

When this distribution pattern was first detected, it presented a puzzle. Because the settling shells are so tiny and light, they should have been carried far away by even slow-moving currents. Some of these shells should have taken perhaps a decade to settle (see Table 5-2). The puzzle was solved when scientists sampled both the water column and the bottom. They determined that, rather than falling as individual shells, they were falling to the bottom as parts of considerably larger **fecal pellets** (see also Box 11-1, page 247), which are the excrement of organisms, in this instance the organisms that feed on the creatures that form the oozes (see Figures 5-18c and 5-18d). When the shells are part of a fecal pellet, the trip from surface water to the deep ocean floor can be as short as 10 days. Thus, the microscopic shells

Figure 5-16 Typical deep-sea ooze. Note the numerous animal tracks and burrows. (Photograph courtesy of Jim Broda, Woods Hole Oceanographic Institution.)

FIGURE 5-17   (a) Drawings of some typical organisms that constitute oozes on the seafloor. (b) Electron micrograph of coccoliths, enlarged 11,000 times. They are actually about one-tenth the diameter of a human hair. (Photograph courtesy of J. C. Hathaway.)

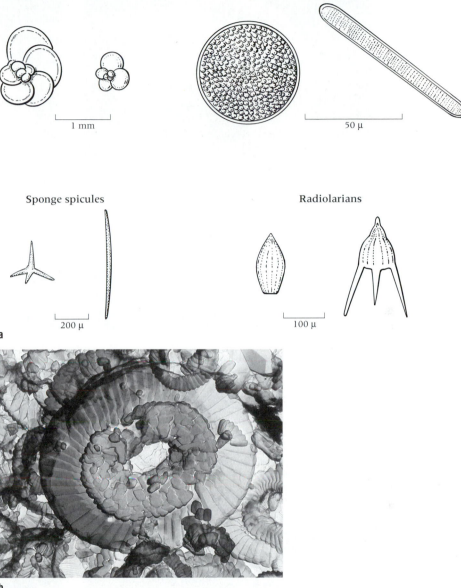

Planktonic Foraminifera

1 mm

Diatoms

50 μ

Sponge spicules

200 μ

Radiolarians

100 μ

a

b

would end up on the ocean bottom beneath where they had lived on the ocean surface. Once on the seafloor, the pellets break apart, leaving the shells to appear as distinct, individual particles.

Calcareous oozes are usually more abundant at shallow depths than in deep water because the calcareous shells are soluble (that is, they can dissolve in seawater). This solubility increases with depth and colder temperature. Thus, calcareous shells are rarely found below 5,000 m (16,000 ft.) because most have been dissolved (see also Box 5-4). The effect is more obvious on the small and delicate coccoliths than on robust foraminiferan shells.

Biogenous sediments are common on the seafloor beneath regions of great biological productivity, such as in equatorial regions and where other sediment types are absent or are deposited at a relatively slow rate (see Figure 5-15). The sedimentation rate of biogenous sediments is about 1 to 5 cm (about 0.4 to 2 in.) per 1,000 years, approximately 10 times faster than the sedimentation rate of clay deposits of nonbiological origin (Table 5-3). The distribution patterns of the sediment on the seafloor can also be linked to the process of seafloor spreading (see Box 5-4).

Another type of biogenous deposit can result from coral reefs that have been exposed to the destructive powers of waves or other erosive forces. If reworking by waves is especially intense, a white mud composed of fine-grained debris may result. The deposits are generally localized around the reef itself.

One of the commonest sedimentary rocks on land is *limestone*, which is composed principally of calcium car-

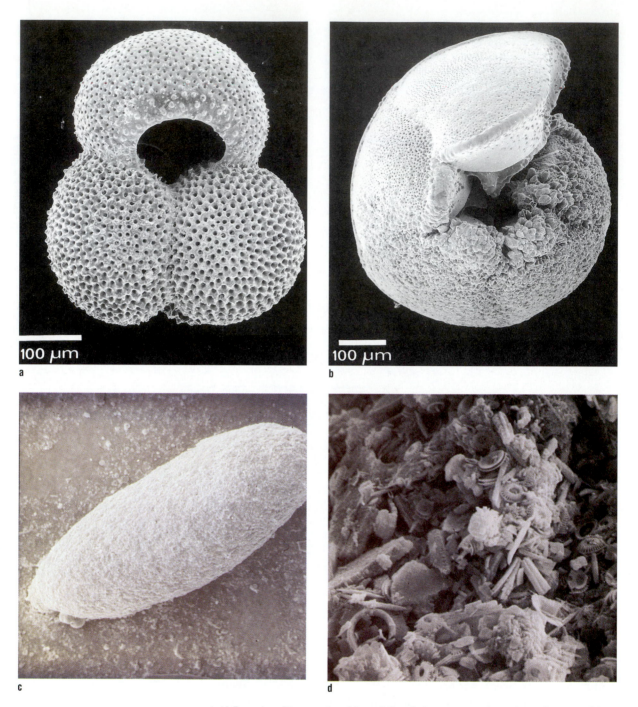

FIGURE 5-18 (a,b) Examples of two species of Foraminifera that are common in marine sediments and fecal pellets: *Globigerinoides ruber* (a) and *Globorotalia truncatulinoides* (b). Scale line is 100 microns or 0.1 mm.
(c) Scanning electron microscope photograph of a fecal pellet produced by a copopod (a zooplankton). The pellet, collected in a sediment trap (see Figure 7-7, page 166), is about 200 microns (0.2 mm) long, or about twice the diameter of a human hair.
(d) Scanning electron microscope photograph of the interior of a fecal pellet. Note the many small particles, many of which are coccoliths (see also Figure 5-17b) or various pieces of animal and plant debris. The pellet is about 50 microns wide, or about half the diameter of a human hair. (Photographs (a) and (b) courtesy of Dr Werner Deuser; (c) and (d) courtesy of Dr. Sus Honjo.)

TABLE 5-2

Settling Velocities of Some Grain Sizes

Particle Name and Diameter	Settling Velocity (cm/second)	Time to Sink 1 kilometer
Sand 100 micrometers	2.5	10.8 hours
Silt 10 micrometers	0.025	46.25 days
Clay 1 micrometers	0.00025	12.25 years

Source: Adapted from Gross, 1990.

bonate. Most limestones result from biological processes, and are composed of plant and animal shells that have solidified into limestone.

Pelagic Clays

Pelagic clay is the term applied to the fine-grained muds occurring in most of the deep ocean basins, far from land. By definition, they contain less than 30 percent biogenous material, so they are **inorganic.** They are generally brown, due to having been oxidized. Very little is known about the origin and source of these clays. The Gobi Desert of Mongolia may be the source for the clays in the northern Pacific. The clay particles are transported by wind and ocean currents. Other possible sources are meteoric dust or volcanic ash.

The sedimentation rate of pelagic clays is extremely low, usually 1 to 2 mm (about 0.04 to 0.08 in.) per 1,000 years, or 1 m per million years. With such a low sedimentation rate, they are diluted and negligible in areas where other types of sediments are forming or accumulating more rapidly. Thus, pelagic clays are significant only in the deep areas of the ocean, far from any land-derived sediments and where solution has removed many of the calcareous shells (see Table 5-3 and Box 5-4). Even under such restrictive conditions, large portions of the ocean, notably in the Pacific, have pelagic clay deposits (Figure 5-15).

Hydrogenous Sediments

Hydrogenous means "water-generated." **Hydrogenous sediments** generally have formed right where they are found, frequently precipitated directly from ocean water. Hydrogenous deposits include nodules of iron and manganese (see Figure 15-11, page 376). These deposits have been found over large areas of the ocean and may be the most important economic deposit of the deep sea (see Chapter 15). They commonly occur as rounded nodules the size of a baseball, but can also be found as slabs or as coatings on rocks. The nodules frequently form around small objects on the seafloor, such as a shark's tooth or

the ear bone of a fish. Since the deposits form by precipitation from seawater, their deposition essentially stops when the nodule or rock is covered with sediment.

The distribution of iron-manganese nodules is usually restricted to areas having an otherwise low sedimentation rate or to areas where strong currents prevent the deposition of other sediments. An example of the latter is the Blake Plateau off Florida, where the swiftly flowing Gulf Stream prevents sediment deposition.

Another hydrogenous sediment also of possible economic significance is phosphorite. **Phosphorite** frequently occurs as pebbles or large grains and is composed mainly of phosphate minerals. Deposits of phosphate on land often are mined for fertilizer. A similar use may be in the future for phosphates occurring in the marine environment. Large deposits of phosphorite are generally restricted to depths of 500 m (about 1,540 ft.) or shallower. This mineral seems to form in basins or areas that contain virtually no oxygen, that is, an anaerobic environment. The absence of oxygen can result when large amounts of organic matter, produced in the surface waters, sink and are oxidized (consuming oxygen) in the deeper waters.

Cosmogenous Sediments

Cosmogenous sediments are particles that quite literally come from outer space (cosmogenous means "generated by the cosmos"). They may be fragments from meteorites, comets, or even parts of large asteroids that may have impacted Earth in the past. Some of these large collisions have been proposed as mechanisms for past worldwide extinctions (see Box 2-4, page 32). Cosmogenous sediments are very rare in the ocean, but sometimes can be detected in the very slowly deposited pelagic clay deposits, where the possibility of dilution is less than with most other marine sediments.

Rocks on the Seafloor

Rocks exposed on the seafloor are generally basaltic rocks rather than the granitic rocks typically found on land. Basalt is found on many oceanic islands, on the mid-ocean ridges, and along some fracture zones (see Figure 5-19). Rocks on the seafloor can also be sediments that have been cemented into rocks or the previously mentioned manganese nodules.

Areas of Nondeposition

There are many places on the ocean floor where no sedimentation is presently occurring. Shallow-water localities where strong bottom currents prevent sediment accumulation are an example. There are regions in the Atlantic and Pacific Oceans where recent sediments are

BOX 5-4

Seafloor Spreading and Deep Sea Sedimentation

Seafloor spreading and the distribution and type of sediment on the deep seafloor are closely related. Prior to the seafloor spreading concept, ocean scientists expected deep seafloor sediment thickness to be 2 to 3 km (1.2 to 1.8 miles) thick, representing material that had accumulated over all geological time. The anticipated distribution pattern was like that shown in Figure 1a, but this expectation was far from what they found.

The start of the Deep Sea Drilling Project in 1968 and the eventual drilling of over 1,000 holes in the deeps of all oceans showed a surprising pattern. First, as discussed in "The Missing Sediments" section (page 116), seafloor sediment thickness was much less than expected, rarely exceeding 600 m (2,000 ft.). Next, the maximum sediment age rarely exceeded 200 million years, only 4 percent of geological time.

Drilling also showed that both the thickness and age of the deepest sediment increased away from the ocean ridge. Remember that the ocean ridge area is the youngest part of the seafloor. These observations are consistent with the concept of seafloor spreading. Indeed, these discoveries were among the key points leading to the general acceptance of the seafloor spreading hypothesis.

One more aspect of seafloor spreading influences deep sea sedimentation. The seafloor is shallower in the ocean ridge area and deepens away toward the flanks (see Figure 4-18, page 94). Because of this, the shallower ridge area may be above the depth where calcium carbonate will dissolve, between 4,500 and 5,000 m (14,760 to 16,400 ft.). The location of this depth, called the *carbonate compensation depth (CCD)*, determines whether falling calcareous shells will be preserved as bottom sediment. Below the CCD, shells are generally dissolved, whereas above it they will be preserved.

Thus there is a tendency for calcareous shells to predominate (assuming they live in the surface waters) near the ridge areas. Further out from the ridge and below the CCD, silicious oozes or pelagic clays tend to be the predominant surface sediment. The carbonate oozes deposited at the ridge areas can be buried under later-deposited silicious oozes or pelagic clays, as the seafloor moves out from the ridge into deeper water. The generally observed sedimentation pattern is shown in Figure 1b. Clearly, seafloor spreading determines the thickness, age, and type of sediment in many parts of the deep sea.

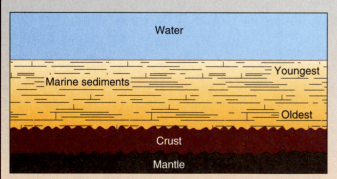

FIGURE 1 (a) The pattern of sediment distribution anticipated before the concept of seafloor spreading had been developed. A thick sequence of sediment was expected to overlie the crust in most of the deep sea.

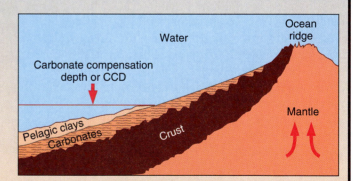

(b) The general distribution of deep sea sediment near an ocean ridge (only one side of the ridge is shown), as determined by drilling and other sampling. As this seafloor moves away from the ocean ridge by means of seafloor spreading, it deepens, eventually going below the CCD. At that time and place, the predominant sediment preserved on the seafloor can change from calcareous oozes to silicious oozes or pelagic clays.

TABLE 5-3

Approximate Time for 1 cm (0.4 inches) of Sediment to Accumulate on the Seafloor

Terrigenous Deposits		Pelagic Deposits	
Near large rivers	1–10 years[1]	Biogenous Sediments	200 to 1,000 years
Continental Shelf	30 years	Pelagic Clay	2,000 to 10,000 years
Continental Rise	100 years		

[1]These numbers are very approximate and are meant only to give an approximation of relative rates.

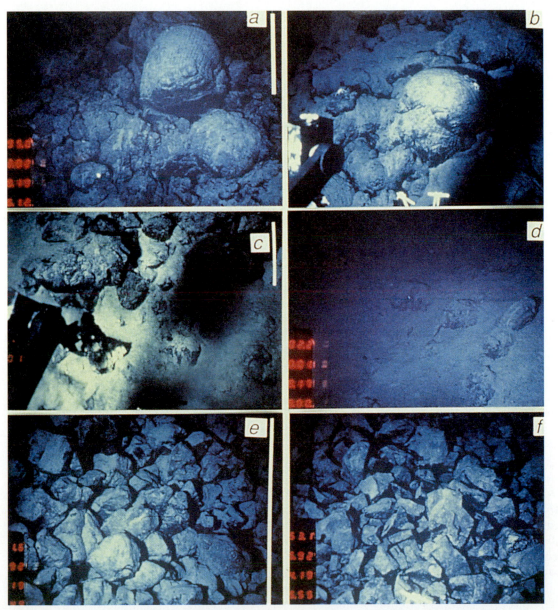

FIGURE 5-19   A series of bottom photographs from the Galápagos spreading center off the western coast of South America in the Pacific Ocean. Note the bulbous pillow lava in a–d and the angular debris in e and f. The arm of the submersible *Alvin* can be seen in band c. The white bar is about 1 m (3.28 ft.) long. The red writing is information relating to the photography. (Photograph courtesy of Dr. Martin C. Kleinrock, Woods Hole Oceanographic Institution.)

absent and the surface sediments are several million years old. Under these conditions, we must assume either that essentially no deposition has occurred for many millions of years, or that any recent deposits have been removed by erosion.

Sediments on the United States Continental Margin

U.S. continental shelf sediments are mainly terrigenous, sand-sized or silt-sized, generally transported by rivers. In the northern areas, glacial material is common. In the south and off the Pacific islands, biogenous sediments prevail. Occasionally, some hydrogenous sediments are found on the continental slope.

Most of the sediments of the continental shelf are **relict,** a geologist's term meaning that they are not from the present-day environment but are relics from a previous one. In this instance, they are relict from the environment before the recent rise of sea level. (An example is a beach deposit on the continental shelf that now is covered by tens of meters of water.) In many areas, continental shelf sediments have been reworked by waves and currents during the period of rising sea level.

Sediments of the continental slope are usually finer-grained than the continental shelf and generally are mud. Much of this mud was carried to the slope during times of lowered sea level. With a lower sea level, material supplied by rivers, wind, and other processes would have been have been carried closer to the continental slope rather than being deposited on the then-exposed shelves. In some areas, especially those off faulted or mountainous coasts like those in Maine or Oregon, rock outcrops are common.

Paleoceanography

Marine scientists have various ways of studying past environmental conditions of the ocean. This field of research is generally called **paleoceanography.** One method of study determines changes in the amount of certain chemical elements in the shells of various organisms. This amount can reflect the prevailing temperature and chemistry of the environment where and when the organism lived. After they die and their remains settle to the bottom, a chemical analysis of their shells, combined with dating by various techniques, can reconstruct the paleoceanographic conditions in which they lived. This can include the timing and duration of previous climatic periods that could have changed the temperature or oceanographic circulation.

You have seen how sediments can vary over horizontal distances. They can also change in character vertically. In other words, there can be variations over time

that form different layers or sequences in the bottom sediment. A study of these sequences is called **stratigraphy** (meaning loosely "graphing of strata"). Its objectives include correlating sediments in one area with those in another, and correlating events in the ocean to events on land. These aims are accomplished by using dating methods such as radioactivity (described in Chapter 7, pages 172–175) or by using the magnetic reversals studied in seafloor spreading research (see Figures 3-6 and 3-7, pages 48 and 50). Stratigraphic studies are more valuable when long cores from the seafloor are obtained by piston coring or drilling.

Dating by Magnetic Reversals

Earth's magnetic field changes in polarity (reverses from north-south to south-north, or vice versa) over time, a phenomenon called **magnetic reversal.** Magnetic-reversal dating is based on the fact that these changes are recorded in the magnetic minerals within rocks and sediments formed at that time. These rocks and sediments can frequently be dated by radioactive methods to give a time for the magnetic reversal.

For example, about 700,000 years ago, a major change or reversal occurred in Earth's magnetic field. It remained essentially reversed until about 1.7 million years ago, when it reversed again to positive, or similar to our present magnetic condition (Figure 3-7, page 50). By taking a long sediment core from the seafloor and determining the magnetic orientation of its minerals, we can identify these reversals. And by relating this reversal pattern to that of a known dated sequence, we can make a good estimate of the sediment's age. A modification of this technique was most valuable in developing the seafloor spreading concept, as we saw in Chapter 3.

The magnetic technique is an extremely good one because the polarity of Earth's magnetic field simultaneously affects all sediments containing magnetic minerals as they are being deposited in all parts of the ocean. Therefore, the magnetic technique can also be used as a correlation device between cores that are geographically far apart (Figure 5-20). Reversals have occurred with surprising speed, perhaps over a hundred years or so.

Dating by Biogenous Sediment

A key fact in stratigraphy is that many organisms that live in the ocean have evolved into different species over periods of time. Therefore, some individual species have time ranges for their existence that can be used both for dating and for correlation. Unfortunately, most marine species have limited geographic ranges—they live either in equatorial, or temperate, or polar regions, but very few live in all three. This means that a single species cannot be used for correlation in all parts of the ocean.

FIGURE 5-20 How the Earth's magnetic-field reversals, sediment type, and fossils can be used to correlate deep-sea cores taken in different parts of the ocean. The magnetic time scale was established by dating the reversals on land, using radioactive isotopes (see Chapter 7, pages 172–175). By measuring the magnetic character of the cores, the cores can be correlated to the time scale, and to each other.

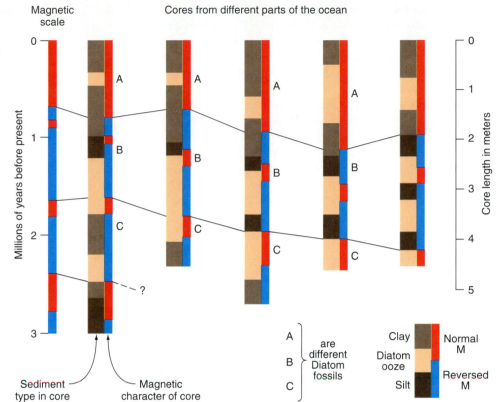

Biogenous deposits are very important to oceanographers because they indicate physical aspects of the environment in which the organisms lived. The organisms best suited for these studies are pelagic Foraminifera, which live mainly in the near-surface waters of the ocean. Certain species are typical of low latitudes (equatorial) and warm water, while others are typical of middle latitudes (temperate) or high latitudes (polar) and cooler water.

Cores of sediments containing foraminiferal ooze mostly show a change in the dominant species with depth. This indicates past changes in surface-water conditions. Some Foraminifera species also experienced a worldwide evolutionary change, which is a time marker, allowing correlation from one core to another. Further, the coiling direction of the shells of some Foraminifera species changed. Apparently this happened in response to temperature changes in the water where they lived.

Another technique for studying the ancient environment is to measure the ratios of various chemical isotopes contained within the shells. (Isotopes are variant forms of the same element that have slightly different properties, explained in Chapter 6.) For example, the uptake of isotopes of oxygen by organisms is influenced by the water temperature where the organisms lived. Thus, the isotope composition can indicate the past water temperature. Results of these techniques are not ab-

solute; in some instances they even conflict. One reason for discrepancy is that Foraminifera may live in different parts of the water column (and therefore at different temperatures) during various life stages.

Corals can also be valuable indicators of past oceanographic conditions. For example, certain species will grow only within a narrow depth range. The method is simple: collect fossil samples of these corals, note their depth relative to present sea level, and date the corals by radioactive techniques. This shows where sea level was when the organism sampled was alive, and how long ago.

Dating Problems

There are three problems in developing the stratigraphic record of the seafloor: (1) sediment does not accumulate everywhere at a constant rate, (2) erosion may remove part of the sediment record in some places, and (3) considerable difference may exist in the sediment types that reach the seafloor. Therefore, in working out the history of the ocean, it is important to date and correlate sediments of different types but of similar ages. A gap in the sedimentary record can result from periods of nondeposition, or it could result from erosion, when bottom currents remove sediments already on the seafloor.

Note that in developing the sedimentary history of a region, you do not directly observe the layers in place, but rather look at small samples the width of a coring tube or drilling pipe (about 6.5 cm, or 2.5 in.). This can often make the unraveling of the past harder than on land, where broad expanses of rock may be viewed directly. Another complicating factor in stratigraphic studies is that organisms living on or in the seafloor often rework the bottom sediments in their normal life activities. In doing so, they can smooth out or mix up sediment characteristics, such as a thin volcanic ash layer that might have existed. Such unique layers can be keys to working out the stratigraphy in a region. For example, if a major volcanic event occurred that deposited ash over a large portion of the ocean basin, that layer can be correlated to others wherever it is found.

The Paleoceanographic Picture

The combination of all these techniques has made possible an acceptable description of recent changes in the ocean. The general picture is that the last glacial period ended about 18,000 years ago, and surface waters of the ocean have been warming since that time. Between 11,000 and 60,000 years ago, the oceans were relatively cold. The warm period that ended 60,000 years ago extended back to 100,000 years ago and was preceded by another glacial period (Figure 5-21).

In summary, sediment studies using radioactive dating, magnetic stratigraphy dating, chemical measurements, and biological variations in fossils have given paleoceanographers a good approximation of past physical characteristics of the ocean. Techniques like these are also being used to speculate on potential variations in ocean circulation and world climate in the future.

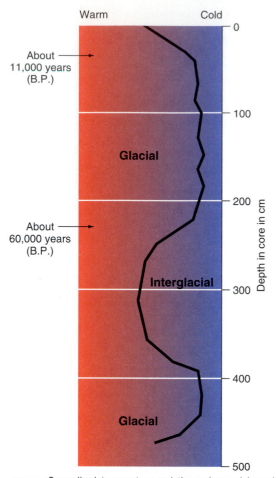

FIGURE 5-21 Generalized temperature variations observed in a deep-sea core. The ancient temperatures are estimated from the relative abundance of pelagic foraminifera. Certain species indicate a warmer temperature of surface waters; other species indicate colder temperature. Major changes occurred about 11,000 and 60,000 years B.P.

SUMMARY

Sediments on the seafloor are divided into four major categories, based on their origin: terrigenous (from previously existing sediment or rock), biogenous (from biological activity), hydrogenous (from seawater), and cosmogenous (from outer space). Those sediments clearly derived from land often dominate the continental margin and can include terrigenous muds, slump deposits, turbidites (from turbidity currents), and glacial deposits.

Pelagic sediments are more typical of the deep ocean and areas far from land. They have settled down through the water column, usually in the absence of strong currents, or have formed directly on the seafloor; biological components often dominate.

Continental sediment may be carried to the ocean by winds, but rivers are usually the main vehicle for transporting terrigenous sediment into the ocean. Some material dissolved in seawater can be precipitated as hydrogenous sediments or used in the biological processes by which animals and plants form shells that eventually accumulate on the seafloor. Another sediment source is volcanic activity, either on land or in the ocean.

The study of past oceanographic conditions is called paleoceanography. An analysis of the chemical change with depth of some sediment aspects can be used to interpret past environmental changes. Changes in shell characteristics or evolution of some species also may indicate previous environmental conditions. Various techniques can be used to correlate sediments, including radioactive dating, magnetic stratigraphy dating, chemical measurements, and biological variations in fossils in the sediments and rocks on the seafloor. A combination of these techniques has helped us understand environmental conditions over Earth's recent history.

QUESTIONS

1. What are the main geological tools for direct sampling of the seafloor?

2. What are the main types of marine sediments? Where does each typically occur?

3. What distinguishes the sediments of the continental margin from those of the deep sea?

4. What factors influence deposition of biogenous sediments?

5. How can marine sediments provide information on past oceanographic conditions?

KEY TERMS

Alvin
ANGUS
ARGO/JASON
bathysphere
biogenous sediment
box corer
calcareous shell
cosmogenous sediment
dredge
drilling ship
Deep Sea Drilling Project (DSDP)
fecal pellet

glacial sediment
Global Positioning System (GPS)
Glomar Challenger
grab sampler
grain size
gravity corer
hydrogenous sediment
inorganic
JOIDES
JOIDES *Resolution*
Loran
magnetic reversal

manganese nodule
marine sediment
mud
navigation
paleoceanography
pelagic clay
pelagic sediment
phosphorite
pinger
piston corer
relict sediment
SCUBA
sedimentation

side-scan sonar
siliceous shell
slump deposit
sorting
stratigraphy
submersible
terrigenous sediment
Trieste
turbidite
volcanic deposit
WASP

FURTHER READING

Anderson, R. A. 1986. *Marine Geology: A Planet Earth Perspective.* New York: John Wiley. A somewhat advanced marine geology textbook.

Hamblin, K. W. 1992. *Earth's Dynamic Systems.* 6th ed. New York, Macmillan Publishing Co. A superb introductory Earth science textbook with excellent illustrations.

Jeffrey, D. 1985. "Fossils: Annals of Life Written in Rock." *National Geographic* 168, no. 2, pp. 182–192. A general article on fossils.

Kennett, J. 1982. *Marine Geology.* Englewood Cliffs, N.J: Prentice-Hall. A detailed and advanced and very comprehensive marine geology textbook.

Libes, S. M. 1992. *An Introduction to Marine Biogeochemistry.* New York: John Wiley and Sons. A slightly advanced textbook covering marine chemistry and how it relates to some aspects of marine geology, biological oceanography, and pollution.

"Marine Geology and Geophysics." *Oceanus* 35, no. 4 (Winter 1992–93). A special issue on recent research in marine geology and geophysics.

Open University. 1989. *Waves, Tides and Shallow-Water Processes.* Oxford: Pergamon Press. Covers most of the aspects of sediment movement in the marine environment.

Shepard, F. P. 1973. *Submarine Geology.* 3d ed. New York: Harper & Row. The third edition of the grandfather of marine geology textbooks, a little dated but still a good general description of many aspects of marine geology.

York, D. 1993. "The Earliest History of the Earth." *Scientific American* 268, no. 1, pp. 90–96. Describes how radioactive techniques are used to learn more about Earth's first 1.5 billion years.

CHAPTER 6

Properties of Water

EARTH IS UNIQUE AMONG THE PLANETS FOR ITS VAST QUANTITIES OF LIQUID WATER. WATER IS ONE OF THE MOST COMMON COMPOUNDS ON THE SURFACE OF OUR PLANET. MANY OF ITS PROPERTIES ARE UNUSUAL WHEN COMPARED TO OTHER SUBSTANCES. WATER HAS THE ABILITY TO ERODE AND REDUCE THE LARGEST MOUNTAINS TO MERE HILLS, YET WATER IS GENTLE ENOUGH TO PRODUCE THE ENVIRONMENT WHERE LIFE BEGAN. WITHOUT WATER, LIFE AS WE KNOW IT WOULD BE IMPOSSIBLE, AND OF COURSE THERE WOULD BE NO WORLD OCEAN. THIS CHAPTER CONSIDERS WATER AND SOME OF ITS CHEMICAL AND PHYSICAL PROPERTIES.

Inside an Arctic ice ridge, viewed during *Fram II* expedition.

(Photograph by Keith von der Heydt, Woods Hole Oceanographic Institution.)

Introduction

About 1,349 million km³ (326 million miles³) of water is spread over Earth's surface. This amount would make a layer 144 km (90 miles) thick if distributed over just the United States. Almost all of this water (97.2 percent) is contained in the ocean (Table 6-1). If the water frozen in glaciers and ice caps were added to the ocean, it then would contain 99.35 percent of the world's water (Figure 6-1). The exchange of water among ocean, atmosphere, and land is called the **hydrologic cycle** (Figure 6-2). Some water continuously moves from one environment to another—for example, from ocean to atmosphere. This movement occurs mainly by evaporation. Other processes in the cycle are **transpiration**, precipitation, and river runoff.

A large portion of animal or plant tissue is composed of water. Without water, most if not all life would be impossible. Because of water's presence in cell tissue, organisms must avoid freezing, because the growth of ice crystals will damage or rupture cell structures. Some marine fish have evolved an antifreezelike compound that allows their survival in near-freezing conditions of the deep ocean (see Box 11-2, page 253).

TABLE 6-1
The World's Water Distribution

Area	Water Volume (miles³)	Percent of Total
Surface Water		
Freshwater lakes	30,000	0.009
Saline lakes and inland seas	25,000	0.008
Rivers and streams	300	0.0001
Total	55,300	0.017
Subsurface water		
Soil moisture	16,000	0.005
Groundwater	2,000,000	0.62
Total	2,016,000	0.625
Ice caps and glaciers	7,000,000	2.15
Atmosphere	3,100	0.001
Oceans	317,000,000	97.2
Total (approx.)	326,000,000	100.0

Source: Adapted from Leopold and Davis, 1966.

The Structure of Water

Early Greek philosophers viewed the world as being composed of four basic components: air, earth, fire, and water. For a long time, water was believed to be a separate and indivisible element, rather than a chemical compound consisting of two different elements (see Box 6-1).

Pure water is an extremely simple compound, yet its properties and behavior are remarkably complex. It was not until 1783 that it was determined that water was composed of hydrogen and oxygen. The chemical composition of a water molecule is simple: two hydrogen atoms to one oxygen atom (H_2O). One of the extraordinary properties of water is that its dissolving power exceeds that of any other liquid. Water also has an exceptional capacity for absorbing heat. It warms or cools more slowly than other liquids. It also warms or cools more slowly than the atmosphere, thereby modifying Earth's surface temperature. The temperature at which pure water freezes has been defined as 0°C (32°F) and the temperature at which it boils has been defined as 100°C (212°F).

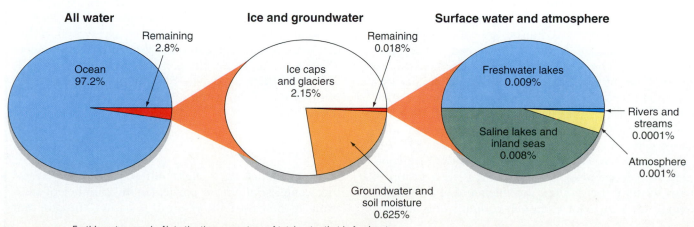

FIGURE 6-1 Earth's water supply. Note the tiny percentage of total water that is freshwater.

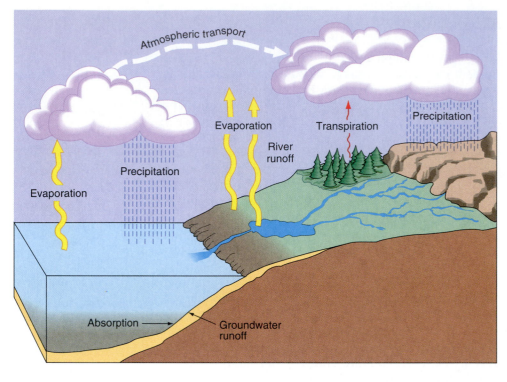

Box 6-1

Chemistry Review

Some basic definitions are necessary for understanding the chemical and biological aspects of water and the ocean. They already may be familiar to you, but review never hurts!

Elements and Compounds

The *atom* is the smallest complete part of an element. Atoms themselves are composed of the subunits *protons, electrons*, and *neutrons*. Protons and neutrons make up the *nucleus* of an atom, and electrons orbit in a ring-like pattern around the nucleus.

Elements are chemical substances comprised of the same atoms and cannot be broken down into simpler substances by chemical means, although they can be broken down by nuclear means. All atoms of a particular element have the same number of electrons in orbit, and protons in the nucleus, when in a neutral state. For example, neutral hydrogen always has 1 electron, neutral oxygen always has 6, and neutral lead has 82.

A *compound* is two or more elements that are combined in a fixed proportion. A *molecule* is a group of atoms forming the smallest possible piece of a chemical compound. Water, for example, contains one atom of oxygen and two atoms of hydrogen. The individual elements oxygen and hydrogen are thus combined to form the molecule H_2O, the compound called water.

When compounds are mixed with seawater they tend to dissolve and break up into individual ions. An *ion* is an atom or group of atoms that is no longer neutral but has an electrical charge. If the charge is positive, the ion is a *cation*. If the charge is negative, the ion is an *anion*. The size of the charge is determined by the number of missing or added electrons.

A chemical or covalent bond can be formed between two atoms when they share electrons, as in the case of the water molecule. An ionic bond occurs when, rather than sharing electrons, one ion loses one or more electrons and the other ion gains them. An example is sodium (written Na^+) when it combines with chlorine (written Cl^-) to form the compound NaCl or common (table) salt.

Isotopes and Radioactivity

Some elements have slight differences in the atoms that comprise them, in particular the number of neutrons in their nucleus. When this occurs, the different

Box 6-1 continued

forms are called **isotopes.** Hydrogen, for example, occurs in three isotopes. The most common has one proton, one electron, and no neutron. The other two isotopes are **deuterium** (doo-TEER-ee-um), which has one neutron in its nucleus, and **tritium** (TRIT-ee-um), which has two neutrons. Both still have only one electron and one proton. Water containing numerous deuterium atoms is called "heavy water" for obvious reasons.

Tritium is *radioactive,* which means that its nucleus is unstable and breaks down or decays to attain the simpler no-neutron form. This releases nuclear energy. The rate of decay is very specific and constant for any isotope, and thus can be used as a method for dating materials that contain the element. Carbon-14 dating is a well-known example of this method.

Temperature

The conversion for Celsius (°C) to Fahrenheit (°F) and vice versa is:

$$°F = (1.8 \times °C) + 32$$

$$°C = (°F - 32) \div 1.8$$

A water molecule is very strong. Considerable energy must be applied to break the covalent bonds that hold together the oxygen atom and the two hydrogen atoms. A **covalent bond** exists between two atoms in a molecule due to the sharing of electrons. The positive charge of the hydrogen atom and the negative charge of the oxygen atom causes them to attract each other. The two positively charged hydrogen atoms, however, repel each other.

Hydrogen has one electron in its outer shell or orbit and space for one more. The outer shell of oxygen has six electrons with space for two more. Unfilled, these electron shells are not stable. When filled, however, they become very stable, such as when a water molecule is formed (Figure 6-3).

When the oxygen and hydrogen atoms are united, they produce a lopsided molecule with the two hydrogen atoms at an angle of 105° to each other (Figure 6-3). This arrangement causes an unequal distribution of electrical charges. The side of the molecule with the two hydrogen atoms has a slight positive charge, and the oxygen side has a negative charge. The net charge, however, is zero. A molecule of this type is called a *dipole,* and its behavior can be compared to that of a magnet. In other words, its positive side is attracted to (or attracts) particles having a negative charge, and its negative side is attracted to (or attracts) positively charged particles.

This distribution of charges allows one water molecule to attract and form a bond with another water molecule (Figure 6-4). The negative charge of one water molecule bonds with the positive side of another water molecule. This is an electrical bond, called a **hydrogen bond.** It is weaker than a covalent bond. An individual water molecule can form a hydrogen bond with four other water molecules (Figure 6-4).

The net effect of hydrogen bonding is that water molecules cling together. This gives water its strong **surface tension,** which causes the surface of water to behave like a thin, elastic film. This is why you can slightly overfill a cup with water, yet not "break" the water's surface. The surface tension is strong enough to allow small insects or even a needle to float on its surface. The hydrogen bonding must be overcome for water molecules to separate, which happens when water vaporizes or ice melts.

The attractive action of water molecules also is important when water comes in contact with the numerous compounds whose atoms are held together by electrical charges (not by covalent bonds). If a water molecule

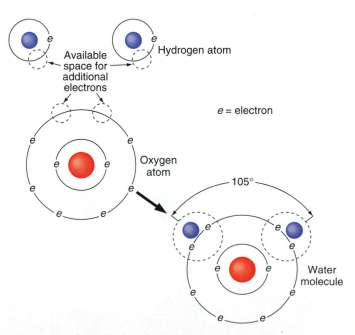

FIGURE 6-3 Approximation of water molecule structure. Water is formed by the joining of an oxygen atom with two hydrogen atoms, in a shape reminiscent of Mickey Mouse ears.

FIGURE 6-4   Clumping together of water molecules by hydrogen bonding. The hydrogen bonds develop between the negative charge of the oxygen ion and the positive charge of the hydrogen ion.

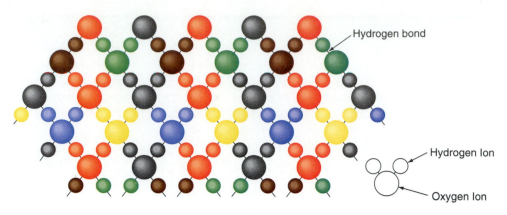

comes between two atoms held together by an electrical charge, its own dipolar electrical charge cancels some of the electrical attraction between the two atoms. As some of this attraction is canceled, the two atoms move farther apart and more water molecules move between the two atoms. Eventually, the attraction between the two atoms becomes insignificant and the atoms remain separated.

The original compound, whose atoms now are surrounded by water, has in this manner been *dissolved* by the water (Figure 6-5). This dipole aspect of water explains why water is an excellent solvent, capable of dissolving many compounds. In summary, water molecules tend to bond with literally anything that crosses their path.

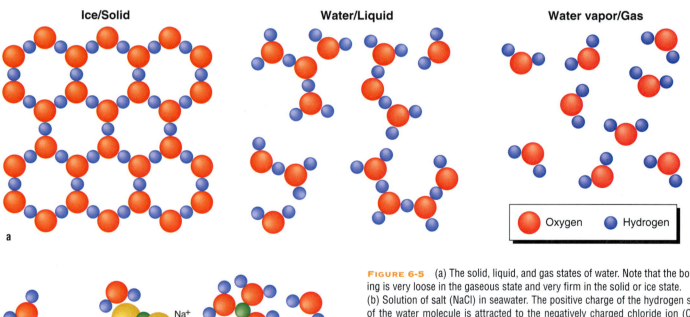

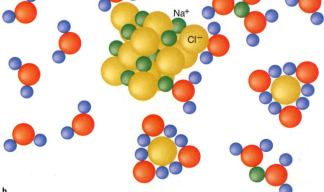

FIGURE 6-5   (a) The solid, liquid, and gas states of water. Note that the bonding is very loose in the gaseous state and very firm in the solid or ice state. (b) Solution of salt (NaCl) in seawater. The positive charge of the hydrogen side of the water molecule is attracted to the negatively charged chloride ion (Cl^-). The negative charge of the oxygen side of the water molecule is attracted to the positively charged sodium ion (Na^+). The salt is fully dissolved when the ions are completely surrounded by water molecules.

Some Properties of Water

Water has several extraordinary properties, summarized in Table 6-2. This section describes some of these properties.

Dissolving Power

Water's dissolving power is unique among fluids. It can dissolve more different liquids, gases, or solids than any other fluid, even strong acids. This dissolving power is reflected in the variety of chemical compounds found in the ocean. Erosion, starting with rainfall on land that slowly dissolves minerals and rocks, ultimately causes most of the ocean's salt content. Evaporation of seawater leaves behind most of the salts, and the water vapor condenses and falls as rain over land, continuing the process (see Figure 6-2). Volcanic processes also add to the ocean's salt content. Chapter 7 presents the sources of the different elements and the chemical processes involved.

Phase

Water occurs naturally in three phases or states: solid, or ice; liquid, or water; and gas, or water vapor (see Figure 6-5a). Changes from one phase to another result from addition or loss of heat. The liquid phase of water dominates Earth for a simple reason: water's high boiling point. At temperatures where some substances have vaporized into gases, water is still liquid. This aspect is very important in the development and maintenance of life. Water's high boiling point results from the dipole structure of the water molecule. As previously mentioned, because of their polarity, individual water molecules cluster together, held by hydrogen bonds. The strength and nature of these bonds determine some physical characteristics of water, such as its boiling temperature.

For water to boil, its hydrogen bonds must be broken. This requires considerable energy, causing the high boiling point. If water consisted of single unclustered and unbound molecules, it would boil at only −80°C (−112°F) and would be a gas under normal conditions. In fact, when water is in the vapor phase, each water molecule is relatively unaffected by any other molecule, because of the distance between the molecules. Other similar hydrogen compounds, such as hydrogen sulfide, have considerably lower boiling points than water, and thus are gases at normal surface temperature.

In the ice phase, water molecules are tightly bound to each other. The structure of ice, however, separates the molecules by relatively large distances (see Figure 6-5a). In the liquid phase, water molecules are less tightly bound to each other, and the lack of structure allows the molecules to be packed closer together. Thus ice is *less dense* than water (whereas most solids are more dense than liquids), explaining why it floats on water.

When seawater freezes, the salts are excluded and the ice is made of relatively pure water. Because of this, and the wider spacing of the water molecules, the less-dense ice floats on seawater. This is important, because if ice sank, the ocean and lakes would freeze from the bottom up, rather than from the top down. If the ocean froze from the bottom up, it would be catastrophic for organisms living on the bottom. If the entire ocean froze

TABLE 6-2

Some Extraordinary Properties of Water

Property	Comparison to other Substances	Comments
Heat capacity	Greatest except for liquid ammonia	Prevents extreme ranges in temperature. Water can store large amounts of heat.
Latent heat of melting/fusion	Greatest except for liquid ammonia	Has a thermostatic effect by absorbing or releasing heat
Latent heat of evaporation/condensation	Greatest of all liquids	Important in heat and water transfer with the atmosphere
Dissolving power	In general greater than any other liquid	Important in erosion and many biologic processes
Surface tension	Greater than any other liquid	Important in cell phenomena and surface water effects
Density	Unique	For freshwater and relatively dilute seawater, maximum density occurs above freezing point. For seawater, maximum density is at its freezing point.

(which is not possible since the ocean freezes from the top down), it would destroy essentially all the animals and plants that live in the ocean. (See section on p. 147 for more about ice in the ocean.)

Density Due to Temperature

The density of a substance is its *mass per unit volume.* Most substances become more dense as they cool, and less dense as they are warmed. This is almost, but not quite, the situation with water. Pure water's maximum density is at 4°C, or about 39°F (Figure 6-6). As temperature rises above 4°C, density decreases, as expected due to warming. As temperature falls below 4°C, however, things are different: water's density *decreases,* due to the way it expands when frozen. At 0°C (32°F), the volume of a block of ice is 10 percent greater than the volume of the same amount of water at 4°C.

This phenomenon is important in the weathering and erosion of rock, concrete, highway surfaces, and tombstones. When water in rock cracks freezes, it expands 10 percent, forcing the rocks or concrete to fracture. The term for this is **frost wedging.**

Another important effect can occur in freshwater lakes in cold climates. If the surface water of the lake is cooled to 4°C, it sinks (because 4°C is the temperature at which freshwater is at maximum density) and is replaced by less dense, warmer bottom water. The process is called overturning or convective sinking.

If these bottom waters lack oxygen and contain hydrogen sulfide, the hydrogen sulfide, with its foul, rotten-egg odor, may be carried to the surface, where it can kill most organisms living there. This type of overturning does not occur once the *salinity* exceeds 24.7 parts per thousand (Figure 6-7). (The concept of salinity is discussed in the next chapter.)

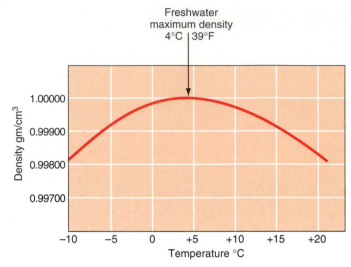

FIGURE 6-6 Density-temperature relation of freshwater. Note that the maximum density occurs at 4°C (about 39°F). This graph makes apparent why ice floats on liquid water, and why colder waters sinks in warm water.

Heat Capacity and Calories

Heat capacity is the amount of heat needed to raise the temperature of 1 gram (0.035 ounce) of a substance by 1°C (1.8°F). Heat capacity is also called *specific heat.* The heat capacity of water is great compared to that of soil, rock, or the atmosphere. Water's temperature changes less for a given amount of heat addition or removal than almost any other naturally occurring material.

Water's great heat capacity has important implications for the ocean. Ocean temperatures range from about −1.9°C (28.4°F) to 30°C (86°F), a range of only 32°C. The land surface has a wider range of about 100°C, and the atmosphere even greater. This high heat capacity of water, combined with the ability to store heat at depths

FIGURE 6-7 Relationship of salinity to temperature of maximum density and to the freezing point. The basic point is that when the salinity is less than 24.7‰, the water can freeze before it has reached its maximum density. If the water has a salinity greater than 24.7‰, the water will sink before freezing. (Adapted from Sverdrup, Johnson, and Fleming, 1942.)

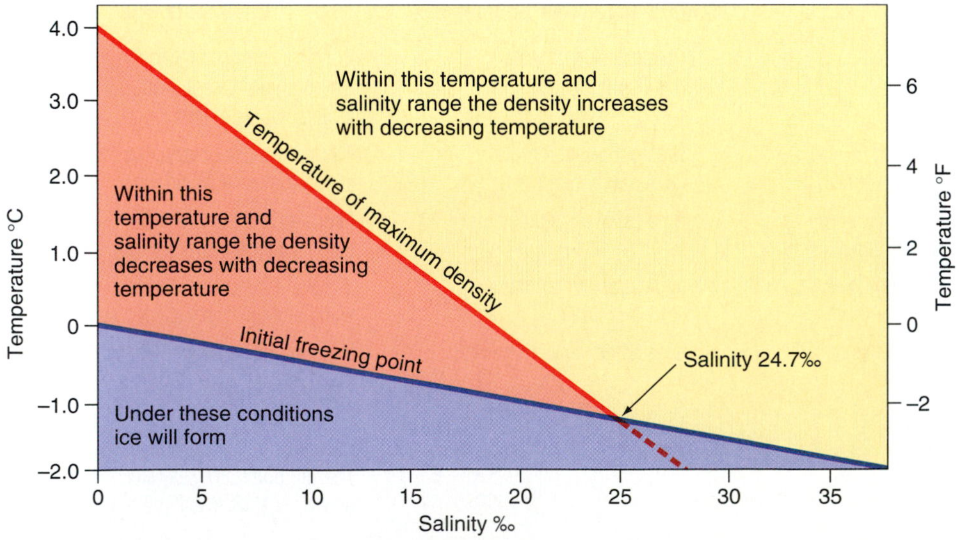

Within this temperature and salinity range the density increases with decreasing temperature

Temperature of maximum density

Within this temperature and salinity range the density decreases with decreasing temperature

Initial freezing point

Salinity 24.7‰

Under these conditions ice will form

below the water surface, moderates the temperature on Earth, preventing wide variations in temperature and making the planet more liveable.

The ability of water to store large amounts of heat is a very important influence on weather phenomena. Evaporation removes considerable heat energy from the ocean and places it into the atmosphere as the evaporated water (water vapor) rises. Here the heat is stored until the water vapor condenses. Condensation releases this stored energy—harmlessly as rain, or destructively as violent storms or hurricanes.

Over the ocean, evaporation is generally greater than precipitation. Over the land, the opposite is true; precipitation is typically greater than evaporation. The evaporation loss from the ocean and precipitation gain by the land is balanced by river inflow from the land into the ocean (Table 6-3).

To change (raise or lower) the temperature of 1 gram of liquid water (about 10 drops) by 1°C requires one **calorie** of heat (added or removed). Considerably more energy is needed if a change of state is involved—to change ice to water, or water to vapor. Melting 1 gram of ice that is already at 0°C (32°F) requires an addition of 80 calories. Evaporating 1 gram of water that is already at 100°C (212°F) requires an additional 540 calories.

Conversely, freezing water already at 0°C requires removal of 80 calories per gram of water to become ice, and condensing water vapor already at 100°C requires removal of 540 calories. These changes and their related additions/removals of heat are shown graphically in Figure 6-8.

Heat Versus Temperature

Be careful not to confuse heat and temperature: they are different! Heat is a *quantity* of energy. Temperature is *how*

TABLE 6-3
Water Input and Output for the Ocean

| | Water Output | Water Input | |
	Evaporation (cm/year)	Precipitation (cm/year)	River Inflow (cm/year)
Ocean			
Atlantic	124	89	23
Pacific	132	133	7
Indian	132	117	9
World ocean average	126	114	12

Source: Data from Budyko, 1974.

rapidly molecules are moving. To appreciate the difference, consider which has a higher *temperature*—a candle flame or a large container of hot water? The candle flame has the higher temperature, of course. Then consider which has more *heat*. Here the answer is the large container of hot water.

The term *calorie* as used here is not the same as the nutritional calorie used in food science.

Heat can be added to water or removed from water in two ways:

1. Heat that causes a temperature change is called **sensible heat** because it can be sensed or felt.

2. Heat that breaks chemical or hydrogen bonds (and eventually may be sufficient to change the phase of water) is called **latent heat,** which means "hidden" heat.

The latent heat of melting is the 80 calories of heat energy that must be absorbed to melt 1 gram of ice (at 0°C)

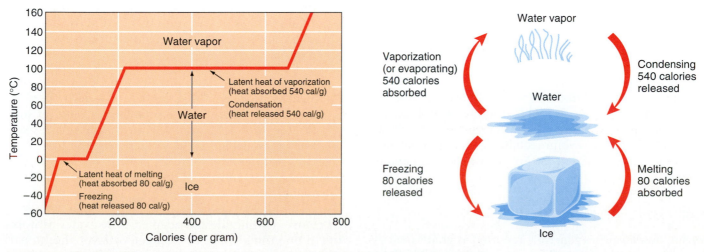

FIGURE 6-8 The three different states of water, and the energy (calories) needed to change from one state to another.

into the liquid state. The latent heat of vaporization is the 540 calories of heat energy that must be absorbed to vaporize (evaporate) 1 gram of water (at 100°C) into the gas state (see Figure 6-8).

Latent heat is released when the process is reversed. In other words, when water is frozen, the water gives up 80 calories of heat to its environment. And when water condenses, it releases 540 calories of heat to its environment. On a large scale in the atmosphere, condensing water releases enormous amounts of heat. As latent heat is released, it can do considerable work such as causing winds, storms, or currents.

The reason for the difference in the amount of heat needed for melting ice (80 calories) and for vaporization (540 calories) is that *all* the hydrogen bonds have to be broken when water evaporates, while only *some* of the bonds must be broken when ice melts.

Note that water does not have to be boiled at 100°C to evaporate. Water can change phase to a gas below 100°C by evaporation. However, this process also requires considerable heat. For example, to evaporate water having a temperature of 20°C requires the addition of 585 calories for each gram of water evaporated. The water molecules left behind (not evaporated) lose their heat energy to the molecules that evaporate. This effect can sometimes be seen as mist rising from warm seawater on a cool day. The process also explains the cooling effect of evaporation, which you can feel by blowing on your wet hand.

Pressure

In the ocean, water pressure increases with depth, due to the increasing mass of the overlying water. Pressure is expressed in the unit *atmosphere*, which is the equivalent of atmospheric pressure at sea level. The pressure of one atmosphere is 1 kilogram per square centimeter (14.7 pounds per square inch). Pressure in the ocean increases one atmosphere for every additional 10 m (32.8 ft.) of water depth. Therefore, in the deepest parts of the ocean, at about 11,000 m (about 36,000 ft.), the pressure is 1,100 atmospheres or 1,100 kg/cm² (over 8 tons/inch²).

Water, with or without salt in it, is almost incompressible—but not quite. Nevertheless, the pressure throughout the ocean does reduce total water thickness by about 53 m (175 ft.). This compressibility generally does not have to be considered except for precise determinations of density.

Density Due to Salinity

Density of seawater is generally reported as grams per cubic centimeter, or g/cm³. The maximum density of *freshwater* (which occurs at 4°C or 39°F), is defined as 1.00

g/cm³. Seawater at the same temperature has an average density of 1.028 g/cm³, due to the salts it contains (see Box 8-3, page 192). As you might expect, the salt ions—sodium, chloride, and others—add their weight to that of the water molecules. Less dense fluids float on heavier fluids (a familiar example is less-dense oil floating on water). Therefore, freshwater tends to float on seawater.

Seawater density ranges from about 1.022 to 1.030 g/cm³, averaging 1.028 g/cm³. Seawater density is sensitive to temperature, salinity (saltiness), and pressure changes (or depth changes, because pressure increases with depth). In seawater, temperature has the greatest effect on density, while pressure has the least. Density is an important property of ocean water because it can determine both the motion of seawater (denser water sinks) and the relative position or layering of water having different temperatures or salinities. Some other properties of water are presented in Box 6-2.

If Water Were Different . . .

It is interesting to consider what the ocean would be like if water properties were different:

- If seawater density were less, ships would have to be much larger to float and to be as efficient as they are now. It is also easier for humans to float in salt water than in fresh water. In some very salty water bodies, like the Great Salt Lake or the Dead Sea, it is so hard to sink that diving into such waters can be very dangerous and is prohibited.

- The high density of seawater is important for many large marine animals. Water supports the great weight of animals like whales, whereas on land, animals have to develop special muscles and a strong skeleton to support their own weight.

- If the surface tension of water were less, rain would not fall as drops but as a disorganized mass. If its surface tension were greater, it would stick to everything like syrup.

- Perhaps most important, if water were less transparent to light, a lesser depth of the ocean could be used by plants for photosynthesis, a process that requires sunlight (see Box 6-3). In such an instance less space would be available for the formation of the basic food of all creatures in the ocean and therefore the total quantity and variety of life in the ocean probably would be less.

Of course, marine life has adapted to seawater as it exists. If one thing could be changed to favor the human species, it would be that many areas of land could use more freshwater.

Box 6-2

Some Other Properties of Water

Cohesion and Surface Tension

These two properties are closely related. As previously mentioned, water molecules can bond with other water molecules in a process known as hydrogen bonding. This structuring holds the water somewhat together because the molecules *cohere* to each other; the bonding is known as *cohesion*.

Surface tension, which results from the cohesion of water molecules, is a measure of how hard it is to separate the molecules of water at its surface. Because of water's cohesive ability, its surface behaves like a weak membrane which can support objects denser than water.

A common illustration of water's surface tension is the floating of a needle or razor blade on the water's surface. In nature, you can observe insects floating or even walking on the water (the common water strider is a familiar example to those who spend time near lakes and streams). Surface tension is increased by the addition of salts, or by decreasing the temperature of the water.

Surface tension applies to all fluids; its degree varies with the bonding, or lack of bonding, among molecules of the fluid. Water has a greater surface tension than any other natural liquid, except the liquid metal mercury. Surface tension can be important in the early stages of wave formation, allowing the wind to get a better grip on the water surface and thus increasing its effect.

Viscosity

Viscosity is the internal resistance to flow or motion.

Its effects are familiar—(syrup, for example, has a higher viscosity than water) and are affected by temperature (think of pouring cold syrup). In the case of water, viscosity also is affected by salt content (salinity). The temperature aspect can be very important to floating marine organisms, because the greater viscosity of colder water makes it easier for them to float. In floating species that have varieties living in both cold-water and warm-water environments, the warm-water varieties frequently have more appendages or feathery limbs than their cold water cousins to aid in floating.

Viscosity can, however, be a handicap for larger swimming organisms since they must overcome the resistance of the water in front of them as they swim forward. The more streamlined an organism, the better it can overcome this slowing effect. In nature, many organisms are very streamlined, especially fish like tuna and porpoises. This streamlining has been applied to ship design to help vessels move faster through the water. Some modern submarines are porpoise-shaped.

Vapor Pressure

Vapor pressure is a measure of how easily water molecules move from the liquid phase into the gaseous phase. Vapor pressure is lowered with increasing salinity because the salts bond to water molecules, making them less available for evaporation. Thus, freshwater evaporates more quickly than seawater.

Box 6-3

Light in the Ocean

Light is a part of the electromagnetic radiation that reaches Earth from the sun. The light that humans can see occupies only a small portion of the total electromagnetic spectrum (Figure 1). The wavelengths longer than visible light include infrared radiation and TV and radio waves. Wavelengths shorter than visible light include X-rays and gamma rays.

Visible light does penetrate the ocean, but about 60 percent of the incoming light energy is absorbed in the uppermost meter of seawater. In very clear water, about 99 percent of the light has been absorbed by a depth of 150 m (492 ft.).

When light travels through water, its intensity decreases as the distance from the ocean surface increases. This loss is not linear, like arithmetic (increments of 1, 2, 3, 4, 5, and so on); instead, it is exponential (increments of 2, 4, 8, 16, 32, and so on). This loss of light is called *attenuation* and has two main

continued

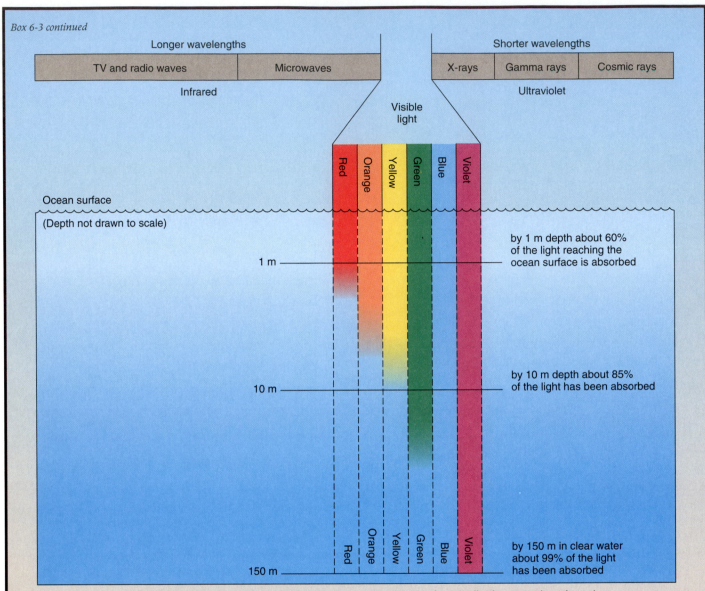

Box 6-3 continued

FIGURE 1   Part of the electromagnetic spectrum. The wavelengths range from smaller than atoms (cosmic rays) to radio waves exceeding 100 km (60 miles) in length. Visible light occupies only a small portion of the spectrum. The key point of this figure is that little light penetrates below the surface of the ocean, and that reds and yellows are absorbed before the blues and violets.

causes. The first is scattering by suspended particles in the water; the more particles present, the greater the scattering. The second cause of attenuation is absorption, in which the light energy is converted into heat or chemical energy. Absorption can be caused by phytoplankton (which use light for photosynthesis), particulate matter in the water, dissolved material in the water, and seawater itself.

Coastal waters tend to have more particles in suspension due to material carried in by rivers and material stirred from the bottom by waves and currents. The coastal region also frequently has high organic growth and a high content of dissolved material. Therefore, light generally penetrates to a lesser depth in coastal regions than in offshore waters, which are relatively clear.

Photosynthesis is the main source of food in the ocean, so the depth of light penetration is important since it controls the depth to which photosynthetic activity can occur. The layer in which photosynthesis takes place is called the **photic zone** and may extend to depths of 200 m (656 ft.) in clear offshore waters, but may be as shallow as 10 m (32.8 ft.) in coastal waters. Plants living on the bottom must be in the photic

Box 6-3 continued

zone to survive. Because only a small portion of the seafloor is shallow enough to be in this zone, the main producers of organic material by photosynthesis are floating plants—the phytoplankton.

Longer wavelengths of light (the reds; see Figure 1) tend to penetrate to less depth than the shorter wavelengths (blues and greens). Our view of the ocean's color is due to what wavelengths are reflected back to our eyes. We generally see the ocean as having a blue-green color since these are the wavelengths that penetrate deeper and are more available for reflection by particles in the water.

In shallow coastal water, seawater may appear more green, yellow, or even red, for these waters generally contain relatively high quantities of suspended sediment and floating organisms. This restricts the penetration of light and reflects more of the longer wavelengths, compared to the clear open ocean where light can penetrate deeper because of the absence of suspended material.

When viewed in deep water, most objects appear black or blue. A dark blue or violet color can sometimes be seen because these wavelengths penetrate deepest and therefore can be reflected by the object being viewed. Black is the result when all other colors are absorbed or filtered out and thus can not be reflected.

Light in the ocean can be measured by light meters or photometers. Either total light or specific wavelengths can be measured. A simpler and cheaper way is with a Secchi disk, a white disk about 30 cm (1 ft.) in diameter that is lowered by line until it disappears from view. This depth is a measure of light attenuation. Light measurements can provide data on biological activity, suspended load, or just general water quality.

Early Studies on the Chemistry of Seawater

The sea and its saltiness have been a puzzle since prehistoric times. Early people learned to obtain salt from seawater by solar evaporation of the water, a technique still used in many parts of the world.

The first scientific study of seawater chemistry is usually attributed to Robert Boyle, an English chemist who published a description of his work in 1670. Near the end of the eighteenth century, the French chemist Antoine Lavoisier discovered that water is a mixture of oxygen and hydrogen. He also devised techniques for analyzing some of the materials dissolved in seawater. Lavoisier and a Swedish chemist, Olaf Bergman, also made the first chemical analysis of seawater. Both men evaporated seawater and then tried to extract different compounds from the residue. The results varied with the techniques and none were very accurate.

Seawater contains over 80 elements and hundreds of different compounds. But by 1819, only chloride, sulfate, calcium, potassium, magnesium, and sodium had been detected in seawater. During the following 50 years, boron, iodine, strontium, silver, lithium, arsenic, and fluorine were discovered in seawater.

Although these early chemists could not accurately estimate the quantity of individual elements in seawater, they were developing a basic understanding of chemical weathering and how eroded material from land was eventually transported into the ocean. For example, chemist Alexander Marcet noted in 1822: "For the ocean having communication with every part of Earth through the rivers, all of which ultimately pour their waters into it . . . I see no reason why the ocean should not be a general receptacle of all bodies which can be held in solution." This idea is not far from the truth, because almost all of the natural chemical elements are found in the ocean.

A major advance in the understanding of the seawater chemistry was made in 1865 by Georg Forchhammer. He noted that, although marked differences may exist in the total salt content among samples of seawater taken from different areas of the ocean, the *ratio* of the major dissolved components in the seawater is essentially constant. This concept is known as **Forchhammer's Principle** or the *Principle of Constant Proportions*.

Forchhammer also observed that silica and calcium were abundant in river water but nearly depleted in seawater; he correctly concluded that this depletion in seawater was due to the action of marine organisms that were absorbing these elements into their shells. Thus he recognized the fact that biological activity plays an extremely important part in the chemistry of the oceans.

The constancy of chemical composition throughout the ocean is perhaps not so surprising when we consider that the ocean is probably several billion years old. If a complete cycle of mixing of the ocean takes 1,000 or 2,000 years, the ocean will have been mixed at least 1 million times during its entire history. It should be fairly uniform.

The Challenger expedition (1872–76) was a major advance in the chemical study of the ocean. William Dittmar, an excellent chemist, analyzed many of the

Challenger's water samples and identified dissolved gases. Dittmar's work confirmed many of Forchhammer's ideas. Dittmar also noted a general decrease in oxygen content with water depth (deepest sample about 1,494 m or 4,900 ft.) and an increase in carbon dioxide in surface water compared with deeper water. However, the importance of these differences, which are due to the processes of photosynthesis and respiration, was not realized at that time.

By the late nineteenth and early twentieth centuries, chemists had established many of the relationships among salinity, density, and chloride content (discussed in more detail in Chapter 7). Determination of salinity was done by chemical reactions with the materials dissolved in seawater. More recently, this has been replaced by new methods, such as procedures for measuring electrical conductivity or devices that can measure salinity and other parameters directly as the instrument is moved through the water (Figure 6-9). Besides being faster, these new techniques are usually more accurate.

By the early twentieth century, chemical oceanographers began to associate the variations of oxygen in the upper parts of the ocean with the biological activity of the microscopic plants (called *phytoplankton*) that float in the water. Other elements also were believed to be involved in the biological processes, especially nutrients such as nitrate, phosphate, silica, iron, and manganese. Many of these nutrients show distinct vertical changes (change in depth) and seasonal changes, which suggests biological activity.

The study of nutrients continues, but emphasis has shifted to studies of vitamins, trace elements (those present in very small quantities), organic compounds including pollutants, and their influence on biological growth and development. These processes are discussed in Chapter 7.

As analytical techniques have improved, more information has been obtained about elements that occur in trace concentrations in the sea. Measurements that are sometimes accurate to one part per billion or even less have shown that more than 80 different elements are present in seawater in measurable quantities. Many elements were actually found in marine organisms before they were discovered in seawater. It is probable that, as measurement techniques improve still further, traces of every naturally occurring element will be found in the sea (see Box 6-4).

FIGURE 6-9   (a) A multiple-use instrument being prepared for lowering from a surface vessel. The instrument package contains a CTD—an electronic device for measuring **c**onductivity (an equivalent of salinity), **t**emperature, and pressure (essentially equivalent to **d**epth)—and at the top a rosette of water sample bottles.

(b) Data from the CTD is transmitted to the ship and monitored on computer terminals. In this process, the water samplers can be electronically closed at appropriate depths by a shipboard observer. (Photograph (a) courtesy of Dr. Peter Wiebe and (b) courtesy of Rod Cantanach, both of Woods Hole Oceanographic Institution.)

Box 6-4

How Do Salts Get Into and Out of the Ocean?

As discussed in Chapter 2, weathering of volcanic rocks was the initial source of many of the cations (positively charged ions) now found in the ocean, including such elements as sodium, magnesium, calcium, and zinc. Erosion and weathering break down and decompose exposed rock and sediment. Eventually various elements and compounds are carried to the ocean by rivers or rainfall.

The anions (negatively charged ions) such as chloride and sulfide came, and still come, from volcanic activity, which releases various gases into the atmosphere, including carbon dioxide, hydrogen sulfide, and chlorine. These eventually become dissolved in water (rainfall or river water) and also reach the ocean.

Initially surprising is that the most abundant elements in present-day rivers are relatively less abundant in the ocean. Likewise, those elements more common in the ocean are relatively rare in river water. The reason for this is simple: the most easily dissolved salts long ago were transported from land into the ocean. The less soluble salts are still being actively eroded and transported, so their concentration is greater in rivers.

The total dissolved material in the ocean is about 5×10^{22} grams (that is 5 followed by 22 zeros). It is estimated that each year an additional 4×10^{15} grams of material (or about 4 billion tons) is added to the ocean by rivers running off the land and by rainfall.

How Salts Are Removed from Seawater

How is this material removed from the sea? Some salts, such as calcium, are involved in biological or geological processes that incorporate them into the sediment on the seafloor. Then, over geological time, tectonic processes will reunite this material with the land, where it can again be eroded and carried into the ocean, in a cycle whose end we cannot foresee.

Another way that salts are removed from the ocean is when seawater evaporates, forming *evaporite deposits,* which can occur in shallow lagoons or estuaries. An excellent example is the evaporite deposits that formed about 5 million years ago in the then-shallow Red Sea. Indeed, the production of common table salt is done by evaporating seawater, or by mining old shallow-sea deposits that resulted from the evaporation of seawater.

Another mechanism for removing elements from the ocean is simply by harvesting the fish or other organisms that inhabit it. In addition, the creatures themselves remove some elements from seawater via their feeding, digestive, and excretion processes. These elements then may be transferred to the seafloor, generally as fecal pellets (see Figures 5-18c and 5-18d, page 124). Animal shells composed of calcium carbonate or silica frequently settle to the bottom and become part of the sediment after the creatures die.

Seafloor spreading and plate tectonics are important in recycling many elements in the ocean. This occurs in several ways. One is by the volcanic processes active along the mid-ocean ridges (see also "Ocean Vents and Recycled Seawater" in Chapter 7, page 169). In addition to elements being discharged by these volcanoes into the seawater, the seawater involved in the venting process removes manganese and some other metals from the seawater to form mineral deposits near the vents (see also "Resources of the Deep Sea" in Chapter 15, page 371).

The other part of this geological cycle occurs when the bottom sediments of the ocean are eventually uplifted as new parts of a continent. Then they can be eroded and again carried to the sea. This may require several hundred million years to complete one cycle.

In general, some elements in the ocean are relatively unreactive and remain in seawater for fairly long periods of time. Others are very reactive and are quickly removed. If the cycles just discussed did not occur, the salinity of the oceans would increase. Indeed, with an annual increase of 4×10^{15} grams of material, it would take little more than 10 million years for the ocean to attain its present salinity. This would mean that the ocean is only 10 million years old. This conflicts with good geological evidence from ancient marine sediments that the oceans have been around for closer to 3 billion years.

How Saltwater Is Different from Freshwater

Seawater, with its numerous elements and compounds, is a very complex solution. Seawater averages about 3.5 percent salt and 96.5 percent water (percent expresses parts per 100, symbolized %). However, because salinity variations as small as 0.1 percent are significant, oceanographers represent salinity in parts per thousand, symbolized ‰. Thus, typical seawater is 35‰ salt.

"Salt" is a term applied to all the elements in seawater, not just to common table salt (sodium chloride, NaCl). Some properties of seawater are similar to those of freshwater, such as its capacity to absorb or give off heat. Other properties, especially those related to biological activity, are strongly influenced by the chemical composition of the water.

The addition of salts to water can alter water's normal properties in some instances. For example, when ions such as sodium and chloride are added to water, they adhere to the water molecules (see Figure 6-5b). The attraction between the ion and the water molecule must then be overcome for saltwater to change state, either to freeze or to boil. This increases the boiling point of seawater compared to freshwater (more heat energy is required to break these stronger bonds). It also decreases the freezing point of seawater compared to freshwater, for essentially the same reason.

Seawater freezes at around −1.9°C (28.6°F). This is an important aspect in the formation of sea ice, discussed later. When seawater boils, freezes, or evaporates, most of the salts remain behind in the liquid phase, increasing its salinity.

Some important changes in the properties of water that occur with the addition of salt are as follows.

1. *Heat capacity* decreases with increasing salinity. (Recall that heat capacity is the amount of heat necessary to raise the temperature of 1 gram of water 1°C.) However, heat capacity also increases with increasing temperature in waters of normal salinity. In other words, as water temperature increases, it becomes harder to remove the last few water molecules from a hydrated ion. Thus the boiling point of seawater increases with increasing salinity.

2. *Density* increases almost linearly with increasing salinity. Pure water has a maximum density at 4°C (about 39°F). The addition of salt lowers the temperature of maximum density. At salinities greater than 20‰, the maximum density occurs at the new freezing point (see Figure 6-7).

3. *The freezing point* is lowered with the addition of salt (see Figure 6-7). Also, seawater becomes denser as temperature drops. The combination of these two characteristics means that the densest ocean water is the coldest and most saline. Conversely, the least-dense seawater has the highest temperature and least salinity. Seawater of normal salinity can exist as a liquid when its temperature is as low as about −1.9°C (28.6°F).

4. *Osmotic pressure* causes a solution to flow or migrate through a semipermeable membrane. This characteristic is very important to organisms, because their cell membranes are semipermeable, through which fluids can move (Figure 6-10). The direction of movement depends on whether the osmotic pressure in the organism is greater or less than that of the ocean. The flow direction is toward the more concentrated medium. It is the water that flows or migrates, not the elements dissolved in solution.

In water having a salinity greater than 24.7‰, there is an important consequence of the relations among salinity, temperature of maximum density, and freezing point: the temperature of maximum density is lower than the freezing point. Thus, as ocean water is cooled, it grows denser (see Figure 6-7). Because cooling works from the surface downward, the surface water becomes denser than the underlying water and sinks. The underlying warmer, less-dense water rises to replace the

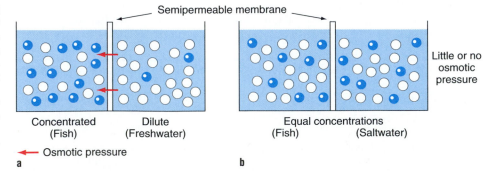

FIGURE 6-10 (a) Osmotic effect on an animal living in a fluid having a higher concentration of elements than its internal fluids. This difference creates an osmotic pressure against which the animal must work to maintain its internal composition. The colored circles represent the different elements.
(b) The concentration of elements in the body fluid and the external fluid is similar, and no significant osmotic pressure results.
Situation (a) is similar to that of a fish in freshwater, situation (b) of a fish in saltwater.

Semipermeable membrane

Concentrated (Fish)   Dilute (Freshwater)

Equal concentrations (Fish)   (Saltwater)

Little or no osmotic pressure

← Osmotic pressure

a   b

cooled water. It in turn becomes cooled and sinks. In this manner, a deep circulation is initiated and maintained, and freezing cannot occur until the entire body of water is cooled to the freezing point. Cooling to this extreme generally does not occur in a large body of water.

However, if the salinity is less than 24.7‰, the temperature of maximum density is reached before the freezing point. In this case, as the surface water is cooled, it attains maximum density and then decreases in density. This means that the water remains near the surface and becomes cooled further. Eventually the freezing point is attained and an ice layer forms on the surface. Thus, the relations of salinity, temperature of maximum density, and freezing point prevent the entire ocean from freezing over.

Ice in the Ocean

Polar ice caps, glaciers, icebergs, and sea ice are major features of our planet (Figure 6-11). Considerable water is tied up within these forms of ice. If it all melted, sea level would rise worldwide by approximately 60 m (about 200 ft.). Ice caps and glaciers contain 29,750,000 km³ (7,000,000 miles³) of water, or about 2.15 percent of the total world's water supply (see Figure 6-1).

Ice, Glaciers, and Climate Change

The distribution of ice on Earth's surface is also a vivid expression of climatic differences on our planet. At times, continental glaciers have covered much larger portions of Earth than they do today. For example, these glaciers reached as far south as New York City only 18,000 years ago. At that time glaciers also covered portions of Australia, New Zealand, South America, Europe, and southern Africa. These "ice ages" are excellent examples of past changes in Earth's climate.

The past advance of the glaciers of course meant lowered temperatures. This also meant lowered sea level, because water in glaciers comes from the ocean. During the glaciation that peaked about 18,000 years ago, worldwide sea level was lowered about 130 m (425 ft.) (see Figure 2-5, page 31). If this happened today, it would leave the world's port cities high and dry and create many millions of square kilometers of new ocean-front real estate.

Note that glaciers and ice sheets not only *result* from climatic change but also help *cause* the change. The reason is that ice, especially snow-covered ice, can reflect almost all incoming solar radiation. Thus very little solar radiation will be absorbed at the surface and converted into heat. The percentage of the amount of radiation *reflected* to the amount of radiation *received* is called

albedo. In the case of snow-covered ice, albedo can approach 98 percent. As the temperature drops, more areas become covered by snow and ice, reflecting even more solar energy back into space. More ice-covered areas reflect more sunlight than the uncovered ground, thus further decreasing the temperature.

So, why doesn't the albedo effect cause the ocean to freeze over completely? Fortunately, some opposing mechanisms counteract the albedo effect. One is seasonal, in that the albedo effect is more important in the summer because more solar radiation is available. In winter, incoming solar radiation from the low-angle sun in polar regions is so slight that albedo effect is almost negligible. Another is that the albedo of dark-colored objects such as rocks is usually very low. Thus the dark-colored body can often attain a higher temperature than the atmosphere and can warm or melt adjacent snow or ice.

As mentioned throughout this text, scientists and others are concerned about a coming warming of global climate caused by human activity. If it occurs, melting of polar ice would surely begin. Consequently, sea level would rise, drowning the world's port cities and moving shorelines inland. Indeed, some feel that one of the first ways to detect any global temperature change is to monitor polar ice for changes. Satellites are especially suited for this task (see Figure 6-11).

Sea Ice

Seawater of average salinity (about 35‰) starts to freeze when it attains a temperature of −1.9°C (28.6°F). During this process, dissolved salts and other compounds in seawater are excluded from the developing crystalline ice structure. This increases the salinity of the remaining unfrozen water, reduces its freezing point, and makes it denser, causing it to sink. This water then is replaced by less-dense seawater (generally of lower salinity), which permits the freezing process to continue. Occasionally, some very salty water (called *brine*) becomes trapped in the ice and generally remains in the liquid state.

Sea ice forms in stages. First, small crystals form. As their numbers increase, they form an icy slush. The slush then can form a thin sheet, unless the wind breaks it into *pancakes*. With further freezing, the pancakes combine into **floes**. Floes can be moved by currents and winds, sometimes colliding with other floes to form large ridges (Figure 6-12). As they continue to coalesce, they form **pack ice**. Pack ice can be moved by winds and currents. When floes stay attached to land, they are called *fast ice*.

Pack ice and fast ice are generally less than 5 m (16 ft.) thick. Both can occasionally break or be penetrated by ice-breaking ships (Figure 6-13). Polar ice, which covers most of the open Arctic Ocean, is much thicker— 50 m (165 ft.) or more.

FIGURE 6-11 Satellite images showing the extent of ice in the Arctic region illustrate the difference in Arctic sea-ice distribution due to seasonal change. The view is looking down on the North Pole (the black circle). The large image was made by *Nimbus 7* in April 1979 and the small one six months later, in September. Sea-ice cover is shown in red, yellow, and purple, with maximum concentration in darkest purple. Light blue areas are open ocean; scattered dark blue patches over the ocean are heavy clouds or rain. The black spot over the North Pole is not traversed by *Nimbus 7* so no data are obtained from this area.

The wider extent of ice in the April figure reflects its growth during the dark winter months. The ice fills most of the Arctic Ocean, Hudson Bay, and the Sea of Okhotsk. It also intrudes into the Bering Sea and along the coast of Greenland. The ice is restructured to a narrow tongue off Labrador (1) due to contact with the warm Gulf Stream water from the South. East of Greenland a huge band of ice has separated from the main ice pack (2). The lower September image was made following the warmer summer months and nearly continual daylight, during which most of the ice pack melted and receded to the confines of the Arctic Ocean. (Images courtesy of National Aeronautics and Space Administration.)

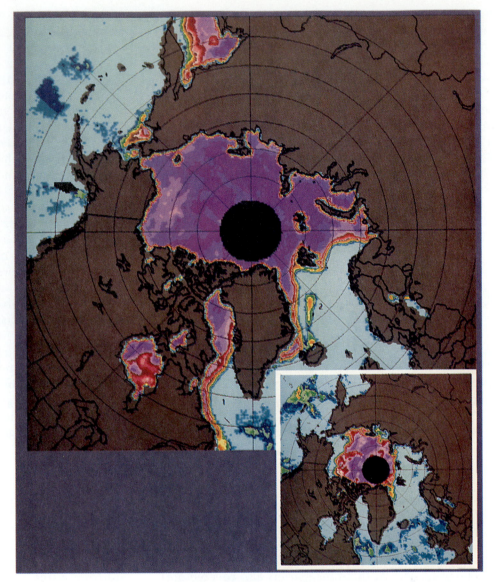

FIGURE 6-12 Two pictures of sea ice ridges in the Arctic Ocean. (Photographs courtesy of Keith von der Heydt, Woods Hole Oceanographic Institution.)

b

c

a

FIGURE 6-13   Three pictures of sea ice.
(a) Edge of the ice sheet. Note polar bear tracks.
(b) The Norwegian icebreaker *Polar Bjørn* during the Marginal Sea Experiment.
(c) Crack in the ice near a scientific camp during the Fram II expedition.
(Photographs courtesy of Keith von der Heydt, Woods Hole Oceanographic Institution.)

Sea ice forms year-round along Antarctica, in the Arctic Ocean, and in other parts of the North Atlantic. As mentioned, an ice layer reflects more incoming solar energy than areas of land or water, because of ice's greater albedo. Thus, once ice becomes established, it can be self-perpetuating. Seasonal and yearly fluctuations in ice distribution can occur at the poles because of weather.

Glaciers and Icebergs

Sea ice is frozen seawater, but glaciers are the product of compacted snow and are formed on land. Icebergs form from glaciers in several ways. In the Antarctic, they typically form by "calving." Parts of ice shelves around the continent break off into the sea, like a cow "dropping" or birthing a calf. In the Arctic, icebergs are more likely to be parts of glaciers that broke off when the glaciers reached the sea (Figure 6-14). Note that much Antarctic ice lies on land, whereas much Arctic ice lies over ocean. Part of the ice sheet over the Arctic Ocean occasionally breaks into large ice islands, some of which have been used for scientific stations.

Only about one-seventh of an iceberg floats above the surface. The famous ship *Titanic* was ruptured on her initial voyage in 1912 by the underwater portion of an iceberg. Following this disaster, which claimed 1,517 lives, the United States initiated an international iceberg patrol to monitor locations of large icebergs and notify ships of their position.

Some icebergs from the Antarctic ice shelf are immense. Recently, one the size of Rhode Island was tracked for several years as it floated around Antarctica. A proposal to use these large icebergs as a source of freshwater is discussed in Chapter 18.

In previous years, the Arctic had military importance because submarines can navigate and hide beneath the ice in the Arctic Ocean. This military aspect, combined with concern about global climatic change, had led to increased study of polar regions (Figure 6-15). The breakup of the Soviet Union has diminished the military aspect, but the concern about global climatic change remains.

FIGURE 6-14 The U.S. Coast Guard icebreaker *Eastward* near an exceptionally attractive iceberg in Arctic waters. (U.S. Coast Guard.)

FIGURE 6-15 (a) Scientific hut used during a research experiment near Prudhoe Bay, Alaska.

(b) An acoustic sound source being lowered under the ice and used for sound transmission experiments during the Fram II expedition. (Photographs courtesy of Keith von der Heydt, Woods Hole Oceanographic Institution.)

SUMMARY

Water is both a very common compound and extraordinary in many of its properties. Without water, life on Earth as we know it would be impossible. About 97.2 percent of Earth's water is in the oceans and 2.15 percent is in ice caps and glaciers. The remainder is in groundwater (0.625 percent), freshwater lakes (0.009 percent), saline lakes and inland seas (0.008 percent), the atmosphere (0.001 percent) and rivers and streams (0.0001 percent).

Although water is a simple compound (H_2O), its unusual properties are due largely to its dipolar structure. The water molecule has a slightly positive charge on its hydrogen side and a slightly negative charge on its oxygen side, which allows water molecules to form hydrogen bonds with each other. This, plus water's dipolar structure, give water many of its remarkable properties.

Water exists in three states: liquid, solid, or gas. It has the greatest dissolving power and greatest surface tension of any liquid, and is among the greatest in heat capacity and latent heat of melting and vaporization. The heat capacity of the ocean is considerably greater than that of the land or atmosphere, a critical factor in moderating major temperature changes on our planet.

The addition of salts to freshwater increases the complexity of water properties. It reduces heat capacity, freezing point, and vapor pressure of water, but increases its density.

Sea ice and its distribution is an important subject of research, especially because the polar regions are critical in determining the effects of any global change in climate.

QUESTIONS

1. What is hydrogen bonding? Why is it so important in determining some of the remarkable properties of water?

2. Why is water extraordinary? Describe its three phases and why they occur.

3. What are the main differences between seawater and freshwater?

4. Why doesn't the ocean freeze from the bottom up?

5. Describe some of the changes that occur in water's properties when salts are added to fresh water?

6. Why do coast waters generally appear to have a more yellow or greenish color than offshore water?

KEY TERMS

albedo	density of water	hydrogen bond	pack ice
anion	deuterium	hydrogen sulfide	phase
atom	dipole	hydrologic cycle	photic zone
attenuation	William Dittmar	ice	pressure
atmospheric pressure	element	iceberg	radioactivity
boiling point	evaporite	ionic bond	salinity
brine	floes	isotope	sea ice
Robert Boyle	frost wedging	latent heat	sensible heat
calorie	Forchhammer's Principle	light	solvent
cation	freezing point	maximum density	surface tension
cohesion	frost wedging	nutrient	transpiration
compound	glacier	osmotic pressure	tritium
compressibility	heat capacity	overturning	vapor pressure
covalent bond	hydrogen	oxygen	viscosity

FURTHER READING

Broecker, W. S. 1983. "The Ocean," *Scientific American* 249 no. 3 pp. 146–60. Some aspects of the chemistry of the ocean.

Broecker, W. S. 1974. *Chemical Oceanography*. New York: Harcourt Brace Jovanovich. A somewhat advanced but solid book on chemical oceanography.

Leeden, F.V.D., F. L. Troise, and O. K. Todd. 1990. *The Water Encyclopedia*. Chelsea, Mich.: Lewis Pub. Co. All the facts about water you might want to know.

Libes, S. M. 1992. *An Introduction to Marine Biogeochemistry*. New York: John Wiley and Sons. A slightly advanced textbook covering marine chemistry and how it relates to some aspects of marine geology, biological oceanography, and pollution.

MacIntyre, F. 1970. "Why the Sea is Salt. " *Scientific American* 223, n. 5, pp. 104–15.

"Marine Chemistry." *Oceanus* 35, no. 1 (Spring 1992). An entire issue on recent research in marine chemistry, including articles on marine organic material, particles in the ocean and hydrothermal activity.

Open University. 1989. *Ocean Chemistry and Deep-Sea Sediments*. Oxford: Pergamon Press. Part of a 5-volume series on the ocean prepared as a University text; sometimes very detailed, but informative.

———. 1989. *Seawater: Its Composition Properties and Behavior*. Oxford: Pergamon Press. Part of the same 5-volume series.

Untersteiner, N. 1986. "Glaciology: A Primer on Ice." *Oceanus* 29, no. 1, pp. 18–23. A discussion of the dynamics of ice and how it forms.

CHAPTER 7

Chemical Processes in the Ocean

Chemical processes in the ocean include both the chemistry of seawater and the chemistry of sediments and rocks of the seafloor. The chemistry of life processes—plants and animals—is also intimately involved.

Chemical oceanographers study the elements and compounds dissolved in seawater and the factors that control their presence. They also survey the distribution of elements and compounds throughout the ocean, including those within organisms and sediments. Distribution is strongly influenced by biological processes, volcanic activity, currents, and human activities. An exciting recent discovery is that hot vents along the ocean's spreading centers contribute significantly to the ocean's chemistry. The full impact of these vents is not yet known and is an important area of present research.

The effects of ocean pollution also are deeply entwined with the chemical, geologic, physical, and biological processes acting within the ocean and among the ocean, atmosphere, and ocean floor.

Water sampling in rough seas from RV *Atlantis II*.
(Photo courtesy of Woods Hole Oceanographic Institution.)

Recent Research in the Chemical Sea

A good way to begin our look at ocean chemistry is to summarize the remarkable variety of research now in progress and its implications for the quality of all lives on the planet.

Increasing **pollution** of the ocean is a main research area for marine chemists. Pollutants include lead from leaded gasoline burned in internal combustion engines, heavy metals such as mercury from various sources, sewage sludge, pesticides such as DDT, radioactive materials, plastics, and petroleum from oil spills. Some of these pollutants, such as leaded gasoline and DDT, have been banned in the United States. However, they still are used in many other countries.

Acid rain has caused considerable ecological damage to lakes, especially downwind of heavily industrialized areas. Fortunately, acid rain has little direct effect on the ocean because seawater is a **buffered solution,** meaning it is able to "adjust" to certain added chemicals. In the case of seawater, buffering results from an excess of carbon dioxide (CO_2) that absorbs or neutralizes large amounts of acidic material without producing a damaging environmental effect. (See Box 7-1 for an explanation of buffering and acidity.)

Scientists concerned about quantities and distribution of pollutants in the ocean have labored to determine where they come from and what harm, if any, they cause to the environment. However, funding cutbacks have caused some marine environmental programs to be reduced or even eliminated in recent years. This has caused marine environmental studies to focus on coastal regions rather than the open ocean. This is reasonable, for environmental problems are relatively minor in the open ocean and the coastal region is where many pollu-

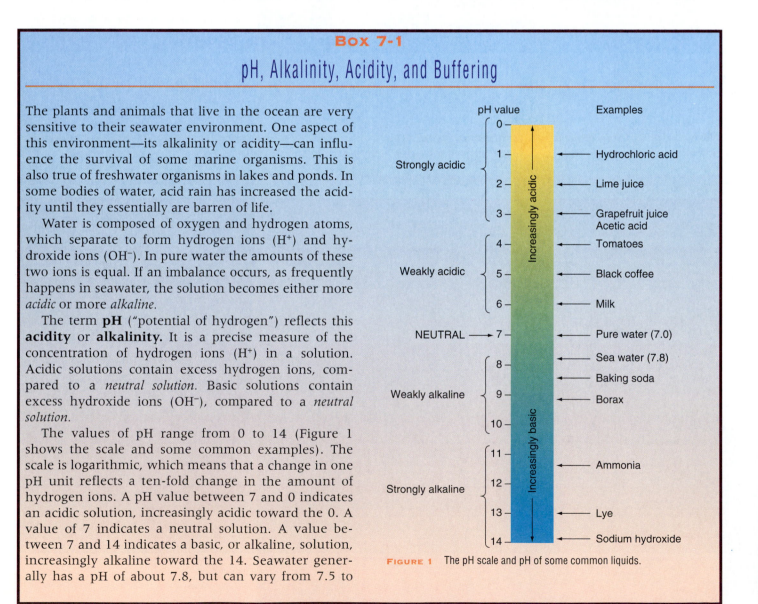

Box 7-1

pH, Alkalinity, Acidity, and Buffering

The plants and animals that live in the ocean are very sensitive to their seawater environment. One aspect of this environment—its alkalinity or acidity—can influence the survival of some marine organisms. This is also true of freshwater organisms in lakes and ponds. In some bodies of water, acid rain has increased the acidity until they essentially are barren of life.

Water is composed of oxygen and hydrogen atoms, which separate to form hydrogen ions (H^+) and hydroxide ions (OH^-). In pure water the amounts of these two ions is equal. If an imbalance occurs, as frequently happens in seawater, the solution becomes either more *acidic* or more *alkaline*.

The term **pH** ("potential of hydrogen") reflects this **acidity** or **alkalinity.** It is a precise measure of the concentration of hydrogen ions (H^+) in a solution. Acidic solutions contain excess hydrogen ions, compared to a *neutral solution.* Basic solutions contain excess hydroxide ions (OH^-), compared to a *neutral solution.*

The values of pH range from 0 to 14 (Figure 1 shows the scale and some common examples). The scale is logarithmic, which means that a change in one pH unit reflects a ten-fold change in the amount of hydrogen ions. A pH value between 7 and 0 indicates an acidic solution, increasingly acidic toward the 0. A value of 7 indicates a neutral solution. A value between 7 and 14 indicates a basic, or alkaline, solution, increasingly alkaline toward the 14. Seawater generally has a pH of about 7.8, but can vary from 7.5 to

pH value — Examples

- 0
- 1 — Hydrochloric acid (Strongly acidic)
- 2 — Lime juice
- 3 — Grapefruit juice / Acetic acid
- 4 — Tomatoes (Weakly acidic)
- 5 — Black coffee
- 6 — Milk
- NEUTRAL → 7 — Pure water (7.0)
- 8 — Sea water (7.8) / Baking soda (Weakly alkaline)
- 9 — Borax
- 10
- 11 — Ammonia (Strongly alkaline)
- 12
- 13 — Lye
- 14 — Sodium hydroxide

Increasingly acidic
Increasingly basic

FIGURE 1 The pH scale and pH of some common liquids.

tion accidents occur, as well as where discharge from rivers, dumped sewage, and wastes first enter the ocean. (Some of these aspects are discussed further in Chapters 14 and 16.)

The effect of plant and animal life processes on the composition of seawater is another important area of research, involving careful assaying of the ocean for elements that have biological connections, especially nutrients and heavy metals. A better understanding could help increase biological productivity (production of organic matter by plants) in the ocean.

One such large-scale assay of the ocean was the GEOSECS program (Geochemical Ocean Sections Study) in the 1970s, a successful multinational project involving several countries. GEOSECS studied chemical properties of the ocean to better understand large-scale oceanographic processes. At many stations in the Atlantic, Pacific, and Indian Oceans, large volumes of water were collected and analyzed, and parts of the samples were stored for further study (Figure 7-1). The data from these stations (Figure 7-2) give a good picture of the general chemical composition of the world's oceans.

The GEOSECS program and others have shown the value of using stable and unstable isotopes to determine the rates and importance of chemical and biological reactions in the ocean. These isotopes are especially useful in studying marine sediments and how seawater mixes. A more recent project, Transient Tracers in the Ocean (TTO), has expanded on this idea by studying the distribution and mixing of several manufactured compounds that have been inadvertently introduced into the environment (for example, tritium from atomic bomb blasts).

Ocean vents are an especially exciting research subject. Discovered during submersible dives along the

---

Box 7-1 continued

8.4. Note that all of these values are mildly alkaline, not acidic.

Liquid water continually breaks up and reforms in solution, in this manner:

$$H_2O \rightleftarrows H^+ + OH^-$$

water   hydrogen   hydroxyl
ion   ion

The arrows indicate that the reaction can go either way.

When this process occurs in pure water, the H^+ and OH^- are of equal concentration. Thus, in this solution, the water has a pH of 7 because no excess of either H^+ or OH^- exists. Pure water is a balanced, neutral solution.

Why, then, do solutions like seawater have pH values different from 7 (neutral)? The reason lies in the different quantities and types of ions in the solution and how they react with each other. The ocean is slightly alkaline because of the reaction of carbon dioxide and other ions with seawater. This reaction is very important and involves this series of possible steps (they vary with conditions):

$$H_2O + CO_2 \rightleftarrows H_2CO_3 \rightleftarrows HCO_3^- + H^+ \rightleftarrows CO_3^{2-} + H^+ + H^+$$

water   carbon   carbonic   bicarbonate hydrogen carbonate hydrogen
dioxide   acid   ion   ion   ion   ions

The arrows indicate that the reactions can go either way. The CO_2, H_2CO_3, HCO_3^-, CO_3^{2-}, and H^+ ions can exist in equilibrium with one other.

The combining of CO_2 and H_2 can produce carbonic acid (H_2CO_3), which may then lose one of its hydrogen ions and form the bicarbonate ion (HCO_3^-). If the bicarbonate ion then loses a hydrogen ion, a carbonate ion with a double negative charge (CO_3^{2-}) results.

If the pH of the seawater were to increase (become more alkaline, less H^+), the reaction could go toward the carbonate (right) side of the equation. The result would be an increase in H^+, lowering the pH. Conversely, if the pH of the seawater were to decrease (become more acidic, more H^+), the reaction would go to the left, removing H^+ and raising the pH.

This process of maintaining the pH is called *buffering*. It prevents sudden changes in the acidity or alkalinity of the seawater. The buffering ability of seawater is critical to many organisms living in the ocean, because they require a fairly constant pH for many of their life processes. If this buffering process did not occur, the removal of CO_2 from seawater during photosynthesis, or its addition during respiration, could result in big swings in the pH. Because of buffering, however, the acidity/alkalinity balance is barely affected.

The slightly alkaline ocean is very beneficial for the animals and plants that develop carbonate shells, including clams, crabs, oysters, foraminifers, and many others. If the ocean were acidic, it could dissolve their carbonate shells, and these life forms would have to exert extra energy to maintain their shells and therefore to live. The slight alkalinity of seawater is also necessary for the formation of many complex organic compounds. If the pH were lower they could not form. In this manner the buffering aspect of seawater facilitates two key necessities for life: maintaining proper pH and providing an appropriate environment for the formation of more complex organic materials necessary for life.

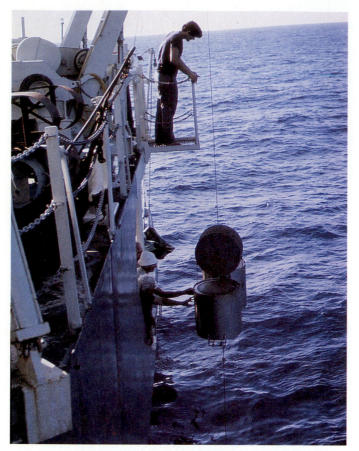

FIGURE 7-1   Lowering of a large-volume water sampler. (Photograph courtesy of Woods Hole Oceanographic Institution.)

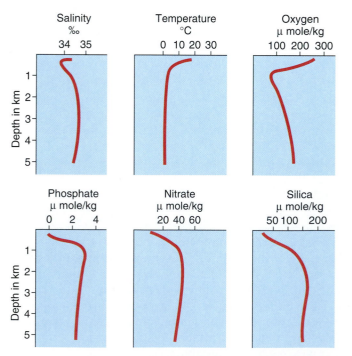

FIGURE 7-2   Some data from a typical GEOSECS station (in this instance, a station between San Diego and Hawaii). (Adapted from Edmond, 1980.)

ocean ridges, these hot-water (**hydrothermal**) vents are related to seafloor spreading and may be a major source of some elements in seawater. One estimate is that, over approximately 8 million years, a volume of water equal to all that in the present ocean cycles through these vents. Most significantly, the water may experience some major chemical alteration. Only a few years ago, most chemical oceanographers thought that nothing could affect the chemistry of the ocean so much in such a relatively short time. The implications of the venting phenomenon are just becoming understood. (See later section on ocean vents, p. 167).

CO_2 and the **greenhouse effect** are a recent and very important area of research for marine chemists. They are studying the effects of increasing CO_2, which is being released into the atmosphere from numerous sources, including natural processes, but mainly from the burning of fossil fuels. Most marine chemists feel that the oceans may absorb much of this increase in CO_2, but perhaps not enough to prevent future climate changes. (This problem is discussed further in Chapter 13.)

Potential mineral resources on the seafloor are another important aspect of chemical oceanography. Deposits of manganese nodules cover over 50 percent of the ocean floor, and cobalt crusts occur on some Pacific seamounts. Other potentially important economic materials—nickel, lead, copper, and zinc—are also concentrated in these deposits. In other regions of the ocean, rocks or sediments contain large quantities of phosphorite that could be mined and used as fertilizer. (More details concerning potential mineral resources are presented in Chapter 15.)

Tools of Chemical Oceanography

Chemical oceanographers are very interested in the distribution of the various chemical components in seawater and the causes of their distribution. Therefore, they need a device for sampling seawater and a technique for measuring the elements or compounds in the seawater. The sampling device:

• Must obtain a sufficient volume of water for analysis

• Must be easily and accurately located according to depth or any other property, such as salinity or temperature

• Must not allow contamination

The volume of water needed depends, of course, on the type of analysis. Salinity determinations now are generally determined electronically, sometimes directly in the ocean without the need for sampling. Some other analyses can require many liters of water, and several such large samples could quickly fill the ship's storage area. Thus, in many instances, chemical analyses are done aboard ship, or a method of concentration is used.

To obtain samples from a specific water depth, a device that can be opened or closed from the ship is required. An excellent instrument for this is the **Nansen bottle** (Figure 7-3), an old, trusted instrument with thermometers that record the water temperature at the time of the sample collection. The device contains two thermometers; one is protected against the effects of pressure, the other is not. When the device is tripped, the temperature values are locked. The temperature difference between the two thermometers is a measure of the pressure on the unprotected thermometer. This difference can then be used to calculate the depth (remember, there is an increase in pressure with depth) from which the water was collected.

To obtain uncontaminated samples, chemical oceanographers may use stainless steel or Teflon-coated samplers. For some research, particular care must be taken to avoid chemical alteration after collection. Refrigeration and preservation to prevent biological growth may be necessary.

Once a sample is collected, numerous chemical techniques may be applied for determination of elements or compounds. The more common shipboard analyses include salinity or chlorinity, oxygen content, and nutrients. In the past, most analyses were done by chemical precipitation. Performed at sea, this is a slow, tedious operation.

Today, better methods are available. Salinity now is determined by the faster and more accurate technique of measuring the electrical conductivity of the water. The more salt in the water, the better it conducts electricity (see Box 7-2). Typical devices make in situ (in-place) measurements while the instrument is lowered through the ocean (see Figure 6-9a, p. 144). Some instruments can electronically measure, record, and transmit the in situ characteristics of the ocean. As the equipment is lowered through the water, data are transmitted through the lowering cable back to the ship and are displayed on a computer (see Figure 6-9b, p. 144). When an interest-

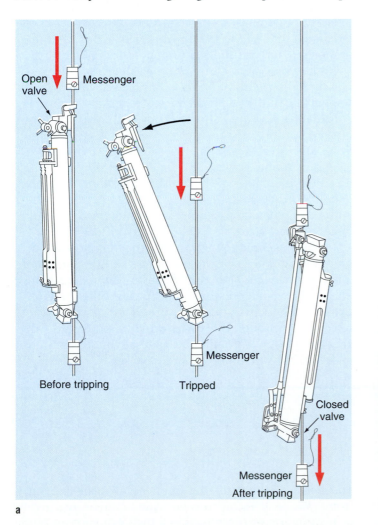

Open valve   Messenger

Messenger

Before tripping   Tripped

Closed valve

Messenger

After tripping

a

b

FIGURE 7-3   Nansen bottle water sampler, named after its inventor, the famous Norwegian explorer of the Arctic Fridtjof Nansen.
(a) The "messenger" (a weight) slides down the wire and hits the Nansen bottle, allowing it to flip over and trap the water inside it by closing two valves. As the bottle flips over, it also releases another messenger, which performs the same procedure to a lower bottle.
(b) Nansen bottle being removed from hydrographic wire. (Image (a) from U.S. Naval Hydrographic Office Publication No. 607.)

Box 7-2

Measuring Salinity

Early attempts to measure the salt content of seawater were limited by chemical techniques and a lack of understanding of the chemistry of seawater. A key breakthrough was made by Georg Forchhammer in 1865. After analyzing numerous seawater samples, he showed that the ratio of the major ions to each other was a constant proportion. This meant that the ratio of the percentage of manganese to the percentage of sodium to the percentage of calcium (and so on) is constant. This is completely independent of the total *amount* of salt content, whether a milligram or a kilogram. This concept is known as **Forchhammer's Principle** or the Principle of Constant Proportions.

Chemist William Dittmar confirmed this observation by analyzing many samples collected during the Challenger expedition (1872–76). Dittmar also suggested that determination of the quantity of a single major element in seawater could be used to determine the total **salinity** of seawater by employing Forchhammer's Principle. This principle is applicable only for seawater collected offshore, because river inflows near land, with their varying amounts of dissolved substances, would affect the ratios before the fresher waters were fully mixed with the seawater.

Also, some constituents in seawater are involved in biological processes, and these are not in constant proportion to the "fixed" constituents of seawater. These elements, such as the nutrients phosphorus, nitrogen, and iron, are called **nonconservative elements;** the elements whose ratios are constant are called **conservative elements.**

In the early days of marine chemistry, major elements in seawater were being detected and measured, but it was a slow process and not always very accurate. Nevertheless, chemists were anxious to develop a standard measure of salinity. A simple definition of salinity is the salt content *in grams* in a *kilogram* of seawater.

Grams per kilogram, or g/kg, is equivalent to parts per thousand, or ‰, which is how salinity generally is expressed.

In 1902, the idea of a relationship was developed between the chloride content of seawater and salinity, commonly is called **chlorinity.** Although differences in chlorinity or salinity can be very small, they are important for determination of seawater density and for the ocean circulation (see Box 8-3, page 192). Therefore, a standard, reproducible sample of seawater salinity is critical for research.

For many years, a standard seawater of known chlorinity was available from the Hydrographic Laboratory in Copenhagen. This sample enabled the comparison of methods and results of chlorinity determination from different laboratories. Since 1975, the Institute of Oceanographic Services in England has maintained the production and distribution of standard seawater.

In the past, salinity, or more typically chlorinity, was measured by chemical titration, a time-consuming and sometimes difficult procedure, especially at sea. Fortunately, however, the presence of the various ions in seawater allows seawater to conduct electricity; the more ions present, the greater the conductivity. Thus, electronic methods were developed to determine salinity. The conductivity method, using a *salinometer,* is very temperature-dependent. Early in its development, bulky equipment was required to maintain a constant water temperature.

Modern techniques for measuring salinity include special sensors to measure conductivity. These are frequently included with other sensors that determine temperature and pressure in a device called a **CTD probe** (conductivity-temperature-depth), which can be lowered through the water rapidly (see Figure 6-9, page 144) to give prompt and generally more accurate measurements than previous techniques.

ing phenomenon is observed, a sample bottle can be "triggered" to collect water at that depth.

Eventually, electronic systems may be developed to measure almost any element in the ocean. These instruments could be lowered or towed through the ocean with the data transmitted to the ship, producing an instantaneous chemical profile of that portion of the ocean. Long-term unattended monitoring systems could be installed on the seafloor or suspended from buoys to measure subtle changes in water chemistry.

What's in Seawater?

The normal salinity of the open ocean usually ranges from 33‰ to 37‰. (This is not true for areas near rivers or melting ice, or those having a high evaporation rate.) The concentration of the six major elements (chlorine, sodium, magnesium, sulfur, calcium, and potassium) constitutes more than 90 percent of the total ions in solution. These and some of the minor elements (strontium, bromine, and boron) have an essentially constant ratio to one another (see the Forchhammer Principle, Box 7-2) and are the "conservative" constituents of seawater. By knowing the amount of one of these elements in seawater, one can calculate the others.

Many of the remaining constituents in seawater, including other elements, dissolved gases, organic compounds, and particulate matter, occur in varying proportions. These variations are due in large part to biological reactions. Indeed, many of the important chemical reactions occurring in the ocean are due to biological processes.

Perhaps surprisingly, the total composition of seawater is not very accurately known. One reason is that some areas of the ocean have not been adequately sampled. Another is that many components are present in extremely small quantities. Furthermore, elements involved in biochemical reactions may vary in concentration by a factor of 1,000 or more from one place to another.

We can divide the chemical composition of the ocean into four parts: (1) dissolved inorganic matter, (2) dissolved gases, (3) dissolved organic matter, and (4) particulate matter. The following four sections describe each of these. A later section, "Life in the Ocean: Biochemical Reactions" (see p. 170), is devoted to the biological effects on the chemical composition of the ocean.

Dissolved Inorganic Matter

Organic material is of biological origin; **inorganic** material is not. By weight, seawater is about 3.5 percent inorganic material dissolved in about 96.5 percent pure water. This dissolved inorganic matter is the major component in seawater. At present, 84 elements have been detected in the ocean. It is important to note that the concentration of many of these elements varies considerably with location, time, season, and especially with biological activity.

Refer to Figure 7-4 in connection with the following:

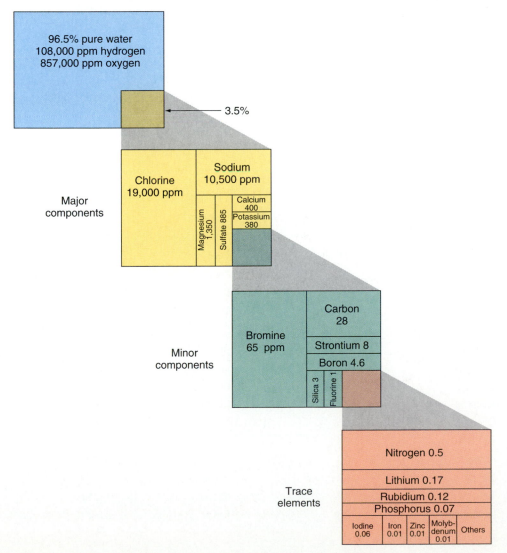

FIGURE 7-4 The main elements in seawater (all values in parts per million, or ppm).

- **Major** inorganic elements are those present in quantities greater than 100 parts per million (ppm), including chlorine, sodium, magnesium, sulfur (usually expressed as sulfate), calcium, and potassium.

- **Minor elements** have concentrations less than 100 ppm but greater than 1 ppm, including bromine, carbon (sometimes expressed as carbonate), strontium, boron, silicon, and fluorine.

- **Trace elements** are those with concentrations less than 1 ppm, including nitrogen, lithium, rubidium, phosphorus, iodine, iron, zinc, and molybdenum. At least 50 other elements are present in quantities less than 10 parts per billion (ppb) (Table 7-1). Eventually, all known naturally occurring elements will possibly be detected in seawater.

In seawater, elements are virtually always present as component parts of a chemical compound (for example, Na and Cl ions instead of combined NaCl), because of the solvent ability of seawater. The relative stability of the various chemical compounds is important in controlling the general composition of the ocean. Apparently, some elements are being concentrated in the ocean, while others are quickly passing through the ocean system and being precipitated onto the seafloor. In other words, the residence time of different elements in the ocean can be extremely variable. (This concept is discussed further in the section "Chemical Reactions and Residence Time," page 166.)

Dissolved Gases

The major gases dissolved in seawater are nitrogen, oxygen, and carbon dioxide. In Table 7-2, note the striking difference in proportion of each gas in the atmosphere and dissolved in the ocean. Oxygen and carbon dioxide are very important in several biological processes and can be produced in the ocean by photosynthesis and respiration processes. Nitrogen gas is rarely involved in re-

TABLE 7-1

Some Minor and Trace Elements in Seawater (all values in parts per billion)

Element	Concentration (PPB)	Element	Concentration (PPB)
Carbon	200–3000	Cobalt	0.2–0.7
Lithium	170	Mercury	0.15–0.27
Rubidium	120	SIlver	0.145
Barium	10–63	Chromium	0.13–0.25
Molybdenum	4–12	Tungsten	0.12
Selenium	4–6	Cadmium	0.11
Arsenic	3.0	Manganese	0.1–8.0
Uranium	3.0	Neon	0.1
Vanadium	2.0	Xenon	0.1
Nickel	2.0	Germanium	0.07
Iron	1.7–150	Thorium	0.05
Zinc	1.5–10	Scandium	0.04
Aluminum	1.0–10	Bismuth	0.02
Lead	0.6–1.5	Titanium	0.02
Copper	0.5–3.5	Gold	0.015–0.4
Antimony	0.5	Niobium	0.01–0.02
Cesium	0.5	Gallium	0.007–0.03
Cerium	0.4	Helium	0.005
Krypton	0.3	Beryllium	0.0005
Yttrium	0.3	Protactinium	2×10^{-6}
Tin	0.3	Radium	1×10^{-7}
Lanthanum	0.3	Radon	0.6×10^{-12}

Source: Data from Goldberg, 1963; Hood, 1963, 1966.

TABLE 7-2

Major Gases of the Atmosphere and the Ocean, by Volume

Gas	Percent of All Gases in Atmosphere	Percent of All Gases Dissolved in Seawater
Nitrogen	78.08	48
Oxygen	20.95	36
Carbon Dioxide[1]	0.035	15

[1]Carbon dioxide in seawater is also found in carbonate, bicarbonate, and carbonic acid ions.

Source: Data from Weilhaupt, 1979; Hill, 1963.

actions except by some bacteria. Inert gases such as helium, neon, argon, krypton, and xenon occur in lesser quantities in seawater. These gases are not known to be involved in any important oceanographic process.

Gases are generally dissolved into the ocean from the atmosphere. However, some very rare gases come from radioactive decay processes within sediment on the ocean bottom. Also, three naturally occurring radioactive isotopic can occur in gaseous forms in the ocean: tritium (^{3}H), carbon-14 (^{14}C), and argon-39 (^{39}A). They are formed in the atmosphere when cosmic rays strike these atoms. The gases ^{3}H and ^{14}C are also formed by the detonation of nuclear bombs. Once in the ocean, these radioactive isotopes continue to decay and can be used to date changes or marine processes. (This is discussed later in the section "Isotopes and Radioactivity in the Ocean," p. 172.)

The solubility of a gas (its ability to go into solution at the sea/air surface) depends on three principal factors:

1. Temperature of the gas and the seawater

2. Atmospheric concentration of the gas

3. Salt content of the seawater (salinity)

The quantity of gases in seawater is mainly determined by these factors (with the notable exceptions of oxygen and carbon dioxide, as discussed below). Below the sea surface, pressure also becomes an important factor. Gases generally are relatively unreactive once in the marine environment. If the quantity of a gas is greater or less than would be expected from these factors, it suggests that something in the marine environment is causing the variation.

OXYGEN. Oxygen concentration in seawater varies with depth. In surface waters, its concentration is related to temperature: the higher the temperature, the lower the solubility of a gas. Somewhere between a few hundred to a thousand meters below the surface, however, an oxygen-minimum or oxygen-poor zone usually occurs (Figure 7-5).

Seawater has two sources of oxygen: the atmosphere and the plants that live in the ocean. Surface waters, because of their contact with the atmosphere, generally contain an expected amount of oxygen. In some instances, a supersaturation (extremely large amount) of oxygen is observed. This is usually due to photosynthesis, which releases O_2 into the water. **Photosynthesis** is the process in which plants use carbon dioxide (CO_2), water (H_2O), nutrients (including some trace elements), and solar energy to produce organic matter—generally expressed as the carbohydrate **glucose** ($C_6H_{12}O_6$)—and oxygen (O_2).

Because the photosynthetic reaction depends on sunlight, it occurs only in the upper layers of the ocean, usually above 200 m (655 ft.) depth. As previously noted, photosynthesis can increase the oxygen content in surface waters. Below and within these upper layers, the oxygen and organic matter are used by organisms, including bacteria. This **respiration** process consumes oxygen, removing it as a gas from the water.

The two complementary reactions can be expressed as follows:

- *Photosynthesis* (by plants) occurs in upper layers of ocean:

$$6CO_2 + 6H_2O + \text{nutrients} + \text{solar energy} \rightarrow \text{organic matter } (C_6H_{12}O_6) + 6O_2$$

- *Respiration* (by plants and animals) occurs throughout the ocean:

$$\text{Organic matter } (C_6H_{12}O_6) + 6O_2 \rightarrow 6CO_2 + 6H_2O$$

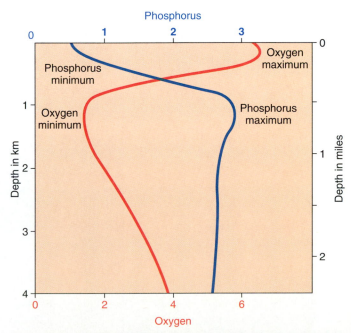

FIGURE 7-5 Vertical distribution of oxygen and phosphorus in the ocean. Oxygen units are milliliters per liter, or parts per million. Each phosphorus unit is equivalent to about 3×10^{-6} g/liter (1/300,000 g/l) of seawater.

An oxygen-minimum zone can result from the respiration of animals and plants and from the bacterial oxidation of organic debris. The presence or absence of the oxygen-minimum zone depends on whether the depletion of oxygen by respiration exceeds the renewal of oxygen by mixing of surface and deeper waters. The increase of oxygen in depths below the oxygen-minimum zone is believed to be due to the influx of oxygen-rich surface waters from the polar regions into the deeper parts of the ocean (see Figures 9-19 and 9-20, page 217).

If an area is isolated from an oxygen source, it is possible for oxygen at depth to be used up. One such isolated area is the Black Sea. Waters devoid of oxygen are called *anaerobic*. However, organic material can still be decomposed in such waters by certain bacteria, using sulfates as an energy source. The process produces sulfide, which can combine with hydrogen to form hydrogen sulfide, a malodorous gas which is lethal to many organisms. If these oxygen-deficient deep waters come to the surface by upwelling or an overturning of the water, mass mortality of animal life in the surface waters usually occurs.

CARBON DIOXIDE.   Carbon dioxide is present in seawater at a concentration 400 times higher than in the atmosphere (see Table 7-2). The total CO_2 in the oceans is more than 60 times that in the atmosphere. One reason is that a cubic meter of water can absorb a larger quantity of carbon dioxide than can a cubic meter of air. A more important reason is that seawater is slightly alkaline and contains available cations such as magnesium and calcium. This allows carbon dioxide to combine with these cations to form carbonates and bicarbonates in seawater, in the manner described below.

Carbon dioxide occurs both as a gas and bound in various compounds. It participates in complicated and interrelated biological and chemical interactions with seawater and plays several important roles in the ocean. It is one of the components in the calcareous shells of organisms. It also serves as a buffer that stabilizes the pH of seawater (Box 7-1). The buffering action of carbon dioxide works like this. When carbon dioxide gas (CO_2) from the atmosphere dissolves in seawater it combines with water to form carbonic acid (H_2CO_3). The carbonic acid then can dissociate (break apart) to form bicarbonate (HCO_3^-) and carbonate (CO_3^{2-}). These forms are in balance (equilibrium) with each other (the $\rightleftarrows$ symbol indicates equilibrium):

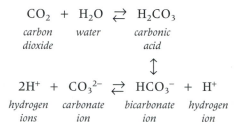

$$CO_2 \;+\; H_2O \;\rightleftarrows\; H_2CO_3$$

carbon dioxide   *water*   *carbonic acid*

$$\updownarrow$$

$$2H^+ \;+\; CO_3^{2-} \;\rightleftarrows\; HCO_3^- \;+\; H^+$$

hydrogen ions   *carbonate ion*   *bicarbonate ion*   *hydrogen ion*

If carbon dioxide is removed from seawater (during photosynthesis by growing plants, for example), some carbonic acid will dissociate to form carbon dioxide and water. This mechanism provides a large reservoir of carbon dioxide for photosynthetic reactions in the ocean. At night, when photosynthesis stops due to lack of sunlight, carbon dioxide continues to be produced by respiration. It too becomes chemically recombined and stored.

These chemical and biological reactions (Figure 7-6) also permit CO_2 to act as a buffer in the ocean, prevent-

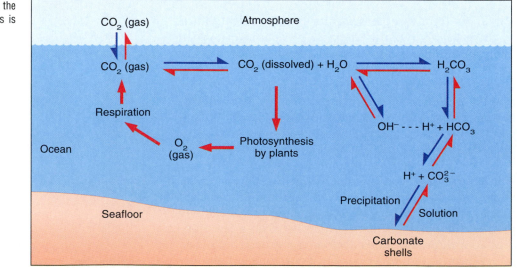

FIGURE 7-6   The carbon dioxide cycle in the ocean. This relatively simple set of reactions is critical to life on our planet.

ing sudden changes in acidity or alkalinity of the water (Box 7-1). The pH level (acidity/alkalinity) can be critical for life processes in many organisms. If the concentration of hydrogen ions (H^+) were to decrease (in other words, if the solution became more basic), the reaction would cause the carbonic acid (H_2CO_3) to produce more bicarbonate (HCO_3^-) and hydrogen ions. On the other hand, if the concentration of hydrogen ions were to increase (if the solution became more acidic), the reaction would go the other way, removing hydrogen ions. In a similar manner, removal of CO_2 (by photosynthesis) or addition (by respiration) has little effect on the pH.

Carbon dioxide in seawater has an interesting relation with pH, temperature, and salinity. If pH remains constant, the total carbon dioxide content increases with increasing salinity and decreasing temperature. However, pH depends in part on the actual amount of carbon dioxide present, and is influenced by water temperature and pressure. The complex, dynamic relation of carbon dioxide with and to the air, sea, and sediments has been one of the most challenging problems in chemical oceanography.

Carbon dioxide also is involved in formation of carbonate shells by many different organisms. (Calcium ions, Ca^{2+}, combine with carbonate ions, CO_3^{2-}, to form calcium carbonate, $CaCO_3$.) In doing so, CO_2 is also removed from seawater. The ability of the ocean to dissolve, store, and remove CO_2 is of major importance in our understanding and prediction of the greenhouse effect (see pages 321–326). This fact may become very important as the quantity of CO_2 in the atmosphere continues to increase, increasing the greenhouse effect, and possibly increasing global warming. Some have suggested that some of the CO_2 could be placed in the ocean. (See Box 7-3 for further discussion of the carbon and carbon dioxide cycle.)

Dissolved Organic Matter

Dissolved organic compounds are of biological origin. In seawater, dissolved organic compounds include carbohydrates, proteins, amino acids, organic acids, vitamins, nitrogen, and phosphorus that are chemically combined into organic compounds. Aside from nitrogen and phosphorus, very little is known about the vertical and horizontal distribution of most dissolved organic material. It is thought that only about 10 percent of the organic compounds present in seawater have been identified. The sources of much of this material are excreta and the decay of dead organisms.

Dissolved organic matter in seawater exists in moderately small, usually variable amounts between 0 to 6 parts per million. Many dissolved organic compounds—especially the nutrients—are critical for the growth and health of plants, animals, and bacteria in the ocean.

Particulate Matter

Particulate matter in seawater (excluding living organisms) includes three basic components: organic debris, various complexes of organic and inorganic material, and mineral particles in the size categories of sand, silt, and clay. The complexes of organic and inorganic material may account for local variations in concentration of some elements in seawater. For example, nearshore waters generally have a high abundance of iron, which may be due to complexes formed by the iron and organic compounds.

Another example of particulate matter in seawater is freshwater diatoms and minerals. These may occur in surface waters thousands of miles from their source, probably transported by winds. In general, particulate matter in the ocean is highly variable and appears to respond to local geography, biological production, atmospheric conditions, and other unknown factors.

The study of how some particulate matter—such as some pollutants—moves through the ocean water to the seafloor is a new and important area in chemical oceanography. *Sediment traps* placed on or near the seafloor and left for long periods (Figure 7-7) have provided quantitative data on particulate matter movement. These traps have revealed two important mechanisms for transporting material from the surface to deeper waters: by fecal matter (waste products, usually in pellet form; see Figure 5-18c and 5-18d, page 124) and by **marine snow** (clumps of fine-grained particles; see Box 11-1, page 247).

In summary, many factors control or influence the chemical composition of the sea:

- Exchange of gases with the atmosphere
- Solubility of different compounds
- Biological and chemical processes by anaerobic bacteria
- Precipitation and exchange with the ocean bottom
- Inflow of freshwater
- Freezing and melting of sea ice
- Chemical reactions that control or influence the concentrations of different elements
- Hydrothermal discharges from deep-sea vents
- Biological processes, including life processes and decomposition of organic matter

Because of their importance, some specific factors are discussed further in the next section.

BOX 7-3

Carbon, Carbon Dioxide, and the Ocean

The importance of carbon dioxide in the living world cannot be overemphasized. Without photosynthesis of CO_2 into oxygen and organic matter, and the reverse process of respiration, most life on Earth would eventually cease. Thus, the amount of CO_2 in the atmosphere and in the ocean is of major concern to scientists and is being monitored. One measure of the planetary flow of CO_2 is the worldwide distribution of carbon.

Carbon, carbon dioxide, and the ocean play a crucial and complex role in global climate change. Somewhere between 5 and 6 billion tons of carbon is added yearly to the atmosphere by human activities, in the form of CO_2. The measured increase of CO_2 in the atmosphere, however, is equal to only about 3 billion tons of carbon. Most of the missing carbon was thought to be absorbed in the ocean. However, recent studies suggest that this is only partially correct. The

ultimate residence of about one billion tons of carbon is unknown.

Carbon and the Ocean

The oceans are the key component for the control of CO_2 and carbon. It is estimated that dissolved within seawater is more than 20 times the carbon that occurs in land plants, animals, and soil. If just 2 percent of the carbon in the ocean were released into the atmosphere as CO_2, it would double atmospheric CO_2 and probably cause a significant temperature increase—global warming.

Fortunately, a balance exists in the exchange and release of CO_2 between the atmosphere and the ocean (Figure 1). However, many factors and processes can affect the solubility of CO_2 in water. Any change in the ocean-atmosphere system could have profound impact

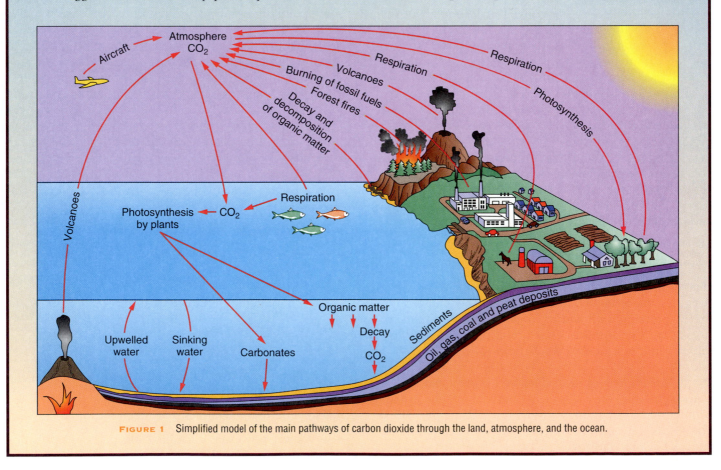

FIGURE 1 Simplified model of the main pathways of carbon dioxide through the land, atmosphere, and the ocean.

Box 7.3 continued

on the total CO_2 distribution. Among the influences on oceanic CO_2 are biological processes, water circulation, pH, pressure, and temperature.

The main biological processes in the ocean involving CO_2 are photosynthesis, which occurs in the upper 100 to 200 m or so of the water column, and respiration, which occurs throughout the ocean. Photosynthesis, which requires sunlight, results in the primary production of organic compounds by plants, combining CO_2 and water. Respiration is the breakdown of the organic matter, releasing the CO_2 back into the environment.

In this biological process, there is a net movement of CO_2 from surface waters into the deeper parts of the ocean. Other factors produce the same effect, including the sinking of dead animals and plants, or parts thereof, which can carry both organic material and calcium carbonate as shell material to the deeper parts of the ocean (Figure 2). Marine snow (discussed in Box 11-1, page 247), falls into this category. The general feeding of animals in the ocean and the excretory materials resulting from this process also move organic material toward the bottom, as does other decomposition processes and the sinking of more dense surface waters.

Carbon dioxide can be returned to the surface by the upwelling of deep water (see pages 214–215). This movement of CO_2 toward deep water and back to the surface again sometimes is called the *biological pump*. As part of this total process, the carbon in CO_2 can also be incorporated into marine sediments as calcium carbonate ($CaCO_3$). Carbonate-rich sediment covers a large portion of the seafloor (see Figure 5-15, page 121).

CO_2 can also be moved laterally by normal seawater movement and currents. Regionally—and only a broad understanding exists—CO_2 tends to be released from the ocean in equatorial regions, especially in the Pacific, and taken up in temperate and subpolar regions. This pattern ties closely to ocean circulation patterns: cooling and sinking prevails in the temperate and subpolar regions (more CO_2 can be absorbed in colder water), whereas upwelling and warming are more common in equatorial regions (less CO_2 can be held in warmer water).

Carbon and World Climates

A better understanding of the carbon cycle is important to evaluating any potential climate change. A major in-

ternational scientific program called **Joint Global Ocean Flux Study (JGOFS)** is considering some of the points raised here. JGOFS has two goals:

1. To determine and understand, on a global scale, the processes controlling the movement over time of carbon and associated biogenic elements in the ocean, and to evaluate the related exchanges among the atmosphere, seafloor, and continental boundaries.
2. To develop a capability to predict, on a global scale, the response of related biological and chemical marine processes to environmental stresses, especially those related to climate change.

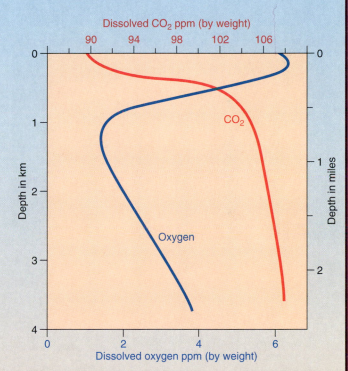

FIGURE 2 The general vertical concentration of oxygen and CO_2 in the ocean. Oxygen concentration is great in surface waters due to exchange with the atmosphere and to photosynthesis, which releases oxygen. Oxygen concentration decreases below the surface because respiration, which uses oxygen, is more prevalent than photosynthesis. Near the bottom, oxygen concentration increases because it is transported there by deep oxygen-rich polar waters. (This process is discussed further in Chapter 9.) CO_2 distribution mirrors oxygen in a way, because it also is involved in the photosynthesis-respiration cycle. However, where oxygen is released, CO_2 is taken up, and vice versa. The CO_2 increase with depth is caused by respiration of animals and bacteria, and by the presence of CO_2 rich water from the Arctic and Antarctic.

FIGURE 7-7 A series of sediment traps being prepared for lowering. (Photograph courtesy of Dr. Sus Honjo.)

Factors Influencing the Ocean's Chemical Composition

The major-element chemical composition of the ocean can be considered constant, in a very general sense. Nevertheless, several ongoing processes influence the ocean's chemistry. These include various chemical reactions, the recycling of seawater through seafloor vents, biochemical reactions, physical processes, and human input.

All of these processes are part of the large, complex geochemical cycle that affects ocean chemistry (Figure 7-8). Many of these processes have been active for over a billion years and have resulted in the present composition of the ocean.

Chemical Reactions and Residence Time

To a large degree, the chemical composition of the ocean is in equilibrium; in other words, in a given volume of the ocean, the quantity of elements stays essentially constant, and so does their proportion. Thus, equilibrium implies that what enters the ocean essentially equals what leaves the ocean (or is deposited on the seafloor). This important point leads to the concept of **residence time,** discussed below.

Large portions of elements in the ocean are erosion products from land, carried to the sea by rivers. At present, rivers carry about 4 billion tons of dissolved material to the ocean each year. Some of this material is recycled—evaporated from the ocean into the atmosphere, precipitated out as rain on land, and carried back into

FIGURE 7-8 A model of the geochemical cycle of the ocean. Although simple, the model does show the main known geochemical processes that occur in or effect the ocean.

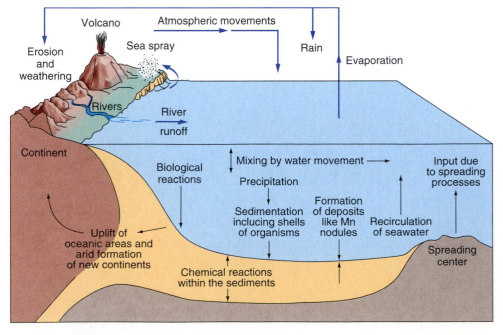

the ocean by rivers. Once in the ocean, different elements have different abilities to react. These differences can be expressed by a simple model (Figure 7-9).

RESIDENCE TIME. The concept of **residence time** assumes that the amount of an element *introduced* into the ocean during a given period of time equals the amount *removed* during the same period of time. The residence time of an element is its amount in the ocean divided by its rate of introduction, or by its rate of removal through precipitation onto the seafloor. In other words, residence time is the time needed to replace the amount of a specific element in the ocean, or the average amount of time that a given atom will stay in the ocean water. It assumes that a steady-state condition exists, or that input equals output. This seems reasonable, for seawater salinity seems to have remained fairly constant for long periods, despite the large volume of salts carried into the ocean by rivers and from deep-sea vents.

Residence time is based on some other assumptions as well. One is that the elements are uniformly and quickly mixed within the ocean. Another is that most elements are introduced by rivers, and in fact, early estimates of residence time just considered the amount of the elements introduced by rivers. Later, scientists emphasized the amount of elements in sediments (Table 7-3). More recently it has been recognized that the input from seafloor vents also must be considered.

Unquestionably, this model provides an oversimplified picture of the ocean. What is important to our understanding of the ocean's composition is the wide range in residence time among elements. For example, aluminum and iron remain in seawater only hundreds of years, while others, such as sodium and magnesium, remain for millions of years (Table 7-3).

IMPORTANCE OF RESIDENCE TIME. The significance of these calculations is that elements having long residence times, such as sodium, are the same elements

Table 7-3

Residence Time of Some Elements in Seawater

Element	Amount in Ocean (grams)	Residence Time (years)
Sodium	147×10^{20}	210,000,000
Magnesium	18×10^{20}	12,000,000
Potassium	5.3×10^{20}	11,000,000
Calcium	5.6×10^{20}	1,000,000
Silicon	5.2×10^{18}	8,000
Manganese	1.4×10^{15}	700
Iron	1.4×10^{16}	140
Aluminum	1.4×10^{16}	100

that are not very reactive in the marine environment and therefore remain in the ocean for long periods. The shorter residence times of silicon, manganese, iron, and aluminum reflect their biological activity. In addition, significant quantities of silicon, iron, and aluminum enter the ocean as particles and can quickly settle to become part of the sediments. These particles include minerals such as quartz, feldspar, or material from volcanic activity.

The great chemical reactivities of manganese, iron, and aluminum are also due to their ability to form mineral deposits on the seafloor, such as manganese nodules. Thus, seawater has fairly high concentrations of some elements that are relatively minor on the land portion of the planet.

Seawater itself has a residence time. Each year an average of 1.26 m (4.1 ft.) of water evaporates from Earth's surface (see Table 6-3, page 139). This evaporated water continues through the hydrologic cycle (see Figure 6-2, page 134) and ultimately returns to the ocean in the form of precipitation or river run-off. If this return ceased, the oceans would evaporate completely in about 3,015 years. (The calculation is 3,800 m divided by 1.26 m/year, or the average depth of the ocean divided by the average evaporation rate.)

Furthermore, because the oceans have been around for at least 3.5 billion years, it is appropriate to say that the average water molecule has gone through the hydrological cycle (evaporation from the ocean and return) more than a million times!

Ocean Vents and Recycled Seawater

One of the most exciting discoveries in oceanography occurred in 1977 when scientists diving in the submersible *Alvin* off the Galápagos Islands (about 1,000 km, or 600 mi. west of Ecuador) discovered a hot spring on the seafloor at about 3 km (almost 2 miles) depth.

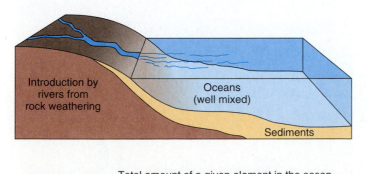

Residence time = $\dfrac{\text{Total amount of a given element in the ocean}}{\text{rate of introduction of element by the rivers}}$

or

Residence time = $\dfrac{\text{Total amount of a given element in the ocean}}{\text{rate of removal as sediment}}$

FIGURE 7-9 Simple residence-time model. The concept does not imply that the different elements are saturated in seawater.

This and later discoveries led to considerable rethinking of ocean chemistry, biology, and ecology.

Discharges of water that are hot (hydrothermal) and salty from the Red Sea and Pacific Ocean have been known since the mid-1960s. However, the findings in the Galápagos and elsewhere are especially impressive and indicate a dynamic process active along many areas of seafloor spreading (Figure 7-10). Not only is a vent discharging hot, salty water, but the water generally contains valuable minerals that are being deposited in the adjacent area.

Most important to science is the fascinating array of animal life around the vents, the question of how these animals live in the vents' unique chemical system (see Box 12-3, page 301), and the chemical implications for the ocean of the water being discharged. A unique biological assemblage of large clams, worms, and crabs is often associated with the vents (see Figures 12-19 to 12-21, pages 300–302).

Since the discovery of the Galápagos vents, similar vents have been found in many other areas in the ocean. These were initially discovered along areas of active seafloor spreading; more recently they have been found in other localities (Box 7-4). In some of the vents associated with seafloor spreading, the water can be nearly 400°C (about 750°F) and literally roars from the vents. Because these waters often contain dark iron sulfides and have a jetlike flow, they are called **black smokers.** The billows of iron sulfides form chimneys of mineralized material through which the hot water continues to be discharged.

These "eruptions," or vents, are found along an oceanic rift where two crustal plates are moving apart, with volcanic rock filling the resulting gap. Among the interesting chemical aspects of these discoveries is that the water from the vents appears to be recycled seawater that has sunk and moved through parts of the ridge crests. In doing so it has picked up relatively large quantities of some elements, as well as heat. Both the elements and the heat are ultimately released to the ocean (Figure 7-11).

Considerable speculation has developed about the value of the minerals near and inside the vents. Generally, each new discovery has been reported in the media, sometimes even in the financial sections of newspapers. However, a realistic evaluation of the worth of any deposit requires more study than has yet been applied to any vent area.

To chemical oceanographers, these deep-sea vents and their discharge of hot, mineral-rich water has led to re-evaluation of several hypotheses concerning the chemistry of seawater and the formation of unique water masses (see Box 9-3, page 218). Taken together, the deep-sea vents can be visualized as a worldwide system for chemically reprocessing the ocean's water and chemicals.

FIGURE 7-10 Different types of ocean vents.
(a) A dramatic example of a "black smoker" photographed along the East Pacific Rise. The black color is due to mineral particles in the water.
(b) A group of shrimp around an active vent in the Atlantic Ocean. Note the shimmering hot water coming from the vent. (Photograph (a) by Dudley Foster; photograph (b) by John Edmond, MIT; both images courtesy of Woods Hole Oceanographic Institution.)

Box 7-4
Vents and More Vents

Subtle hints of the discharge of hot, salty water from the seafloor came from research in the early 1960s along the East Pacific Rise and the Red Sea, and in a 1972–73 study of the Mid-Atlantic Ridge by a French-American team. However, it remained until the 1977 findings of *Alvin* at the Galápagos Ridge for the actual discovery to be made. Both the variety and size of the organisms around the vents were a surprise. To this day, many questions concerning their origin and distribution remain somewhat of a mystery.

Although the initial discovery of vents releasing hot water along ocean ridges was quite startling, more amazement was in store. This included finding exotic fauna that lived on chemically produced food, rather than that produced by photosynthesis (see Box 12-3, page 301). If this doesn't seem earth-shattering, consider that our entire concept of life was based on the photosynthesis-respiration cycle. Now that we have seen life forms that base their existence on chemistry other than photosynthesis, what does this imply for the possibility of life on other planets? The discovery of new life forms at the vents continues.

Further exploration led to other discoveries in the Pacific and later in the Atlantic Ocean, regions that often feature the dramatic **black smokers.** The reason that water so hot can exist without turning to steam is due to the pressure of the overlying seawater. Recall that each 10 m of depth adds 14.7 lb./in.^{2}, or one atmosphere, of pressure. One vent area in the Pacific off the coast of Oregon discharges a huge, distinct body of water called a *megaplume*. This body is so distinctive that it has been mapped (see Box 9-3, page 218).

Vents in the Gulf of Mexico

For a while, it was thought that the discovery of vents and their unique biological communities would be confined to ocean ridge areas. Then came another surprise in 1984, when a study of a continental slope region along the Florida coast in the Gulf of Mexico found some organisms similar to those of the deep-sea vents. Water of greater than normal salinity, 46‰, was flowing from cracks in a limestone outcrop. These waters were rich in hydrogen sulfide and were apparently supporting the growth of bacterial mats similar to those found in the deep-sea vents. More study and drilling into the limestone found waters having temperatures up to 115°C (239°F) and salinities approaching 250‰.

Apparently this hot, salty water migrates along cracks and joints in the rock before reaching the ocean. The origin of these waters could be related to the processes that form the many hydrocarbon deposits in the region (the Gulf coast is the site of numerous offshore oil-and-gas deposits). Indeed, that year, another area with a fauna similar to deep-sea vents was observed in the Gulf of Mexico, south of New Orleans. The Florida discharge was at a depth of about 3,200 m (10,496 feet); the one south of New Orleans was found at about 650 m (2,132 feet). These types of vents may be common near locations where hydrocarbons are being formed.

Later, large tube worms, mussels, and clams were found at several more sites in the Gulf of Mexico. Their occurrence with oil and gas seeps seems to be common. Dr. James Brooks, a Texas A&M oceanographer, has noted prolific oases of tube worms and mussels at oil and gas seeps. The gills of mussels collected from some of the sites have bacteria that use methane (the main component of natural gas) as a source of carbon. The mussels in turn eat the compounds produced by the bacteria. This process is similar to that on ocean vents, except that deep-sea vents involve hydrogen sulfide gas.

Vents in Subduction Zones and a Lake

Another region where vents have been found is in subduction zones, the areas where sediments and rocks on one plate are being thrust under another plate (see Figure 3-9, page 52). So far they have been discovered on the continental slope off Oregon and along portions of the Japan trench and the Peru-Chile trench.

In the early 1990s, a study was made of Lake Baikal in Russia, the world's deepest (1,637 m, or 5,369 ft.) and possibly oldest (25 million years) freshwater lake. The expedition discovered several areas of high heat flow on the lake floor. Then, using an ROV (Remotely Operated Vehicle), scientists found a vent area with several unique species of life. One interpretation is that the lake bed is slowly cracking apart, creating a rift valley, and that the process has led to the discharge of warm water. It seems reasonable to expect that discoveries of vents and their associated living communities will continue.

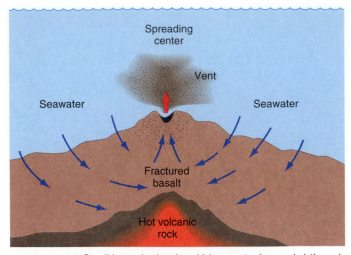

FIGURE 7-11   Possible mechanism by which seawater is recycled through spreading center areas.

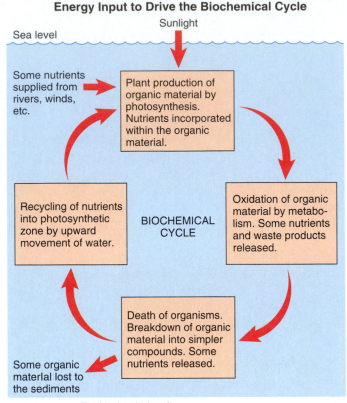

FIGURE 7-12   The biochemical cycle.

The volume of certain elements being added to the ocean by vent processes may exceed the volume from any other process. The rate of water input is still poorly known. Some scientists have suggested, however, that the movement of dissolved material by vent water may actually be comparable even to the volume of material contributed to the ocean from weathering on land.

One estimate is that about 6,000 m³ (22,400 ft.³) of water per second are entering the sea from all the vents together. This is about one-third of the Mississippi River's flow at New Orleans. At this rate, vent water would fill the entire ocean in about 7 million years. This could mean that the oceans have been recycled through the vent system about 500 times since they were formed. If this number is correct, it would clearly change many ideas about ocean chemistry and would necessitate revision of the geochemical model of the ocean.

Life and the Ocean: Biochemical Reactions

Probably the most important chemical reactions in the ocean are those caused by life processes. Collectively they form the **biochemical cycle.** These processes cause a flow of nutrients, which travel from the organic material formed by plants during photosynthesis, through the life cycle of the organisms in the ocean, and eventually back into the photosynthetic zone (Figure 7-12). (See further discussion in Chapter 12.)

We already have examined the effects of photosynthesis and respiration on the oxygen and carbon dioxide content of surface waters (the upper 100–200 m, or about 330–660 ft.). From this, it is clear that plants and animals in the ocean greatly influence seawater composition. Other elements are also involved in the biochemical cycle, such as nitrogen, carbon, phosphorus, and trace elements like silicon and iron. These elements are removed from seawater by the formation of organic ma-

terial during the photosynthetic growth of marine plants. The elements are later returned to seawater as waste and decomposition products from the organic material.

You might expect organic material to pile up on the seafloor, but little generally accumulates. This is because the organic material either decays or is consumed while falling through the water or when it reaches the bottom, where it is digested by bottom-dwelling organisms. When the organic matter decays (which requires oxygen), its nutrients are released back into the environment. Certain areas, such as the Black Sea, accumulate organic matter because oxygen is absent in the bottom waters.

The consumption of oxygen in the oxidation of organic material helps to explain the relative absence of oxygen at intermediate depths of the ocean (see Figure 7-5). Oxidation (or respiration), unlike photosynthesis, is not powered by light energy and can occur at any depth and at any time of day or night.

FOUR LAYERS OF NUTRIENTS.   We can classify the distribution of nutrients in seawater into four layers, based on their abundance:

1. *Low-nutrient surface layer* (photosynthetic zone); usually about 100 to 200 m thick (about 330 to 660 ft.). Nutrient concentration is low because it is utilized during photosynthesis and incorporated within the organic material.

2. *Increasing nutrient layer;* several hundred meters thick. Nutrient concentration increases very rapidly due to its release from organic material by oxidation processes.

3. *Maximum nutrient concentration layer;* usually about 500 to 1,000 m thick (about 1,650 to 3,300 ft.).

4. *Uniform nutrient layer;* usually extends to the bottom. Nutrient concentration is uniform.

In some instances, the maximum concentration of nutrients does not coincide with the minimum concentration of oxygen. This is due either to the presence of nutrients released during a previous biological cycle, or to a different source for oxygen.

VERTICAL DIFFERENCES.   The oxidation and subsequent return of the nutrients in the second layer results in a large and important reservoir of nutrient-rich water below the photosynthetic zone (upper 100–200 m). These nutrients can return to the photosynthetic zone only by the physical circulation or movement of the water. Here they will become reincorporated into organic material.

This movement takes place in several ways: by worldwide circulation of the oceans; by upwelling (a type of vertical mixing) in coastal, offshore, and equatorial regions; and by annual vertical mixing in temperate and high-latitude areas. (Chapter 9 describes how these movements occur.)

The process of nutrient use results in a downward movement of nutrients due to biochemical reactions. This is compensated for by an upward movement of these same elements as a result of water circulation. The organic material that is deposited in the sediments is removed from the system, but this can be balanced by material that is added from rivers and other sources (see Figure 7-12).

HORIZONTAL DIFFERENCES.   In addition to these vertical differences in nutrient concentration, ocean-wide differences and seasonal variations also occur. The deep waters of the Pacific and Indian Oceans, for example, contain more phosphorus and less oxygen than do the deep waters of the Atlantic and Arctic Oceans. This is caused by the source of the waters, their composition, and their subsequent modification by biological and physical factors.

Seasonal changes in nutrient content are most evident in temperate areas, where phytoplankton have two growth periods each year—the spring and early autumn. Studies in the English Channel that have analyzed the animals and plants living in the water, and the water itself, show seasonal patterns and changes in the various forms of phosphorus (Figure 7-13). They also indicate two annual plankton growth periods, each accompanied by an increase in dissolved organic phosphorus.

OTHER ELEMENTS.   We have focused on the biochemical reactions that affect nutrients. However, many other elements are concentrated by organisms. This occurs either through photosynthesis or through animal consumption of organic material and eventually liberating it, either as waste or through death and decomposition. For example, calcium and silicon are used by some organisms to form their shells. Extensive deposits of these shells exist on the seafloor (see Figures 5-15 to 5-18, pages 121–124). Several crustacean species concentrate copper, and in some marine organisms, trace elements occur in greater concentration than in the seawater.

In considering the distribution of reactive elements in seawater, we must also evaluate the water dynamics. For example, if the movement or flow of water is sufficiently large, it can overwhelm or obscure differences in chemical composition caused by biological processes, evaporation, heating, or any other phenomena.

Physical Processes

Changes in the salt content of ocean waters are mainly caused by differences in evaporation and precipitation (rainfall), and melting or freezing of ice. In some polar regions, greater precipitation dilutes surface water, causing lower salinity. In subtropical waters, relatively high salinity can occur where evaporation is high.

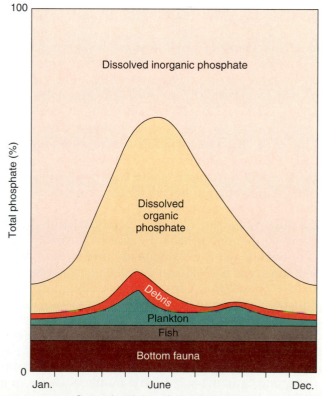

FIGURE 7-13 Seasonal variation of phosphate in waters of the English Channel. (After Harvey, 1950.)

Currents, waves, and other turbulence usually keep the upper parts of the ocean moderately well mixed. Near land, warm surface water may be blown away from the coast by strong offshore winds and replaced by up-welling of cooler, nutrient-rich subsurface water. This process can create large amounts of nutrient-rich water and, in turn, high levels of photosynthesis, making these areas rich in biological resources.

Human Input

Human input to the composition of the ocean is small, but observable. It can be seen, for example, in the distribution of the element lead. Every nation on Earth uses the internal combustion engine, which burns fuel that often contains lead compounds. This has caused a significant increase in lead concentration in the atmosphere and consequently in the ocean.

Worldwide, more than 250,000 tons of lead enter the ocean each year. Near industrialized regions, this input can be seen in surface layers of the ocean, where lead content can be 5 to 50 times that of deeper waters. Some countries, such as the United States, have restricted the use of lead in automotive gasoline. Other examples of human input to the ocean are given in Chapter 16, which deals with the subject of pollution.

Isotopes and Radioactivity in the Ocean

Atoms of elements occur in several forms called **isotopes.** Recall from Box 6-1, page 134, that isotopes of an element are identical except for the number of neutrons in the nucleus. This gives rise to variants like hydrogen ^1H, deuterium ^2H, and tritium ^3H. Typically, for any element, there is an extremely common isotope (for example, ^1H), whereas its other isotopes are rare (^2H, ^3H). Generally, all isotopes of an element have chemical properties similar to those of the common form of the element. Thus, all isotopes of an element behave the same in chemical reactions.

The physical properties, however, differ slightly. Small differences in atomic weight are evident as small variations in boiling point, freezing point, and rates of diffusion of the rare isotope compared with the common form. These small variations can be used to study some oceanographic processes.

Some isotopes are radioactive and change (decay) by losing pieces of their atomic structure to become a different isotope or even a different element. For example, carbon occurs in three forms: carbon-12 (^{12}C), carbon-13 (^{13}C), and carbon-14 (^{14}C). The numbers 12, 13, and 14 indicate the different atomic weights of the element.

These three forms are chemically similar, but, ^{14}C is radioactive and decays to nitrogen-14 (^{14}N). The original radioactive isotope is known as the *parent* and the resulting form or forms are called the *daughters*. If the daughters are radioactive, they too will decay and change.

The rate at which radioactive isotopes decay is specific for each isotope (Table 7-4). The range of decay rates is spectacular, from 4.5 billion years down to microseconds, depending on the isotope. The **half-life** is how long it takes for half of the original quantity of a radioactive material to decay into another isotope. Therefore, by measuring the relative amounts of the parent and daughter that are present in a sample, one can determine how long the decay has been going on, and thus the age of the sample (see Figure 7-14).

Potassium-40 (^{40}K), for example, decays to argon-40 (^{40}Ar). If the ratio of potassium-40 to argon-40 in a rock sample is, say, a million parts to one part, that means the sample was formed very recently, for very little of the ^{40}K has had time to decay. If the ratio is 1 to 1, that means that half of the original ^{40}K has decayed to ^{40}Ar. Therefore, the age of the rock is 1.3 billion years, because that figure is the half-life of ^{40}K, or the amount of time necessary for half of the ^{40}K to decay to ^{40}Ar. If the ratio is 1 to 3, the age of the rock is 2.6 billion years (three-fourths of the original ^{40}K has decayed to ^{40}Ar).

It is important to note that the age obtained indicates when the rock *last solidified from a molten state*. The reliability of this technique is based on two requirements:

1. No daughter isotope was originally present in the molten rock as it solidified.

2. No parent or daughter isotope has been *added or removed* during the subsequent aging of the rock.

It is usually difficult to prove that a single parent-daughter pair in a sample meets these requirements. Contamination, addition, and removal happen easily over time. If, however, several isotope pairs are present in the sample, and if more than one of these parent-daughter

---

TABLE 7-4

Some Important Radioactive Isotopes and Their Half-Lives

Parent	Stable Daughter	Half-Life in Years
Thorium-232 (^{232}Th)	Lead-208 (^{208}Pb)	1.4 billion
Uranium-238 (^{238}U)	Lead-206 (^{206}Pb)	4.5 billion
Potassium-40 (^{40}K)	Argon-40 (^{40}Ar)	1.3 billion
Uranium-235 (^{235}U)	Lead-207 (^{207}Pb)	0.7 billion
Carbon-14 (^{14}C)	Nitrogen-14 (^{14}N)	5,680

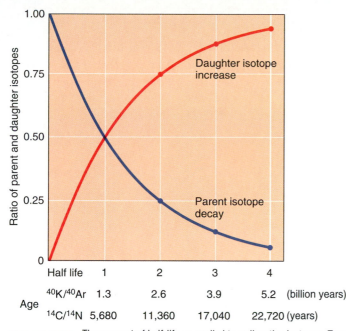

Age

	Half life	1	2	3	4	
^{40}K/^{40}Ar		1.3	2.6	3.9	5.2	(billion years)
^{14}C/^{14}N		5,680	11,360	17,040	22,720	(years)

FIGURE 7-14 The concept of half-life as applied to radioactive isotopes. Two parent/daughter examples are shown: potassium-40 (^{40}K), which decays to argon-40 (^{40}Ar) and has a half-life of 1.3 billion years, and carbon-14 (^{14}C), which decays to nitrogen-14 (^{14}N) and has a half-life of 5,680 years. See the text for additional discussion.

pairs are measured, they can provide independent checks of one another. In other words, two different radioactive isotopes will produce similar results only if both meet the two requirements. (Another possibility is that both are incorrect in exactly the right proportions, but this is highly unlikely.)

Isotopes that have half-lives of billions of years are best suited for evaluating the age of Earth, because measurable amounts of the parent will still be present after such long periods of time. Isotopes with relatively short half-lives, like the 5,680 years of ^{14}C, are used for measuring more recent events of shorter duration, such as dating recent sediments or archaeological events.

Slight differences in behavior of some stable isotopes can be used to interpret some chemical and physical oceanographic processes, such as evaporation and temperature changes. Radioactive isotopes, both those occurring naturally and those created by nuclear bomb explosions, can be used to study and date the history of the ocean. They can also date presently active dynamic processes, such as circulation and ocean mixing. Radioactive isotopes are useful in determining rates of accumulation of sediments. By dating sediments, past ocean events can be compared with past events on the continents.

We will now look at the two fundamental types of isotopes—stable and radioactive—and see how their behavior is valuable in oceanographic research.

Stable Isotopes

The value of studying stable isotopes is well illustrated by water, which contains these isotopes:

- Hydrogen occurs in three forms: hydrogen-1, hydrogen-2, and hydrogen-3. The common hydrogen atom (^{1}H) constitutes more than 99.9 percent of all hydrogen atoms. The *deuterium* atom (^{2}H) is heavier and much less abundant. The *tritium* atom (^{3}H) is a very rare hydrogen isotope and is radioactive (unstable).

- Oxygen occurs in three forms: oxygen-16, -17, and -18. The lightest and most abundant isotope is ^{16}O, making up over 99.7 percent of all oxygen atoms. ^{17}O and ^{18}O are heavier and rare.

A typical water molecule is composed of the lighter isotopes ^{1}H and ^{16}O, so it weighs less than it would if any of the heavier isotopes (^{2}H, ^{3}H, ^{17}O, ^{18}O) were substituted into the molecule. Therefore, the common "light" water molecule is more active physically. More of these light molecules go into the air above a water surface, creating vapor pressure. Light-water vapor pressure is greater than the vapor pressure of "heavier" waters. Therefore, it is easier for molecules containing lighter isotopes to evaporate from surface waters of the ocean. An important consequence is that when evaporation occurs, surface waters are *decreased* in their lighter isotope components and *increased* in their heavier isotopes.

Latitude can also be important, because more evaporation occurs in hot equatorial areas than in cooler high-latitude areas. Knowing these facts, one can measure the hydrogen and oxygen isotopic composition of seawater and determine some history of the seawater, such as its past temperature.

Shell-forming animals and plants that live in the surface waters also reflect the conditions of the water by the ratio of ^{18}O to ^{16}O within their shells. Since these organisms build their shells from whatever isotopes are at hand, this automatically creates a record of the isotope ratio at the time of shell formation. After death, the shells settle to the ocean floor and become incorporated into the bottom sediments.

Thus a drill core of sediments can provide a continuous record of historical isotope ratios, and therefore of any temperature changes that occurred over time in the surface waters. If radioactive isotopes are also present and measured, the sediment record can be specifically dated in years and compared with events on land.

Other variations of isotopes within the biological environment are useful. For example, the heavier isotopes of oxygen are more abundant in the oxygen-minimum zone. Apparently, marine plankton and other organisms that use organic material in their oxidation processes selectively remove the lighter isotopes of oxygen, leaving the heavier isotopes behind.

Radioactive (Unstable) Isotopes

Radioactive isotope analysis also is valuable to the oceanographer. The rates of radioactive decay can be correlated with the rates of chemical reactions in the ocean. These isotopes are especially useful in the study of deep-sea sediments, which are deposited very slowly. Marine geochemists use radioactive isotopes to correlate events over large areas of the ocean, and even to correlate them with events on land.

Radioactive isotopes in the ocean and sediment have three possible origins:

1. Isotopes formed when Earth was formed

2. Isotopes from cosmic sources and solar reactions with the atmosphere

3. Isotopes from nuclear reactions produced by humans

The first group of "primordial isotopes" is generally used for dating extremely slow processes, such as sedimentation rates of very slowly deposited deep-sea sediments. These radioactive elements have extremely long half-lives and therefore are still present in measurable quantities, even after about 4.6 billion years of Earth history.

Isotopes produced by cosmic and solar reactions with the atmosphere are valuable to oceanographers because of their relatively short half-lives (Table 7-5). Short half-lives are better for measuring short-term events, such as water circulation and ages of sediments that were deposited relatively quickly.

CARBON-14. Probably the most useful of these isotopes is ^{14}C—generally called **carbon-14**—which is formed by the interaction of cosmic rays with the atmosphere. This produces higher-energy neutrons, most of which are captured by nitrogen molecules in the atmosphere (^{14}N) to form ^{14}C. The ^{14}C combines with oxygen, producing carbon dioxide. By exchange with the atmosphere, this radioactive carbon dioxide then enters the ocean, where it is used in life processes. The ^{14}C decays at a constant rate; its half-life is 5,680 years. Because carbon atoms are part of all living things, carbon-14 can be used to date things that were once alive. It is commonly used in archeology and similar fields.

The ^{14}C in organisms or surface waters is in equilibrium with the environment, but when the organisms die, photosynthesis and respiration stop, no additional ^{14}C is added, and the remains sink. The ^{14}C will continue to decay, however, at the steady rate of a 50 percent decrease every 5,680 years. Therefore, measuring ^{14}C content indicates how long the material *has ceased to be in contact with the surface waters*. For water, the resulting age can also reflect the degree of mixing with waters of differing ^{14}C ages.

TABLE 7-5

Some Radioactive Isotopes That Are Produced by Cosmic and Solar Reactions, and Their Half-Lives

Isotope	Half-Life in Years	Source
Carbon-14 (^{14}C)	5,680	Cosmic rays and nuclear bombs
Silicon-32 (^{32}Si)	500	Cosmic rays
Tritium (^3H)	12	Cosmic rays and nuclear bombs
Strontium-90 (^{90}Sr)	28	Nuclear bombs
Cesium-137 (^{137}Cs)	30	Nuclear bombs

In general, ^{14}C can be used to date only material less than about 100,000 years old. This is due to its relatively short half-life and the difficulty of detecting the small amounts remaining beyond this period of time. Two corrections must be made to the ratio of ^{14}C to stable carbon before it can be used for dating. It is sobering to note that both corrections are necessitated by human activity. One correction is for the recent increase in carbon in the atmosphere due to the burning of fossil fuels; the other correction is for the ^{14}C produced by nuclear explosions.

An important issue in oceanography is to understand the motion of water and how it mixes, both horizontally and vertically. Such information is needed to determine the fate of pollutants in the oceans, to understand the biological and geochemical cycles, and to understand climatic changes. Carbon-14 can be measured in various water masses and used to calculate their age, or at least how long the water has been away from the surface. These calculations are based on numerous assumptions, and they usually depend on models of differing complexity. ^{14}C dating of marine sediments is an especially valuable technique for understanding the recent climatic history of the ocean.

ARTIFICIAL RADIOACTIVE ISOTOPES. Radioactive isotopes are considered *artificial* when they are produced by testing of atmospheric nuclear weapons or discharges of nuclear waste. Although they are pollutants, they can be valuable tools for studying marine processes. Because human nuclear activity is recent (the past 50 years), the sources of these pollutants and their time of input into the environment generally are known. Thus, these radioactive isotopes can be used to date recent marine phenomena are especially useful in establishing the rate of water movement by determining the distribution of such materials in surface and deep waters. The time when these substances reached the surface of the ocean and their relative amounts can usually be documented.

If they are then found below the surface or far from the point of injection, oceanographers can infer how they traveled, and how fast.

One research program in the late 1980s that used this concept was called **Transient Tracers in the Ocean (TTO).** Among its objectives was to better understand the role of the ocean in global climate and how the marine environment responds to inputs of substances made by humans, including ^3H and ^{14}C produced from nuclear bomb blasts and the deliberate discharge of nuclear waste.

A key discovery in this program came following GEOSECS studies in which the presence of ^3H (tritium) was noted in the deep waters of the North Atlantic (Figure 7-15). Because ^3H has a half-life of only about 12 years (it decays to helium, ^3He), it was surprising to see it penetrate so far into the ocean. In general, ^3H, the heaviest isotope of hydrogen, is relatively rare.

The answer lay in major input from nuclear testing in the 1950s and 1960s. This input into the atmosphere eventually reached the surface waters of the ocean and, by mixing, the deeper waters. By measuring the helium/tritium ($^3He/^3H$) ratio in different areas, an evaluation of mixing and its rate could be made. Without the input from manufactured bombs, this analysis would have been impossible because the small quantities of ^3H that occur naturally are too minute to measure.

Another surprising discovery, detected during a TTO cruise, was that the deeper waters of the North Atlantic were colder and slightly less salty than measured in a cruise nine years earlier. Although the differences were slight—0.15°C colder and 0.02‰ less saline—they are very significant. The reasons for the change seem linked to atmospheric effects because these determine many characteristics of surface water, which sinks in a few specific areas to form the deep waters of the ocean. It is premature to say whether these changes reflect a worldwide climatic change, but clearly these changes in the ocean will be intensely studied in coming years.

In 1986, a major accident occurred in the Chernobyl nuclear power plant in the Ukraine (part of the former Soviet Union). Large amounts of radioactive material were discharged into the atmosphere and some subsequently settled into the ocean, especially the nearby Black Sea. This disaster provided an opportunity to study how the ocean can absorb and transport material.

Scientists quickly placed a series of sediment traps in the Black Sea (similar to those shown in Figure 7-7) to catch the settling particles. The data are still being studied, but some results are evident. It appears that the largest flux of material to the bottom occurs during periods of high growth by plankton. This suggests that the radioactive material is first absorbed by phytoplankton (small ocean plants), which then are eaten by zooplankton (small ocean animals). The material reaching the bottom probably is aggregate fecal material from the zooplankton.

FIGURE 7-15 Distribution of tritium in the waters of the western Atlantic Ocean. The figure is a vertical profile from the north (right side of figure) to the south, extending almost the entire length of the Atlantic. At the bottom of the figure are familiar locations on land. Note how tritium has penetrated to a considerable depth in the waters of the North Atlantic. Tritium units (0.2, 5, and so on) are number of tritium atoms in 10^{18} atoms of hydrogen. Data from numerous GEOSECS cruises.

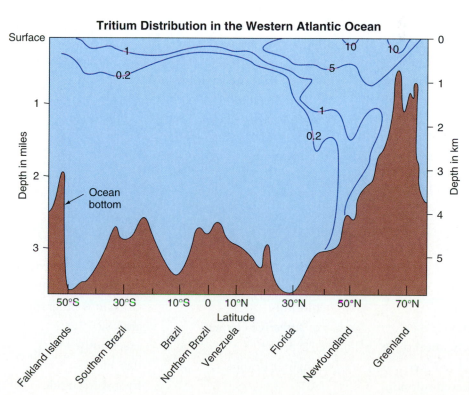

SUMMARY

Chemical oceanographers principally study the chemical aspects of the ocean, which in many instances are intimately involved with biological activities. Early studies in this field stressed cataloging what was present in seawater. More recent efforts have focused on the processes that control the composition of the ocean, including especially the discharge of material from deep-sea vents, and rates of reactions (radioactive isotopes are especially useful in this category).

Seawater is a mixture of dissolved inorganic matter, dissolved organic matter, dissolved gases, particulate matter, and, of course, water. The amount of many elements in seawater can be estimated by Forchhammer's Principle if the concentration of one element not involved in biological reactions is known. Biological processes, especially photosynthesis, can strongly influence the concentrations of some elements (especially nutrients) and some gases (especially oxygen and carbon dioxide). The role of carbon and carbon dioxide, although not fully understood, is clearly important in climate problems.

The key factors that influence the ocean's chemical composition are: (1) exchange with the atmosphere; (2) solubility of compounds; (3) reduction of organic matter by anaerobic bacteria; (4) precipitation to and exchange with the ocean bottom; (5) inflow of freshwater; (6) freezing and melting of sea ice; and (7) chemical and biological reactions and processes, including the discharge of hydrothermal solutions from deep-sea vents.

Hydrothermal activity along spreading centers is an example of a process that can strongly influence the chemical composition of the ocean. Recently, vent areas have been found in regions other than along spreading centers. Considerable understanding exists concerning some biochemical aspects and their seasonal variations and influences on the ocean; however, much is yet to be learned. Elements that are unreactive in seawater tend to have a long residence time in seawater.

Isotopes, especially some radioactive ones, can be used to introduce the measurement of time in the assessment of oceanographic and chemical processes. Some stable isotopes, because of slight differences in their physical properties, can be used to learn about past changes in certain oceanographic processes, such as temperature and evaporation.

QUESTIONS

1. How has the discovery of hydrothermal vents along ocean spreading centers changed our view of the chemistry of seawater?

2. What is the Forchhammer Principle and why is it important?

3. What are the major processes affecting the distribution of oxygen and carbon dioxide in the ocean?

4. Discuss why carbon dioxide tends to move toward deeper parts of the ocean.

5. Discuss the processes of photosynthesis and respiration and why they are critical to the ocean and to life in general.

6. Discuss the key factors that influence the chemical composition of the ocean.

7. Describe the relation among carbon dioxide, pH, and buffering of seawater. Why is buffering of seawater important?

8. What are the key nutrients in the ocean? What is their general distribution pattern?

9. What are isotopes? How can they be used in oceanographic research?

KEY TERMS

acid rain	carbon dioxide	electrical conductivity	hydrothermal water
acidity	carbon-14	Forchhammer's Principle	in situ
alkalinity	chlorinity	GEOSECS Program	inorganic
bacteria	conservative element	glucose	isotope
biochemical cycle	CTD probe	greenhouse effect	Joint Global Ocean Flux Study (JGOFS)
black smoker	daughter isotope	half-life	
buffered solution	DDT	hydrogen sulfide	major element

manganese nodule	ocean vent	photosynthesis	stable isotope
marine snow	organic matter	pollution	trace element
minor element	oxygen	pressure	Transient Tracers in the Ocean (TTO)
Nansen bottle	oxygen-minimum zone	radioactivity	
nitrogen	parent isotope	residence time	tritium
nonconservative element	particulate matter	respiration	
nutrient	pH	salinity	

FURTHER READING

Berner, E. A., and R. A. Berner. 1987. *The Global Water Cycle.* Englewood Cliffs, N.J.: Prentice Hall. A somewhat advanced text on the chemistry of water and the environment.

Broecker, W. S. 1974. *Chemical Oceanography.* New York: Harcourt Brace Jovanovich. A somewhat advanced but solid book on chemical oceanography.

———. 1983. "The Ocean." *Scientific American* 249, no. 3, pp. 146–60. Some aspects of the chemistry of the ocean.

Holm, N. G., ed. 1992. *Marine Hydrothermal Systems and the Origin of Life.* Norwell, Mass.: Kluwer. Several papers on various aspects of hydrothermal vents, focusing on how they might be important in the origin and evolution of early life.

Jannasch, H. W. 1984. "Chemosynthesis: The Nutritional Basis for Life at Deep-Sea Vents." *Oceanus* 27, no. 3, pp. 73–78. A description of some of the chemical and biological processes working at ocean vents.

Libes, S. M. 1992. *An Introduction to Marine Biogeochemistry.* New York: John Wiley and Sons. A slightly advanced text-book covering marine chemistry and how it relates to some aspects of marine geology, biological oceanography, and pollution.

"Marine Chemistry." *Oceanus* 35, no. 1 (Spring 1992). An entire issue on recent research in marine chemistry, including articles on marine organic material, particles in the ocean, and hydrothermal activity.

Macintyre, F. 1977 "Why the Sea Is Salt." *Scientific American* 223, no. 5, pp. 104–15.

Open University. 1989. *Ocean Chemistry and Deep-Sea Sediments.* Oxford: Pergamon Press. Part of a 5-volume series on the ocean prepared as a university text; sometimes very detailed, but informative.

———. 1989. *Seawater: Its Composition Properties and Behavior.* Oxford: Pergamon Press. Part of the same 5-volume series.

Untersteiner, N. 1986. "Glaciology: A Primer on Ice." *Oceanus,* 29, n. 1, pp. 18–23. A discussion of the dynamics of ice and how it forms.

CHAPTER 8

Physical Aspects of the Ocean

PHYSICAL OCEANOGRAPHY IS THE MEASUREMENT AND STUDY OF THE PHYSICAL PROPERTIES OF THE SEA AND HOW THEY VARY OVER TIME AND OCEAN SPACE, COMBINED WITH A THEORETICAL EVALUATION OF THE PROCESSES THAT CONTROL THESE PHYSICAL PROPERTIES. THE MEASUREMENT-AND-STUDIES PORTION MAINLY DESCRIBES THE PROPERTIES AND MOTION OF THE OCEAN. THE THEORETICAL PART HAS BEEN EMPHASIZED IN RECENT RESEARCH, WITH FOCUS ON THE CAUSES FOR THE OBSERVED DISTRIBUTION OF SEAWATER PROPERTIES, CIRCULATION PATTERNS, AND OTHER PHENOMENA. THIS CHAPTER DISCUSSES FOUR KEY PHYSICAL PROPERTIES OF SEAWATER: TEMPERATURE, SALINITY, DENSITY, AND THE VELOCITY OF SOUND IN THE OCEAN.

The lowering of a water sampling device. Various characteristics of the ocean, such as temperature, are measured by electronic sensors on the device and transmitted to the surface ship. Individual water sampling bottles can be triggered at the desired study depths.

(Photograph courtesy of Peter Wiebe, Woods Hole Oceanographic Institution.)

Introduction

Physical oceanography is perhaps the most independent of the subdisciplines in oceanography, though it cannot be wholly separated. For example, the critical physical property of seawater density is partly determined by water chemistry, which is the province of chemical oceanography. Variations in density affect the ocean circulation and currents. Likewise, the discovery of some currents and their extent was first determined by biological oceanographers, who noted the distribution and range of marine organisms.

The distribution of nutrients (and therefore the areas of biological activity) in the ocean is, to a large degree, controlled by the physical conditions of the sea. The bottom topography of the ocean, which is the province of marine geology, can affect large-scale oceanic circulation. Finally, the interaction of the ocean and the **atmosphere** is an extremely important influence on Earth's weather and climate and ultimately global change. Thus, the relatively independent field of physical oceanography is not really so independent.

During World War II, our lack of knowledge about the ocean's physical aspects propelled studies of subjects such as how sound travels through water (mainly to detect submarines) and how to predict wave height in coastal areas (for safe troop landings). More recently, programs that are often large in scale and worldwide in scope have proven effective for studying the very important but often subtle aspects of physical oceanography.

Physical oceanographers are especially interested in the interrelationships of the atmosphere and the ocean. These are important controls of both atmospheric and oceanic circulation. Recently, the interaction of the atmosphere and the ocean has also become a dominant issue with oceanographers who are interested in global change. Understanding and predicting weather, both oceanic and terrestrial, awaits better understanding of air-sea interactions. This includes special phenomena like El Niño (see page 326). These topics are treated further in Chapters 10 and 13.

The ability to predict specific characteristics of the ocean is very important. At present, the only phenomena we can determine in advance are tides, surface waves, and tsunamis (large ocean waves produced by submarine earthquakes). Predictability of the characteristics of the upper layers of the ocean would be especially useful to predict **sound propagation** and to understand many aspects of marine biology, such as the distribution of fish species.

Despite recent discoveries, solutions to many key problems of oceanography are limited by a critical lack of knowledge of general ocean circulation. One difficulty in acquiring this knowledge is the problem of sampling and observing a three-dimensional ocean over time and throughout its area and depth. The ocean is turbulent, with considerable variability over time and space. Conventional observations by ships provide important data, but they cannot provide the large scale and timeliness of observations needed to answer essential questions concerning the ocean and climate or the ocean's effect on biological processes.

To obtain this data, a new set of technological developments is necessary. These new technologies include remote-sensing satellites, drifting buoys that can monitor and radio-transmit oceanographic data, and sophisticated computers to assimilate the data and test models of ocean circulation. The development of these and other new instruments permits accurate in-place measurements of temperature, salinity, currents, and other variables (see opening figure to this chapter). **Satellites** have become an important tool for physical oceanographers and will become even more so (Box 8-1). Advances in computer data processing now allow quick results, in some instances even while the vessel is still in the study area.

Problems such as large-scale oceanic circulation can best be solved by **synoptic measurements** (many measurements taken at the same time over a large area). In the past, synoptic observations required numerous and expensive surface ships. Now, however, recently developed fixed or floating buoys (Figure 8-1) and **oceanographic platforms** used in conjunction with airplanes, submarines, satellites, and surface vessels make the procedure more efficient.

Instruments of the Physical Oceanographer

A prime objective of the physical oceanographer is to measure the important physical properties of seawater, such as its temperature, salinity, and density. These measurements then can be used to deduce evaporation and heat exchange, currents and water movements, and other physical processes occurring in the ocean.

Temperature and Depth Measurements

There are three common ways of measuring the temperature of ocean water. One is by using a very accurate recording thermometer, which is usually attached to a **Nansen bottle** (see Figure 7-3, page 157) and lowered into the ocean. When the Nansen bottle is at the desired depth, a weight (called a messenger) is dropped down the wire; when it hits a trigger on the bottle, the bottle overturns, releasing another messenger to perform the

Box 8-1

Eyes in the Sky: Satellites in Oceanography

As the science of oceanography advances, it has become necessary to observe large parts of the ocean simultaneously, as well as to monitor certain areas over long periods. In both instances, surface ships cannot do the job, because the slow speed of ships combined with their limited range permit, at best, only a very small portion of the ocean to be observed at once. Satellites overcome these limitations in a spectacular manner.

One key advantage of satellites is their ability to obtain data from a very large area in a very short time. Another is that they can observe areas such as the Arctic, which surface ships have difficulty reaching. The point about coverage becomes clear when you realize that an oceanographic research ship can cover about 460 km (250 miles) a day, whereas satellites fly at about 29,000 km (18,000 miles) per hour. By the time a surface ship has covered about 100 km, the conditions of the water that were measured when the ship started may already have changed.

Satellites are routinely used for navigation and weather observation, and are also obtaining widely varied information about Earth's surface. An example of oceanographic information being collected by satellites is the change in salinity where river water flows into the ocean, which can be detected from variations in how seawater reflects light (reflectivity). Similarly, infrared techniques are used to detect temperature variations (see Figure 1). The color of the ocean surface can be related to the amount of plankton (microscopic floating plants) in the water (see Figure 2). Images

from the coastal zone and nearshore regions can be used to detect how currents are moving sediment and to observe some effects of pollution. Fishing boats can use satellite information to locate fish. All ships can use satellite data to avoid storms or icebergs, and to improve their speed by traveling in certain currents.

Satellites with specialized sensors and measuring instruments can obtain data on a global scale (the temperature data used to prepare Figure 1 is a good example). The satellite data are transmitted to receiving stations where they are processed by computer. The data volume is immense and requires considerable storage and computer facilities.

Just like setting up a camera for best results, a satellite is placed in a specific type of orbit for the best view of whatever it is to collect data about. A satellite can be placed in a high orbit, so as to remain synchronized above a specific spot on Earth's surface, allowing 24-hour observation of the area. It also can be placed in a relatively lower orbit and at various inclinations to Earth's equatorial plane, so it circles Earth but covers a different area with each orbit. In this manner it is possible to obtain a complete coverage of Earth in about 12 hours.

From the data, charts can be prepared showing daily, weekly, monthly, or yearly variations in ocean temperature, wind speed, or current. Indeed, data similar to weather maps for land can be developed for the ocean. This type of information is valuable for many reasons: weather prediction, monitoring or detecting pollution,

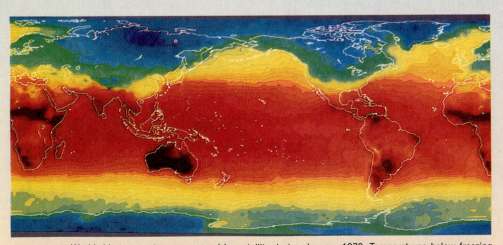

FIGURE 1 Worldwide temperatures measured by satellite during January 1979. Temperatures below freezing (0°C or 32°F) are shown in green and blue. In Siberia and Canada, temperatures as low as −30°C (−22°F) were measured. Temperatures above freezing are shown in yellow, red, and brown. (Photograph courtesy of National Aeronautics and Space Administration.)

Box 8-1 continued

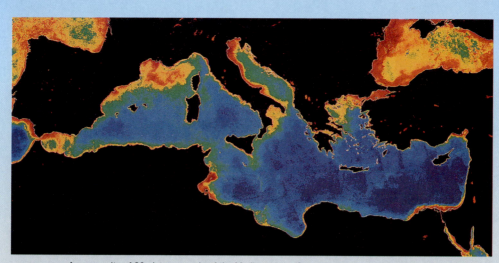

FIGURE 2   A composite of 30 pictures made of the Mediterranean Sea showing relative biological productivity, as determined by the color of the water. Blue indicates relatively clear, plankton-free water; orange and red are plankton-rich. (Note especially the Atlantic Ocean and the Black Sea, in the upper left and upper right, respectively.) In some areas along the coast, the high productivity is due to pollution from human activities. (Photograph from NOAA.)

oceanographic research, fishing, and shipping (ships can be rerouted to avoid ice or storms).

The *LANDSAT* satellites revolve around Earth, providing systematic and repetitive coverage of the land and ocean. Such coverage can be extremely valuable in detecting changes over time in environmental conditions. *SEASAT A*, launched in 1978 especially for oceanographic studies, produced some remarkable pictures and data before it failed 3 months later. Following *SEASAT A*, other satellites were able to measure surface winds, ocean color, sea temperature, sea ice distribution, and other measurements important to science, shipping, waste disposal, and national security.

A major series of U.S. and foreign satellites for oceanographic research is planned for the coming years (see Figure 18-17, page 458). These satellites will provide data on surface winds, sea state, water temperature, wave height, currents, chlorophyll distribution, and other aspects of the sea. These data are critical to evaluating worldwide ocean phenomena, especially those pertinent to research on global change. They also have practical value for ship routing, search-and-rescue operations, global weather forecasting, predicting El Niños, and the like. It is very likely that major advances in the understanding of our planet will come from satellites.

same task on other bottles farther down the line. When the bottle overturns, it traps a sample of seawater that can be used later for a salinity determination or other measurements.

The overturning of the bottle also breaks the mercury column inside the thermometer, preserving the temperature measurement made at that time. If this did not happen, the temperature would rise as the bottle passed through warmer surface waters as it was lifted back to the ship, thus giving an inaccurate reading.

Usually two thermometers are used, not only to measure temperature, but depth. How can thermometers measure depth? One is exposed to the pressure of the overlying seawater, whereas the other is protected from this pressure. The tremendous pressure of seawater at depth squeezes the unprotected thermometer and its mercury column, causing an inaccurate temperature reading, but one that is proportional to the pressure.

Thus, the difference in the readings between the two thermometers reflects the pressure at depth. This in turn indicates the depth, because pressure corresponds to depth.

These temperature measurements should be accurate to better than 0.02C° (about 0.03F°). This high degree of accuracy is necessary for two reasons: (1) temperature has a very important effect on density and other physical properties, and (2) temperatures in the deep sea are relatively constant and have such extremely small variations that only a very accurate thermometer can record them.

The lowering of several Nansen bottles or similar instruments at one locality constitutes a **hydrographic station.** Hydrographic stations obtain measurements at a few discrete depths, which is more efficient than taking measurements throughout the entire **water column** (full depth of the ocean at that place).

a

b

c

FIGURE 8-1   (a) A group of buoys used in a large-scale physical oceanography study. These buoys are 3 m (9.8 ft.) in diameter and are filled with a special noncompressible material, giving them the necessary buoyancy to float on the surface. The towers on the buoy are used to attach meteorological (weather) sensors. The buoys then are anchored to the ocean bottom with wire rope and nylon line; other sensors can be attached to this line.
(b) Launching the buoy.
(c) The buoy in position. (Photographs (a) and (b) by Nancy Brink and photograph (c) by Jerry Dean, both of the Woods Hole Oceanographic Institution.)

Continuous Measurements

In recent years, techniques have been developed that continuously measure temperature, salinity, and other parameters. One simple device, developed during World War II for measuring temperature, is the **bathythermograph (BT),** which can be quickly lowered from a vessel even while the ship is moving. The device contains sensors to detect pressure and temperature that produce a continuous plot of temperature versus depth on a coated glass slide. The BT, which is simple to use, is not as accurate as a good thermometer but has the advantage of producing a continuous picture. An expendable version, the XBT, transmits its data directly to a surface vessel or an airplane (Figure 8-2).

Other new instruments combine the accuracy of thermometers with the speed and continuous measurement capabilities of the BT. They use various sensors and either transmit the data directly to the ship by

radiotelemetry or store it until the instrument is retrieved. Because they also can measure several variables and are very accurate and sensitive, these devices can be used to examine small-scale details of the ocean's physical properties (Figure 8-3). These instruments have several important advantages: they are quick, give a continuous reading, can be used with electronically triggered water samplers (see Figure 6-9, page 144), and are very accurate. Disadvantages include their cost and the need for supporting electronic cables and winches.

Other instruments can be put on the seafloor or suspended from floating or fixed platforms to monitor various ocean parameters for long periods of months to perhaps a year (see Box 8-2). Some devices (Figure 8-4) can be recalled from the seafloor by an acoustic signal. The signal instructs the device to drop its attached weights, allowing it to float to the surface. Once found (which is not as hard as it sounds—the device is

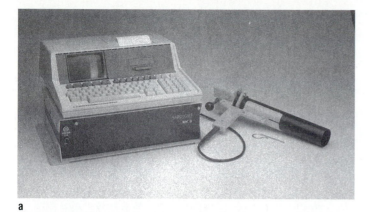

a

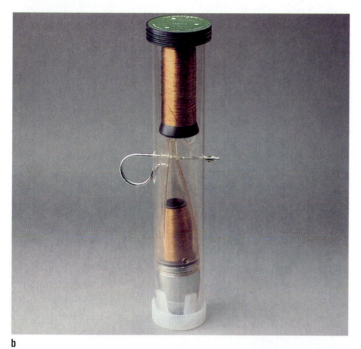

b

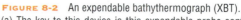

FIGURE 8-2   An expendable bathythermograph (XBT).
(a) The key to this device is this expendable probe containing a thermistor or temperature-measuring device connected to a spool of fine wire.
(b) A transparent view of the probe. After it is launched into the ocean, wire is unreeled as the probe falls through the water to the bottom. An electrical signal representing changes in temperature is transmitted by the wire to a recorder. Because the rate of fall of the probe is known, depth can be read directly from the recorder.
(c) The launching of an XBT. This device can be launched while the ship is moving, or even from an airplane, in which case a radio transmits the data to the plane. (Photographs (a) and (b) courtesy of Sippican Corporation, Marion, Mass.)

c

FIGURE 8-3   High-Resolution Profiler developed at the Woods Hole Oceanographic Institution. This instrument selectively samples the water column as it free-falls through it. On deployment, the negatively buoyant profiler sinks to a programmed depth, at which it releases ballast weights and floats back to the surface for recovery. The profiler measures temperature, salinity, pressure, and horizontal and vertical flow of water, allowing a "profile" of the water column to be constructed. (Photograph courtesy of Raymond W. Schmitt, Woods Hole Oceanographic Institution.)

BOX 8-2

Observation Stations: *Fram*, Ice Islands, and Buoys

Some important contributions in the early history of oceanography were made by vessels that were frozen into the Arctic polar ice and allowed to drift with it for several years. One especially interesting expedition was that by Fridtjof Nansen to study how Arctic winds and currents affect pack ice. Nansen built a specially strengthened ship, *Fram,* that would not be crushed by expanding ice when it froze around the ship. The *Fram* became frozen into an ice floe and drifted from 1893 to 1896, almost reaching the North Pole. *Fram* was eventually freed from the ice and returned to its home port in Norway. These expeditions clearly showed the value of permanent or semipermanent observational stations on the ocean.

A drifting **ice station** was first established in the Arctic Ocean by the former Soviet Union in 1937. The United States started its first scientific camp in the Arctic in 1952 on Fletcher's Ice Island, more commonly known as T-3. The drift track of this ice island was observed, either by its temporary occupants or before that by airplane, from 1947 to 1964.

One advantage of floating ice islands, besides permitting normal oceanographic observations in the polar region, is the large geographic coverage they afford. Many prominent Arctic phenomena, such as auroras, magnetic conditions, and ice drift, can be conveniently studied from these floating oceanographic platforms.

Permanently fixed platforms that are attached to the seafloor can be used to measure various oceanographic parameters such as temperature and currents. These platforms can be radar and navigational towers, piers,

FIGURE 1 FLIP, a Floating Instrument Platform. The 108 m (355 ft.) long platform was developed by the Marine Physical Laboratory of the Scripps Institution of Oceanography, University of California, San Diego. In the horizontal position (a), it is towed to a research site. Once on-station, its ballast tanks are flooded, and the platform "flips" (b), eventually becoming vertical (c). In the vertical position, FLIP is about 17 m (55 ft.) above the level of the water and is extremely stable. FLIP is used for various types of research. When work is completed, FLIP resumes its horizontal position by forcing water from its tanks. (Photographs courtesy of Scripps Institution of Oceanography.)

Box 8-2 continued

or weather and lighthouse ships. Advantages of fixed platforms are their relatively small cost in comparison with ships and their stability for long-term, uninterrupted measurements.

A different type of oceanographic instrument is the **Floating Instrument Platform (FLIP)** from the Scripps Institution of Oceanography. In its "flipped" position (see Figure 1), FLIP is very stable. Its up-and-down motion is only a small fraction of the wave motion around it. Because of its stability, FLIP has been used successfully in studies of wave movements and sound transmission in the ocean.

A type of platform commonly used in modern oceanography is the anchored buoy system (see Figure 2). Buoy systems may drift or be attached to the ocean bottom. Often one or more buoys are left at the surface as markers and to store information obtained from underwater sensors attached to the buoy line. These surface buoys can also record weather conditions, surface currents, and waves (see also Figure 8-1).

Subsurface sensors can measure currents, temperature, or other parameters. Sometimes floats or glass spheres are attached to keep the mooring line taut and to reduce the horizontal and vertical motion of the system. Subsurface buoys may contain a power supply and recorders for other sensing devices. The mooring line often ends at a release mechanism near the bottom that may be triggered on command, allowing recovery of the buoy system.

Information obtained by buoys may be collected and recorded within the system until retrieved by a surface vessel. More sophisticated buoy systems transmit data to an onshore station or relay them to an orbiting satellite.

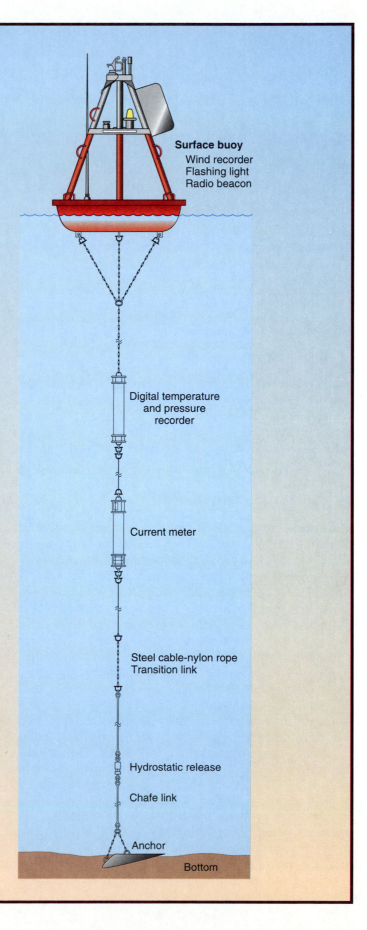

FIGURE 2 Schematic of an oceanographic buoy system (see also Figure 8-1).

FIGURE 8-4 Popup Profiler. This tripod device is lowered to the seafloor, where it releases acoustic beacons (sound-transmitting floats) that slowly rise to the surface. The position of the beacons is tracked by the profiler on the seafloor. The trajectory of the beacons as they rise is an indication of currents in the water that they pass through. Eventually the profiler is recalled electronically from the seafloor to the surface and the data are retrieved. (Photograph courtesy of Al Bradley, Woods Hole Oceanographic Institution.)

equipped with flashing lights and electronic signals), the data are recovered and the instrument is ready for reuse. With more sophisticated systems, the data can be transmitted to a satellite and from the satellite to any location, such as an office, while the instrument remains on the ocean surface (Figure 8-5).

Acoustical Measurements

A new and promising technique for measuring physical properties in the ocean is **acoustical tomography.** The name comes from the Greek word *tomos,* meaning section or slice. The principle is similar to CAT scan technology used in medicine (CAT stands for Computerized Axial Tomography). In the ocean, several transmitters (sound sources) and receivers are distributed throughout the water (Figure 8-6). Sound is then emitted and received. Numerous pathways of sound travel are possible, and the speed of sound in the water is influenced by various oceanographic characteristics, especially temperature, salinity, and pressure.

One of the first experiments with this technique used four sound sources and five receivers around a 300 km² (186 miles²) area of the ocean, thus resulting in 20 different horizontal paths for the signals to travel. By using sophisticated computer techniques and long-term observations, changes in arrival time of the sound at a particular receiver may indicate some change in physical characteristics. An example is increased flow of an ocean current, which has a different temperature and salinity than the surrounding water and hence influences the velocity of sound in the area where the acoustical tomography measurements are being made.

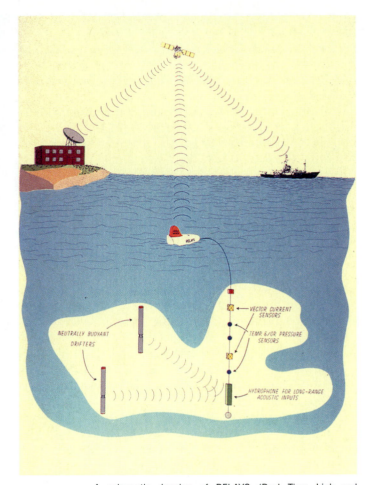

FIGURE 8-5 A schematic drawing of RELAYS (Real Time Link and Acquisition Yare System), a multisensor system. This device is used to measure several oceanographic parameters over a period of time. The data can be telemetered via satellite to a land station. The surface buoy contains a telemetry transmitter for sending the data, a data processing computer, batteries, and solar cells. Attached to the cable may be current meters and sensors to measure pressure (or depth), temperature, and conductivity (to determine salinity). The hydrophone can be used to receive or transmit sound, detect background noise, and receive data from other instruments. A satellite can track the surface buoy geographically. (Image courtesy of Woods Hole Oceanographic Institution.)

The acoustical tomography technique is new. If successful, it can avoid many limitations of conventional measurements and be relatively quick (a large area can be "sensed" in minutes). The ability to examine large areas of the ocean can be very useful in detecting ocean variations that are possibly due to global change.

General Characteristics of the Ocean

The ocean has numerous physical characteristics, but the three most important are salinity, temperature, and density. Further, these three characteristics interact. We will now look at each.

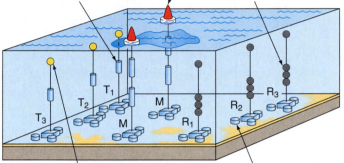

Conventional moorings (see Figure 8-1 and Box 8-2) are used to measure the general characteristics of seawater.

The transmitter sends out sound signals at a predetermined frequency, communicating with instruments at other locations.

The receiver is a hydrophone and the sound information is stored in the device.

A radio and strobe light permit easy location and recovery of the transmitters, receivers, and moorings.

Three transponders are placed at the base of each device. They transmit acoustical signals that are used to accurately measure the position of the mooring.

a

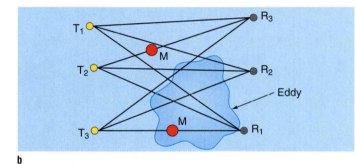

b

FIGURE 8-6 A side (a) and top (b) view from the surface of a possible acoustical tomography configuration. Sound is sent from each transmitter (T) and received by each receiver (R). The distance between the transmitters and receivers can be hundreds of kilometers. The "conventional moorings" (M) provide reference data on temperature, salinity, and currents at specific points to calibrate the tomographic measurements. The "eddy" represents a water mass having different temperature, salinity, or both from the surrounding water.

Salinity

Salinity is defined as the total amount in grams of dissolved salts in 1 kg (about 2.2 lbs.) of seawater, expressed in parts per thousand (‰) by weight. It generally is impractical to analyze for every component dissolved in seawater, so usually just a single element related to salinity is measured; chloride is the one used. The concentration of chloride in parts per thousand is measured and multiplied by a constant; the result is the water's salinity. This technique works because seawater's major components have a ratio essentially constant to each other. (Recall the Forchhammer Principle, page

158; also reexamine Figure 7-4 on page 159, which shows the main elements in seawater.)

Salinity is more frequently determined by measuring the **electrical conductivity** of seawater, a method accurate to 0.002‰. The conductivity of seawater is a measure of its ability to transmit an electrical current and is dependent on the concentration of ions (charged atoms) in the water. By comparing the conductivity of a sample of seawater with that of a seawater standard, salinity can be estimated. (Box 7-2, page 158, describes the technique for determining the salinity of seawater.)

The principle factors influencing salinity are **evaporation** and **precipitation.** This makes sense, because the more water evaporates, the more the salt in it becomes concentrated. Likewise, the more precipitation that is added to seawater, the more the salt in it is diluted. Figure 8-7 illustrates this relationship.

SALINITY AND GEOGRAPHY. A less important factor is freezing or melting of ice. When seawater freezes, salts are left behind, concentrating them in the remaining liq-

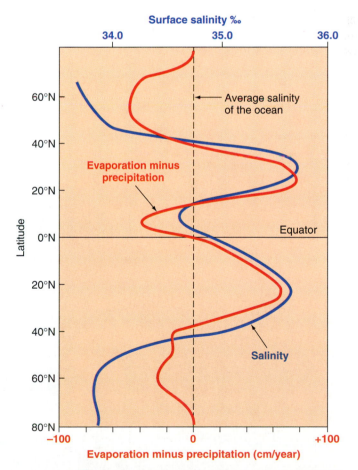

FIGURE 8-7 The vertical axis shows Earth's latitude, north and south. Plotted by latitude are the average values of surface salinity and the annual difference between evaporation and precipitation. Note how closely the average surface salinity corresponds to evaporation minus precipitation. (Data from various sources, including Wüst and others, 1954.)

uid water. In polar regions, salinities are greater in winter when ice is being formed and less in summer when the ice (mainly fresh water) melts.

Another factor is runoff from land. Salinity tends to be greater off desert regions at about 25°N and 25°S, where evaporation is high, concentrating the salt. Salinity tends to be less in rainy areas around 40° to 50°N, and in rainy parts of the tropics, because rainwater dilutes the salt (see Figures 8-7 and 8-8). The same is true in areas where large rivers enter estuaries and then the ocean. The salinity of the Gulf of Mexico, for example, is influenced by the large inflow of freshwater from the Mississippi River.

In local areas that have high evaporation, such as the Red Sea and Persian Gulf, salinity can be concentrated to as high as 40‰. In most parts of the ocean, however, the salinity range is from 33 to 37‰, with a median value of about 34.7‰. Higher salinity values occur near the arid equatorial areas and lower values near the polar regions (Figure 8-8). Actually, about 75 percent of the water in the ocean has a salinity range between 34.50 and 35.00‰, so accurate measurements are necessary to resolve small differences. Differences of a few hundredths of a part per thousand can be important for some oceanographic processes.

SALINITY LAYERS IN THE OCEAN. So far we have looked at the horizontal distribution of salinity across the ocean. Now, looking at salinity vertically, you can see that the distribution of salinity with depth occurs in four zones:

1. The surface zone, a well-mixed layer of generally uniform salinity, 50 to 100 m thick (165 to 330 ft.)

2. The **halocline,** a zone with a rapid change in salinity

3. A thick zone of relatively uniform salinity, extending to the ocean bottom

4. An occasional zone of minimum salinity in some areas, at a depth of 600 to 1,000 m (1,900 to 3,300 ft.)

These different zones are shown in Figure 8-9. Also shown are the thermocline, halocline, and pycnocline. The **thermocline** is the zone where the temperature changes more rapidly than the water above or below it. As noted, the halocline is a zone with a rapid change in salinity. The **pycnocline** is the zone where water density changes rapidly, usually due to changes in temperature and salinity. Figure 8-10 shows a typical vertical pattern of temperature and salinity, in this case across the North Atlantic.

Temperature

Temperature is probably the most commonly measured oceanographic variable. It can be measured with mercury thermometers mounted on water sampling devices, such as Nansen bottles, or by electronic devices. Sophisticated

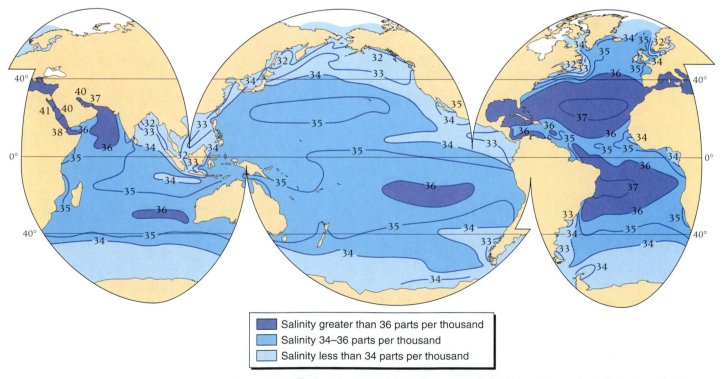

Salinity greater than 36 parts per thousand
Salinity 34–36 parts per thousand
Salinity less than 34 parts per thousand

FIGURE 8-8 Surface seawater salinity during summer in the Northern Hemisphere. Note the relatively higher values of salinity north and south of the equator and the relatively low values in the polar regions. (Adapted from Sverdrup, Johnson, and Fleming, 1942.)

FIGURE 8-9 The ocean's temperature, salinity, and density are far from uniform, and vary with depth. These variations are shown in the three graphs. The thermocline is caused by a rapid change in temperature with depth. The halocline is due to a rapid change in salinity with depth. The pycnocline can be caused either by unrelated variations in temperature or salinity, or by variations that occur in them together.

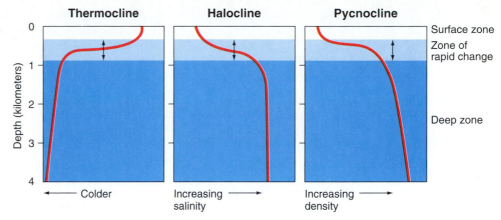

instrumentation using temperature sensors and transmitting devices can permit continuous measurements of temperature as the device is lowered through the ocean (see Figure 6-9, page 144). Other devices, such as infrared radiometers, when used from an airplane or satellite, can give instantaneous readings of the surface temperature of the sea (see Figure 1 in Box 8-1).

HEATING AND COOLING THE OCEAN. The ultimate source of heat for the ocean is the sun. However, this

FIGURE 8-10 A salinity section (a) and temperature section (b). The section runs west to east across the North Atlantic Ocean at about 24°N. Note the broadening in the scale from the top to the bottom of the figure. It is obvious from these sections that most of the variability in temperature and salinity occurs in upper parts of the ocean. In deeper water, these properties are relatively constant. The black area is the ocean bottom, and the relatively high and irregular region in the center is the Mid-Atlantic Ridge. (Adapted from Fuglister, 1960.)

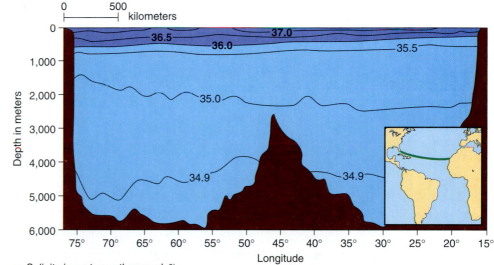

a   Salinity in parts per thousand, ‰

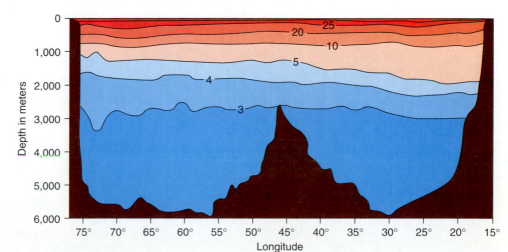

b   Temperature in degrees C

heat arrives at the ocean both directly and indirectly in three ways:

1. Direct radiation from the sun and sky

2. Conduction of heat from the atmosphere

3. Condensation of water vapor, which releases latent heat stored in the water vapor when it evaporated

The surface of the ocean is cooled in three ways:

1. Radiation from the surface back to the atmosphere

2. Conduction of heat back to the atmosphere

3. Evaporation, which removes heat from the ocean to power the evaporation process

Ocean currents can transfer heat from one area to another by bringing into contact bodies of water having different temperatures.

OCEAN TEMPERATURES. The surface temperature of the ocean is closely related to latitude and time of year. Surface temperature is related to latitude because more heat per unit area is received at the equator—where the sun is higher in the sky—than at the poles, where the sun is low in the sky or even below the horizon during winter (see Figure 9-6, page 205). Surface temperature is related to time of year because the sun is higher in the sky and there is longer daylight in the summer than in the winter.

The usual temperature pattern with depth in the ocean (Figures 8-9 and 8-10) consists of three principal layers:

1. A warm, well-mixed surface layer, from 10 to perhaps 500 m (about 33 to 1,640 ft.) thick

2. A transition layer, below the surface layer, called the main thermocline, where the temperature decreases rapidly with depth. The transition layer can be 500 to 1,000 m (about 1,600 to 3,300 ft.) thick

3. A layer up to several kilometers thick that is cold and relatively homogeneous

The character of the main thermocline varies with latitude (Figure 8-11). It is virtually absent in polar regions, where most of the ocean surface is covered with ice in winter and solar radiation is small in summer. In the tropics the thermocline may be close to the surface. Areas having a strong seasonal warming also may have a temporary or seasonal thermocline in the surface layer.

Ocean surface temperatures show a relatively straightforward pattern, with warmer surface waters in the equatorial regions and cooler waters near the poles (Figure 8-12). There are deviations from this pattern due to surface currents as well as some seasonal effects. The seasonal effects are driven mainly by variations in solar radiation (discussed further in Chapter 9).

As we shall see in Chapter 9, the ocean is constantly transferring heat from equatorial regions toward the colder poles. On the surface, this transfer is accomplished by currents that move warm water toward the poles, such as by the Gulf Stream in the Atlantic Ocean. In polar regions, relatively denser waters slowly sink and move toward the equator, forming the deeper water layers of the ocean.

A large portion of the seawater in the ocean is relatively cold; about 75 percent has a temperature below 4°C (39.2°F). Again, this is due to solar radiation heating only the upper parts of the water column, and denser colder water sinking to the bottom.

Below the main thermocline, temperature and salinity values are usually closely related. This relationship can be used to define different water types or water masses. Temperature-salinity relationships can also indicate the source and mixing of water masses and are especially important in studies of deep-ocean circulation.

Seawater has a very high heat capacity, meaning that considerable heat must be added to water before its temperature rises. Therefore, seawater does not have rapid temperature changes. This is one of the reasons why it takes so long in the summer before the ocean becomes warm enough for swimming. As a result, the ocean with its high heat capacity has a moderating influence on coastal climates.

Density

The third important physical property of seawater is **density.** Density of seawater is its mass per unit volume, for example, grams/cubic centimeter (g/cm^3). Seawater density has a narrow range, only from about 1.022 to 1.03 g/cm^3. Thus it is slightly denser than pure water,

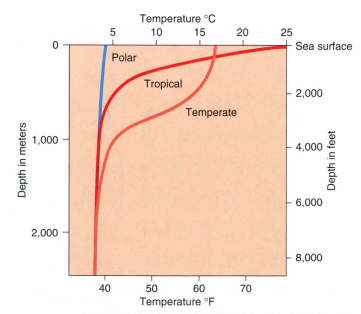

FIGURE 8-11 Temperature profiles and position of the thermocline in polar, temperate (middle latitudes), and tropical areas. (Adapted from Charnock, 1971.)

Ocean Surface Temperature (in C°) in February

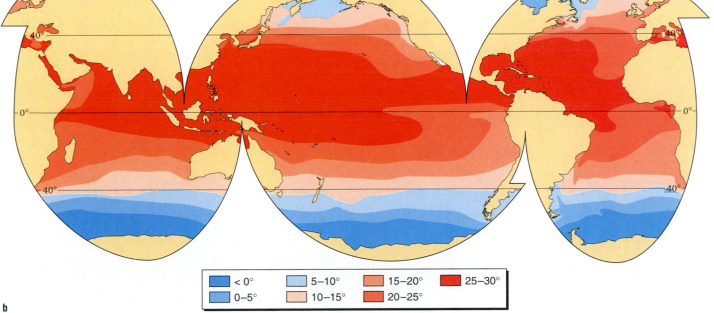

Ocean Surface Temperature (in C°) in August

FIGURE 8-12   General sea surface temperatures in February (a) and in August (b). Note that the temperature contours tend to parallel the latitude and that when it is winter in the northern hemisphere, it is summer in the southern hemisphere. (Adapted from Sverdrup, Johnson, and Fleming, 1942.)

which is the density reference at 1.000 g/cm³. Seawater density is controlled by three variables that have complex interactions: salinity, temperature, and pressure. In general:

- Density increases with increasing salinity (more salt molecules add mass).

- Density increases with increasing pressure, which is to say that density increases with increasing depth (more molecules are compressed closer together).

- Density increases with decreasing temperature (water "contracts"—its molecules move closer together at lower temperatures).

Thus, colder, deeper, more saline water is usually denser water. Seawater density can be calculated if these three variables are known precisely (see Box 8-3).

Denser water tends to sink. Less-dense water tends to rise or be pushed up to the ocean's surface. This vertical sorting or movement by density is caused by a combination of two interacting forces, buoyancy and gravity. The movement eventually forms stable density strata in the ocean (increasing density with increasing depth).

The upper 100 m or so (330 ft.) of the ocean is strongly influenced by wind and waves and therefore is well mixed and relatively uniform. Below this surface layer, large changes in temperature (the thermocline in Figure 8-9) and salinity (the halocline in Figure 8-9) produce a corresponding rapid increase in density (the pycnocline in Figure 8-9). The deep denser waters of the ocean occur below the pycnocline.

An inversion can develop in which denser water overlies less-dense water. In this situation, the water layering is unstable and overturning generally occurs. This instability can result from high evaporation of surface waters, which increases their density. The denser water eventually sinks (to be replaced by less-dense deeper water) until it reaches a zone of similar density.

The relatively large density differences that produce the stable pycnocline effectively isolate surface waters from deep waters. Polar regions are the exception, where limited solar radiation and the ice cover cause the thermocline, halocline, and consequently the pycnocline to be absent or poorly defined. In polar waters, strong density stratification does not exist. The absence of this stratification allows relatively easy mixing of the surface waters with deeper waters.

It is this easy mixing that is believed to be responsible for oxygen being carried to ocean depths. Surface waters absorb oxygen from the atmosphere in polar areas (cold waters can hold more oxygen than warm waters). This

---

Box 8-3

More about Seawater Density

Density is an extremely important parameter of seawater, because a very small variation in density can lead to important changes in ocean circulation. The density of seawater determines the position of the various water masses in the ocean (see, for example, Figure 9-20, page 217).

Density is defined as the mass divided by the volume (g/cm³). The density of pure water at 4°C (39.2°F) is 1.00 g/cm³. A typical rock has a density of about 3 g/cm³ and thus is three times denser than water.

Temperature. If seawater is warmed, its density decreases, because warming increases the distance between individual water molecules, resulting in fewer of them in a given volume. Decreasing the temperature removes energy, so the water molecules come closer together, thus increasing the density. If the salinity is kept constant, colder seawater is more dense than warmer seawater and sinks. Temperature is the main control on seawater density.

Salinity. Seawater is denser (generally averaging about 1.0289 g/cm³) than pure water because of the various salts it contains. Liquids of lesser densities float on liquids of higher densities; therefore, fresh water floats on seawater until they become mixed together. As seawater gets saltier it becomes more dense and sinks until it reaches a level having a similar density.

Pressure. Seawater density is also influenced by pressure, although this effect is smaller than the effect of salinity or temperature. Pressure due to the weight of the overlying seawater compresses water molecules, thus increasing their density. However, the difference in the pressure effect on water at the ocean surface compared to that of the deepest part of the ocean is only about 5 percent.

The pycnocline, the region of rapidly changing density with depth, is generally a barrier to the mixing of deeper high-density water with that of the lower-density surface water. (Surface water is less dense because it has been warmed by solar radiation.) The position of the pycnocline is clearly related to the zones of rapid change in salinity (halocline) and temperature (thermocline), as shown in Figure 8-9.

cold, denser, oxygen-rich water sinks and mixes with the deep water (see Figures 9-19 and 9-20, page 217).

A general summary of the surface water pattern of temperature, salinity, and density is shown in Figure 8-13.

Underwater Sound

Underwater sound is an important tool for oceanographers. It is used to measure ocean depth and to examine the character and thickness of Earth's crust. Biological oceanographers can use sound to detect and study organisms (see Figures 12-14, page 295, and 15-17, page 384). Military use of sound for submarine detection and locating objects on the seafloor has also encouraged study of underwater sound.

Sound Velocity

The velocity of sound in the ocean directly depends on temperature, salinity, and pressure (depth). Sound velocity in seawater ranges from 1,400 to 1,570 meters (4,593 to 5,151 ft.) per second. For perspective, this is very roughly 1.5 kilometers per second, or just under a mile per second, about four times faster than sound travels in the air. Thus water depth can be estimated based on sound velocity. For example, if the sound from a "pinger" on the seafloor takes 1 second to reach the surface, you know that the depth is about 1,500 m.

The velocity of sound in water increases with increasing salinity, temperature, and pressure (depth). The rates of increase are:

- *Salinity:* 1.3 m/sec. faster for each part-per-thousand increase in salinity (about 4.2 ft./sec. for each 1‰)

- *Temperature:* 4.5 m/sec. faster (approximately) for each 1°C increase in temperature (14.8 ft./sec. for each 1.8F°)

- *Depth:* 1.7 m/sec. faster for each 100 m increase in depth (5.6 ft./sec. for each 328 ft.)

As you can see, temperature has the greatest influence over the rate of sound travel in water. Although each factor causes only a small change, these changes significantly affect estimates of water depth as determined by sound velocity.

Corrections for change in salinity, temperature, and pressure must be applied to obtain accurate estimates of sound velocity. For this purpose, general correction factors have been developed for most areas of the ocean. More accurate estimates of sound velocity are possible with a **sound velocimeter,** a device that is lowered into the ocean to measure sound velocity directly, eliminating the need for corrections. Sound is also used in a new technique called *acoustic tomography* (see Figure 8-6) and to measure changes in parameters such as temperature or salinity (see Box 8-4).

The changes of sound velocity in the ocean can be divided into three zones, shown in Figure 8-14a:

1. The surface zone (about 50–100 m or 165–330 ft. thick), where the waters are well mixed. The sound velocity increases with depth in this zone mainly due to the pressure (depth) effect (because temperature and salinity stay about the same).

2. A zone where the sound velocity decreases because of rapid temperature decreases (thermocline). The slowest velocities are at about 600 m (2,000 ft.) in the Pacific and 1,200 m (4,000 ft.) in the Atlantic. The difference in the depth of these zones in the two oceans is due to subtle differences in their temperature and salinity values with depth.

3. A zone where the sound velocity increases with increasing pressure and the temperature is relatively constant

Like ocean waves, sound waves can be refracted or bent. Consequently, sound waves bend toward areas of

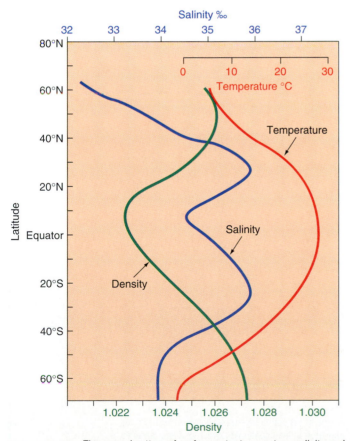

FIGURE 8-13 The general pattern of surface water temperature, salinity, and density, plotted against latitude. (Illustration adapted from Charnock, 1971, and Packard, 1964.)

Box 8-4

A Shot Heard Around the World

All observers agree that gases such as carbon dioxide are increasing in the atmosphere. The question is what the global impact of this increase will be. One of the scientific challenges in evaluating various global-change scenarios is to determine if Earth is actually warming. The obvious approach is to measure long-term temperatures worldwide, but measuring temperatures in land locations is difficult because of daily and seasonal variations and the effects of urbanization on the measuring sites.

An innovative experiment to ascertain whether the planet is warming was recently proposed by Dr. Walter Munk of the Scripps Institution of Oceanography and some colleagues. The idea is based on two points: (1) the atmosphere and the oceans are closely linked to each other, so an increase in the temperature of one would be reflected by a temperature increase in the other; and (2) the warmer the water, the faster sound travels in it.

The experiment Munk and others organized was to transmit sound from a specific location, **Heard Island**

in the Indian Ocean, to receivers at various other locations around the world. The time that the sound arrives at each station is accurately recorded and compared with similar transmissions at later times (a 10-year duration is planned for the experiment). If the overall long-term time-of-travel from all points decreases over the years (that is, the sound travels faster, on average), it could mean that the ocean is warming.

Heard Island (see Figure 1) was chosen because of its strategic location: sound transmitted from this region can directly reach receiving stations in all oceans. The distances between the transmission point and reception points are accurately known, essential to the precision of this experiment. The sound will travel in the sound channel (see Figure 8-14) and should be detectable as far away as San Francisco, over 18,000 km (11,200 miles) distant. It will take the sound slightly over 3 hours to travel from Heard Island to San Francisco. It should be emphasized that the subtle changes predicted—averaging about 0.005°C (0.009°F) per year

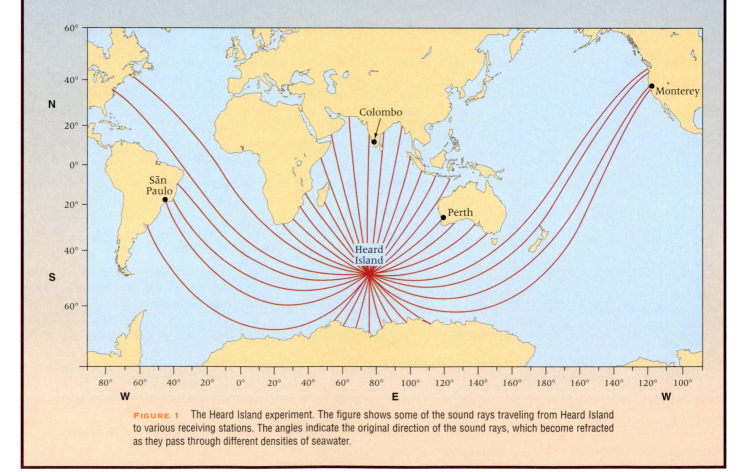

FIGURE 1 The Heard Island experiment. The figure shows some of the sound rays traveling from Heard Island to various receiving stations. The angles indicate the original direction of the sound rays, which become refracted as they pass through different densities of seawater.

Box 8-4 continued

over the ocean at depths of about 1,000 m (3300 ft.)—are very difficult to confirm with conventional ways of measuring temperature.

If the oceans are warming, as some think, the travel time from Heard Island to San Francisco could be reduced by about 0.25 seconds per year. Similar increases in travel rate should be detected at other stations. It is an interesting experiment that might help determine whether Earth is warming.

The Heard Island experiment did, in one surprising respect, create considerable controversy. This concerned whether the high level of sound would affect the mammals (especially whales and seals) in the region. A special permit is needed from the U.S. Government for experiments that might affect marine mammals. This was not initially realized by the Heard Island scientific team and led to some last-minute filing of forms. (The two government regulations involved are the Marine Mammal Protection Act and the Endangered Species Act, both concerned with injury to or harassment of organisms.)

Underwater speakers were placed at a depth of 200 m (660 ft.). The sound was pulsed on and off for one hour, then quiet for two hours, then repeated. The sound produced at the transmission site was low in frequency, similar to a foghorn, but *much* louder. At a distance of 100 m (330 ft.) from the speakers, it was estimated to be about the same as that from a jet engine at high speed (similar, perhaps, to the sound level at a very serious heavy-metal concert).

It is hard to know whether the sound bothered marine animals, but it is quite possible since many species, such as porpoises and some whales, depend on sound to navigate and to communicate with each other.

The Heard Island team eventually got their permit in early 1991 and made several modifications to their plan, but there were some hard feelings among environmental groups that special treatment was involved. This was ironic because the experiment focused on one of the main environmental issues of our time: Is Earth experiencing a global warming?

For various reasons, including environmental concerns, the next phase of the experiment, in 1994, was to involve a more modest sound level. The plan was to place loudspeakers off Hawaii and California at 1,828 m (6,000 ft.) depth and generate sound levels more like those of a chamber music group. The objective remains the same, and it will still take a decade or so to complete the experiment.

The new proposed experiment, however, drew numerous protests from various environmental groups that were concerned about the impact of the sound on marine animals (especially whales, dolphins, and turtles). Opponents of the project say that about 700,000 marine animals could be affected, some deafened. Proponents say that the sound level and frequencies should not be damaging and that the project will be closely watched and changed if there are problems. The debate has involved numerous meetings, congressional hearings and the media. The National Marine Fisheries Service, part of NOAA, is the agency that oversees the government regulations. As of early 1995, no decision has been reached.

lower sound velocity. This **refraction,** combined with vertical variation of sound velocity in the ocean, produces shadow zones and sound channels. A **shadow zone** is an area where relatively little sound penetrates. It occurs in the upper part of the ocean where the surface zone, an area of increasing sound velocity, overlies the area of decreasing sound velocity and the sound is transmitted only in the surface zone (Figure 8-14b). The reason is that sound is refracted upward in the increasing-velocity zone and downward in the decreasing-velocity zone. In both instances, sound is deflected toward an area of lower sound velocity, producing the shadow zone. The existence of the shadow zone is important for military reasons: it is very difficult to detect a submarine in a shadow zone (Figure 8-14c).

A **sound channel** can occur where the velocity of sound reaches a minimum value (Figure 8-14b). Sound traveling in this minimum-velocity zone is refracted up-ward or downward back into the area of lower velocity and thus is maintained in the minimum-velocity zone. There is little energy loss in this zone and sound can be transmitted for thousands of kilometers. The sound channel is known as **SOFAR** (for SOund Fixing And Ranging). Experimentally, sound has been transmitted through it over 25,000 km (about 15,500 mi.), or about sixth-tenths of the way around the world! This transmitting ability is the basis for the Heard Island experiment described in Box 8-4.

The SOFAR channel has a practical use for ships in distress. An explosive charge detonated in this channel by a vessel can be detected at coastal receiving stations, and the position of the vessel can be calculated from the different times at which the sound arrives at different stations.

Remember that sound is really a variety of mechanical energy that vibrates the water it moves through. As

FIGURE 8-14 (a) Sound velocity profiles in the ocean. Arrows indicate direction of increasing temperature or sound velocity. Zones 1, 2, and 3 are discussed in the text.
(b) Shadow zone and sound channel.
(c) Shadow zone formed by the refraction of sound due to changes in the velocity of sound with depth in the ocean. A submarine in the shadow zone would be very difficult to detect.

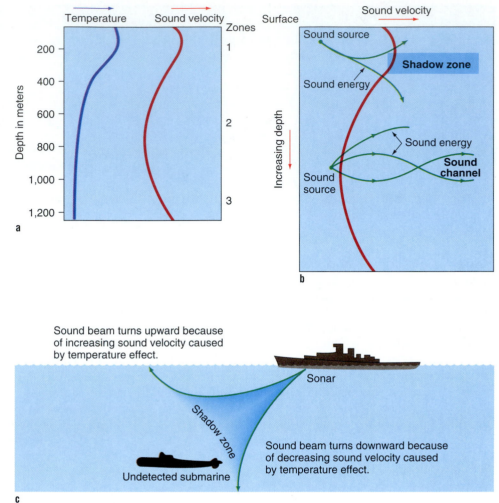

sound energy travels through water, it decreases in energy because it spreads out, is absorbed, or is scattered:

- Sound energy loss due to spreading is proportional to the square of the distance traveled (sound traveling 3 km has only one-ninth of its original energy).

- Sound energy is absorbed by the water and converted to heat. Absorption is proportional to the square of the sound frequency: the higher the frequency, the greater the absorption.

- Sound can be scattered by particles, marine organisms, gas bubbles, and the ocean bottom itself (sound is reflected from the ocean bottom).

Many potential uses of sound in the ocean will be tried in future years. For example, sound might be able to detect specific types of fish (by their size or swimming pattern), to track fish migration, or even to attract fish to an area where harvesting is easier.

SUMMARY

Physical oceanography is the measurement and study of the physical properties of the sea and how they vary over time and ocean space, combined with a theoretical evaluation of the processes that control these physical properties. As such, it probably is the most independent field of oceanography.

In spite of some recent discoveries, physical oceanographers are still limited in their knowledge of the general circulation of the ocean. Several large-scale international programs using various observational techniques, especially satellites, have been started to study the basic physical characteristics of the ocean and its general circulation. This knowledge is very important in understanding the ocean's role in any future global change.

Three basic properties of seawater influence many oceanic processes: salinity, temperature, and density.

These properties have distinctive patterns with depth and can be used to distinguish between different water masses or bodies of water. Many of the patterns of temperature and salinity are the result of processes occurring at the air-sea interface, such as heating and evaporation.

Stratification of ocean water is due to density differences. These in turn are related to salinity, temperature, and pressure (or depth). Underwater sound, an important present and perhaps future tool for marine scientists, is also influenced by temperature, salinity, and pressure.

QUESTIONS

1. What are the key physical properties of seawater, and how are they measured?

2. Why are satellites destined to play an increasing role in physical oceanographic research?

3. Describe the general characteristics of ocean water temperature and how this pattern has developed.

4. Describe the general characteristics of ocean water salinity and what factors influence it.

5. How are the halocline, thermocline, and pycnocline formed?

6. How does salinity and temperature influence density? Why is the density of seawater an important parameter?

7. Describe the pattern of sound velocity in the ocean. What factors cause this pattern?

8. What is the SOFAR channel? Why does it exist? What are some of its uses?

KEY TERMS

acoustic tomography

atmosphere

bathythermograph (BT)

density

electrical conductivity

evaporation

Floating Instrument
 Platform (FLIP)

Fram

halocline

Heard Island

hydrographic station

ice station

LANDSAT

Nansen bottle

oceanographic platform

precipitation

pycnocline

radiotelemetry

refraction

salinity

satellite

SEASAT A

shadow zone

SOFAR

sound channel

sound propagation

sound velocimeter

sound velocity

synoptic measurement

temperature

thermocline

transmitter

underwater sound

water column

XBT

FURTHER READING

Baker, J. D. 1991. "Towards a Global Ocean Observing System." *Oceanus* 34, no. 1, pp. 76–83. Describes how worldwide data obtained from satellites would be used to update weather and monitor the health of the ocean.

Baker, J. D., and W. S. Wilson. 1986. "Spaceborne Observations in Support of Earth Science." *Oceanus* 29, no. 4, p. 76–85. A general article on how satellites are used in some fields of science.

Charnock, H. 1971. "Physics of the Ocean." In *Deep Oceans*, ed. P. Herring and M. Clark. London: Praeger Press, pp. 82–120. An excellent discussion of many of the physical aspects of the ocean.

Knauss, J. A. 1978. *Introduction to Physical Oceanography*. Englewood Cliffs, N.J: Prentice-Hall. A general but slightly advanced text on physical oceanography.

Open University. 1989. *Seawater: Its Composition Properties and Behavior*. Oxford: Pergamon Press. Slightly advanced but good discussion of seawater and its properties.

"Physical Oceanography." *Oceanus* 35, no. 2 (Summer 1992). A special issue on recent research in physical oceanography.

Price, J. F. 1992. "Overflows: The Source of New Abyssal Waters." *Oceanus* 35, no. 2, pp. 28–34. A discussion of how deep water "originates" at the poles and travels throughout the ocean.

Spindel, R. C., and Peter F. Worcester. 1990. "Ocean Acoustic Tomography." *Scientific American* 263, no. 4, pp. 94–98. A good general article on tomography in the ocean.

Sverdrup, H. U., M. W. Johnson, and R. H. Fleming, 1942. *The Oceans: Their Physics, Chemistry, and General Biology*. Englewood Cliffs, N.J: Prentice Hall. Generally considered the "bible" of books on the ocean. It is complex but contains a wealth of information about the ocean.

Van Der Leeden, Frits, Fred L. Troise, and David Keith Todd. 1990. *The Water Encyclopedia*. 2d ed. Chelsea, Mich.: Lewis Publishers. All you would want to know about water on this planet.

Von Arx, W. S. 1977. *An Introduction to Physical Oceanography*. Reading, Mass.: Addison-Wesley. A solid and readable text on physical oceanography.

CHAPTER 9

Ocean Circulation

INTERACTIONS BETWEEN THE OCEAN AND THE ATMOSPHERE ARE MANY, INFLUENCING MANY BASIC CHARACTERISTICS OF BOTH THE OCEAN AND THE ATMOSPHERE AND PRODUCING A DISTINCT CIRCULATION PATTERN IN EACH. IN SOME INSTANCES IT IS DIFFICULT TO ESTABLISH WHICH IS CAUSE AND WHICH IS EFFECT. THE OCEAN AND THE ATMOSPHERE ARE ESSENTIALLY IN CONSTANT MOTION, BEING MOVED BY DENSITY CHANGES, WIND, AND EARTH'S GRAVITY. THE MAIN DRIVING FORCE FOR THE DENSITY CHANGES AND THE WIND IS HEAT FROM THE SUN.

OCEAN CIRCULATION IS IMPORTANT FOR MANY REASONS. HORIZONTAL OCEAN CURRENTS MOVE WATER AND HEAT FROM EQUATORIAL REGIONS TOWARD THE POLES. THIS HAS CONSIDERABLE INFLUENCE ON THE CLIMATE AND WEATHER OF MUCH OF OUR PLANET. VERTICAL OCEAN CIRCULATION CAN MOVE OXYGEN AND OTHER IMPORTANT GASES TO THE OCEAN DEPTHS. IT ALSO TRANSPORTS NUTRIENTS FROM DEEPER WATERS TO THE SURFACE, WHERE THEY CAN BE USED BY PLANTS IN PHOTOSYNTHESIS.

The launching of a sophisticated Swallow float in the equatorial Atlantic from the research vessel *Oceanus*. This device has glass spheres (top of instrument) that hold batteries and electronics. This Swallow float has drifted with the current in the SOFAR channel at a depth of about 3,300 m (10,800 ft.) and transmitted data to a series of listening stations for over four years (see also Figure 9-1d).

(Photograph courtesy of Philip L. Richardson, Woods Hole Oceanographic Institution.)

Introduction

When sailing ships dominated the ocean, water currents that moved toward one's destination formed the fast lanes for ship movement. Currents moving the opposite direction amounted to breakdown lanes, causing great lost time in travel. Thus, knowledge of the ocean's currents meant financial advantage and power. Benjamin Franklin made an early attempt at mapping ocean currents—recall his map of the Gulf Stream (see Figure 1-3, page 7). Later, Matthew Fontaine Maury examined numerous ship logbooks and deduced that oceanic currents are related to the wind. His observations (1855), and those of the later Challenger expedition (1872–76), provided a start toward understanding the dynamic structure of the ocean.

It was not until the German Meteor expedition (1925–27) that oceanographic studies changed from a worldwide and general focus to more localized and detailed work. Scientists on the Meteor expedition primarily studied the circulation of the South Atlantic Ocean and deduced the movement of water in the deep ocean by taking numerous measurements of temperature and salinity and observing patterns in their data.

As noted in various chapters, lack of knowledge about general ocean circulation is a fundamental impediment to understanding many aspects of the climate, biology, and chemistry of the ocean. The principal difficulty in getting this knowledge lies in observing the ocean's three vast dimensions (length, width, and depth) over sufficiently long periods. This problem was addressed in the mid-1980s when physical oceanographers developed a program called the **World Ocean Circulation Experiment,** or WOCE (see Box 9-1). Although its main results are still to be reported, considerable insight has already been gained about the complexities of ocean circulation.

Box 9-1
The World Ocean Circulation Experiment (WOCE)

The **World Ocean Circulation Experiment (WOCE)** is the largest study ever made of the ocean; its design alone required five years and involved scientists from about 40 countries. The observational phase of this experiment occurs during 1990–97. Information is being obtained from all the oceans using ships, buoys, seafloor instruments, and satellites, and the large amounts of data collected requires very sophisticated computers and data reduction schemes. Measurements being made include temperature, salinity, various radioactive tracers, current velocity at various depths, winds, air-sea interactions, and sea-level changes. The overall goals of the World Ocean Circulation Experiment are both scientific and practical:

- To understand the general circulation of the global ocean well enough to be able to model its present state and predict its evolution in relation to long-term changes in the atmosphere
- To provide the scientific background for designing an observing system for long-term measurement of the large-scale circulation of the ocean

An understanding of circulation requires knowledge of how the physical properties of seawater are changed by heating and cooling, how the atmosphere interacts with the ocean, and how Earth's rotation affects the ocean. Obtaining this knowledge requires a combination of mathematical studies and extensive and elaborate field measurements.

WOCE's results should improve our models of ocean circulation, allowing us to predict aspects of global change. A better understanding of ocean circulation and the other data being collected can be very valuable in studying ocean processes related to biology, chemistry, and geology. WOCE should show how human activities are influencing the ocean and our planet. For example, are our activities causing sea level slowly to rise, or are they causing the ocean gradually to warm?

Another large and related program is **Tropical Ocean Global Atmosphere (TOGA) program,** which as its name indicates, focuses on tropical regions and problems such as El Niño. TOGA has two main objectives:

- To investigate the tropical ocean-global atmosphere system—its variability, predictability, interactions, and the mechanisms responsible
- To investigate the feasibility of modeling ocean-atmosphere interaction, to allow prediction of climate change

With the conclusion of these programs, our understanding of ocean circulation should be considerably improved.

Measurement of Currents

The ways that ocean **currents** are measured illustrate how oceanographers use both very simple methods and advanced technology. We will examine direct and indirect methods of measuring surface and subsurface currents.

Simple Measurement of Currents

SURFACE CURRENTS. **Ship drift** is one direct method of measuring speed and direction of surface currents that requires no electronics or large expenditures. In this method, the difference between the anticipated arrival point and the actual arrival point of a ship is assumed to be due to currents (Figure 9-1a). In reality, other factors, such as the wind, also can influence the actual arrival point.

Another simple, direct method of measuring currents is to put bottles into the water and note their movement. Or they may be left to drift freely unobserved (Figure 9-1b). Commonly called a **drift bottle**, a card is placed inside it offering a small reward to any finder who returns the card and tells where and when the bottle was found. Floating devices like drift bottles have certain disadvantages. For example, neither their exact route, the time it took them to get to where they are found, nor the influence of the wind or other factors on their drift can be known. Nevertheless, by plotting the launchings and recoveries of many drift bottles, a general notion of large-scale surface oceanic circulation can be gained.

DEEPER CURRENTS. A slightly more elaborate device is a **drogue,** which floats below the surface and can be moved by currents. The drogue is placed slightly below the surface to eliminate the direct effects of the wind (Figure 9-1c). Usually attached to some kind of float, it can be tracked from a surface ship by radar, radio direction finder, or even satellite if the drogue is equipped with a transmitter. A parachute can also be used as a drogue.

A very sophisticated type of drogue is the **Swallow float,** named for its inventor, British oceanographer John Swallow. The Swallow float is a sealed aluminum tube that is dropped into the ocean (Figure 9-1c). It is slightly heavier than seawater, so it sinks. However, it is less compressible than seawater, so at a certain depth its density equals that of the seawater. At that depth, which can be predetermined, the device floats and moves with any current that is present. (In other words, the density of seawater increases with depth more than the density of the Swallow float increases with depth.) If a sound-emitting device is attached to the Swallow float, the float's movement can be monitored by tracking the sound. A Swallow float can be *delicately* balanced to float in the **SOFAR channel** (see opening figure to this chapter), thus allowing its signals to be received at land stations. These floats are frequently called SOFAR floats.

If a surface ship is doing the monitoring, the ship itself may drift with the current being monitored. Thus the ship must have some method of locating its own position relative to the current measuring device. In the past, one way to solve this problem was to anchor a buoy to the seafloor to provide a reference location for the ship. Today, navigation satellites are commonly used to determine the ship's position.

Measurements from Swallow and SOFAR floats show that the deep ocean, once thought to be relatively motionless, can have currents moving as fast as almost 50 cm (20 in.) per second, or about 1 nautical mile per hour (1 knot). However, considerably slower speeds, around 1–2 cm/sec., are more typical of the deep ocean (Figure 9-1d).

A rather clever way of indirectly measuring surface currents is to note the position and drift patterns of abandoned derelict sailing vessels and other drifting objects. Philip L. Richardson, a physical oceanographer, has examined old records (Figure 9-2a shows one from 1888) and plotted a trajectory of drifting derelicts during the period 1883–1902 (Figure 9-2b). These plots show the general circular pattern of surface circulation in the North Atlantic. They also add some historical insight to oceanography.

Modern Techniques for Measuring Currents

More modern types of current-measuring devices are **telemetering buoys, free-fall devices,** and **current meters** (Figure 9-3). Current-measuring devices can be suspended from fixed objects such as buoys or lightships and can measure the speed and direction of the passing currents over time. More sophisticated devices transmit the data to shore-based laboratories or computers.

Current-measuring instruments can record either the average current velocity over a long period of time or the velocity at a given moment. This is important because currents vary. A current-averaging device may show an average flow northward at 50 cm/sec., whereas the current may actually have been moving at various speeds part of the time, and even in other directions.

To get a better picture of the variability of the ocean, more sophisticated devices are combining different measurement techniques with remote sensing and telemetry. These devices include neutrally buoyant drifters (Figure 9-4, similar to Swallow floats), moored devices with current meters or other sensors, surface ships, satellites, and the electronic telemetering of data to a shore laboratory (see Figure 8-5, page 186). Such systems can give a multidimensional picture of ocean variables over a period of time. The use of several of these systems will give even more extensive coverage.

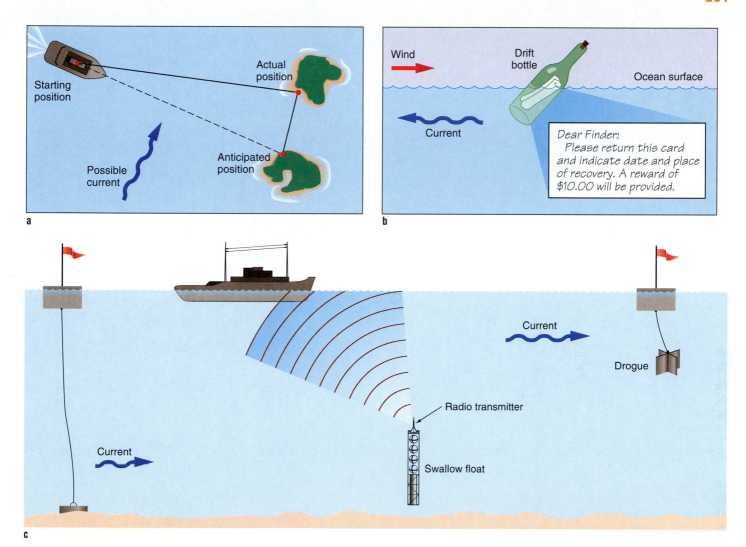

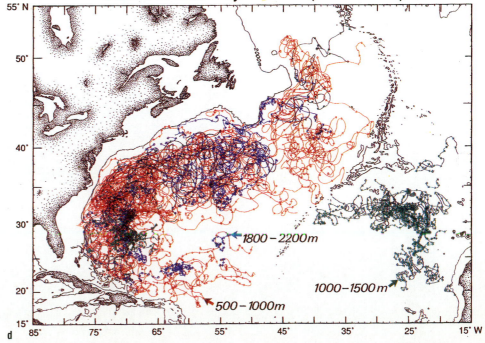

FIGURE 9-1 Ship drift (a), drift bottle (b), drogue and Swallow float (c), are some direct methods of measuring currents. In the last method, the ship maintains its position relative to a buoy and observes the movement of the drogue (visually) or the Swallow float (electronically). The opening figure to this chapter shows the launching of a sophisticated type of Swallow float.

The confusing pattern of lines (d) shows the trajectories of Swallow floats released into the North Atlantic during the period 1972–89. The colors indicate the depth of the floats. The arrowheads show the float position every 30 days. Most of the trajectories are concentrated around the Gulf Stream and show the varieties of meanders and eddies typical of that region. The pattern on the eastern side of the Atlantic (right-hand side of figure) represents water flowing from the Mediterranean Sea into the Atlantic. (Illustration courtesy of Philip L. Richardson, Woods Hole Oceanographic Institution.)

FIGURE 9-2 (a) Part of an 1888 U.S. Navy Pilot Chart for the North Atlantic Ocean. The purpose of the chart was to help sailors avoid collision with various wrecks and drifting material. Shown are the locations and dates at which derelict ships were sighted. The condition of the ships is shown schematically: stern down, bottom up, and so on. Drifting buoys are shown as small circles with a dot. Steamship routes are depicted, as are fog belts (stippled area extending east of Cape Cod).

(b) A composite of the trajectories of drifting derelict ships and buoys noted from monthly charts published during the interval 1883–1902. Study of this maze reveals the general clockwise circulation pattern of the North Atlantic. (Both photographs courtesy of Philip L. Richardson.)

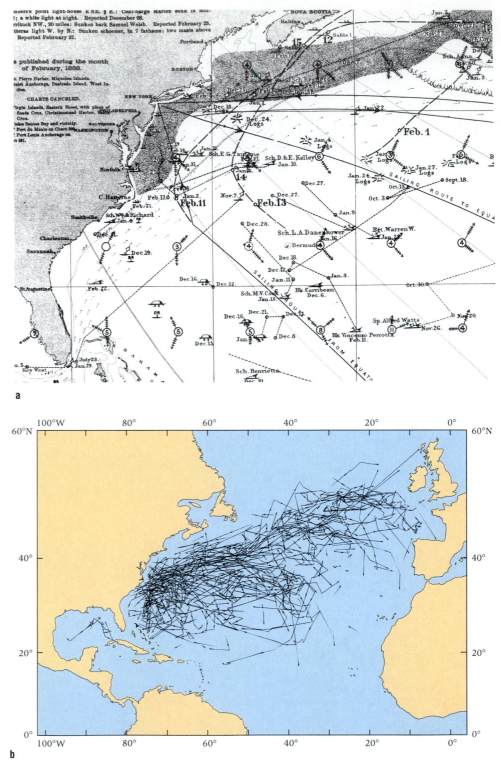

a

b

c

FIGURE 9-3   (a) A lineup of vector-averaging current meters. They can be attached to a mooring line of a buoy at selected depths to measure and record current speed and direction.

(b) The measuring part of the current meter. Water moving by the rotor causes it to turn; the number of turns per unit time is a measure of the current flow rate, which is recorded inside the instrument.

(c) Biological growth on a current meter left on the New England continental shelf for 6 months. (Photographs (a) and (b) courtesy of Woods Hole Oceanographic Institution; photograph (c) courtesy of Shelley Lauzon, Woods Hole Oceanographic Institution.)

FIGURE 9-4 A vertical current meter, a neutrally buoyant, free-floating instrument to which ballast is added or removed to make it sink or rise to a predetermined depth and return. As it sinks and returns, the instrument measures pressure, temperature, and vertical current velocity. The picture shows the device just before release. (Photograph courtesy of Nancy Brink, Woods Hole Oceanographic Institution.)

Of course, there are several problems—both environmental and human—with leaving instruments unattended in the ocean. Biological growth can reduce the capabilities of the instrument (Figure 9-3c). Storm damage, wave damage, and corrosion can take their toll on delicate scientific instruments exposed to the elements. And, unfortunately, theft and vandalism occur.

Indirect Measurement of Currents

Indirect methods of determining currents include measuring temperature, salinity, oxygen content, or several other properties of seawater. If these properties are strongly influenced or determined by surface phenomena, their character or quantity at depth or along a distance from a source can be an indication of oceanic mixing and currents.

The distribution of certain organisms can also indicate current directions. The presence of floating organisms in an area far from their usual habitat suggests that they were transported there by currents.

Physical oceanographers, like other oceanographers, use modern instruments as well as the more conventional devices just described. Satellites, aircraft, and large buoys laden with electronic instruments provide data about various oceanic parameters. Analysis of these data is commonly done by high-speed computers, in many cases directly on research vessels.

Now we turn our attention to the energy that our planet receives from the sun. This energy ultimately drives the circulation of both the ocean and the atmosphere.

Solar Energy

Radiant energy from the sun is absolutely crucial to Earth's physical systems and to life on Earth. As this **solar energy** enters our atmosphere, some is reflected, while some is absorbed and changed into other forms of energy (Figure 9-5). Because the ocean covers 71 percent of Earth, it receives the major amount of incoming solar energy (sunlight).

In spite of the numerous complex processes affecting incoming **solar radiation,** our planet's temperature has remained relatively stable over long periods. This indicates that the incoming solar radiation and the outgoing radiation from Earth are in balance. Without this balance, Earth would either heat up or cool off, resulting in rather dramatic effects to the ecology of our planet.

It is very important that incoming solar energy does not reach Earth's surface uniformly. The equatorial areas receive more solar energy per unit area than the poles. The reason is clear from Figure 9-6. Solar energy strikes Earth nearly head-on at the equator, concentrating its energy over a smaller area. Toward the poles, however, the same energy is spread over a larger area.

The incoming energy from the sun is also affected by the atmosphere, which both absorbs and reflects portions of the solar energy received. In polar regions, the energy must travel through a greater thickness of atmosphere, causing less to reach the surface (Figure 9-6). These two facts—angle of solar radiation and thickness of atmosphere—ensure that the equatorial regions receive the most heat per unit area, the temperate regions somewhat less, and the poles the least.

Earth's Heat Budget

Considering the entire planet, the amount of heat received is balanced by the amount of heat eventually lost back into space. Without such a balance, Earth would

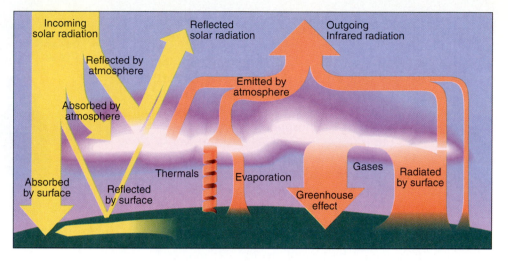

FIGURE 9-5 The heat budget for our planet. This illustration shows the fate of incoming solar radiation. Some of this energy is absorbed by the atmosphere, land, and oceans, and some is also reflected back into space. Greenhouse gases in the atmosphere (discussed in more detail in Chapter 12) are especially efficient in trapping radiation and may be heating our atmosphere. The widths of the arrows indicate the relative amounts of energy. (Illustration from Mix, Farber and King, 1992, *Biology: The Network of Life* [New York: HarperCollins].)

become either cooler or warmer. The various gains and losses in heat are viewed by scientists as a **heat budget,** with gains in one place balanced by losses elsewhere.

In the ocean, the heat budget works like this:

1. Heat is *added* to the ocean by radiation from the sun and the atmosphere, by conduction from the atmosphere, and by the condensation of water vapor (recall the *latent heat of condensation* from Figure 6-8, page 139).

2. Heat is *lost* when it is absorbed by water evaporating from the ocean (recall the *latent heat of evaporation* from page 139).

3. Heat is *lost* when it is both radiated and conducted from the sea surface back into the atmosphere.

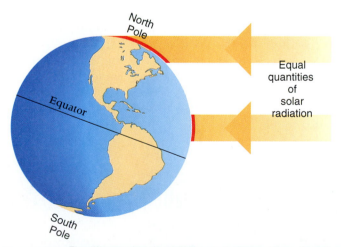

FIGURE 9-6 The difference in solar radiation received per unit area for polar and equatorial regions. In the equatorial region the sunlight falls perpendicular onto the surface, thus concentrating its warmth in a relatively small area, compared to the polar regions, where a similar amount of sunlight falls over a much larger area. In the simplest terms, this is why the equator is hot and the poles are cold.

These processes all occur in the upper parts of the ocean. Heat can be moved from one part of the ocean to another by currents and other forms of water motion. An excellent example is the **Gulf Stream,** which carries warm water along the East Coast of the United States across the northern Atlantic Ocean, where it warms areas of northern Europe, including the British Isles.

The rate of these processes varies with the solar energy input, which in turn varies with Earth's movements relative to the sun—day/night and the seasons. It is common sense that more heat is lost than gained at the polar regions, and that the converse is true for the equatorial regions. Satellite observations have confirmed this. Currents, winds, and general ocean circulation move heat from the equator toward the poles and maintain the present surface temperature pattern.

Water has one of the greatest **heat capacities** of any substance, meaning that it can absorb and hold considerably more heat than can land or the atmosphere. Seawater, for example, holds about 4,000 times more heat than the atmosphere. Indeed, the upper 3 m (about 10 ft.) of the ocean holds more heat than all the atmosphere. This great heat capacity means that water is slow to heat and slow to cool, making the ocean's temperature relatively stable in a world where temperatures fluctuate rapidly with day and night.

The temperature of the ocean, therefore, is relatively stable day to day, whereas land temperatures generally change considerably over a 24-hour period. Hence the heat-storing capacity of the ocean is very important in modifying and influencing continental climate, an influence that can be observed on the west coasts of land in intermediate latitudes of the Northern Hemisphere, such as California and England. Here, the dominantly onshore winds transport warm air from the sea to the land.

Atmospheric Circulation

As noted, the atmosphere receives a larger quantity of solar radiation per unit area in the equatorial regions than in the polar regions. To maintain Earth's heat balance, some of this heat is transferred to the higher latitudes by movements of the atmosphere.

If Earth were not rotating, a simple circulation would probably exist between the equator and the poles (Figure 9-7). The air at the equator would be heated and would expand; this expansion would lower its density, so it would rise. The rising would create a low-pressure area at the surface. Cooler air from the surrounding area would move toward the low-pressure area at the equator and in turn would be heated, would expand, and rise. At the poles, the air would be cooled, would contract, and sink, creating a high-pressure area. The cold, dense polar air would move toward the equator (from high to low pressure). This would result in a steady wind from the north in the Northern Hemisphere, and from the south in the Southern Hemisphere.

You can visualize high-pressure and low-pressure areas, both for water or air, as something like a hill. The high-pressure area is at the "hilltop." The steeper the hill, the faster is the motion (air or water) from the hilltop to the bottom. The steeper the hill, the greater the pressure gradient. Likewise, the steeper the hill, the stronger is the wind that created the pressure "hill" in the first place.

This simple "one-cell" model of atmospheric circulation would be fine if Earth did not rotate. But it does, so the model must be expanded into a more realistic and complex three-cell version, because Earth's rotation deflects circulating air. This deflection—which results in the air turning toward the right in the Northern Hemisphere and toward the left in the Southern Hemisphere—is called the **Coriolis effect,** named after the French mathematician Gaspard Gustave de Coriolis, who described it. The Coriolis effect is an important and complex aspect of ocean circulation, as described in Box 9-2.

Thus, the atmosphere does not simply circulate in a straight line along the pressure gradient from a high-pressure area (poles) to a low-pressure area (equator). Instead, air circulation is deflected by Earth's rotation (Coriolis effect) into curved paths (Figure 9-8). It is important to remember that the wind is not *caused* by the Coriolis effect; it just is *deflected* by it.

In the Northern Hemisphere, rising warm air from the equator starts to flow north toward the pole. As it leaves the equator, it is deflected toward the right, cools, and eventually descends at about 30°N latitude. Part of this descending air completes the cell by heading southward, toward the equator (causing the so-called

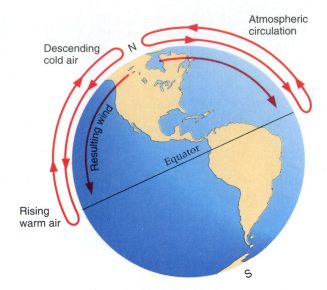

FIGURE 9-7 If Earth did *not* rotate on its axis, this is the probable pattern of atmospheric circulation and resulting winds.

trade winds). The remaining air continues northward toward the pole, causing the *westerly winds* or **westerlies.** (When meteorologists refer to a westerly wind, they mean that it blows *from* the west; when oceanographers refer to a western-flowing current, they mean that it goes *toward* the west.) This air is eventually warmed, rises, and is also deflected to the right, forming a second cell.

In the north polar region, the air is cooled, becomes denser, sinks, and travels southward until it is heated sufficiently to rise again at about 60°N, causing the **polar easterlies.** The resulting winds and circulation cells are shown in Figure 9-8. Areas of rising or sinking air generally have calm winds, such as the doldrums around the equator and the horse latitudes at about 30°N and 30°S. Areas where air travels along Earth's surface generally have steady and intense winds, such as the westerlies and the trades. A synoptic view of worldwide wind patterns, obtained by satellites' observations, is shown in Figure 9-9.

Atmospheric Circulation Drives Ocean Circulation

The wind system just described is key in generating surface ocean currents. The winds stress the ocean's surface and the water responds by moving. The result is the **wind-driven circulation** of the ocean (Figure 9-10). As we shall see, however, the direction of currents is in-

Box 9-2

The Coriolis Effect

The **Coriolis effect** (also called the Coriolis deflection) is a very important concept in understanding the circulation of the atmosphere and ocean water. It can be a complex concept to understand, but it results simply from the fact that the planet is rotating. Earth rotates in an easterly direction, counterclockwise if viewed from above the North Pole. The Coriolis effect is a deflection or change in direction of a freely moving object such as the atmosphere, the ocean, aircraft, rockets, birds, baseballs, and so on. The magnitude of this deflection increases with the velocity of the object and with latitude. The effect is zero at the equator, increasing to a maximum at the poles.

What is observed is determined by the viewer's frame of reference. To an observer in the Northern Hemisphere, standing on the ground, a freely moving object appears to be deflected to the right of its direction of motion. (In the Southern Hemisphere, the deflection is to the left.) As the object travels, Earth is rotating beneath it. By contrast, to an observer in space, the object appears to travel in a straight line.

Have you ever been in a car next to another one, when one starts slowly to move? Without an additional reference point, it may be very hard to tell whether you are in the moving vehicle or the station-ary one. Let's examine this further. (For the rest of this discussion, assume that we are in the Northern Hemisphere with the right-curving situation.) Look at Figure 1. Imagine that you are at the North Pole and shoot a harmless rocket to land at the equator. Imagine also that the trip will take two hours. If Earth did not rotate, the rocket would simply travel to its destination.

However, as the rocket travels toward the equator, Earth is rotating. It rotates completely (360°) in 24 hours. Therefore, in one hour, it rotates 15° (360° divided by 24 hours = 15°). So, in two hours, Earth rotates 30° to the east. When the rocket reaches the equator, *it will arrive 30° west of where it would have hit if Earth were not rotating.* Earth has turned beneath your rocket as it traveled.

To our observer on the ground, it appears the rocket has been deflected to the right. Actually, however, it traveled in a straight line, and it is Earth's turning that produced the apparent deflection. If the rocket were fired the other direction, toward the North Pole from the equator, a similar thing would happen.

This is the Coriolis effect. It has a similar, if less dramatic, impact on slower-moving atmosphere and ocean currents. The effect, however, is extremely important.

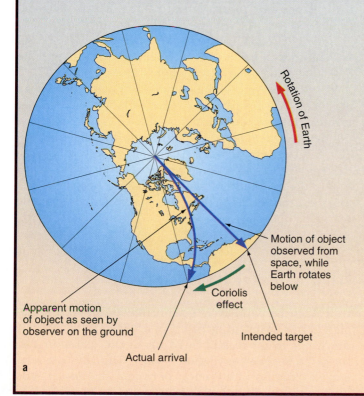

Rotation of Earth

Motion of object observed from space, while Earth rotates below

Coriolis effect

Apparent motion of object as seen by observer on the ground

Intended target

Actual arrival

a

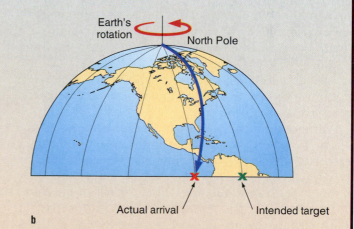

Earth's rotation

North Pole

Actual arrival

Intended target

b

FIGURE 1   The Coriolis effect or deflection is caused by Earth's rotation. The deflection makes any freely moving object travel in a curved path relative to the Earth's surface. [This example shows a polar view (a) and a side view (b).] In this example, a rocket is shot from the North Pole toward the equator. If the trip takes two hours, the rocket will land 30° off target (see text).

60° N

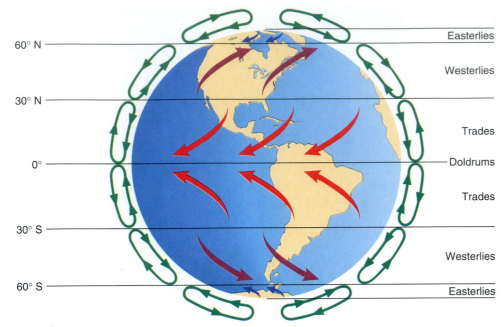

fluenced by other factors. For the following description, compare the patterns of atmospheric circulation in Figure 9-9 and ocean circulation in Figure 9-10.

The easterly trade winds form **equatorial currents,** which are common to all oceans. In the Atlantic and Pacific Oceans these currents are intersected by land, which complicates the pattern, deflecting currents to the north and south. These deflected currents travel along the western parts of the oceans. Called **western boundary currents,** they are among the largest and strongest ocean currents. One, the Gulf Stream, transports water at the astonishing rate of about 100 million m³/sec., more than 100 times the combined outflow of all rivers of the world. Table 9-1 shows the velocity and volume of some major ocean currents.

Western boundary currents are driven across the ocean by westerly winds. They form **eastern boundary currents,** which flow back into the equatorial region, thus completing large **gyres,** or circles. Large-scale gyres of this type occur in the subtropical regions of the North and South Pacific, the North and South Atlantic, and the south Indian Oceans (obvious in Figure 9-9). Smaller and weaker gyres exist in the northern subpolar regions of the Atlantic and Pacific Oceans.

No gyres occur in the southern subpolar region, probably because no land barriers obstruct water flow to create a gyre. Therefore, the **Antarctic Circumpolar Current,** also called the **West Wind Drift,** flows completely around the world (see Figure 9-9).

An interesting gyre occurs in the north Indian Ocean. Its rotation changes direction every 6 months due to half-yearly reversals in the atmospheric circulation pat-

tern, called **monsoons** (Figure 9-11). In fall and winter, the northwest trade winds prevail. The pattern reverses in spring and summer and prevailing winds are from the south and southwest. This results from a change in the location of the doldrums: northward in winter and southward in summer. In turn, this shift is caused by the difference in heat capacity of land and water. In spring, the land becomes relatively warmer; in fall, the reverse is true. These relative changes in temperature shift the atmospheric circulation cell between the equator and the 30° latitudes, North and South. This in turn changes the direction of the ocean circulation.

The interaction of the atmosphere with the ocean produces two distinct types of circulation. One is the **wind-driven circulation** just discussed. The other is the density circulation, or **thermohaline circulation.** The wind-driven circulation is stronger, and its significance is restricted to about the upper 1,000 m (3,300 ft.) of the ocean. Thermohaline circulation, on the other hand, extends into the deep sea. We now shall examine each type of circulation.

Wind-Driven Circulation

How does the wind drive ocean circulation? In the simplest terms, wind drags on the water surface, causing the water to move and build up in the direction that the wind is blowing (Figure 9-12a). This, in turn, creates a pressure difference between the lower and higher areas, with the higher pressure at the downwind

FIGURE 9-9 The sea-surface wind field over (a) the Pacific Ocean and (b) the Atlantic and Indian Oceans. The data from which these images were generated were obtained by a radar scatterometer flown on the *SEASAT A* satellite. Three days of data are shown. The arrows point in the direction the wind is blowing; the longer the arrow, the greater the wind speed. Light winds (less than 4 m/sec., or 9 miles/hour) are shown in blue. Strong winds (greater than 14 m/sec., or 31 miles/hour) are shown in yellow. (Illustrations courtesy of NASA.)

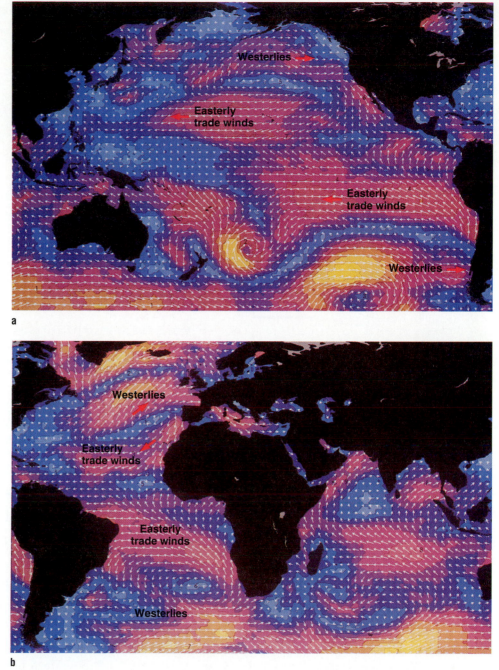

end, where the water has "piled up." This water-pressure difference wants to push the surface water back down toward the region of lower pressure. In other words, the water has been forced by the wind to "flow uphill," but gravity pulls it back down the slope, as shown in the illustration.

The Coriolis effect applies to moving water, just as it applies to moving air. In the open ocean, surface water is deflected by the Coriolis effect and moves at an angle of 45° to the wind (Figure 9-12b). This deflection is to the right in the Northern Hemisphere. Oceanwide, a balance generally exists between the effect of the pressure gradient (or gravity) on the moving water and the deflection of the Coriolis effect.

When this balance exists, no further deflection of the moving water occurs, and the current flows around the high-pressure area, forming a gyre (Figure 9-12c). (Remember that water cannot flow uphill; and it will only flow to the right in our Northern Hemisphere example, due to the Coriolis effect.) When a balance

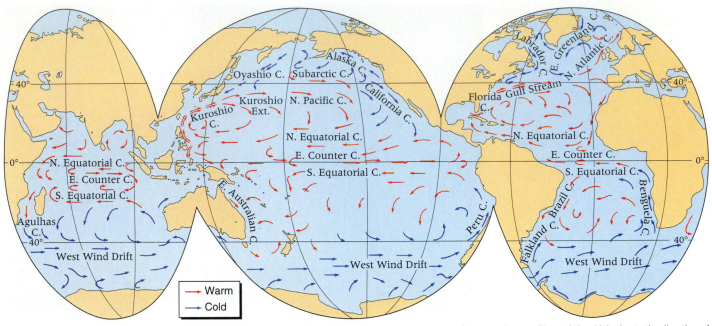

FIGURE 9-10 Major surface currents of the ocean. Compare these to Figure 9-9, which shows the direction of surface winds over the ocean. Is the relation between atmospheric circulation and ocean circulation evident? Note that the Indian Ocean gyre is shown in its winter condition; in summer, it rotates in the opposite direction. (Adopted from Sverdrup and others, 1942.)

exists between pressure-gradient forces and the Coriolis effect, the resulting currents are called **geostrophic currents.** The continents form the outer boundaries of these ocean-circling currents, or oceanwide gyres. The gyres occur only north and south of the equator. They cannot exist at the equator, where the Coriolis effect is zero.

Each of the five existing ocean-wide gyres is dominated by a relatively strong and narrow western boundary current, with a relatively weak and broad eastern boundary current on the other side. The gyres and boundary currents are:

Gyre	Western/Eastern Current Pair
North Atlantic	Gulf Stream and Canary Current
South Atlantic	Brazil Current and Benguela Current
North Pacific	Japan or Kuroshio Current and California Current
South Pacific	East Australian Current and Peru or Humboldt Current
Indian Ocean	Agulhas Current and West Australian Current

The western boundary currents are stronger for two main reasons: (1) the driving force of the strong trade winds, and (2) the fact that the central high-pressure area tends to be offset toward the western part of the ocean basin, due to Earth's easterly rotation. The water flowing around the high-pressure area therefore is squeezed closer to the western edge of the ocean basin.

Ekman Spiral

When the surface water molecules are moved by wind, they in turn move, by friction, water molecules below them, and so on to a depth of about 100 m (330 ft.) or slightly deeper. The frictional forces are such that the

TABLE 9-1

Velocity and Transport of Some Major Ocean Currents

Current	Maximum Velocity (cm/sec)	Transport (millions of m³/sec.)
Antarctic Circumpolar (West Wind Drift)	—	100
Gulf Stream	200–300	100
Kuroshio	200	50
North Equatorial Pacific	20	45
Equatorial Undercurrent	100–150	40
Peru or Humboldt	—	20
Brazil	—	10

Source: Data from Warren, 1966, and others.

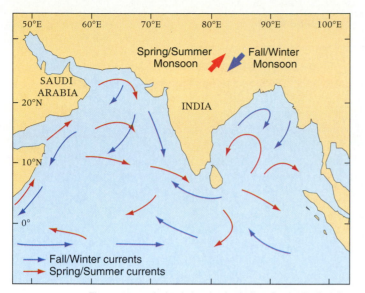

FIGURE 9-11 The monsoonal circulation in the Indian Ocean. During the winter and fall winds from the northeast dominate. The pattern reverses in the spring and summer when the winds are from the southwest. This changing wind pattern causes a corresponding reversal in the oceanic circulation.

motion becomes considerably slower with depth, generally ceasing by about 100 m depth. The water below the surface, however, also is deflected by the Coriolis effect. The net result is a spiral, in which each successively lower layer moves more to the right than the layer above it (or to the left in the Southern Hemisphere).

The net effect is that the slow deeper current is "twisted around" to where it may actually flow opposite to the surface current (180°). This spiraling current is called the **Ekman spiral** (Figure 9-13), named after the

Swedish scientist who theorized it. Because of this spiraling, the *net* or average flow or transport of water is about 90° to the direction of the wind (to the right or left, depending on whether you are in the Northern or Southern Hemisphere). This net flow of water is called the **Ekman transport.**

In summary, the surface waters move at 45° to the wind, the net transport is at about 90° to the wind, and at some depth, not exceeding 100 m, the current may actually be moving 180° opposite that of the surface current, although considerably slower.

Countercurrents and Undercurrents

Between the great northern and southern gyres of the oceans, **countercurrents** flow near the equator. They flow easterly, opposite to the westerly flowing equatorial currents of the gyre on either side of them. Some of these countercurrents flow slightly below the surface and are called **undercurrents.** One of these, the Pacific Equatorial Undercurrent, flows from west to east at the equator 100 m (330 ft.) or deeper below the ocean surface. Also called the Cromwell Current, this current is about 300 km (185 miles) wide, only a few hundred meters thick, and has velocities as great as 3 knots (150 cm/sec.). The origin of undercurrents and their relationships to the overlying water are not fully understood.

Eddies and Rings

Eddies are large circular flows of water that can occur along the edge of an ocean current (Figure 9-14). Some

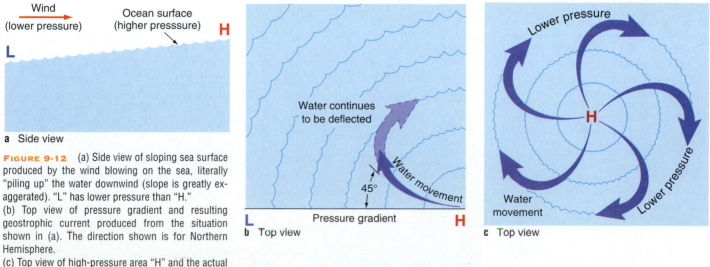

FIGURE 9-12 (a) Side view of sloping sea surface produced by the wind blowing on the sea, literally "piling up" the water downwind (slope is greatly exaggerated). "L" has lower pressure than "H."
(b) Top view of pressure gradient and resulting geostrophic current produced from the situation shown in (a). The direction shown is for Northern Hemisphere.
(c) Top view of high-pressure area "H" and the actual motion of the surface water being driven by the wind (in the Northern Hemisphere). The net result is for the water to flow around the high-pressure area forming a gyre.

FIGURE 9-13 The Ekman spiral, produced by wind blowing on the ocean surface. Surface waters will move at 45° to the wind. The motion is to the right in the Northern Hemisphere and to the left in the Southern Hemisphere. Deeper layers also are progressively deflected relative to the layers above them, although they move more slowly. Eventually the flow direction at depth can become 180° to the flow of the surface water. The average or net transport is 90° to the wind direction.

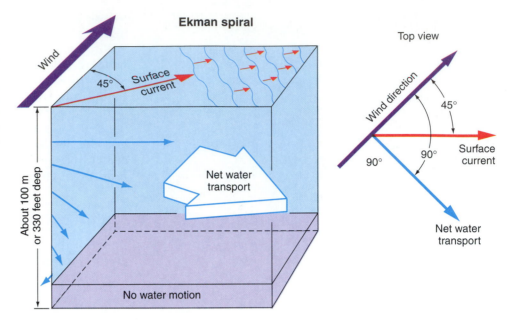

FIGURE 9-14 Satellite picture of eddies in the central Mediterranean Sea taken from the space shuttle *Challenger*. This picture was taken by a civilian oceanographer from a height of 45 km (28 miles), and covers an area 210 km (130 miles) on a side. The sunlight was reflected toward the camera, producing a surprising sunglint pattern. In the central portion of the photograph the smooth surface of the sea reflects light like a mirror and appears bright. In regions where surface waves roughen the surface, sunlight is scattered, so those areas appear darker. Away from the strongly illuminated central portion of the figure, however, the pattern reverses, with smooth regions appearing darker and rougher regions appearing lighter. In regions, 1, 2, and 3, spiral eddies are clearly visible. The wakes of several ships (4) can be seen as horizontal lines crossing the surface. In some instances the wakes are displaced (5), indicating the presence of currents. (Photograph by Paul Scully-Power; courtesy of National Aeronautics and Space Administration.)

can be about 200 km wide (about 125 miles) and last about 2 months. Several types have been observed, including ones that occur only below the ocean surface. In the Pacific, eddies have been identified up to 1,000 km (600 miles) in diameter.

These water masses are similar to weather systems in the atmosphere. They have been compared to underwater storms with "water winds" of several tens of centimeters per second. They can affect large areas for periods of several months. In doing so, they are moving water, essentially unmixed with the surrounding water over large horizontal distances. One key question, still unan-

swered, is how these eddies interact with other large-scale oceanic circulation patterns.

Somewhat similar to oceanic eddies are large ringlike features first discovered along the Gulf Stream. These are large loops or meanders that pinch off from the Gulf Stream and form circular **rings** of water having strong currents within them (Figure 9-15). As a ring forms, a column of water is captured from one side of the current and carried to the other side. The water in the ring then has qualities of salinity, temperature, oxygen content, and other properties from its original side that are different from its new surroundings. Like eddies, these rings

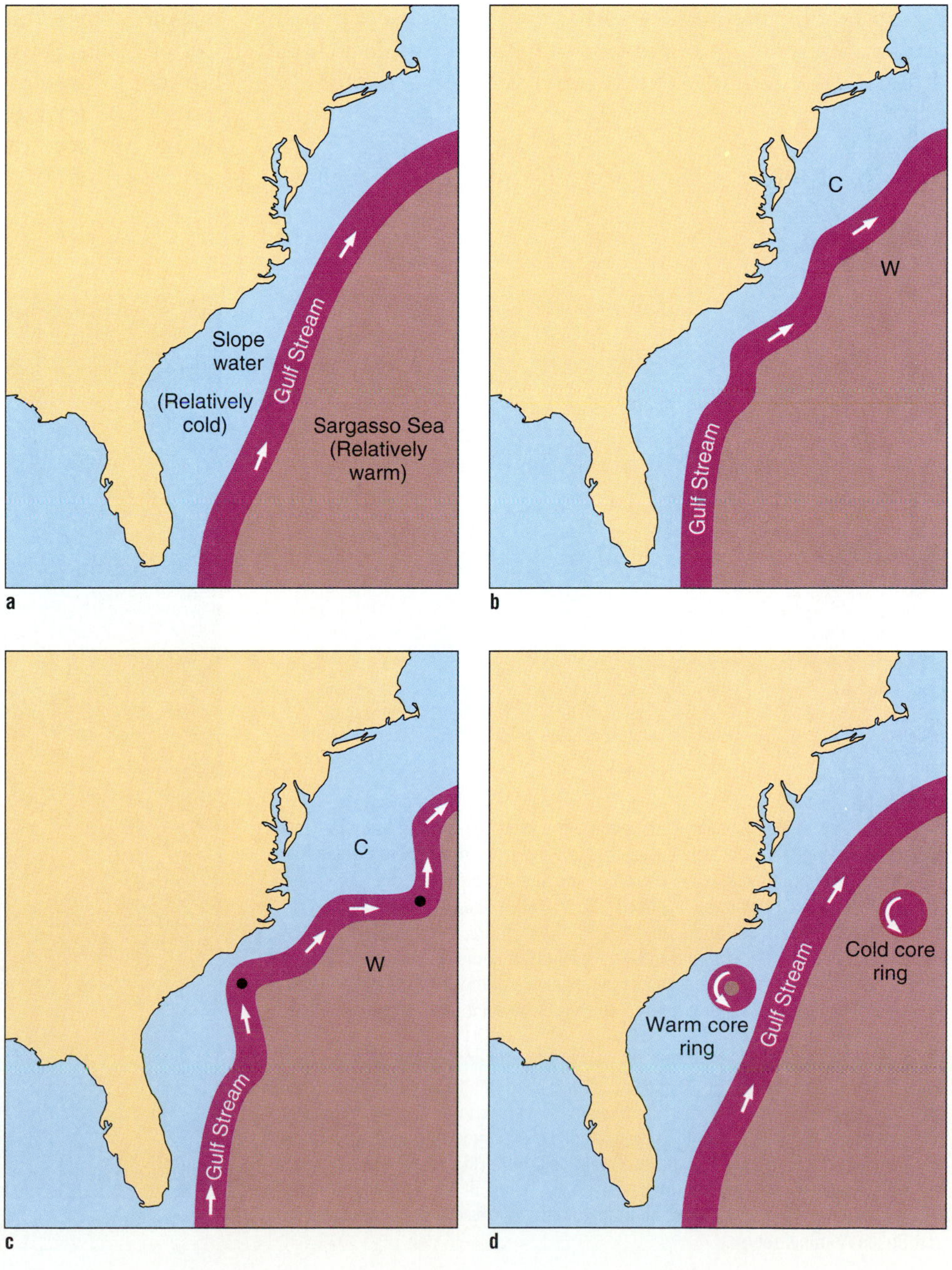

FIGURE 9-15 When Benjamin Franklin published his 1770 map of the Gulf Stream, little did he realize its complexity. Shown here are the formation of rings from the Gulf Stream. The sequence a, b, c and d occurs over a period of weeks. Those rings that incorporate Sargasso Sea water and that end up north of the Gulf Stream are called *warm-core rings* because their core temperature is warm relative to the surrounding water. Those found to the south and enclosing water from over the continental slope are *cold-core rings*. Figure 9-16 shows the movement of a cold-core ring.

can move large masses of water, transporting organisms or pollutants to other areas in the process.

Rings often are about 150 to 300 km wide (about 100 to 200 miles) and may extend to depths of 3,500 m (about 11,500 ft.). Several of these rings frequently exist in an area, and some have been tracked for months (Figure 9-16). Some rings endure for 2 to 3 years (see Figure 9-1d). Similar rings also have been found in the Japan or Kuroshio Current region, which is the Pacific analog of the Atlantic's Gulf Stream, and along the West Wind Drift or Antarctic Circumpolar Current.

The area between a ring and the surrounding seawater has disturbed or distorted physical characteristics. Interestingly, submarines might be harder to detect in this area because they are often detected by careful determination of sound velocity, and such observations are hard to make in regions having varied physical conditions. (See sound velocity discussion in Chapter 8, page 193.)

As in the case of eddies, little is known about how these rings affect larger aspects of the ocean. The rings will be among the more important topics considered by physical oceanographers in the coming years.

Upwelling

Besides causing a horizontal movement of water, the action of the wind on the sea surface may also produce vertical motion. For example, when one wind-driven current collides with another it causes a convergence and sinking. When a wind-driven current collides with land, it also produces sinking. When currents move away from each other or away from land, it produces a divergence, and water rises from depth to the surface.

One type of divergence is **upwelling,** which can occur when prevailing winds blow parallel to a coast. In many instances the motion of the water is offshore, due to the Coriolis effect (Figure 9-17a). When subsurface waters are brought to the surface they often have high nutrient contents; thus an area of high biological productivity may result. Surface water can sink by essentially the same process if the water flows toward the land (Figure 9-17b).

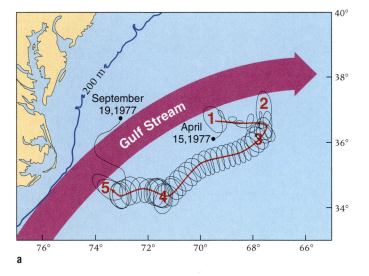

a

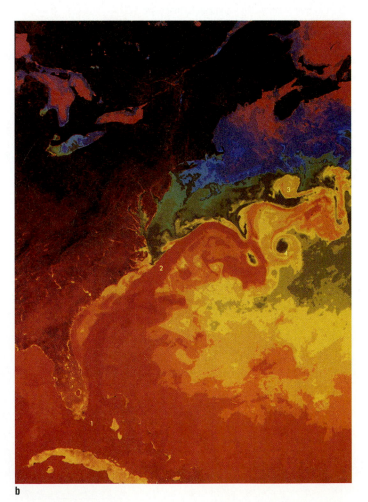

b

FIGURE 9-16 (a) Trajectory of a satellite-tracked, free-drifting buoy looping around the center of a cold-core Gulf Stream ring from April 15 to September 15, 1977. The ring movement is shown by the shifting center of the loops. The numbers refer to positions: 1 = April 15; 2 = April 26–29; 3 = May 15–17; 4 = July 31–August 2; 5 = September 13–15. During April, the ring became connected with the Gulf Stream and moved rapidly (0.5 knots) eastward. In May it separated from the Gulf Stream and began its slow southward drift at 0.1 knots. In September, the ring rapidly and completely coalesced with the Gulf Stream and was lost.
(b) Image showing temperature of the seawater in the area of the Gulf Stream. Data is from a composite of 35 satellite passes made in 1984. The red and orange correspond to warm water, the blues and purples to cold water. Note the riverlike appearance to the Gulf Stream and the presence of warm-core rings (yellow surrounded by green, labeled 3) and cold-core rings (green surrounded by yellow, labeled 4). (Image (a) based on the work of Dr. Phil Richardson, Woods Hole Oceanographic Institution. Image (b) courtesy National Aeronautics and Space Administration.)

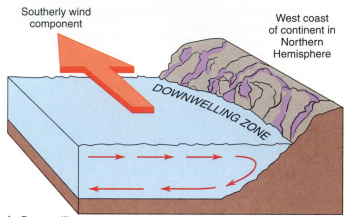

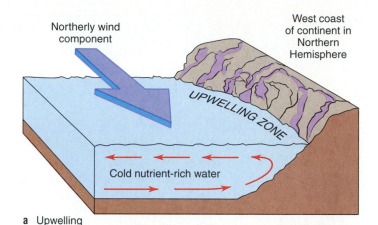

Northerly wind component

West coast of continent in Northern Hemisphere

UPWELLING ZONE

Cold nutrient-rich water

a Upwelling

Southerly wind component

West coast of continent in Northern Hemisphere

DOWNWELLING ZONE

b Downwelling

FIGURE 9-17 (a) Upwelling or rising of surface waters due to northerly nearshore winds, and (b) sinking of surface waters due to southerly nearshore winds. These phenomena occur along the western coasts of continents. (c) Satellite images taken along the West Coast of the United States that show coastal upwelling. The right-hand figure shows sea-surface temperature. Upwelled cold water (8–9°C) is shown in violet and purple, intermediate-temperature water by blues and greens, and warmer (14–15°C) water by yellows and reds. Warm California Current water is in yellows and reds. Three upwelling areas are indicated (north to south) at Cape Blanco (1), Cape Mendocino (2), and Point Arena (3). Meanders in the California Current are also evident (4, 5, and 6), as well as long filaments of upwelled water (7). The left-hand figure shows plankton chlorophyll pigment concentration. The highest levels are red, intermediate levels are yellow and green, and lowest levels are purple. The upwelled water is rich in nutrients and provides an excellent environment for high phytoplankton growth, as seen by the correlation of colors indicating upwelled water in the nearshore region and high phytoplankton growth (8). Some of this high growth is carried offshore (9). (Photograph (c) courtesy of National Aeronautics and Space Administration.)

Upwelling can have a profound influence on the biological productivity of an area (Figure 9-17c). It has been estimated that 50 percent of the world's fish catch comes from upwelling areas, although these areas in themselves comprise only about 1 percent of the total area of the ocean. The reason is simple: upwelling brings cool, nutrient-rich waters to the surface, which in turn permits increased growth by phytoplankton and thus provides more available food for other organisms (like fish) higher up in the food chain.

A program called **Coastal Upwelling Ecosystems Analysis (CUEA)** made an 8-year study of upwelling and showed how weather systems hundreds to thousands of kilometers away can cause local currents and upwelling (Figure 9-18). Local conditions such as submarine canyons or other coastal configurations can influence the strength of the upwelling process. Other processes, such as the meanders of large currents like the Gulf Stream (see Figure 9-16), may also force nutrient-rich waters toward the surface.

FIGURE 9-18   The major coastal upwelling areas of the world (shaded) and the weather circulation patterns that drive them. (Adapted from Hartline, 1980.)

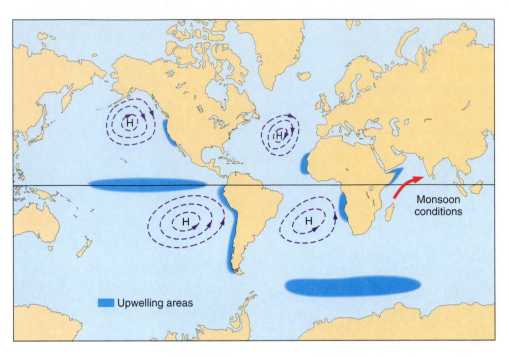

Upwelling areas

Monsoon conditions

Thermohaline Circulation

The thermohaline circulation is mainly a deep-water process, caused primarily by variations in water density. Because density is controlled mostly by temperature and salinity, this density circulation is called **thermohaline circulation** (*thermo-* for temperature, plus *haline* for salt). Density differences that drive this circulation usually develop at the sea-air interface. Thus, the wind-driven and thermohaline circulation systems are related.

Observations of thermohaline circulation at or near the surface are difficult because the circulation rate is very slow (a few cm/sec. or even less). Further, thermohaline circulation is obscured by the faster-moving wind-driven circulation (10s to 100s of cm/sec.). Most information concerning thermohaline circulation comes from detailed subsurface measurements of temperature, salinity, and dissolved oxygen.

New insight into thermohaline processes comes from studies that measure human-introduced substances, which include pollutants and carbon-14 from nuclear explosions. The studies observe how far these substances have moved into the ocean depths. One surprising result, from the Transient Tracers in the Ocean (TTO) program, was that deeper waters of the North Atlantic have grown colder and less salty over a 10-year period. Although the differences are small, they are considered significant. Some believe they may reflect worldwide climatic change.

Thermohaline circulation is a convection process. Dense cold water formed in high latitudes (polar re-

gions) sinks and slowly flows toward the equator. Much of the deep water of the ocean has acquired its characteristics in this manner. This process occurs principally in two places, the North Atlantic and the Antarctic (Figure 9-19).

In the North Atlantic, the cold, more dense water sinks to near the bottom and moves south across the equator (Figure 9-20). This water, called the **North Atlantic Deep Water,** is defined by its temperature, salinity, and oxygen and is easily distinguished from other water masses present in the Atlantic Ocean (Figure 9-21). (A **water mass** is a body of water having a distinct pattern of salinity and temperature throughout its range.)

In the Antarctic region, an Antarctic Bottom Water and an Antarctic Intermediate Water are formed. The former, one of the densest bodies of water in the ocean, travels north along the bottom under the equator. The Intermediate Water also travels north, but at a depth of about 1 km (about 0.6 miles) below the surface, because it is less dense.

Thermohaline circulation obviously is extremely important in establishing deep ocean conditions. Because deep water has its origins in polar or near-polar regions, it is easy to understand why deep waters are so cold. Their relatively high oxygen content compared to shallow tropical waters also reflects their surface polar origin. Without this surface polar source of oxygen, deep ocean waters could become oxygen-depleted by oxidation of the organic matter that falls through them.

Bottom-flowing water masses can also be influenced by seafloor topography. Dense waters formed in the

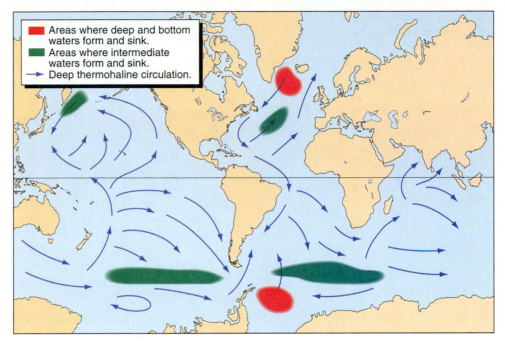

FIGURE 9-19 The deep thermohaline circulation of the ocean. The two red ovals indicate areas of sinking and formation of deep or bottom water, the green ovals indicate areas where intermediate waters form and sink. Arrows show the general flow of water in the deeper parts of the ocean. Note the intensified currents along the western sides of the oceans. The hypothesis for this flow plan was developed by Dr. Henry M. Stommel. (Adapted from Stommel, 1958.)

Arctic Ocean are prevented from reaching the Atlantic by a submarine ridge. The Mid-Atlantic Ridge also is a barrier to flow between basins of the western and eastern Atlantic.

Thermohaline circulation is fairly slow. The bottom waters of the ocean sometimes eventually come back in contact with the surface through this circulation, but the process may require centuries.

Some recent observations have found that hydrothermal events associated with seafloor spreading may cause small and unique water masses (see Box 9-3). Little, however, is known about them.

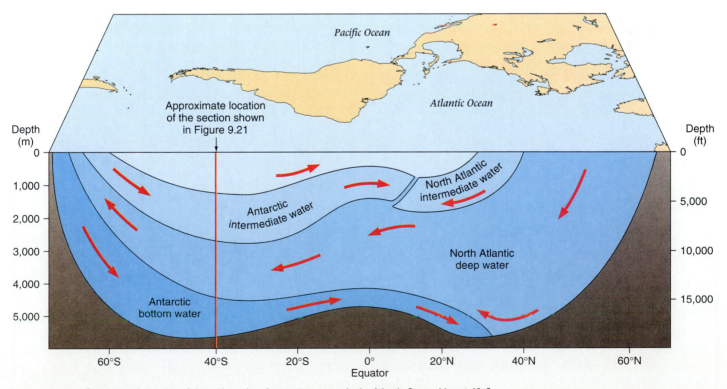

FIGURE 9-20 Diagrammatic section of the major subsurface water masses in the Atlantic Ocean. Line at 40°S indicates approximate location of the section shown in Figure 9-21.

FIGURE 9-21 Salinity, temperature, and oxygen measurements made at GEOSECS Station 60 in the South Atlantic (35°58' S and 42°30' W). The distinct variations in the measured parameters easily define the three main water masses (see also Figure 9-20).

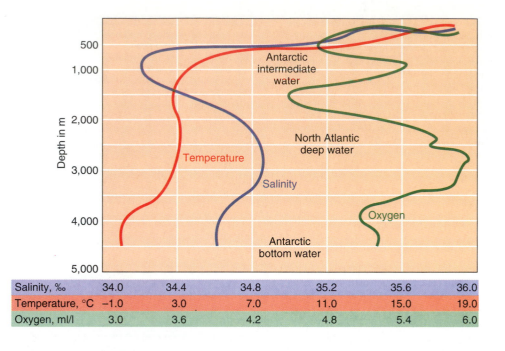

Salinity, ‰	34.0	34.4	34.8	35.2	35.6	36.0
Temperature, °C	−1.0	3.0	7.0	11.0	15.0	19.0
Oxygen, ml/l	3.0	3.6	4.2	4.8	5.4	6.0

Box 9-3

Megaplumes

A remarkable set of scientific discoveries was the finding of vents in all major oceans and even in one freshwater lake (see Box 7-4, page 169). Though these discoveries have dramatically influenced the thinking of marine geologists and geophysicists, chemical oceanographers, and biological oceanographers, they had rarely interested physical oceanographers, who generally felt that, although vents are interesting, they had little impact on the physical condition of the ocean.

This view changed with a 1986 discovery by scientists from the Pacific Marine Environmental Laboratory of NOAA (National Oceanic and Atmospheric Administration). While making routine hydrographic measurements about 450 km (260 mi.) off the coast of Oregon along the Juan de Fuca Ridge, they detected an anomalous body of water about 0.25°C (about 0.5°F) warmer than the surrounding water. The warm-water body was found to be about 700 m (2,000 ft.) thick and about 20 km (12 miles) in diameter. The top of the water body was about 1,000 m (3,300 ft.) above the seafloor and resembled a frisbee in its appearance. The size of this water body was so immense that it was dubbed a **megaplume** (Figure 1). In 1987 a second megaplume was discovered somewhat north of the 1986 find.

One of the NOAA scientists who discovered megaplume, Dr. Edward T. Baker, estimated that the heat causing the megaplume was equal to the yearly discharge of 200 to 2,000 black smoker-type vents (see Figure 7-10, page 168). He also concluded that the discharge probably came from a fairly large crack or rift on the seafloor, rather than from many individual vents. Water from these vents or cracks was creating a unique water mass, somewhat like a volcanic cloud being injected into the atmosphere.

Apparently, the rising hot water mixes with the surrounding cold water until a density equilibrium is reached. The resulting plume starts to move horizontally. The actual distance these plumes can move is unclear, but about 5 years earlier a helium-rich body of water was tracked by Drs. John Lupton and Harmon Craig more than 2,000 km (1,250 miles) from its presumed source.

If these plumes are common, it may explain one mystery concerning vents: how do their unique fauna move from one area to another? One hypothesis is that larval forms may be carried to new sites in the plumes, analogous to winds carrying and dropping seeds.

A further surprise came from this region when a team of NOAA scientists, led by Robert Embley, compared some seafloor surveys made in 1989 with those made in 1981. They observed changes both in the character and elevation over a 16 km (about 10 miles) stretch of the seafloor in the region, a potential source

Box 9-3 continued

area for one of the megaplumes. In one place at least 25 m (80 ft.) of rock had been added. It was proposed that when the hot water was emitted it may have been accompanied by large flows of volcanic material. In any case, both processes are due to seafloor spreading.

Underwater pictures (Figure 2) later confirmed the volcanic flows; an estimate of the total flow volume of new volcanic material is about 50 million m³, equal to that of the 1977 eruption of Kilauea volcano in Hawaii. Perhaps more important was that this was the first well-documented example of a deep-water volcanic eruption on an ocean ridge. Baker estimated that about 100 million m³ of water at about 350°C (662°F) was discharged in a few days or less to produce the plumes.

Reference

Embley, R. W., W. Chadwick, M. R. Perfit, and E. T. Baker. 1991. "Geology of the Northern Cleft Segment, Juan de Fuca Ridge: Recent Lava Flows, Seafloor Spreading and the Formation of Megaplumes." *Geology* 19, no. 8, pp. 771–75.

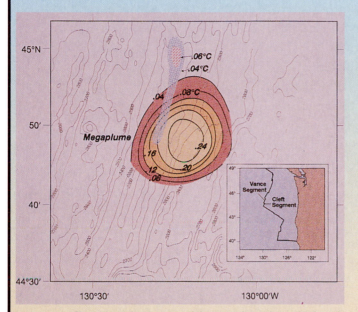

FIGURE 1   Plan view of the area around the megaplume site on the Juan de Fuca Ridge. The figure shows the size and intensity of the megaplume relative to an idealized view of a normal (smaller and cooler) plume from this region. Temperature is indicated in degrees Celsius; the megaplume is about 20 km (12 miles) wide. The figure also displays the bottom contours, in meters. The insert shows the Vance and Cleft segments of the Juan de Fuca Ridge; the megaplume comes from the Cleft segment. (Illustration courtesy of Dr. Edward T. Baker, NOAA.)

FIGURE 2   Underwater photograph of the seafloor in the megaplume area. Note the very recent lava (darker material) that was discharged in the 1980s, overlying the older volcanic rock (lighter material). The photograph shows an area about 2–3 m across. (Photograph from Dr. Robert W. Embley, courtesy of NOAA/PMEL VENTS Program.)

SUMMARY

The major oceanic current systems result from the combined influence of the wind blowing on the ocean (surface currents) and variations in seawater density (deep currents). Both movements result directly from solar energy. Thermohaline circulation, mainly a deep-water process, is caused by density variations that result from temperature (differences in heat received over the surface of Earth) and salinity (the effects of dilution and evaporation on the concentration of salts).

Surface currents mainly result from winds that cause high-pressure areas. These currents are deflected by Earth's rotation (Coriolis effect). The presence and shape of continents and the ocean floor also influence circulation patterns.

Under certain circumstances, upwelling of subsurface nutrient-rich waters to the surface can create conditions favorable for increasing biological growth.

QUESTIONS

1. How does solar energy affect the ocean?
2. What is the ocean's heat budget?
3. How are currents measured?
4. What causes surface currents in the ocean?
5. How are Western Boundary currents formed and why are they so strong?
6. Why is the Coriolis effect important to understanding ocean currents?
7. What is thermohaline circulation? How does it develop?
8. Describe the processes that cause upwelling. Why is it important?

KEY TERMS

Antarctic Circumpolar Current

Coriolis effect

countercurrent

Coastal Upwelling Ecosystems Analysis (CUEA)

current meter

currents

drift bottle

drogue

eastern boundary current

eddies

Ekman spiral

Ekman transport

equatorial current

free-fall device

geostrophic current

Gulf Stream

gyre

heat budget

heat capacity

megaplume

monsoon

North Atlantic Deep Water

polar easterlies

rings

ship drift

SOFAR channel

solar radiation

solar energy

Swallow float

telemetering buoy

thermohaline circulation

Tropical Ocean Global Atmosphere (TOGA) program

trade wind

undercurrent

upwelling

water mass

West Wind Drift

westerlies

western boundary current

wind-driven circulation

World Ocean Circulation Experiment (WOCE)

FURTHER READING

Baker, E. T. 1991–92. "Megaplumes." *Oceanus* 34, no. 7, pp. 84–91. Information on the megaplume discovery on the Juan de Fuca Ridge.

Eagleman, J. R. 1985. *Meteorology: The Atmosphere in Action.* 2d ed. Belmont, Calif.: Wadsworth Publishing Co. A good textbook on the atmosphere.

McLeish, W. H. 1989. "Painting a Portrait of the Gulf Stream." *Smithsonian* 19, no. 2, pp. 42–55. A well-written story of the Gulf Stream.

Open University. 1989. *Ocean Circulation.* Oxford: Pergamon Press. An advanced but good discussion of ocean circulation.

"Physical Oceanography." *Oceanus* 35, no. 2 (Summer 1992). A special issue on recent research in physical oceanography.

Pickard, G. L., and W. J. Emery. 1982. *Descriptive Physical Oceanography: An Introduction.* 4th ed. Oxford: Pergamon Press. An advanced but good textbook on general physical oceanography.

Price, J. F. 1992. "Overflows: The Source of New Abyssal Waters." *Oceanus* 35, no. 2, pp. 28–34. A discussion of how deep water "originates" at the poles and travels throughout the ocean.

Richardson, P. L. 1985. "Derelicts and Drifters." *Natural History,* June, pp. 43–49. A discussion of some of the data presented in this chapter.

Richardson, Philip, L. "Tracking Ocean Eddies." *American Scientist,* 81, no. 3, May–June 1993, pp. 261–71.

Spindel, R. C., and Peter F. Worcester. 1990. "Ocean Acoustic Tomography," *Scientific American.* 263, no. 4, pp. 94–98. A good general article on tomography in the ocean.

Stommel, H. 1958. "The Circulation of the Abyss." *Scientific American* 199, no. 1, pp. 85–90. A classic describing the deep circulation of the ocean.

Sverdrup, H. U., M. W. Johnson, and R. H. Fleming. 1942. *The Oceans: Their Physics, Chemistry, and General Biology.* Englewood Cliffs, N.J: Prentice Hall. Generally is considered the "bible" of books on the ocean; complex, but contains a wealth of information about the ocean.

Webster, P. J. 1981. "Monsoons." *Scientific American* 245, no. 2, pp. 109–18. A discussion of the processes that form monsoons, and computer simulations that can be used to predict dry and rainy phases.

Wiebe, P. H. 1982. "Rings of the Gulf Stream." *Scientific American* 246, no. 3, pp. 60–70. A good discussion on the origin of Gulf Stream Rings.

Wunsch, Carl. 1992. "Observing Ocean Circulation from Space." *Oceanus* 35, no. 2, pp. 9–17. How satellites can provide new insights into global and regional oceanography.

CHAPTER 10

Waves and Tides

WAVES IN THE OCEAN COME IN VARIOUS SIZES AND ARE CAUSED BY DIFFERENT PROCESSES, FROM THE COMMON TYPES CAUSED BY WINDS TO THE RARE ONES (TSUNAMI) CAUSED BY EARTHQUAKES, MARINE SLUMPING, OR FAULTING. ANOTHER WAVE (TIDE) IS CAUSED BY THE GRAVITATIONAL ATTRACTION OF THE MOON AND SUN COMBINED WITH EARTH'S ROTATION. WAVES ARE FREQUENTLY BENEFICIAL BECAUSE THEY MIX OCEAN WATER IN THE COASTAL ZONE, AND THEY PROVIDE THE KEY COMPONENT TO A POPULAR SPORT—SURFING. UNDER SOME CIRCUMSTANCES, HOWEVER, WAVES CAN BE VERY DANGEROUS, CAUSING CONSIDERABLE DAMAGE AND LOSS OF LIFE. THIS CHAPTER CONSIDERS THE VARIOUS TYPES OF OCEAN WAVES, THEIR CAUSES, AND THE FACTORS THAT CHANGE THEM AS THEY TRAVEL ACROSS THE OCEAN.

The NOAA research vessel *Oceanographer* riding a very large wave.

(Photograph courtesy of National Oceanic and Atmospheric Administration.)

Introduction

In the last chapter we saw how the wind generates ocean currents. A smaller-scale result of wind blowing on the ocean surface is the formation of waves. Later in this chapter we will consider waves formed by processes other than the wind. But the various characteristics of waves described here are generally the same, regardless of how the waves form.

Waves: Energy in Motion

Waves are mechanical energy traveling through water. This is an important point—it is the *energy* that is moving with the speed of the wave, not the water. The water may also move, but at a much slower speed and generally in a circular or back-and-forth motion, except in shallow water where the waves break. Thus, what we see in a wave is the shape or **waveform** that this energy forces the water to take (Figure 10-1). Every waveform has certain properties:

- **Wavelength** (L) is the horizontal distance between any two similar points on the waveform, such as two successive crests (as shown in the figure) or troughs. Wavelength is measured parallel to the travel direction of the wave.

- **Period** (T) is the time interval that two successive wave crests take to pass a reference point.

- **Wave height** (H) is the vertical distance measured from the wave crest to the wave trough.

- **Water depth** (d) is from the bottom of the wave trough to the sea bottom.

- **Wave velocity** (C) is the wavelength divided by the period (velocity C = wavelength L/period T).

Wavelengths of waves formed by the wind generally range from about 60 to 150 m (about 200 to 450 ft.). The maximum height of a water wave is controversial, because it is hard to make a precise measurement when you are surrounded by very large waves. However, a height of 34 m (112 ft.) has been observed. Many suspect that larger waves are possible. The period of wind-generated waves generally does not exceed 20 seconds, and shorter periods of 12 seconds or less are more typical.

As mentioned, it appears that ocean waves are moving water along, but in the open sea this is not so. It is the *wave energy* (made visible by the waveform) that is rapidly advancing. The water itself moves forward only slightly. The proof of this is readily visible: watch a floating object bob up and down with each passing wave. The wave energy moves horizontally and rapidly through the water surface, but the object itself—and the water it is floating in—just moves up and down, with only a slight forward movement (Figure 10-2a). Therefore, when talking about wave motion, we must distinguish between the motion of the *wave energy* (waveform) and the motion of the *water particles*.

When talking about waves and "deep water," we mean that the depth of the water is at least half the wave's wavelength. (If the wavelength is 60 m, for example, then "deep water" is at least 30 m to the bottom.) In deep water, the motion of individual water particles at the surface is orbital, as shown in Figure 10-2b. An individual water molecule actually makes a circular trip on the wave energy, returning nearly to its original position as one waveform passes.

With depth below the surface, however, the size of the orbit quickly decreases (Figure 10-2b). The velocity of the water particles, in addition to the size of the circle, also decreases rapidly with depth. By a depth of about one-half the wavelength, the orbital motion is essentially zero. In other words, a surface wave having a wavelength of 60 m (about 200 ft.) would not be noticeable by a diver at a depth of 30 m (about 100 ft.).

FIGURE 10-1 Important characteristics and morphology of a wave in deep water.

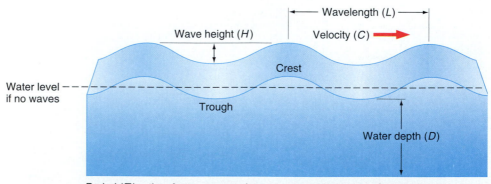

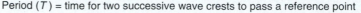

Period (T) = time for two successive wave crests to pass a reference point

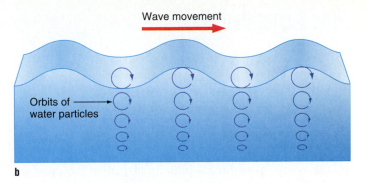

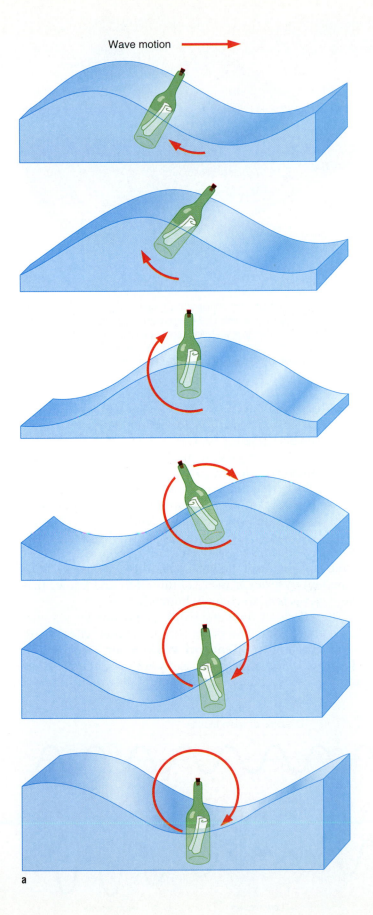

FIGURE 10-2   (a) An object (a drift bottle) floating on the surface as a wave passes (in deep water) clearly shows the orbital motion associated with the waveform. Note that the waveform moves forward, but the water particles only move in an orbit or circular pattern.

(b) This diagram shows how the orbital motion of the wave decreases with depth. At a depth of about half the wavelength (in deep water) the orbital motion is essentially zero.

Wind-Generated Waves

Wind-generated waves form on the sea surface by the transfer of energy from moving air (wind) to the water. Exactly how energy is transferred from the wind to the ocean is a very complicated process that is not completely understood. The smallest waves formed by the wind, called **capillary waves,** have a period of less than $\frac{1}{10}$ of a second and can be observed on a very calm sea or lake when its surface is initially disturbed by a puff of wind. Their wavelength is less than 2 cm (less than an inch) and they are short-lived. With increasing wind speed, however, the properties of these waves—such as their height, period, and length—will increase.

The height and period of wind-generated waves are functions of three factors: (1) wind velocity, (2) wind duration, and (3) **fetch,** or the distance over which the wind blows (Figure 10-3). The individual effect of these factors still is being researched, but some relationships are clear:

- Wave height and wavelength generally increase (up to a definite maximum) with increasing wind velocity and duration.

- Fetch is important in determining wavelength. Wavelengths of only a few meters are common in lakes, where the fetch is short; wavelengths of hundreds of meters are typical for oceanic waves because the fetch is long.

- In shallow waters such as lakes, water depth can become another important factor influencing the waves.

As noted, the maximum possible height of wind-generated waves is not known, because of the difficulty of measuring wave height during storms. Waves have been reported up to 800 m (2,600 ft.) in length, which indicates a period of 23 seconds. Such large waves can result from the intersection of waves coming from different directions or having different wavelengths. Where they intersect, the wave patterns interfere with one another,

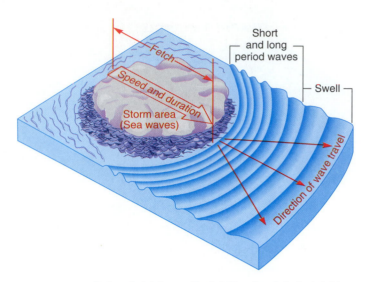

FIGURE 10-3 Factors that influence the height and period of wind-driven waves: fetch (distance over which the wind blows), wind velocity, and wind duration. Wave size will increase with wind speed, wind duration, and fetch. Waves advance across the ocean as "swell," no longer powered by the wind that spawned them but by the energy they acquired from the wind.

resulting in reinforcement (increase) in the wave pattern in one place and reduction (decrease) in the wave pattern elsewhere (Figure 10-4).

The main effect of this interference is that there can be a dramatic change (increase or decrease) in the wave height. Such a phenomenon can explain so-called freak or *rogue waves* that seem to appear from nowhere and have caused extensive damage to ships at sea. This phenomenon of interference also explains why the sea surface may be calm for a while and then quickly exhibit some large waves, followed by another calm period. The experienced surfer is aware of this effect and knows that large waves tend to come in bunches.

The effects of different wind speeds on the sea surface—the levels of waves energized, and their effects—are summarized in the **Beaufort Wind Scale** (Table 10-1). Accomplished mariners can easily estimate the Beaufort number by observing the condition of the ocean surface.

FIGURE 10-4 Two or more wave patterns of different wavelength can interfere with each other, forming a new wave pattern. In some instances, the interference may cause extremely high, or "rogue" waves.

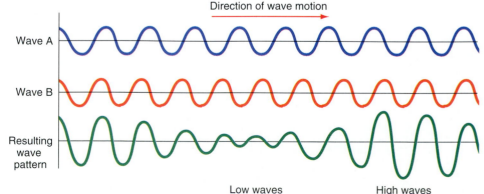

TABLE 10-1
The Beaufort Wind Scale

Beaufort Number	General Description of Wind	Condition of Sea	Wind Speed (miles/hour)	(km/hour)	Wave Height (ft.)	(m)	Conditions on Land
0	Calm	Sea smooth as mirror	Less than 1	Less than 1	0	0	No leaf motion
1	Light air	Small waveletlike scales; no foam crests	1–3	2–5	0.5	0.15	Smoke drifts; slight leaf motion
2	Light breeze	Waves short; crests begin to break	4–7	6–11	1	0.3	Wind felt; leaves rustle; wind vanes move
3	Gentle breeze	Foam has glassy appearance not yet white	8–12	12–20	2	0.6	Leaves and twigs move; small flags extend
4	Moderate breeze	Waves now longer; many white areas	13–18	21–29	5	1.6	Small branches move; movement of paper litter, dead leaves, and dust
5	Fresh breeze	Waves pronounced and long; white foam crests	19–24	30–39	10	3.1	Small trees sway; wavelets on streams and ponds
6	Strong breeze	Larger waves form; white foam crests all over	25–31	40–50	15	4.7	Large branches moving; umbrellas blown around
7	Moderate gale	Sea heaps up; wind blows foam in streaks	32–38	51–61	20	6.2	Trees swaying; hard to walk
8	Fresh gale	Height of waves and crests increasing	39–46	62–74	25	7.8	Small branches break; car steering affected
9	Strong gale	Foam is blown in dense streaks	47–54	75–87	30	9.3	Branches on ground; loose shingles removed
10	Whole gale	High waves with long overhanging crests; large foam patches	55–63	88–101	35	10.8	Trees uprooted; some damage to houses and buildings
11	Storm	High waves; ships in sight hidden in troughs	64–75	102–120	—	—	Considerable damage to structures and trees
12	Hurricane	Sea covered with streaky foam; air filled with spray	Above 75	Above 120	—	—	Major damage; entire towns may be devastated

Wind-Generated Waves: Sea, Swell, and Surf

Wind-generated waves play out their lives in three stages: sea, swell, and surf (Figure 10-5). Waves generated in the area directly affected by the wind, frequently where a storm is occurring, are called **sea.** Sea waves are irregular, with no systematic pattern. Since each is the product of an independent wind-dragging-water incident, these waves have different periods and heights, and travel in various directions.

As these waves leave the windy or stormy region where they were generated, the longer waves of similar period outdistance the shorter waves because their velocity is greater. This is evident from the wave velocity equation: velocity C = wavelength L/wave period T.

The waves eventually assume a more uniform pattern because waves of similar dimensions travel at similar speed and tend to "fall in" together. Waves in this regular pattern are called **swell** (Figure 10-5b). As swell travels still farther from the generating area, their pattern tends to remain constant in length, but they decrease in height because they are losing energy.

Ultimately, the wave energy dissipates entirely, gradually converted to heat as the wave performs the work of propelling countless water molecules in orbital motions as it travels (recall Figure 10-2) and eventually reaches shallow water. However, the ocean is such an efficient fluid for conducting wave energy that sometimes a wave pattern is capable of traveling across an entire ocean.

The third stage of wind-generated waves, called **surf** (Figure 10-5c), occurs near shore when the wave height increases and the wave breaks. Breaking surf waves differ from sea and swell waves because the water particles are no longer traveling in an orbital motion but are moving toward the beach. This results in a large amount of energy released by the breaking wave against the beach

a

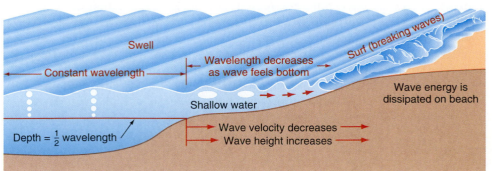

b

c

FIGURE 10-5 Wind-generated waves.
(a) Sea waves in the generation (storm) area.
(b) Swell. Note how the wave pattern has become relatively uniform.
(c) Surf. The waves are breaking. (Photographs (a) and (b) courtesy of Woods Hole Oceanographic Institution, photograph (c) courtesy of Dr. David Aubrey.)

on impact. The work done by this energy is no mystery: it is called beach erosion, and it grinds down massive areas of beach worldwide every year.

This is a good point at which to view how Earth's systems tie together: energy from the sun causes the atmosphere to move (wind), which imparts some of this energy to the ocean surface, generating waves. These waves can travel hundreds to thousands of kilometers to dissipate their energy against the shore, eroding it. This has been going on for billions of years and will continue for billions more, as long as there is an ocean, and a sun to provide the energy.

Waves in the Coastal Region

When waves reach shallow water, only their periods remain the same; all other characteristics change, including wavelength, height, and velocity (Figure 10-6).

FIGURE 10-6 Waves entering the surf zone. Note the decrease in their wavelength and increase in wave height when the wave form reaches shallow water (water depth less than half the wavelength). The period remains the same. After the wave breaks its energy will be spent on the beach, sometimes causing erosion.

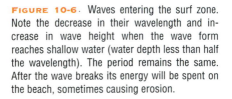

Swell

Surf (breaking waves)

Constant wavelength

Wavelength decreases as wave feels bottom

Shallow water

Wave energy is dissipated on beach

Depth = $\frac{1}{2}$ wavelength

Wave velocity decreases

Wave height increases

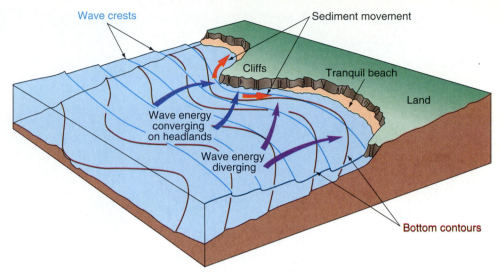

FIGURE 10-7 Wave refraction in the nearshore region. As the waves enter shallow water, they "feel" the bottom and are slowed relative to those parts of the wave that are still farther out in deeper water. This results in a bending or refraction of the wave pattern. Wave energy becomes focused or converges on topographically high areas. The wave energy spreads out or diverges over open bay-like areas.

Wavelength and velocity decrease in shallower depths. This change is small until the depth of the water (*D*) equals one half the wavelength of the wave. At this depth the wave is said to "feel" bottom, and this makes the wave rapidly increase in height.

In shallow water, the orbital motion of the water molecules becomes compressed and the patterns become flattened. More energy is being forced into less space, so the waves start to increase in height. When a wave becomes so high that the particle velocity at the wave crest exceeds the overall wave velocity, the wave tumbles forward, or breaks. On a gently sloping beach, this usually occurs when the ratio *H/D* is between 0.8 and 0.6.

If waves enter shallow water perpendicular (head-on) to the shore, they will dissipate their energy directly against the beach. However, when waves enter shallow water at some other angle to the beach, or if they encounter irregular changes in the nearshore bottom topography, their direction of travel changes (Figure 10-7). This change, called **refraction,** occurs when one part of the wave reaches shallow water before another part of the wave does. This first part to reach shallow water slows, causing the entire wave to turn toward the shallow water. Thus the wave crests tend to parallel the bottom topography or contours (Figures 10-7 and 10-8) and break nearly parallel to the coastline. Refraction can also occur around objects in the water, such as icebergs (see Figure 6-14, page 150).

Refraction depends on bottom topography, wavelength, and direction of approach. On irregular coasts, refraction can cause a concentration of wave energy, or **convergence,** on topographically high areas such as submerged ridges, headlands, or elevated points. A dispersion of wave energy, or **divergence,** occurs over submarine canyons or in bays. These are generally areas of calmer waters, compared to regions of convergence. Erosion, as you might expect, is greater in areas of convergence and deposition is greater in areas of divergence.

After waves break, water is carried into the surf zone and transported toward the beach, which is a barrier. Some or most of the water can return seaward along the bottom of the surf zone. With the usual angular approach of the waves to the beach, however, water also is transported along the beach, generally with considerable sediment (Figure 10-9). The movement of water along the beach is called **longshore current.** It increases until it can overcome the incoming waves, at which time the water flows seaward in what is called a **rip current** (see also Figure 14-5, page 340). Rip currents, because of their seaward flow, can be dangerous to swimmers.* The positions of rip currents depend on submarine topography, beach slope, wave direction, and the wave height and period. Rip currents often are inaccurately called "rip tides."

Langmuir Cells

When strong winds blow along the ocean surface, they can produce a series of convection cells, typically 15–30 m (50–100 ft.) in width and less than 10 m (33 ft.) in depth. Called **Langmuir cells** for the scientist who first described them, these cells have alternating left and right rotating patterns, with long axes parallel to the wind direction (Figure 10-10). This pattern causes a succession of surface areas of divergence and convergence. Part of the cell may be made visible by floating debris, which frequently accumulates over the convergence (or sinking) part of the cell. Langmuir cells can be important in transporting nutrients and gases and in mixing of the surface waters.

---

*If caught in a rip current, it generally is best to let the current carry you seaward. Do not try to swim against it to reach shore. As the rip current diminishes in strength, swim parallel to the beach until out of the effect of the rip current, then start to swim ashore.

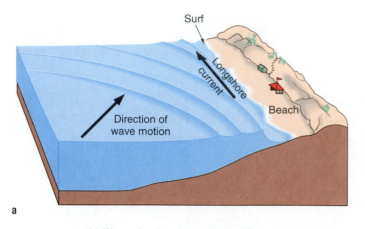

a

FIGURE 10-8 (a) Wave refraction along a beach. When waves approach a straight shoreline at an angle, the portion of the wave in shallow water will slow relative to that part of the wave in deeper water. This effect produces a refraction or change in direction of the waves. This process will also produce a longshore current and ultimately a rip current (see Figure 10-9a).
(b) Wave refraction along a beach near Santa Barbara, California.
(c) Wave refraction around an island (left center). White features at top are clouds with their shadows below. Picture is taken from a height of 16 km (10 miles). (Photograph (b) courtesy of Dr. David Aubrey; photograph (c) courtesy of U.S. Air Force, Cambridge Research Laboratories.)

c

b

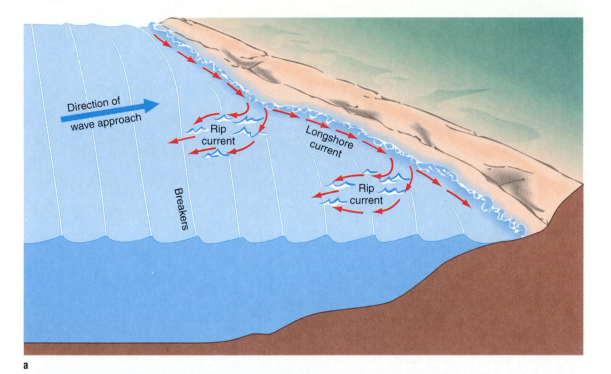

a

b

FIGURE 10-9   (a) Formation of longshore current and rip currents in the nearshore region. The rip current results from the buildup of water in the nearshore area that eventually returns seaward as a narrow and swift current.
The longshore current constantly moves sand along the beach, some of which returns seaward in the rip current.
Note also that a longshore current generally forms a rip current.
(b) Air photo of a series of small rip currents along the coast of Cape Cod. The rip currents are the small "puffs" off
the lower part of the land.

FIGURE 10-10 Langmuir cells. Note the alternating left-hand and right-hand circulation, which forms divergent and convergent zones. The convergent zones are marked by the accumulation of debris and foam. (Adapted from Open University, Ocean Circulation, 1989.)

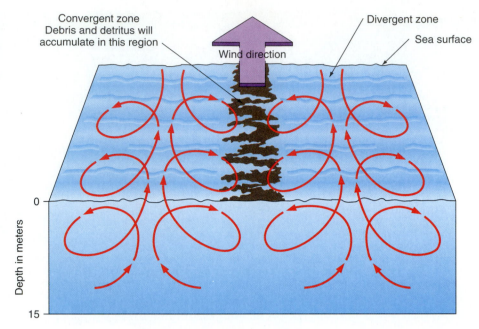

Internal Waves

An **internal wave** can occur between fluids of slightly different densities. In the ocean such waves generally cannot be seen directly, but can be detected by systematic and closely spaced temperature observations. Such measurements have shown that internal waves are common in the ocean. They travel beneath the air-sea surface, and may have a height greater than 30 m (100 ft.).

The periods of internal waves are 5 minutes or more. They can break like surface waves, but below the surface. Their presence sometimes is hinted at by slow-moving surface slicks composed of plankton, fine-grained sediment, or surface-water contamination (Figure 10-11) that accumulate over the trough of the internal wave.

Any condition causing waters of different density to come in contact with each other can cause internal waves, for example, an outflow of freshwater from a river or mixing of waters of different salinity or temperature. What initiates an internal wave still is uncertain, but they probably are related to other forms of energy in the ocean, such as currents, winds, or tides. Internal waves can cause the thermocline to move up and down slowly with the wave motion.

Catastrophic Waves

Set a bowl filled with water on a table. Blow on the water very hard, bump the bowl sharply, then drop a pebble or an ice cube into the bowl. By doing these various things, you have just simulated what Earth's atmosphere and lithosphere can force the ocean to do under extreme conditions of energy release: generate **catastrophic waves.**

These waves are generated by energy from large atmospheric disturbances such as hurricanes and typhoons (storm surges), by energy from seafloor disturbances such as earthquakes and slumping (tsunami), and by energy from surface landslides and ice breaking off glaciers into the water (some examples are given in the following paragraphs). Catastrophic waves often cause property damage and loss of life.

Storm Surges

Strong onshore winds, frequently associated with hurricanes, can push water up onto a coast, causing an exceptionally high sea level called a **storm surge.** Storm surges can be dangerous, especially if they coincide with high tides in low coastal regions. In the Gulf Coast area of the United States, storm surges have been known to raise the water level as much as 7 m (about 23 ft.). In 1900, over 6,000 people drowned during such a storm in Galveston, Texas, killed by the storm surge rather than by high winds or collapsing buildings.

Storm surges differ from other waves because the water level rises gradually rather than in a quick, rhythmic rise and fall. This is why hurricane forecasters constantly warn people to prepare for the surge and evacuate low-lying areas before the water level rises. A further discussion of hurricanes and their impacts is presented in Chapter 13.

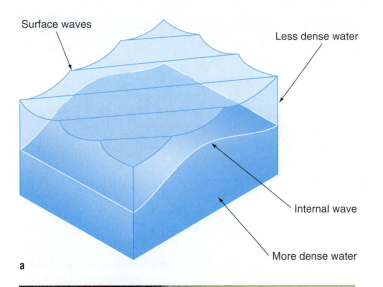

a

b

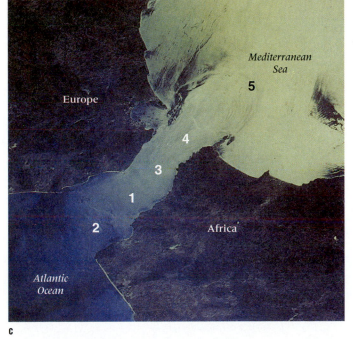

c

FIGURE 10-11 (a) A simple illustration of how an internal wave can be formed between waters of differing densities.
(b) Sea-surface expression of internal waves. The surface slicks, composed of fine-grained sediment, are related to the troughs of the internal waves.
(c) Satellite view of a remarkable sequence of waveforms in the Strait of Gibraltar region. These different forms result from the large volume of water flowing from the Atlantic Ocean (lower left) into the Mediterranean (upper right). Europe is above the Strait, Africa below. A corresponding outflow occurs from the Mediterranean into the Atlantic, but this happens below the surface. Numbers 1 and 2 indicate features due to turbulence of the inflowing Atlantic water (because the flow is constricted at the narrow strait). Features 3 and 4 are the surface expression of internal waves formed at the interface between the incoming and outgoing flow. Feature 5 also indicates internal waves, in this instance formed by the "sloshing" back of the tides between the Atlantic and the Mediterranean. These patterns are visible because of the reflection of sunlight glittering off the sea surface. (Photograph (b) courtesy of E. C. LaFond, U.S. Naval Undersea Center. Photograph (c) taken from the Space Shuttle *Challenger*, which was named for the research vessel *Challenger*; courtesy of National Aeronautics and Space Administration.)

Landslide Surges

In a **landslide surge,** the movement of large quantities of rock or ice (due to earthquakes or glacial movements) into the ocean or especially into an enclosed bay generates immense waves. An exceptionally large wave occurred in Lituya Bay, Alaska, in 1958. Triggered by an earthquake, a volume of rock estimated at 30,000,000 m³ (about 40 million yards³) fell directly into the bay from a height of about 1,000 m (3,300 ft.) This caused a wave that rose over 500 m (1,600 ft.) onto the mountainside on the other side of the bay (Figure 10-12). In another dramatic example, over 15,000 people drowned in a similarly formed wave on the Japanese island of Kyushu in 1792.

Tsunami

Tsunami (*soo NAH mee*) is a Japanese word meaning "harbor wave," indicating where these massive waves often cause their destruction. Tsunami are commonly but mistakenly called "tidal waves" because they come up like an extremely rapid tide, but in fact they have nothing to do with the ocean's daily tides. Tsunamis are caused by submarine movements in the lithosphere, produced by the jolts from earthquakes, slumping, or volcanic eruptions (Figure 10-13).

In deep water, a tsunami may have a wavelength up to 700 km (435 miles), a period from 5 to 60 minutes, travel at speeds over 640 km (400 miles) per hour, and yet have a wave height (amplitude) of only a few centimeters

a

b

c

FIGURE 10-12 (a) Lituya Bay, Alaska in 1954, before the giant wave.
(b) In 1958 wave damage is apparent.
(c) Wave damage on the north shore of Lituya Bay. View is about 3 km (about 2 miles) from the bay entrance. Width of the zone of destruction is about 600 m (about 2,000 ft.) at right margin of photograph. Note trees with limbs and bark removed. (All photographs courtesy of U.S. Geological Survey.)

FIGURE 10-13 A schematic model of the development of a tsunami generated by a fault on the seafloor. The wave produced is essentially undetectable when it travels in deep water, but can be awesome once it reaches the coast.

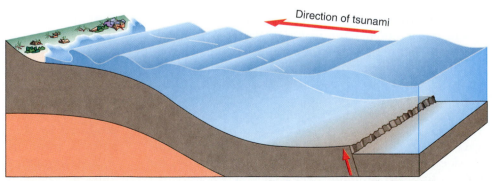

Direction of tsunami

Earthquake causes a displacement of the sea floor, initializing a tsunami (about an inch), too small to notice in the open sea. However, when a tsunami reaches shallow water and breaks against the coast, its massive size and energy can produce a wave higher than any wind-generated wave.

The destructive effect of a tsunami is strongly controlled by submarine topography. A sloping beach or land near a submarine ridge will experience high and damaging waves, while a wide and shallow continental shelf can absorb much of the wave energy and diminish the damage from a tsunami. Figure 10-14 shows a tsunami coming ashore in Japan.

One tsunami may be the source of the story of the so-called "lost continent of Atlantis" (see Box 10-1). Numerous other catastrophic tsunamis have occurred, including one that entered the Tagus estuary, which forms the harbor of Lisbon, Portugal. In 1755, thousands of curious people foolishly ventured out onto the exposed estuary seafloor when sea level surprisingly dropped by several meters. In this unfortunate instance, a trough of a major tsunami had reached the area. The subsequent crest and later waves killed more than 50,000 people and caused immense damage.

Another notable tsunami occurred in 1883 with the eruption and collapse of the volcano Krakatoa in Indonesia. The collapse of large sections of the island volcano into the ocean produced waves exceeding 30 m (100 ft.) in height, which destroyed over a hundred villages and killed more than 33,000 people.

More recent earthquakes have produced tsunamis that caused many deaths and much property damage. A 1960 tsunami that struck Chile in 1960 killed more than 1,000 people. In July 1993 a large earthquake off Japan caused a tsunami that struck northern Japan. Together, the earthquake and tsunami killed more than 100 people and caused considerable property damage. In the U.S., since 1946, six major tsunami have hit Hawaii, Alaska, or the West Coast, causing over 350 fatalities and nearly a billion dollars in damage. Figure 10-15 shows

a

b

c

FIGURE 10-14   Three views of a tsunami, generated by an offshore earthquake, that struck Japan in 1983. The tsunami is shown hitting the Oga Aquarium in Akita, Japan. In the first figure (a), the water is drawing away from the shore in the early stages of the tsunami. In the second figure (b) the inundation is at its maximum. The third figure (c) shows the area after the wave has withdrawn. The tsunami caused about $800 million in property damage and killed more than 100 people. (Photograph by Takaaki Uda, Public Works Research Institute, Japan, distributed by the National Geophysical Data Center, NOAA.)

Box 10-1

A Historic Tsunami

One of the more interesting myths about the ocean concerns the so-called "lost continent" of Atlantis. The legend comes from the writings of Plato, who reported the story of a prosperous and intellectually advanced island nation that was destroyed by earthquakes and storms and disappeared beneath the sea. Over the years, adventurers and scientists have searched for such a lost island in the Atlantic Ocean, the Caribbean Sea, and the Mediterranean Sea. Books and movies have been done on the subject, but convincing data remained elusive, except for one possibility in the Mediterranean Sea.

Recent research by oceanographers, geologists, and archeologists has found evidence of a major tsunami that came ashore on the island of Crete around 1450 B.C. The tsunami was due to a volcanic eruption that destroyed much of the nearby Greek island of Thíra, sometimes also known as Santorini (Figure 1). Before the eruption, Thíra was inhabited by the Minoans, who developed a very sophisticated civilization.

Besides causing a catastrophic tsunami, the eruption also resulted in the collapse of the volcanic island. This undoubtedly brought the Minoan civilization to a close. Archeological studies of the remains of the Minoan culture on Thíra indicate that this island may indeed have been the Atlantis of Plato's story.

A good general description of the destruction of Thíra and its implications is contained in a 1985 article by J. Mavor, "Atlantis and the Catastrophe Theory," *Oceanus* 28, no 1, pp. 44–51. Also see F. W. McCoy and G. Herkian, "Anatomy of an Explosion," *Archaeology*, May/June 1990, pp. 42–49.

Figure 1   An engraving of an 1866 eruption on the Greek island of Santorini. Some aspects are probably exaggerated. However, it may resemble the 1500 B.C. eruption and subsequent tsunami that destroyed the Minoan civilization on this island. The sinking of the central portion of the island is thought by some to be the source of the legend of the lost continent of Atlantis. (Photograph courtesy of P. Hedervari, National Geophysical Data Center, NOAA.)

the pattern of a tsunami received at three widespread points in the Pacific following a 1964 Alaskan quake.

The tsunami threat is ever-present in the tectonically active Pacific basin. In response, an international early warning system was created to provide advance warning of tsunami following significant earthquakes. The warning system allows people time to take preventive action following an earthquake but before the tsunami approaches. The warnings must be given quickly, for tsunami travel at speeds rivaling those of a commercial airliner.

In the early warning system, sensitive seismographs located at stations around the Pacific record the shock waves from earthquakes. Observers quickly determine the earthquake's epicenter and calculate when a resulting tsunami should arrive (Figure 10-16). People in areas

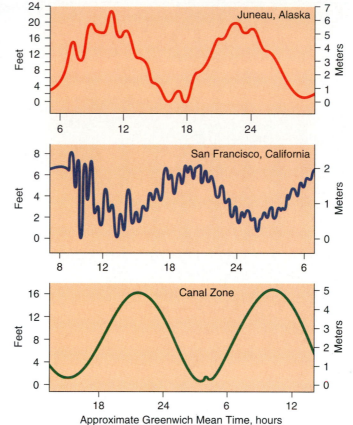

FIGURE 10-15 Tidal gauge records of the 1964 Alaskan tsunami from three different locations along the Pacific Basin. Note that the tsunami arrived later at the cities farther from earthquake's epicenter. These records do not show individual waves but rather the broad up-and-down movement of the water. (After Spaeth and Berkman, 1967.)

that could be hit by the wave then can be forewarned. The system seems to have been successful, but only in areas where adequate communication systems exist.

Most tsunami in the Pacific are caused by submarine movements and earthquakes along the plate boundaries that somewhat discontinuously encircle the Pacific (see Figures 3-12 and 3-13, pages 55 and 56). Tsunami generated in these areas travel outward and reach most other areas of the Pacific, as is quite evident from Table 10-2.

Fortunately, the U.S. West Coast has received little damage from tsunami compared to other coastal areas around the Pacific. The reasons are interesting:

1. As a source of earthquakes, the U.S. West Coast, although tectonically active, is actually relatively stable compared with other coastal areas of the Pacific.

2. Tsunamis initiated in the Aleutian and South American areas approach the U.S. West Coast diagonally. This causes their energy to be delivered in a glancing blow, rather than head-on, causing less destruction.

3. The relatively wide continental shelf on the U.S. West Coast (compared to the other shelves around

the Pacific) causes the waves to lose considerable energy before reaching shore.

4. The U.S. West Coast has many high-cliffed or hard-rock coastal regions that are less susceptible to damage than low-lying beaches.

Stationary Waves

A wave type common to many enclosed bodies of water—such as bays, lakes, and bathtubs—is the **stationary wave,** also called a **standing wave** or **seiche** (*saysh*). In a stationary wave, the waveform does not move forward; instead, the water surface moves up and down. The motion is similar to that of soup in a bowl that has been tilted and then put down on a flat surface, sort of a slow and gentle back-and-forth sloshing (Figure 10-17). The water surface remains stationary at certain locations, called nodes, while the rest of the surface moves up and down.

Stationary waves can be generated by storms, rapid changes in atmospheric conditions, or sudden disturbances to the water surface. These events temporarily pile up water in one part of the enclosed water body.

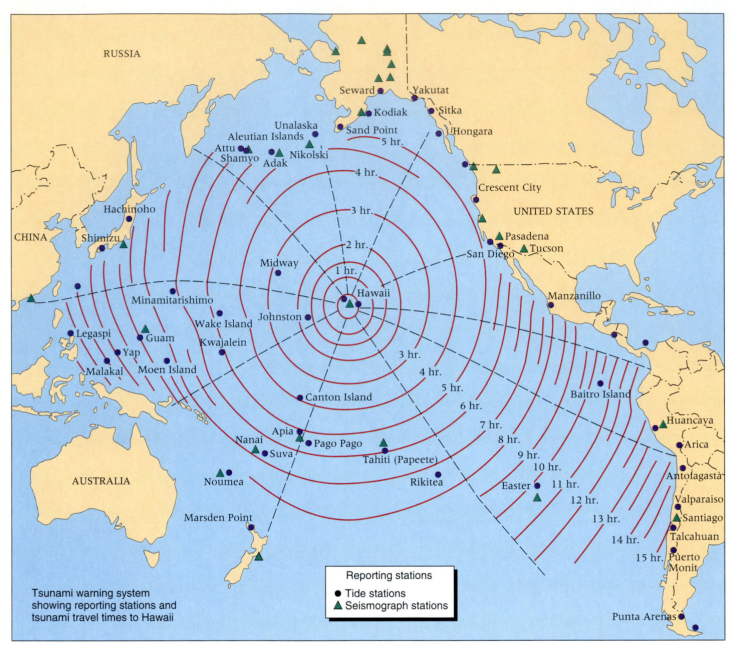

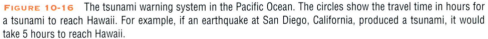

FIGURE 10-16 The tsunami warning system in the Pacific Ocean. The circles show the travel time in hours for a tsunami to reach Hawaii. For example, if an earthquake at San Diego, California, produced a tsunami, it would take 5 hours to reach Hawaii.

When the force is removed, the water flows back, forming a wave. Once this wave is generated, water in the lake or bay continues to oscillate at a rate controlled by the length and depth of the basin.

Although generally not dangerous, seiches have on occasion been responsible for property damage and even loss of life. A series of winter storms in 1987 caused a seiche on Lake Michigan that raised the water level several meters, causing considerable erosion and flood damage in Chicago.

Ocean Tides

Tides are the rhythmic rise and fall of sea level familiar to anyone who has lived or vacationed at the seashore. One of humanity's earliest scientific ventures was to explain and predict the tides. Tidal movements were observed, measured, and recorded by early people, who noted their relationship to the moon. Many hypotheses of tidal cause and techniques of tidal prediction were de-

TABLE 10-2

Events Associated with the Earthquake at Prince William Sound, Alaska, 1964

Time	Event
03:36 A.M.	Earthquake occurs off the northern shore of Prince William Sound, Alaska.
03:44	Seismic sea-wave warning alarm rings in Tsunami Warning Center, Honolulu, Hawaii.
04:35	Kodiak, Alaska, experiences tsunami wave 3–4 m (10–13 ft.) above the mean sea level.
05:02	First warning issued.
05:55	Kodiak confirms existence of tsunami.
07:00	Tsunami reaches Tofino, British Columbia, Canada.
07:08	Kodiak reports waves of 11 m (36 ft.) at 05:40; 12 m (39 ft.) at 06:30; 10 m (33 ft.); seas diminishing.
07:39	1 m (3 ft.) wave arrives in Crescent City, California; some evacuees foolishly return thereafter.
07:50	Four persons drown in DePoe Bay, Oregon.
09:00	Tsunami reaches Hilo, Hawaiian Islands.
09:20	4 m (13 ft.) wave (probably the fourth) sweeps into Crescent City, causing great damage.
10:20	Tsunami reaches east coast of Hokkaido, Japan.
10:38	Tsunami reaches northeast coast of Honshu, Japan.
01:55 P.M.	Tsunami reaches Kwajalein, Marshall Islands.
07:10	Tsunami reaches La Punta, Peru.

Source: *Tsunami Warning System in the Pacific*, Intergovernmental Oceanographic Commission publication.

veloped in the eighteenth and nineteenth centuries. Prediction has recently been improved, mainly through the use of high-speed computers.

Causes of the Tides

Gravity and Inertia. Tides are caused by two forces, gravity and inertia (gravity was discussed in Chapter 4, page 71). To refresh your memory the gravitational attraction between two bodies is directly proportional to their masses and inversely proportional to the square of their distance apart. (In mathematical terms, gravity is proportional to mass/distance^2.)

The bodies of interest here are the Earth, sun, and moon. Earth and the moon are about 400,000 km

(nearly a quarter of a million miles) apart. Earth and the sun are about 150 million km (about 93 million miles) distant. The mass of the sun is about 27 million times that of the moon. The moon, however, is very roughly 400 times closer to Earth than the sun, so its nearness gives it a greater effect. The bottom line is that the gravitational effect between the moon and Earth is about twice that of the sun.

The gravitational attraction among Earth, the moon, and the sun helps keep these bodies in their orbital relations to one another. However, without some counterbalancing force to gravity, Earth, the sun, and the moon would all be pulled against each other. This counterbalancing force is **inertia,** which is the tendency of a moving object to continue moving in a straight line. It is this

FIGURE 10-17 A simple stationary (standing) wave, or seiche. After the water has been put in motion, it continues to move (slosh) back and forth, forming the seiche. Arrow indicate direction of wave motion.

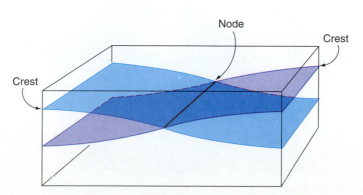

force (sometimes called *centrifugal force*) that holds water in a bucket when you swing the bucket in an overhead arc. It is inertia that makes an automobile tend to go straight when you are trying to make a tight turn at high speed.

Without inertia, Earth, the moon, and the sun would crash together quickly. Without gravity, they would fly apart into the Universe. Because these forces are in balance, these bodies have maintained their present orbits for millions of years. This is an overall balance, for either gravity or inertia may prevail briefly at various positions of Earth and the moon in their orbits.

Gravitational attraction affects everything on Earth— solid earth, atmosphere, and water—but the results on the first two cannot be observed by the unaided eye. The effect on the oceans, however, is obvious: the daily tides, which are low, Earth-spanning wave forms that vary widely in height when they reach a coast, depending on the nature and location of the coastline.

THE MOON. Because the moon is so much closer to Earth than the sun, its gravitational effect on the tides is about twice that of the sun. Let us first consider the sim-ple case of the gravitational attraction between Earth and moon (Figure 10-18). The gravitational attraction is strongest on the side of Earth that happens to be facing the moon, simply because it is closer. This attraction causes the water on this "near side" of Earth (*N*) to be pulled toward the moon. Inertia is also active here, at-tempting to keep the water in place. But the gravitational force exceeds it and the water is pulled toward the moon, causing a "bulge" of water on the near side toward the moon.

On the opposite side of Earth ("far side," *F*), the gravitational attraction of the moon is less because it is farther away. Here, the inertial force exceeds the gravitational force, and the water tries to keep going in a straight line and thus moves away from Earth, also forming a bulge.

Thus, the combination of gravity and inertia creates two bulges of water. One forms where Earth and moon are closest, and the other forms where they are farthest apart. Over the rest of the globe the two forces are relatively in balance.

Because water is fluid, the two bulges stay essentially aligned with the moon as Earth rotates. Thus, a coastal area on Earth may experience two tidal highs when it

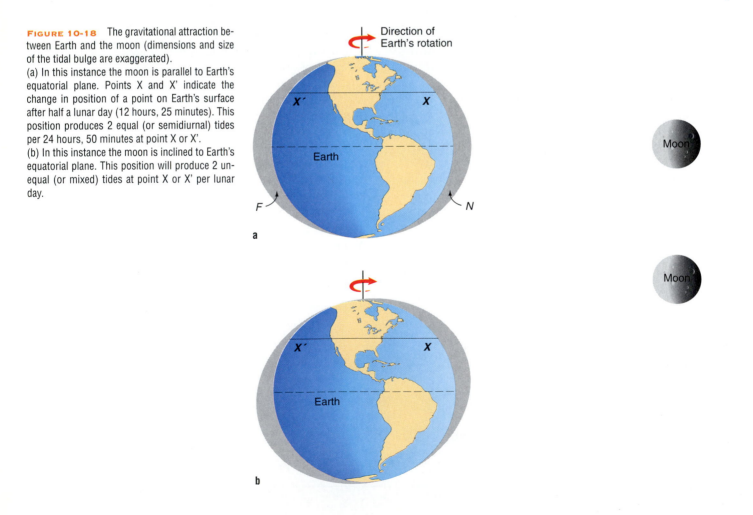

FIGURE 10-18 The gravitational attraction be-tween Earth and the moon (dimensions and size of the tidal bulge are exaggerated).
(a) In this instance the moon is parallel to Earth's equatorial plane. Points X and X' indicate the change in position of a point on Earth's surface after half a lunar day (12 hours, 25 minutes). This position produces 2 equal (or semidiurnal) tides per 24 hours, 50 minutes at point X or X'.
(b) In this instance the moon is inclined to Earth's equatorial plane. This position will produce 2 un-equal (or mixed) tides at point X or X' per lunar day.

passes through the bulges (the moon side and the opposite side) within this time period. Of course, two lows also are experienced when the shore is outside of the bulge area. As we shall see shortly, three possible tidal patterns actually exist.

It takes Earth about 24 hours to rotate once, relative to the sun. But, because the moon is moving with respect to Earth as Earth is spinning, it takes Earth a little longer to complete a rotation relative to the moon—24 hours and 50 minutes. Thus, two daily tides occur 12 hours, 25 minutes apart.

THE SUN. The gravity of the sun also exerts a strong tidal influence on the ocean. The tidal bulge produced by the sun is only 46 percent of that produced by the moon, due to the sun's greater distance from Earth. Remember that although the gravitational force between two bodies varies directly with their mass, it also diminishes inversely with the square of the distance between them.

The effect of the sun becomes especially important when the sun and moon happen to align with Earth. The combined gravitational attraction of the two bodies produces a very strong tide that "springs forth" onto the coast, and thus is called a **spring tide.** (Spring tides have nothing to do with the season of spring.) Spring tides occur roughly every 14 days, at the times of new moon and full moon (Figure 10-19).

Relatively weak tides, called **neap tides,** occur when the sun and the moon form a right angle with the Earth and essentially are opposed to each other. This also occurs about every 14 days, at first-quarter moon and last-quarter moon. The **tidal range** (the maximum water height at high tide, minus the minimum water height at low tide) is greater than average during spring tides and less than average during neap tides.

The orbits of the moon and Earth are not perfectly circular but rather are ellipses. This causes their distance apart to vary; sometimes they are closer together and sometimes they are farther apart, and the same is true for their distance to the sun. Under certain conditions, this can cause extreme tidal conditions (see Box 10-2).

Tidal Currents

The tides are basically large, low wave forms that cause currents, called **tidal currents.** Tidal currents in the open ocean are relatively weak; near land, however, they can attain speeds of several kilometers per hour. Tidal currents in shallow water and estuaries can be very important geologically. They can move large amounts of sediment that can eventually shoal or block harbors and must be removed by dredging. In some estuaries during high tide, a large wave forms and travels upstream. Called a **tidal bore,** this wave can be as high as 3.3 m

(10 ft.) or more and have speeds over 15 km (about 9.3 miles) per hour.

The tidal range worldwide averages between 1 and 3 m (3.3 to 3.9 ft.), but can attain 20 m (about 65 ft.) in some areas, such as the Bay of Fundy in Nova Scotia or the Gulf of California. Such exceptional tides are generally due to the geographic position and geometry of an area. A long V-shaped or U-shaped basin (such as the two examples cited), facing into the direction of the incoming tide, can generally compress and focus the incoming water to form higher tides.

When the tide is rising—water is coming in toward the shore—the current is called a **flood tide.** Eventually the incoming water attains high-tide level and starts to withdraw or fall. The current produced by the falling current is called the **ebb tide.**

There are three basic tidal patterns. Most areas have two high tides and two low tides a day. When the two highs and the two lows are about the same amplitude, the pattern is called a semidaily or **semidiurnal tide** (Figure 10-20). If the highs as well as the lows each differ in height, the pattern is called a **mixed tide.** Some areas, such as the Gulf of Mexico, have only one high and one low tide each day, which is called a **diurnal tide.** The U.S. West Coast tends to have mixed tides, whereas a semidiurnal pattern is more typical of the East Coast. Figure 10-18 shows why some areas have semidiurnal tides and others have mixed tides.

Because the time interval between high tides is about 12 hours, 25 minutes (half a lunar day), high tide occurs about 50 minutes later every day (2×25 minutes). This simple fact shows that tides are primarily influenced by the moon. If they were mainly controlled by the sun, they would occur at the same time every day, based on our normal 24-hour solar day.

The Importance of the Tides

The tides are important for several reasons. Tidal mixing of nearshore waters removes pollutants and recirculates nutrients. Tidal currents also move floating animals and plants to and from their usual breeding areas in estuaries to deeper waters. People who fish frequently follow tidal cycles to improve their catch, because strong tidal currents concentrate bait and smaller fish, thus attracting larger fish. When sailing ships were more common, departures or arrivals in a harbor had to be closely linked to the tidal cycle.

The rising and falling motion of the tides has potential for generating electrical power. This has been applied in only a few areas. As you might expect, the greater the tidal range in an area, the greater the amount of energy that is available. Therefore, the Bay of Fundy is a prime candidate. A tidal power plant already exists along the

FIGURE 10-19   The four phases of the moon and how they and the sun produce the spring and neap tides. Neap tides occur when the trough of the solar tide is aligned with the crest of the lunar tide or when the trough of the lunar tide is aligned with the crest of the solar tide. Spring tides occur when the crests of both lunar and solar tides are aligned. The view is from the North Pole and the tides are greatly exaggerated. The red arrow indicates the direction of Earth's rotation.

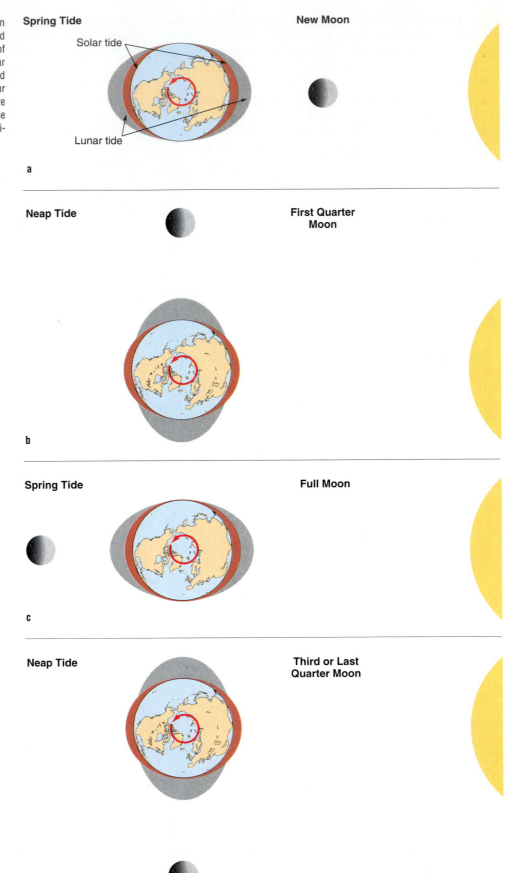

Box 10-2

High Tides

In late 1990, a series of exceptionally high tides occurred. Newspapers called them "the high tides of a lifetime." That description was a bit extravagant, because considerably higher water can occur when normal high tides and storms coincide. Nevertheless, in a statistical sense, and eliminating any effect of storms or onshore winds, these tides indeed were higher than any predicted for the next 60 years.

These high tides were due to an unusual alignment of Earth, the moon, and the sun. As shown in Figure 10-19, high tides occur twice a month at full moon and new moon, when these bodies align with Earth and the gravitational force of the sun and moon are in the same direction. Astronomers call this situation a syzygy (*SIH-zih-jee*), a Greek word meaning "yoked together."

Several other things happened in late 1990 to make the tides so significant. One was that the high tides occurred when Earth was closest to the sun. The other two things were the positions of the moon and sun relative to Earth's equator, complications that had the effect of increasing the tides. The net result was to produce a tide slightly higher than normal, by a few centimeters (an inch or two).

The coincidence of these alignments led certain scientists—who generally were not oceanographers—to predict that the strong tidal pull would also affect land and cause major earthquakes. Some predictions especially emphasized this possibility for the central United States. Most earth scientists and oceanographers rejected such predictions, and indeed the period passed uneventfully, with no unusual earthquakes.

Diurnal Tide   New Orleans, Louisiana

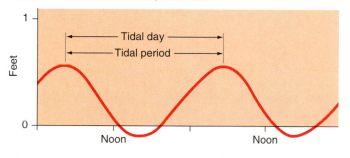

Semidiurnal Tide   New York City (The Battery, New York Harbor)

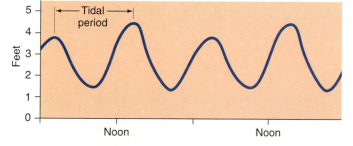

Mixed Tide   Los Angeles (outer harbor)

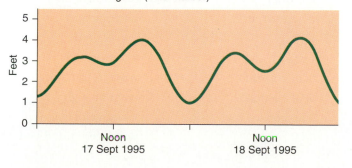

Rance River in France (see Figure 18-4, page 446). This use of the ocean is discussed further in Chapter 18.

Tidal Friction

You might think that Earth rotates smoothly beneath its tidal bulges, but research shows that this is not so. The tide causes friction between Earth and its ocean. This **tidal friction** actually slows Earth's rotation, although the slowing rate is extremely small—it has increased the time between sunrise and sunset by just 0.001 second in the last 100 years!

Nevertheless, this small rate can become significant over billions of years of geologic time. About 400 million years ago, the number of days in a year was close to 400, and the length of these days would have been about 22 hours, because the time it took Earth to make one complete revolution around the sun has not changed significantly. (Scientists are able to determine these figures by counting the daily growth rings preserved in ancient coral, similar to the way trees have annual growth rings.)

There were other effects of a change in the speed of Earth's rotation. It would have affected the moon, because these two bodies are closely linked by gravity. If Earth spun faster, the moon did, too. For the forces of gravity and inertia to balance, Earth and the moon must have been closer together. With the moon closer to

FIGURE 10-20   The three principal types of tides: diurnal, semidiurnal and mixed.

Earth, tides would have been much stronger than today, because of stronger gravitational attraction.

These stronger tidal currents would have created biologic and geologic conditions different from those common today. Some scientists have suggested that the increased tides due to the proximity of the moon may have provided the impetus for the evolution of hard-shelled organisms. Soft-shelled organisms, living in shallow-

water conditions, probably would have had difficulty surviving in such a rigorous environment. Geologically, vast inland seas, flushed once or twice daily by high tides, would have covered many of the low-lying areas of the world.

Putting this all together indicates that, in the geological past, there were more—but shorter—days in the year, the moon was closer to Earth, and the tides were higher.

SUMMARY

Winds blowing on the ocean surface can form waves. Many characteristics of wind-formed waves are related to wind velocity, duration of time that the wind blows, and the fetch, or extent of water over which the wind blows. Characteristics of waves remain fairly constant once they leave the generating area, and they travel away as swell. When they reach the coastal zone the shallowness of the water causes changes in many characteristics and ultimately results in breaking waves that cause erosion or deposition of sediment. The breaking waves usually cause longshore and rip currents.

Among the other categories of waves are tsunami and storm surges, both of which can be dangerous and can result in the loss of life and property. Other wave types are seiches (stationary waves) and internal waves.

Tides, the daily or twice-daily rhythmic rising and falling of sea level, are caused by the gravitational attraction among Earth, the moon, and the sun. Tides will cause coastal currents that can mix coastal waters and move floating plants and animals. In areas having a high tidal range, there is potential for using tidal movement as an energy source.

QUESTIONS

1. What factors influence the character of surface ocean waves?

2. Describe the development of a wind-generated wave from the area where it forms until it breaks on a beach.

3. Describe a rip current, how it forms, why it is dangerous, and what to do if caught in one.

4. What is a tsunami? How is it formed?

5. Why are there tides? How are they caused?

6. Why do tides alternate in strength twice a month?

KEY TERMS

Beaufort Wind Scale
capillary wave
catastrophic wave
convergence
diurnal tide
divergence
ebb tide
fetch
flood tide
gravity
inertia

internal wave
landslide surge
Langmuir cell
longshore current
mixed tide
neap tide
period refraction
rip current
rogue wave
sea
seiche

semidiurnal tide
spring tide
standing wave
stationary wave
storm surge
surf
swell
tidal current
tidal friction
tidal bore
tidal range

tsunami
wave crest
water depth
waveform
wave height
wavelength
wave period
wave refraction
wave trough
wave velocity
wind-generated wave

FURTHER READING

Fox, W. T. 1991. *At the Sea's Edge: An Introduction to Coastal Oceanography for the Amateur Naturalist.* New York, Prentice-Hall. A good and not very complex introduction to coastal marine science.

Garrett, C., and L.R.M. Maas. 1993. "Tides and Their Effects." *Oceanus* 36, no. 1, pp. 27–37. Describes how the tides can impact many aspects of coastal mixing and circulation.

Kampion, Drew. 1989. *The Book of Waves: Form and Beauty on the Oceans,* Santa Barbara, Calif.: Arpel Graphics. A beautifully illustrated book on waves.

Knauss, J. A. 1978. *Introduction to Physical Oceanography.* Englewood Cliffs, N.J: Prentice-Hall. A general but slightly advanced text on physical oceanography.

Open University. 1989. *Ocean Circulation.* Oxford: Pergamon Press. An advanced but good description of various forms of ocean circulation.

———. 1989. *Waves, Tides and Shallow-Water Processes.* Oxford: Pergamon Press. An advanced but solid description of tides and waves.

"Physical Oceanography." *Oceanus* 35, no. 2 (Summer 1992). A special issue on recent research in physical oceanography.

Pickard, G. L., and W. J. Emery. 1982. *Descriptive Physical Oceanography: An Introduction.* 4th ed. Oxford: Pergamon Press. An advanced but good textbook on general physical oceanography.

Sverdrup, H. U., M. W. Johnson, and R. H. Fleming. 1942. *The Oceans: Their Physics, Chemistry, and General Biology.* Englewood Cliffs, N.J: Prentice Hall. This book generally is considered the "Bible" of books on the ocean. It is complex, but contains a wealth of information about the ocean.

CHAPTER 11

Life Forms of the Ocean

A GREAT MANY OF THE WORLD'S LARGEST, MOST BEAUTIFUL, AND MOST EXOTIC ORGANISMS LIVE IN THE OCEAN. INDEED, AFTER MORE THAN A CENTURY OF STUDY, NEW MARINE CREATURES ARE STILL BEING DISCOVERED. THE OCEAN OFFERS A WIDE RANGE OF ENVIRONMENTAL CONDITIONS FOR LIFE. TWO DISTINCT ENVIRONMENTS, THE OCEAN WATER AND THE SEAFLOOR, PROVIDE NUMEROUS AND VARIED HABITATS FOR MARINE PLANTS AND ANIMALS. IN THIS CHAPTER WE WILL STUDY OCEAN ORGANISMS: THEIR HABITATS, MODES OF LOCOMOTION, AND GENERAL BIOLOGICAL CLASSIFICATIONS.

A jellyfish. Note the small fish (about 10 cm or 4 in. long) living within the jellyfish.

(Photograph courtesy of Dr. Laurence P. Madin, Woods Hole Oceanographic Institution.)

Introduction

Based on biology, we can divide the marine environment into two realms: the **benthic,** which applies to the seafloor or ocean bottom, and the **pelagic,** which refers to the overlying water (Figure 11-1). The pelagic realm can be further subdivided into a **neritic system,** meaning the water overlying the continental shelf, and an **oceanic system,** meaning the water overlying deeper parts of the ocean. The benthic realm is usually divided into a **littoral system,** or the tidal zone, out to a depth of 200 m (660 ft.), and a **deep-sea system.**

Further depth divisions are possible in both the benthic and pelagic realms (Table 11-1). Note that some similar terms are used in the table for both the benthic and pelagic realms. Also note that the depth divisions are not absolute figures.

Another way of dividing the ocean is based on the lowest depth at which sufficient light for photosynthesis can usually penetrate the water—about 200 m (about 660 ft.). The upper, lighted area is called the **photic zone;** it is within this region that photosynthesis by plants occurs. The deeper, darker region is the **aphotic zone.**

These environments, which are defined by physical aspects such as depth and light penetration, frequently represent environmental boundaries for organisms. In other words, assemblages of organisms (called **communities**) coincide with the physical environment. Examples include organisms that are restricted to a certain depth range, or plants that can photosynthesize only where sufficient light is available. Such an environmental region is called a **biotope.**

TABLE 11-1

Depth Ranges of the Biological Environments of the Ocean

Depth	Benthic Environment	Pelagic Environment
	Littoral system	
Above high tide	Supratidal	
High tide to low tide	Intertidal	
0 to 200 m (0 to 660 ft.)	Subtidal	*Neritic system*
	Deep-sea system	*Oceanic system*
200 to 2,000 m (660 to 6,600 ft.)	Bathyal	Bathyal
2,000 to 6,000 m (6,600 to 20,000 ft.)	Abyssal	Abyssal
More than 6,000 m (more than 20,000 ft.)	Hadal	Hadal

Benthic Environments

The **benthic environment** is the ocean bottom. It covers a wide range of oceanographic conditions, from the exposed shoreline areas to the deep hadal environment (see Table 11-1 and Figure 11-1). Benthic organisms (those that live on the ocean sediments, or in them) vary according to the different conditions of the environment. Thus, benthic plants and animals of the deep sea are different from those of coastal waters.

FIGURE 11-1 Divisions of the marine environment. The two large divisions are the benthic realm (seafloor) and pelagic realm (water). Each is divided into subsystems.

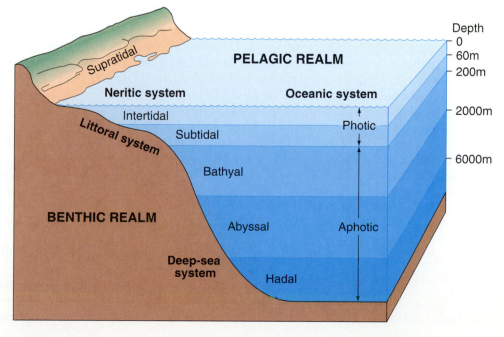

Littoral System

Littoral is from a Latin word meaning "of the shore." Thus the **littoral system** is the shore system, and it is divided into three subsystems. The supratidal ("above the tide") is usually exposed to air. The intertidal ("between the tides") is exposed to air or covered by water, depending on the tides. The subtidal ("below the tide") is nearly always submerged.

The **supratidal** environment is extremely stressful for marine organisms. The term "fish out of water" loosely applies to these creatures, for animals and plants living there are exposed to air almost continuously, being immersed only during periods of extremely high tides and storms and by the spray from breaking waves. Living forms in this environment—generally small gastropods (snails and slugs) and lichens on rocks, and crabs and amphipods (shrimplike animals) on beaches—are similar the world over, and are generally resistant to desiccation.

The **intertidal** environment includes the region periodically exposed at low tide. The width of the intertidal region depends on the tidal range and the slope of the ocean bottom. Animals living in this environment must also withstand the effect of breaking waves. Some do this by attaching themselves to rocks; others by burrowing into the bottom, which also lessens some of the harmful effects of exposure at low tide.

The outer edge of the intertidal environment is near the depth limit at which most attached plants can grow on the bottom; they cannot grow in deeper water because there is insufficient light for photosynthesis. The kelp beds off the California coast are an exception; these plants attached to the bottom can grow up to the sea surface and receive their needed sunlight. Actually, only a small portion of the seafloor is available for the growth of attached plants, and even within this small area many portions cannot be used because the bottom may be muddy or otherwise unsuitable for plants. The intertidal region may be the best-studied marine biological environment, especially because it can be observed just by walking or swimming. The life forms in this environment are very numerous and varied.

The outer part of the littoral system is the **subtidal** division. Its depth can extend to 200 m (660 ft.) or more. Organisms living here are continuously underwater. Photosynthetic producers decrease with depth because light needed for photosynthesis decreases. The outer part of the subtidal division generally conforms to the edge of the continental shelf.

Deep-Sea System

The deep-sea system is composed of the *bathyal, abyssal,* and *hadal* divisions. Understanding the origin of the words helps with remembering these terms. They basically mean "deep," "deeper," and "deepest": *bathy-* means deep; *a-byss* means "without bottom," and *Hades* refers to the underworld of Greek mythology. The deep-sea system is not as well known as the shallower littoral system. The deep-sea system is devoid of plant life because there is no light for photosynthesis; however, certain animals and bacteria can live at these depths.

Oceanographic conditions of the deeper parts of the ocean (below depths of 200 m, or 660 ft.) are relatively uniform. Temperature decreases slowly with depth, salinity is essentially constant, and pressure increases 1 atmosphere with each 10 m (33 ft.) of depth.

Relatively little is known about the biological environment of the deep sea. Most of our information about the deep sea is derived from underwater photographs, deep-sea dredgings, and more recently from bottom observations made in submersibles and ROVs.

Because most organisms in this environment are composed primarily of water, with few if any air spaces, and because water is not very compressible, deep-sea pressure itself is not an excluding factor for life there. Based on laboratory studies, however, pressure can impact specific life processes, such as the ability to produce proteins. Thus organisms that live in the deep sea must have adapted to overcome such problems.

Food in the deep sea is not as abundant as in shallower waters. Animals living in the deep sea are thought to receive most of their food from organic material that settles from the near-surface waters to the ocean bottom (see Figure 11-24b, for example). The production of organic matter in surface waters is usually greater in coastal waters and decreases with distance from land, becoming much less over the abyssal plains. This quantity is further reduced by its disintegration and decay while sinking through the water (see Box 11-1). Thus the amount of food reaching the deep-ocean bottom generally is small. It follows, therefore, that the animals of the deep sea should be scavengers rather than large predators. Many animals of the deep sea are bizarre in appearance (Figure 11-2) and are often small.

The density of life in the ocean can be expressed by a measure called **biomass,** which refers to the amount of living organisms in grams per square meter of the ocean bottom. Biomass typically shows high values in the littoral region and low values in deep water (Table 11-2).

Pelagic Environments

The **pelagic environment** is the water above the seafloor. It can be divided into a nearshore *neritic system* and an offshore *oceanic system.* The border between the

Box 11-1

Marine Snow

During shallow-water diving, oceanographers long have noted that the water often contains numerous fine-grained particles or clumps of particles. When scientists were able to dive into deeper parts of the ocean with submersibles, they found that these same types of particles were present and were frequently very visible because they reflected the powerful lights from the submersible. Generally called **marine snow,** these particles have been examined microscopically and photographed (see Figure 1).

Marine snow aggregates can attain a few centimeters in diameter, but are generally only a few millimeters wide. Subsequent collection of seawater and filtering showed the particles to be a mixture of living organisms (including bacteria), detritus, fecal material, and possibly some inorganic material such as clay particles, often in a gelatinous mix. The very fragile aggregates can rarely be sampled with conventional techniques because they break apart easily. They have been collected using sediment traps (see Figure 7-7, page 166) suspended at various water depths or on the seafloor and described as the oceanographer's equivalent of a rain gauge.

Regardless of how carefully they are collected samples can still break, and thus representative samples are rare. Each aggregate is essentially a small ecosystem, complete with nutrients. Recent studies using microelectrodes have detected oxygen and pH gradients around marine snow particles. This suggests that both photosynthesis and respiration are occurring at rates different from the surrounding water. These properties can change on the time scale of hours as the microecosystem changes.

Marine snow, as expected, is generally more abundant in surface waters. As particles settle toward the seafloor, they may lose or gain components. An important biological and chemical aspect of marine snow is that it is a mechanism for particles to reach the ocean bottom quickly, where they can provide food or become part of the bottom sediment. In some regions, the seafloor has a thin layering of fluff—the collection from a marine snowfall—a few weeks after a surface plankton bloom.

It also appears that a large amount of marine snow may be eaten in the upper parts of the water column and "repackaged" as fecal pellets that then sink to the seafloor.

FIGURE 1   Marine snow photographed at 800 m (about 2,600 ft.) depth off Panama in the Pacific Ocean. The distance along the bottom of the figure is about 10 cm (about 4 in.). (Photograph courtesy of Dr. Susumu Honjo, Woods Hole Oceanographic Institution.)

two is not very definite and is often set at the edge of the continental shelf, similar to the border between the benthic littoral and benthic deep-sea systems. Figure 11-3 summarizes important characteristics of the marine environment.

Neritic System

The **neritic system** of the pelagic environment generally has very diverse conditions, especially at or near the freshwater discharge from a river. Organisms living in this environment must be able to tolerate a wide range in salinity. Nutrients can enter the nearshore neritic environment by upwelling due to coastal winds (see Figure 9-17, page 215) or from rivers.

A good nutrient supply will sustain an abundant growth of phytoplankton, the basic food of the sea. This, in turn, attracts other forms of life, developing complex food interactions. As a result, the neritic area is generally the ocean's most biologically productive region, and is where most of the seafood (fish and shellfish) consumed by people is harvested.

FIGURE 11-2 (a) Some deep-sea fish. They look ferocious, but are usually very small. Lengths are in centimeters (2.54 cm = 1 in.).
(b) Professor Bob Griffith of the University of Massachusetts at Dartmouth examines a viper fish.
(c) One of the bizarre fishes of the deep sea, commonly called a tripod fish (*Bathypterois bigelow*). (Illustration (a) adapted from R. V. Tait and R. S. DeSanto, *Elements of Marine Ecology* (New York: Springer-Verlag, 1972). Figure (b) courtesy of Prof. Bob Griffith; (c) courtesy of Dr. C. D. Hollister, Woods Hole Oceanographic Institution.)

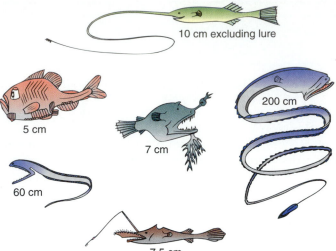

a

b

c

TABLE 11-2
Average Biomass in Different Parts of the Ocean

Area	Biomass (grams/m²)
Coastal zone	100–5,000
50 to 200 m depth (about 165 to 660 ft.)	200
about 4,000 m (about 13,100 ft.)	About 5
Kuril-Kamchatka Trench	
about 6,000 m (about 20,000 ft.)	1.2
about 8,500 m (about 27,900 ft.)	0.3
Tonga Trench, 10,500 m (34,440 ft.)	0.001

Oceanic System

The **oceanic system** of the pelagic environment can be divided as shown in Table 11-1 or, as previously mentioned, into light (photic) and dark (aphotic) zones, with the boundary at about 200 m (660 ft.) or sometimes less (Figure 11-3). Some of the surface-water conditions of the neritic environment, such as its salinity, can vary somewhat due to river discharge. Temperature decreases with depth, with the greatest vertical change occurring in the thermocline. The temperature of surface waters varies with latitude (see Figure 8-12, page 191). Nutrients are usually low in the surface waters and increase with depth.

The deep ocean is an area of almost complete darkness. Currents on the whole are relatively slow, a few centimeters a second. There are very subtle changes in temperature or salinity, if any. The abyssal-pelagic area is

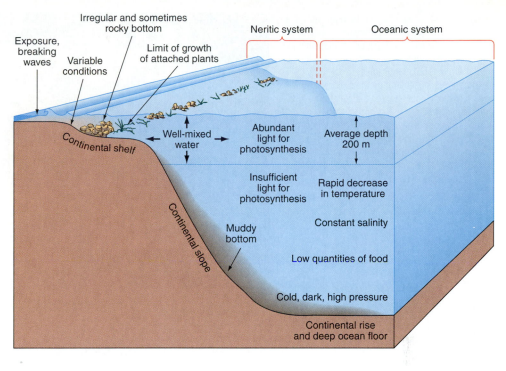

FIGURE 11-3 General characteristics of the marine environment. Note the variable conditions typical of the upper part of the ocean compared to the relatively constant conditions in the deeper part.

one of the world's largest ecological units, enclosing about three-quarters of the total volume of the ocean.

Relatively little is known about the deep hadal-pelagic zone, mainly because of the difficulty of sampling the animals that live in these waters. The hadal-benthic life has been better sampled.

For decades it was generally believed that the pelagic realm of the ocean was something of a biological desert. However, recent studies using improved techniques—particularly for measuring extremely small plankton—have shown that the productivity may have been underestimated by a factor of two or three. Hence categorizing the pelagic realm as a desert is perhaps unfair, but it is still considerably less productive than coastal regions. Notable exceptions are the deep-sea vent communities, discussed in Chapters 7 and 12.

Classifying the Ocean Population

In biology, several ways exist to classify living things, including physical characteristics, habitat, and whether or not they produce their own food. One very general way of classifying ocean organisms is by their mode of locomotion or movement, which relates to their main habitat. In this manner we can divide ocean organisms into three large groups: plankton (floaters with limited movement), nekton (swimmers), and benthos (seafloor dwellers).

Plankton

Plankton (from the Greek word for "wandering") is a general term for organisms with very weak or limited locomotion. They are moved mainly by ocean currents (Figures 11-4 and 11-5; see also Figure 11-12, page 260). Plankton can be either animals (zooplankton), plants (phytoplankton), or bacteria (bacterioplankton). Most plankton are microscopic, but the group also includes some large floating forms, such as jellyfish and sargassum weed. In terms of biomass, plankton comprise the largest group of organisms in the ocean.

PHYTOPLANKTON. **Phytoplankton** are the base of the oceanic food chain and thus may be the most important individual group of life forms in the sea. These organisms convert water and carbon dioxide into organic material through photosynthesis. As plants, phytoplankton must live in the photic zone. Their distribution and growth, however, can show pronounced vertical and seasonal variations. The vertical distribution is due to the depth to which sufficient light for photosynthesis can penetrate, a depth that may be only 1 m (a little over 3 ft.) in nearshore sediment-laden waters, to perhaps as much as 200 m (660 ft.) in the clear open ocean (see Figure 12-6, page 287).

Seasonal changes in phytoplankton distribution are due to various relationships among light, nutrient supply, temperature, herbivores (organisms that eat

FIGURE 11-4   (a) This photograph shows a variety of plankton. The large organism slightly to the right of center is a copepod, a zooplankton. The dimensions of the illustration are 1.7 by 1.0 mm (1.0 mm = 0.039 inches).
(b) Diagrammatic illustration of some representative marine phytoplankton. 1–6 are diatoms, 7–10 are dinoflagellates, 11 is a coccolithophore, and 12 is a blue-green alga. All are microscopic. (Photograph (a) courtesy Dr. Don Anderson, Woods Hole Oceanographic Institution. Illustration (b) from Nybakken, *Marine Biology*, 3d ed. (New York: HarperCollins, 1993).)

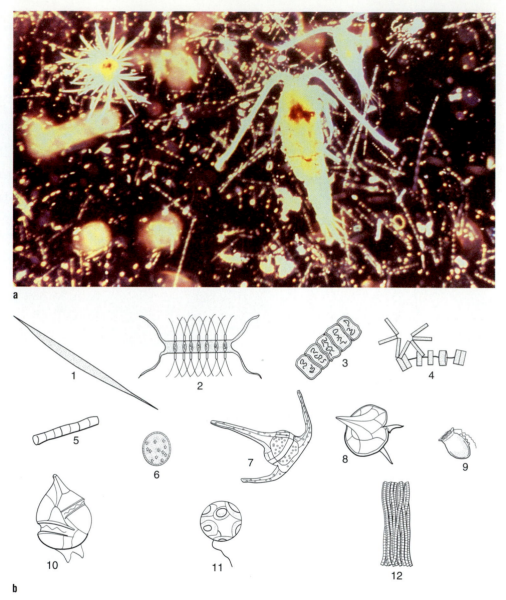

plants), and other factors. Phytoplankton growth can be very rapid, sometimes as much as six cell divisions per day. Because each cell division doubles the number of organisms, 500 organisms could bloom into 32,000 in one day.

ZOOPLANKTON.   Many sea animals have a planktonic stage, usually beginning at birth, during which they float freely in the ocean, relocating and absorbing nutrients as they go. **Zooplankton** include representatives from almost every group of marine animal.

Zooplankton are of two types: **holoplankton** (permanent plankton), which spend their entire lives as plankton, or **meroplankton** (temporary plankton), which spend only a portion of their lives (usually the lar-

val stage) as plankton. At other life stages, meroplankton "grow up" to become nektonic (swimming) or benthic (bottom-dwelling) forms.

Zooplankton are also very important in the economy of the sea because they eat phytoplankton. This makes phytoplankton-derived energy, in the form of food, available for higher life forms when they in turn eat the zooplankton. Zooplankton that feed on phytoplankton are frequently called grazers or herbivores.

Plankton also can be classified by size. The smaller ones are:

Ultraplankton (less than 0.005 mm in diameter) can include bacteria (bacterioplankton). Because of their small size, this group is difficult to study.

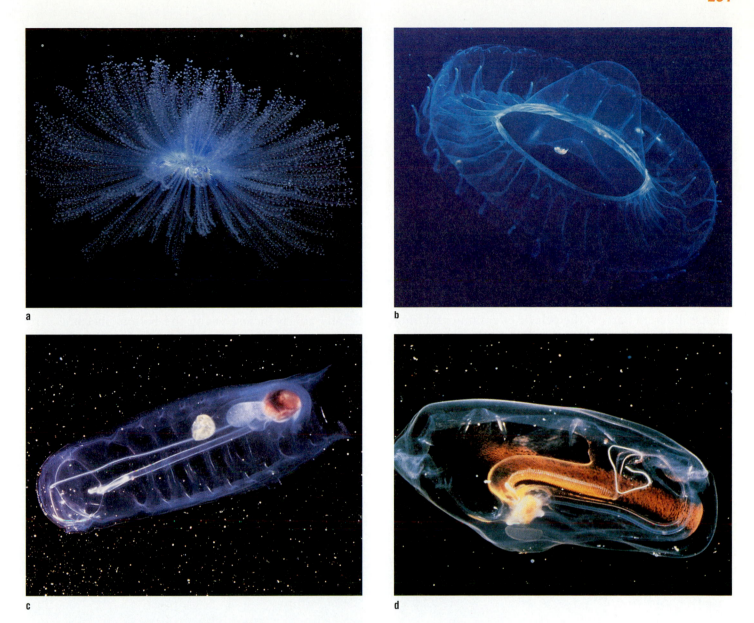

a

b

c

d

FIGURE 11-5 Examples of some types of plankton: (a) and (b) are jellyfish; (c) and (d) are salps, which are tunicates. In a, note the small fish (about 10 cm, or 4 in. long) living within the jellyfish. (Photographs courtesy of Dr. Laurence P. Madin, Woods Hole Oceanographic Institution.)

Nanoplankton (0.005 to 0.07 mm) are also hard to study since they generally pass through the openings of collecting nets.

Microplankton (0.07 to 1 mm) are large enough to be captured in nets and can be studied with conventional microscopic techniques.

On the other end of the scale are plankton such as jellyfish, some of which can attain 15 m (49 ft.), including tentacles. These are called **macroplankton.**

Some plankton are especially valuable to oceanographers because certain species are characteristic of particular bodies of water. These species, called **indicator species,** can be used to trace the origin and movement of the water body. Many areas of the seafloor are covered with the shells, or **tests,** of planktonic organisms.

A geologically important group of organisms is the protozoan Foraminifera (see Figure 5-18a and 18b, page 124) whose shells, and even the coiling direction of chambers in the shell, provide valuable information about the history and climate of the ocean when these organisms lived.

Nekton

Nekton (from the Greek word for "swimming") include animals that are able to swim freely, independent of ocean currents (Figure 11-6). This group excludes plants, which may float but do not actively swim. Nekton encompass many advanced animal forms, such as fish and whales and other sea-dwelling mammals. Nekton have the mobility to search actively for food and to avoid predators. These animals also can migrate across the ocean. Most exist either on or near the ocean surface or at depths on the ocean bottom (see Figure 11-2).

Nekton are commercially important to humans and, because of their feeding habits, strongly affect other forms of life. Many nekton feed mainly on plankton, limiting and controlling the phytoplankton population.

Nekton are generally the least restricted form of animal life in the sea (see Box 11-2). Although they may inhabit different parts of the pelagic environment during their lives, their distribution is still limited somewhat by temperature and pressure. The influence of these and other environmental factors is not as well understood as it is for the benthic and planktonic life forms.

Some animals (and plants) have **gas bladders** to adjust their buoyancy. The gas is similar in composition to atmospheric gases. For some plants, like the large brown kelp (see Figure 11-11, page 259), a gas bladder allows part of the plant to float near the surface while the rest remains attached to the seafloor. For animals, like some jellyfish, gas bladders allow them to float at or near the surface. Gas bladders have a disadvantage, too, since organisms that depend on them—especially fish—cannot move up and down rapidly (go into areas of different pressure) without first changing their buoyancy by adjusting their internal gas content.

Benthos

The term **benthos** comes from the Greek word for "deep" or "deep sea." Benthos (Figure 11-7), also called **benthic life,** are organisms that live on the ocean bottom, or in it. Some organisms, such as barnacles and oysters, have planktonic larval forms that eventually settle and attach themselves to the bottom (become benthos) for their adult lives. Other types of benthic life, such as worms and clams, may burrow into the bottom. Still others, such as starfish and echinoids, may creep slowly across the bottom. In water shallow enough to allow photosynthesis, some plants can form part of the benthic life.

a

b

FIGURE 11-6 Examples of some types of nekton.
(a) Porpoises swimming and playing in front of a research vessel.
(b) Yellowtail surgeonfish swimming near the Galápagos Islands. (Photograph (a) courtesy of Eben Franks; photograph (b) courtesy of T. M. Rioux; both of Woods Hole Oceanographic Institution.)

Box 11-2

Fish Antifreeze

Because of its salt content, seawater can remain a liquid at temperatures a few degrees below freezing (0°C or 32°F). As seawater approaches this temperature, it becomes denser (see Figure 6-7, page 138) and sinks. Thus, deeper water in the ocean is generally the coldest. An interesting aspect of this phenomenon is that animals, including fish, live successfully in such an environment without freezing. This is all the more remarkable because the body fluids within most organisms freeze at temperatures higher than the freezing point of seawater.

To accomplish this, the organism must prevent the growth of ice crystals within its blood and other body fluids. If such ice crystals were to grow, they would kill the animal by expanding and rupturing its cell walls. Recent studies have shown that some fish and probably other organisms can produce a type of antifreeze that prevents the growth of the life-threatening ice crystals.

Research on fish living in cold water has shown that their blood contains a variety of proteins or protein-sugar compounds that attach to any ice crystal as it begins to form, thus arresting further growth. Further studies undoubtedly will detect similar compounds in the other creatures that live in such cold conditions.

In the deep sea, benthic organisms live in perpetual darkness where food generally is in short supply. The environment is cold and uniform. This is very different from the shallow-water benthic environment, where conditions are variable and food is more plentiful.

Benthic organisms can have geologic importance because they can modify some physical and chemical properties of sediment during their life on the seafloor. In addition, many of the fossils in sediments are the remains of benthic organisms. Bacteria also are part of the benthic life. A small group of bacteria use *chemosynthetic* processes to produce organic material from inorganic substances and thus are an important food source in the deep sea.

Classifying Organisms by Their Physical Characteristics

The number and variety of life forms on Earth is impressive. Biologists estimate that more than 10 million species presently exist, although less than 30 percent have been fully described. They also believe that over 70 million species have lived and become extinct over geological time. To develop a comprehensive classification scheme for this many organisms is therefore quite a challenge.

Taxonomy is a system for classifying and naming organisms. There are various taxonomic systems, but in

FIGURE 11-7 Examples of some types of benthic (seafloor) life forms. (a) In this coral reef environment, animal in the center is *Tridacna*, a giant clam.

(b) A deeper-water environment showing a skatelike animal partially buried in bottom sediment. (Photograph (a) courtesy of Dr. Phillip Lobel.)

every one the highest-ranking category is the **kingdom.** Each kingdom (for example, Animalia) is subdivided by characteristics (backbone or not, feathers or not, and so on) all the way down to the species level (for example, *Larus atricilla,* the laughing gull).

Here is the taxonomic pattern, using as an example Kingdom Animalia and ourselves:

Taxonomic Level	Classification of Humans
Kingdom	Animalia
Phylum	Chordata (subphylum Vertebrata)
Class	Mammalia
Order	Primates
Family	Hominoidea
Genus	*Homo*
species	*sapiens*

Note that the last two levels, genus and species, are used for the scientific names of organisms, such as *Canis familiaris* (common dog) or *Acer saccharum* (sugar maple). These names are necessary because common names are too inconsistent for scientific use. Scientific names are in Latin because the system was devised in 1758, at a time when Latin was widely used by educated people.

The Five Kingdoms

This book uses a widely recognized taxonomic system that classifies all living things into five kingdoms:

Kingdom Monera Monera are primitive one-celled organisms that lack a cell nucleus. The two phyla are bacteria and blue-green algae.

Kingdom Protista Protista include one-celled organisms not included in Monera, and phytoplankton. Examples of Protista are algae, protozoans, slime molds, and some parasitic organisms. Important marine protists are foraminifera and radiolarians.

Kingdom Fungi Fungi are organisms that absorb organic food, such as mushrooms, molds, and lichens. These are not common in the deep sea, but can occur in intertidal zones as lichens, which are fungi living in a mutual relationship with algae. The algae provide the food and the lichens provide cover that retains water during exposure to the air.

Kingdom Plantae Plantae mainly comprise the common land plants, most of which have roots, stems, and leaves. These types of plants are relatively rare in the ocean, and if present are found only in coastal areas.

Kingdom Animalia Animalia includes all the multicellular animals, such as mollusks, fish, amphibians, reptiles, birds, and mammals.

Table 11-3 shows where some marine organisms fit into the five-kingdom scheme. Figure 11-8 shows the grouping and possible evolution of the five kingdoms.

The **species,** usually the smallest unit in a systematic classification, is generally reserved for organisms that can breed among themselves and produce fertile young, but cannot breed with different organisms. The five-kingdom approach used in this chapter is one of several ways to define organisms in an evolutionary scheme. (If you are interested in pursuing this subject further, see "Further Reading" at the end of this chapter for texts on biology or marine biology.)

TABLE 11-3

Common Marine Organisms and How They Fit into the Five-Kingdom Classification

Kingdom Monera	Kingdom Protista	Kingdom Fungi	Kingdom Plantae	Kingdom Animalia
Bacteria	Green algae	Lichens	Angiosperms	Porifera, sponges
Blue-green algae	Brown algae			Cnidaria, corals
	Red algae			Annelida, worms
	Yellow-green algae			Arthropoda, crustaceans
	diatoms			Mollusca, clams, squid
	dinoflagellates			Echinodermata, starfish
	coccolithophoridal			Chordata, fish, mammals
	Foraminifera			
	Radiolaria			

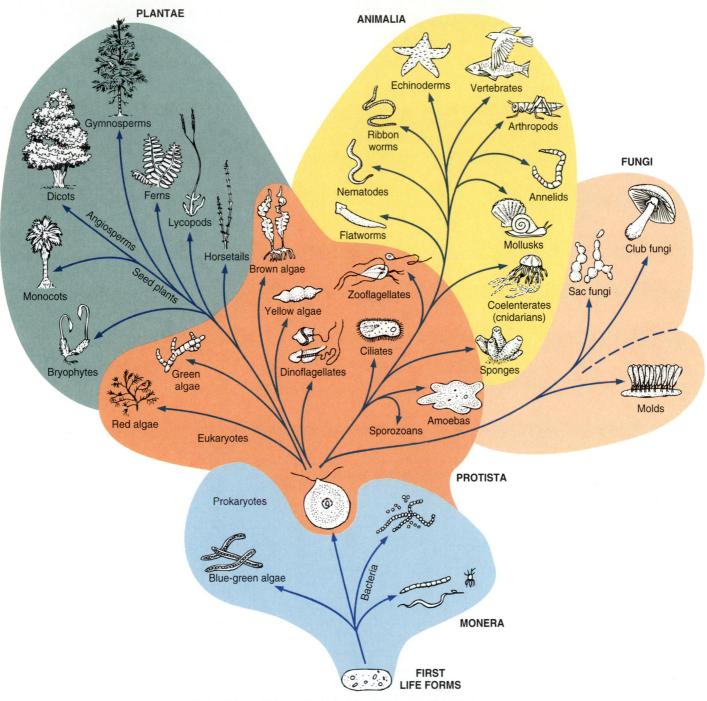

FIGURE 11-8 The five-kingdom scheme of living things, widely used by biologists and oceanographers. Some common life forms are shown. (Illustration adapted from Beck, Lien, and Simpson, *Life: An Introduction to Biology,* 3d edition (New York: HarperCollins, 1991).)

The cells of all species can be divided into two basic types. A cell having a nucleus is *eukaryotic*. A cell lacking a nucleus is *prokaryotic*. Prokaryotic cells are simpler and evolved first. Today only Kingdom Monera has these cells. The first organisms to evolve are thought to have been bacteria that did not need oxygen for their metabolic processes (anaerobic). The cells of organisms in the other four kingdoms are eukaryotic. Virtually all biologists concur that Kingdom Monera contains the simplest known organisms and is the kingdom from which the other four kingdoms evolved (Table 11-4).

TABLE 11-4

Some Characteristics of the Five Kingdoms

Kingdom	Examples	Major Aspects	Form	Ability to Photosynthesize	Ability to Move
Monera	Bacteria	Single prokaryotic cells, no nuclear membrane, little internal organization; includes heterotrophs, autotrophs	One-celled	Some do	Some can
Protista	Algae, protozoans	Eukaryotic cells, single-celled or as colonies, nuclear membrane	One-celled	Some do	Some can
Fungi	Molds, mushrooms, fungus	Eukaryotic cells, multicellular, heterotrophic	Molds and yeasts; one-celled and multicellular	No	Generally no
Plantae	Ferns, mosses, flowering plants	Eukaryotic cells, multicellular, well-defined and developed tissues	Multicellular	Yes	Generally no
Animalia	All invertebrates, vertebrates, fish, humans, etc.	Eukaryotic cells, multicellular, well-developed tissues and organs	Multicellular	No	Yes

There are about 50 different phyla in the five-kingdom scheme, and all have at least one species that lives in the marine environment. Many of the important ones are described in the following pages. We begin our tour with the smallest and simplest: the bacteria.

Bacteria and Blue-Green Algae

Bacteria and blue-green algae are both members of Kingdom Monera, one-celled prokaryotes. Although poorly understood, bacteria are important ocean organisms because of their role in producing organic material.

Bacteria occur in almost every environment and their variety appears almost infinite. On land, bacteria are known to grow in hot, salty springs where temperatures approach boiling (100°C, or 212°F). They have also been found around the hot vents associated with mid-ocean ridges (see Figure 12-19c, page 300). Bacteria likewise can dwell in near-frozen brines having temperatures of −2°C (about 28°F). Some bacteria live in oxygen-free conditions and others in oxygen-saturated conditions, whether these exist in surface waters or at the greatest depth of the ocean.

Bacteria can be divided into two groups: heterotrophs and autotrophs. The **heterotrophs**—probably the more common group—use organic material obtained from other organisms for their nutrition. **Autotrophs** use inorganic material and carbon dioxide to form organic material. In this respect, they are similar to photosynthetic plants. The amount of organic material produced by bacteria in this manner is small compared to that produced by plants, but this small quantity may be important in the deep sea, where other sources of food are scarce.

Blue-green algae are abundant in rivers and lakes but are not too common in the ocean, except near some rivers and in some tropical regions. These organisms are generally very small and poorly developed; some are planktonic. Even though they are called blue-green algae, some may be other colors due to accessory pigments within the plant. A dramatic example of this is the floating form *Trichodesmium*, which has a red hue. It is this bacterium that often gives the Red Sea its color and thus its name.

Plants of the Sea

The fundamental difference between plants and animals is that most plants are *producers*—they have chlorophyll and can photosynthesize their they own food. Animals do not have chlorophyll and therefore must eat plants or other animals to obtain their food. This is equally true on both land and in the ocean.

On land, plants are immersed in air, which provides a medium for oxygen and carbon dioxide exchange. Plants are rooted in the soil not only for support but also to contact soil water, which contains vital nutrients.

Seawater can be compared to soil water because it carries the nutrients necessary for plant life. The similarity between land plants and marine plants, however, ends there. On land, plants have adapted with extensive root systems to obtain water and food, and with leaves to obtain oxygen and carbon dioxide. In the ocean, plants are completely surrounded by water that provides nutrients, oxygen, and carbon dioxide. Thus, no real root systems are necessary, except to attach some plants to the seafloor in shallow water.

Insufficient light for photosynthesis penetrates the great depths of much of the world's oceans, thus making large portions of the seafloor unfavorable for plant growth. Some plants in shallow water receive sufficient light and are attached to the seafloor, but only a small portion of the seafloor (less than about 2 percent) is both shallow and solid enough for attached plants. If the water is very shallow, the plant can be exposed and may dry out at low tide, or can be damaged by breaking waves. Large temperature changes are common in shallow water, and such changes can affect the plant's ability to photosynthesize or reproduce.

Most plants in the ocean are planktonic. These usually microscopic, single-celled plants produce the bulk of organic material in the ocean and can occur in unbeliev-

ably large quantities. Essentially all the ocean's animals feed on these plants, either directly, or indirectly by eating other animals that feed on plants. Remember that phytoplankton are restricted to the upper layers of the ocean, because of their dependence on light, while consumers of the organic matter produced by plants occur throughout the ocean.

Marine plants are classified into two kingdoms, Plantae and Protista. Kingdom Protista contains all of the phytoplankton, a vast group. Kingdom Plantae, however, has few representatives, for reasons that will become clear. Let us now look at these two kingdoms.

Kingdom Plantae

Kingdom Plantae has many members on land but relatively few in the ocean. Of the seven phyla in Plantae, only one has marine representatives—the Angiospermophyta, or flowering plants. Of more than a quarter of a million flowering plant species worldwide, only about 200 marine species thrive. Only about a quarter of these 200 can survive fully submerged in seawater.

Some of the Angiospermophyta are familiar because they are visible to us. Among the more common are salt-marsh plants, mangroves, and sea grasses. The salt-marsh plants include *Spartina* and *Salicornia*, which occur in salt marshes around the world (Figure 11-9a). Because they can not tolerate total submergence, these plants occur in the higher parts of the marsh.

Spartina

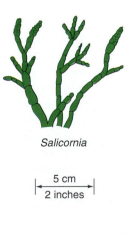

Salicornia

5 cm
2 inches

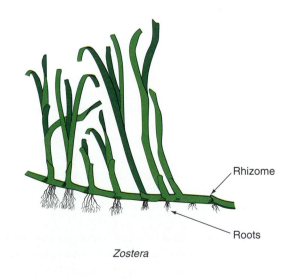

Rhizome

Roots

Zostera

b

FIGURE 11-9 (a) *Spartina* and *Salicornia*, two plants typical of salt marshes (see also Figure 1-1a, page 3). (b) *Zostera* (eel grass), a shallow-water seagrass. Note the rhizome from which new leaves and plants grow.

a

The term **mangrove** actually refers to a group of trees whose roots can tolerate submergence in seawater (see Figures 14-18b and 14-18c, page 350). Mangroves are tropical-to-semitropical trees limited in distribution to 30 degrees north or south of the equator. Both salt-marsh plants and mangroves are capable of trapping sediment and contributing to the seaward growth of coastal areas.

One sea grass is *Zostera*, commonly called eelgrass. This plant (Figure 11-9b) possesses true roots attached to a stem, or rhizome. It can reproduce either by seeds or by sending up new leaves from its rhizome. Eelgrass grows in water usually less than 5 m (about 16 ft.) deep, where wave action is not very strong. When the plant dies and is broken, its pieces may be carried out to sea and become an important source of food in offshore waters.

Where *Zostera* grows thickly, it protects many forms of animal life. An entire biological community, from small diatoms to nearshore animals, depends on the eelgrass environment. In the early 1930s a disease killed most of the eelgrass on the Atlantic coast and caused considerable damage to the scallop population because scallops depend on eelgrass for refuge. Fortunately for the scallop fisheries, the eelgrass recovered and has repopulated most of its original habitat.

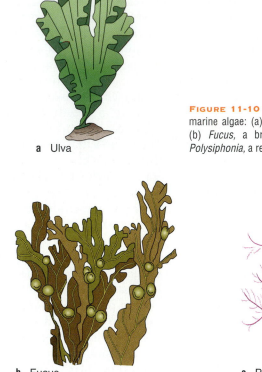

a Ulva

FIGURE 11-10 Common forms of marine algae: (a) *Ulva*, a green alga; (b) *Fucus*, a brown alga; and (c) *Polysiphonia*, a red alga.

b Fucus

c Polysiphonia

Kingdom Protista

Kingdom Protista presents a problem in taxonomy since it is a hodgepodge kingdom of single-celled life forms, some plant-like (algae) and some animal-like (foraminiferans). The more common marine producers, including all phytoplankton, are in Kingdom Protista. They have no true roots, stems, or leaves.

The most important and abundant producers are algae, which are generally divided into groups, mainly by color (Figure 11-10): green algae, brown algae, red algae, and yellow-green algae. The first three are usually attached and are commonly called seaweed. Yellow-green algae are mainly planktonic. Over 90 percent of the photosynthesis in the ocean is done by phytoplankton.

GREEN ALGAE. Green algae occur mainly in freshwater, in areas where freshwater and saltwater are mixed, and in the shallower parts of the littoral system. They rarely occur below 10 m (33 ft.) depth and thus are restricted to the well-lighted upper part of the ocean. A common form is the alga *Ulva*, or sea lettuce (Figure 11-10a). Green algae sometimes impart a distinct green color to the water. They can also form an algal slime on boats and other submerged objects.

BROWN ALGAE. Brown algae, which include the large plants collectively called **kelp,** occur mainly in the marine environment. One of the largest, *Nereocystis*, may grow to be 60 m (200 ft.) long and forms kelp beds common along some coasts. These algae are sometimes harvested from boats by cutting off the top layers of kelp. They are used mainly for algin, a food additive, and as a source of iodine, potassium, and iodine-based drugs.

The large forms that make up the kelp beds are attached to the bottom by a **holdfast,** a branched structure that attaches to a rock (Figure 11-11). A long tube called a **stipe** extends from the holdfast and terminates in one or more blades or fronds, often with gas-filled bladders for flotation. The gas gives the plant buoyancy, permitting it to float.

There are other forms of brown algae. Some are small, delicate, branching plants; others are big and broad like *Fucus* (see Figure 11-10b). Brown algae tend to be best developed in cooler parts of the ocean, especially in nearshore, rocky-bottomed areas.

One brown alga, *Sargassum*, extends its range in an interesting fashion. *Sargassum* actually grows in tropical areas, but when torn loose by waves, it drifts with the currents. It also grows and multiplies while drifting. Large quantities of these *Sargassum* algae accumulate in an area of the North Atlantic that is named for them: the Sargasso Sea. Before the plant dies and sinks, it forms a unique environment that shelters many kinds of animals, including some small floating crabs and fish.

RED ALGAE.   Red algae are among the prettiest organisms in the sea (see Figure 11-10c). Their red color comes from their abundance of red pigment (although some "red" algae actually are purple, brown, or green). Like other producers, red algae contain green chlorophyll pigment, but the green color is masked by other pigments. Red algae extend farther out to sea than the other forms. Some red algae that live on coral reefs (such as *Lithothamnion*) have the ability to precipitate calcium carbonate, which helps form and cement the reef into a structure with sufficient strength to withstand wave erosion.

Red algae can be important producers of organic matter, especially in the outer areas of the continental shelf. Some species are commercially valuable because they are a source of *agar,* which is used, among other things, as a thickening agent for ice cream and other products.

YELLOW-GREEN ALGAE.   The yellow-green algae include several types of organisms and are a very important group of protists in the ocean. They are mainly planktonic organisms containing chlorophyll and are therefore capable of photosynthesis. They live in the surface waters of all the oceans. Because they float, yellow-green algae are not restricted to the nearshore areas, as are the other species of algae.

The three most notable groups of yellow-green algae are diatoms, dinoflagellates, and coccolithophoridae. Diatoms are more common in colder waters; dinoflagellates are more abundant in warmer waters.

Diatoms are perhaps most important in this group. These microscopic organisms occur both in saltwater and freshwater (Figures 11-12a and 11-12b). Single-celled diatoms can combine with other individuals to form long chains or groups. In some areas of the ocean, production and growth of diatoms is especially rapid, since diatoms can multiply in a geometric manner (2 diatoms become 4; 4 become 8; 8 become 16; and so on). In this manner, their numbers can become very large extremely quickly—within days. Diatoms have siliceous shells, and if enough of them accumulate on the ocean bottom after death, they can form a deposit called diatom ooze (see Figure 5-15, page 121).

Diatoms have the ability to form *resting spores,* where the organism goes into a form of hibernation. They do so when environmental conditions become unfavorable and can remain in this configuration for periods of months or years. When conditions improve, they resume reproducing.

Dinoflagellates are other important members of the yellow-green algae group (Figure 11-12c). Some of these organisms are animal-like (they consume organic matter rather than produce it), whereas others are definitely plants. Dinoflagellates have flagella that resemble tails, which they use for locomotion to seek nutrients or better environmental conditions. Many dinoflagellates are luminescent. When excited, they impart a glowing color to the water. The purpose of this is not understood.

A large sudden growth, or bloom, of dinoflagellates will discolor the water, causing a "red tide" (see Figure 16-9, page 407). Humans can be poisoned—and may possibly, but rarely, die—if they eat mussels or clams that have fed on certain species of dinoflagellates and diatoms during a red tide period (this subject is discussed further in Chapter 16).

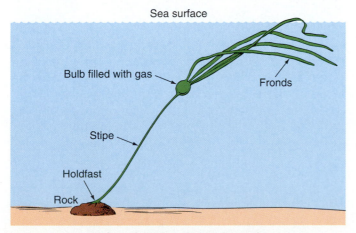

FIGURE 11-11   (a) General structure of a large brown alga.

(b) An alga that has washed ashore. Note the holdfast, which still is attached to a small rock.

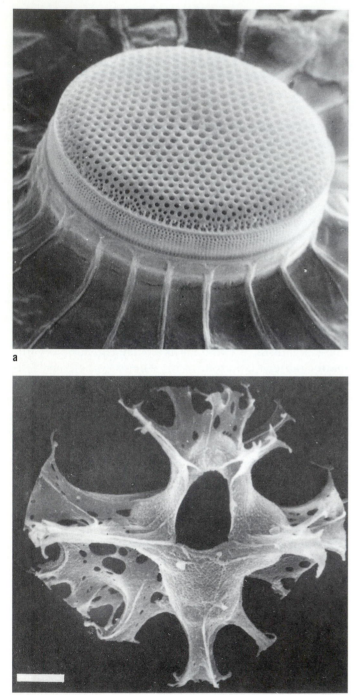

FIGURE 11-12 Electron microscope photographs of some yellow-green algae.
(a and b) Diatoms, about 20 to 80 microns (about 0.0008 to 0.0031 in.) across.;
(b) is a colony of diatoms.
(c) dinoflagellate; the white bar in (c) is 10 microns long (about 0.0004 in.).
(d) Coccolithophoridae, about 10 microns in diameter. The other objects in the
figure are debris and parts of other organisms. (Photographs courtesy of Dr.
Susumu Honjo.)

The tests (shells) of dinoflagellates, when they have them, are easily destroyed after death. Thus these organisms do not form extensive deposits on the seafloor.

Coccolithophoridae are another group of yellow-green algae, consisting of smaller organisms, about 5 to 10 μ in diameter (Figure 11-12d). These organisms were first recognized by geological oceanographers who noted their calcareous shells in deep-sea sediments. They were too small to be collected in plankton nets. Though not fully understood, the contribution by the Coccolithophoridae and other small yellow-green algae to the biological economy of the sea appears to be considerable, especially in the open ocean where nutrients are low.

In summary, plants of the sea are very different from those on land. In the ocean, few areas exist where attached plants can grow and receive enough light for photosynthesis. Therefore, the organic producers in the ocean are mostly microscopic floaters that can use the light available in the upper 200 m (660 ft.) of the ocean. Even though much of the plant life is microscopic, its numbers are sufficient to feed the remaining population of the ocean.

Animals of the Sea

Again, the fundamental difference between plants and animals is that most plants have chlorophyll and can photosynthesize their own food. Animals do not have chlorophyll and therefore must eat plants or other animals to obtain their food.

All animal phyla have marine representatives, whereas some do not have a terrestrial (land) representative. The important marine animals are described here in the order of their increasing biological complexity. With the exception of the Protista, all other animallike organisms mentioned are from the Kingdom Animalia.

Kingdom Protista

Kingdom Protista consists mostly of single-celled, microscopic, animal-like organisms, including the orders Foraminifera and Radiolaria (see Figures 5-17 and 5-18, pages 123 and 124). Many of these organisms live in the surface waters of the ocean. When they die, their shells settle to the bottom, covering large areas (see Figure 5-15, page 121). Because certain species are characteristic of certain geologic time periods, Radiolaria and Foraminifera can be used to determine the age of sediments and to correlate one sediment deposit to another. Most Foraminifera shells are composed of calcium carbonate; Radiolaria shells are mainly silica.

Kingdom Animalia

PHYLUM PORIFERA. Porifera, or sponges (Figure 11-13a and b), are multicellular benthic animals. Over 10,000 different species are known. Sponges are classified by the composition of their internal skeletons, which may be calcareous, siliceous, or spongin material (the component of natural commercial sponges). Porifera lack organs and well-defined tissues.

PHYLUM CNIDARIA. The phylum Cnidaria is more complex because its members have developed tissues and a high degree of polymorphism (individual species may occur in a variety of forms). The three classes of this phylum are important. The class Hydrozoa (see opening figure to this chapter and Figures 11-13c and 11-13d) includes some forms of jellyfish. One common variety is *Physalia*, known as the Portuguese man-of-war.

Large jellyfish, sometimes as much as 2 m (almost 6½ ft.) in diameter, belong to the class Scyphozoa. The third class, Anthozoa, contains most corals and sea anemones (Figure 11-14). Corals are important because their calcareous skeletons can form the core of large coral reefs (see Figures 12-16 and 12-17, pages 298 and 299).

PHYLA PLATYHELMINTHES AND NEMATODA. Platyhelminthes and Nematoda (*helmis* is the Greek word for "worms") include the unsegmented varieties of worms. Platyhelminthes, the flatworms, include *Planaria*, the "cross-eyed" worm commonly studied in high school or college biology, and the tapeworms found in humans, which are only one of many parasitic genera in this phylum. Adult individuals in this phylum range widely in size, from under 1 cm (0.4 in.) to over 20 m (65 ft.).

Some larger forms have been called "sea monsters." However, even though they are long, they are very thin and not quite up to the standards of a Loch Ness–class sea monster. Nematodes are round, or thread, worms. One land variety, hookworm, or *Trichinella*, is a parasite dangerous to humans.

PHYLUM CHAETOGNATHA. Chaetognatha (pronounced KEY-tog-natha) is a relatively small group with few known genera. Chaetognaths are small, transparent creatures that resemble worms and are commonly called arrow worms or glass worms. They are very voracious and can quickly consume large numbers of smaller organisms.

PHYLUM ANNELIDA. Annelida (pronounced a-NEL-i-da) are earthworms, leeches, and sandworms (Figure 11-15a). They have segmented, elongated bodies. Most marine annelids are benthic, with many living in burrows. The burrowing forms are important because their

a

d

FIGURE 11-13 Some sponges and jellyfish: (a) red sponge; (b) glass sponge; (c) deep-sea jellyfish generally found below 500 m (1640 ft.); (d) jellyfish commonly called a "sea nettle." (Photograph (b) by Dr. Robert Ballard; (c) by Dr. L. P. Madin; all courtesy of the Woods Hole Oceanographic Institution.)

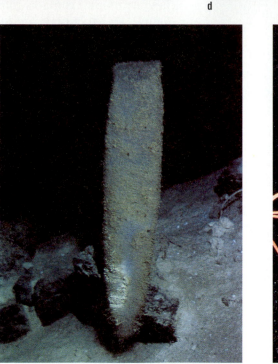

b

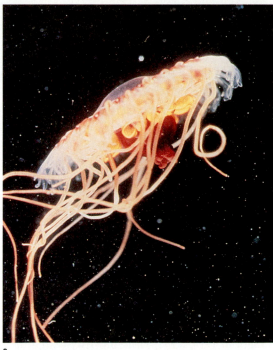

c

incessant churning mixes the upper 10 to 20 cm (4 to 8 in.) of bottom sediment. These worms, along with some mollusks and crustaceans, are the dominant animals in deep-sea sediments.

PHYLUM ARTHROPODA. Arthropoda is the largest phylum on land and sea, both in numbers and total mass. Arthropods have an external skeleton and numerous jointed appendages. This phylum includes three very important groups:

Uniramia (pronounced you-nuh-RAIM-ee-uh) are the insects.

Chelicerata (kuh-LISS-er-a-ta) includes spiders, horseshoe crabs, and sea spiders.

Crustacea includes crabs, lobsters, and shrimp.

Insects, which form a vast and important class of land animals, are almost completely absent from the sea, although some spend part of the early stage, generally as

FIGURE 11-14 Sea anemones: (a) a shallow-water anemone found in a coastal pond;

(b) sea anemone more typical of the open ocean. (Photograph (a) courtesy of Woods Hole Oceanographic Institution. Photograph (b) courtesy of Dr. Phillip Lobel.)

larvae, in the marine environment. Only *Halobates,* the water strider, spends its entire life in the sea, literally walking on the water surface.

The marine Chelicerata include horseshoe crabs and sea spiders (Figure 11-16a), or pycnogonids. The latter are not true spiders. Like insects, the Chelicerata form a relatively minor group in the ocean.

The Crustacea are divided into several classes and subclasses. One subclass, Cirripedia, includes *barnacles* (Figures 11-16c and 11-16d). Adult barnacles generally have hard shells and live attached to the ocean bottom, to fixed objects (pier pilings and undersides of ships), or to other animals. They have a pelagic larval stage that accounts for their growth on deep-sea objects such as

FIGURE 11-15 (a) Polycheate worms;

(b) flatworm. (Photographs courtesy of Woods Hole Oceanographic Institution.)

a

b

c

d

FIGURE 11-16   Some members of the phylum Arthropoda: (a) sea spider; (b) variety of shrimp (order Decapoda); (c) acorn barnacle, commonly found in intertidal regions; (d) barnacle fully extended in feeding position. (Photograph (b) courtesy of Dr. Laurence P. Madin, Woods Hole Oceanographic Institution; (c) and (d) courtesy of P. J. Oldham, Marine Biological Laboratory.)

whales and ships. Their occurrence on the hull of ships causes friction that reduces a ship's speed, so they must be removed periodically from most vessels.

The most numerous of the crustaceans are members of the subclass Copepoda (Figure 11-17). Copepods are important because they eat phytoplankton and concentrate it in their bodies for other larger animals to eat. In this manner they form an essential link in the food chain of the sea. Most copepods are pelagic zooplankton (see Figure 11-4).

Probably the best known crustacean order is Decapoda, which includes shrimp, crabs, and lobsters (Figure 11-18). Most of these animals are benthic, and only a few are pelagic forms. This group of animals is important to those who fish for and trap them.

Another order, Euphausiacea, are also common zooplankton. These animals are somewhat larger than copepods and are more advanced in their development. Euphausids, along with some copepods, are capable of extensive vertical migration, sometimes traveling

FIGURE 11-17 Most growth stages of the copepod *Labidooera aestiva,* drawn to the same scale. Some parts of the head appendages are not shown until stage XI, an adult stage. It generally takes less than 3 weeks for a copepod to go from stage I to the fully developed adult stage. (Photograph courtesy of Dr. George D. Grice.)

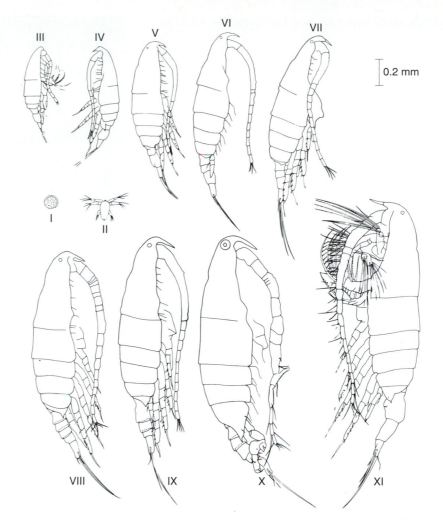

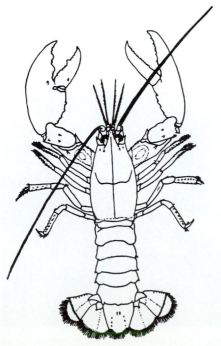

FIGURE 11-18 (a) Drawing of the American lobster, *Homarus americanus,* and (b) the real thing. It takes about 7 years for a lobster to reach the size at which it can be legally harvested.

several hundreds of meters a day. Euphausids—krill, for example (Figure 11-19)—are the favorite food of some whales.

The order Amphipoda consists mainly of benthic organisms, with some pelagic forms. Figure 11-20 shows some large amphipods feeding at a bait can set at a depth of over 5,000 m (over 17,000 ft.) in the North Pacific Ocean.

PHYLUM MOLLUSCA. Mollusca is a major ocean phylum containing many familiar species. In many genera, mollusks have a soft body covered by a hard shell, which may have one or two parts or eight pieces. Three classes of Mollusca are noteworthy: Gastropoda, Bivalvia, and Cephalopoda (Figure 11-21).

Gastropods include terrestrial and benthic-living snails, slugs, and floating forms such as pteropods. These animals have a "foot" that is used to move on the bottom. In many forms, the foot is attached to a hard, spiral-shaped shell, which can be absent from some planktonic species. Pteropods, one of these planktonic forms, may settle on the bottom and form extensive seafloor deposits after death.

Bivalves, which include clams, oysters, scallops, mussels, and the like, have a two-piece shell. Most marine animals of this order live either attached to the bottom or burrowed into it. Some forms burrow into wood, such as boat docks and piers, causing extensive damage. Two genera, *Teredo* and *Bankia,* can bore holes into wood up to 30 cm (about 1 ft.) long.

Bivalves constitute an important human food source. The settlements of early people often can be identified by large piles of empty shells, mainly bivalves.

Cephalopods include squid, octopus (Figure 11-22a), and nautiloids. In these animals, the solid foot typical of the other classes is divided into arms or tenta-

FIGURE 11-20   A very large group of amphipods swarming around a bed of mussels at a deep-sea vent in the Pacific Ocean. The density of this group is estimated at more than 1,000 individuals per liter (about a quart) of seawater. They are about 5 mm (0.2 in.) in length and live in complete darkness. The swarm was found downstream from where hydrothermal water discharges through seafloor cracks. The reason for their swarming behavior is unknown. (Photograph courtesy Dr. Cindy Lee Van Dover, Woods Hole Oceanographic Institution.)

cles. Squid are very common in the ocean and serve as an important food source to many other creatures. The so-called giant squid, which may be as long as 18 m (about 60 ft.), is the largest living invertebrate animal known (Figure 11-22b).

PHYLUM ECHINODERMATA. Echinodermata is an exclusively marine phylum. It includes sea stars (Figure 11-23a), brittle stars, sea urchins (Figures 11-23b and 11-23c), and sea cucumbers. These animals have a five-

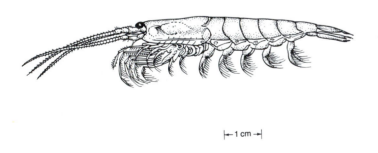

|← 1 cm →|

a

FIGURE 11-19   (a) Drawing of the common Antarctic krill, *Euphausia superba,* and (b) the real thing. Its length is about 5 cm (2 in.).

b

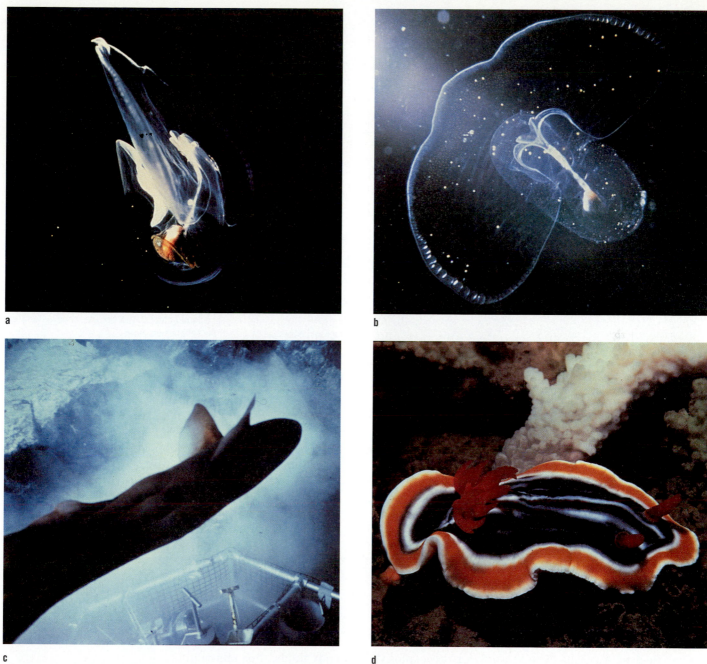

a

b

c

d

FIGURE 11-21   Some representatives of the Mollusca phylum: (a) pteropod; (b) mollusc larvae; (c) cephalopod, called an octopod (those are not ears by its head, but fins); (d) nudibranch (larger animal) and pteropod. (Photographs (a) and (b) courtesy Dr. Laurence P. Madin, Woods Hole Oceanographic Institution; (c) courtesy of Dr. Robert Ballard; (d) courtesy of Great Barrier Reef Marine Park Authority.)

sided symmetry and an internal skeleton. Most of the forms are benthic. Two classes are worth further mention: Holothuroidea and Asteroidea.

Holothuroidea includes sea cucumbers (Figure 11-24), which can live in all depths of the ocean. Some sea cucumbers have a very peculiar ability: they can eviscerate, which means they can literally turn them-

selves inside out. In times of danger, they can discharge their internal organs, leaving a meal for their attackers, and later regenerate another set of organs.

Asteroidea is well represented by sea stars the world over. They are frequently seen in littoral regions but also occur in deeper parts of the ocean. Starfish, as they are commonly called (they are not fish, of course), feed on

a

b

FIGURE 11-22   (a) A pugnacious octopus who became annoyed and attacked the mechanical arm of the submersible *Alvin*.
(b) A giant squid that washed up on Plum Island, north of Boston, Massachusetts. The main part of this creature was about 2 m (7 ft.) long, with arms about 2.5 m (8 ft.). Its estimated weight was 200 kg (440 lb.). (Photograph (a) courtesy of Woods Hole Oceanographic Institution. Photograph (b) courtesy of William B. Coltin.)

oysters, clams, and sometimes even coral. In doing so, they can considerably damage these valuable resources. In the past, when people fishing caught starfish they would often cut them in half, thinking they had killed the animals, and throw them back into the ocean. Starfish, however, can regenerate their lost arms, so by cutting them in half, people were actually doubling their numbers!

Several years ago a dramatic population increase of the starfish *Acanthaster* (Crown of Thorns) was noted on some Pacific coral reefs (see Figure 11-23d). This creature feeds on coral and has been responsible for large-scale destruction of reefs in places such as the Great Barrier Reef of Australia. The reasons for the explosive growth of this starfish variety are unknown. In some areas, the rapid growth quickly decreased to more normal conditions, again for unknown reasons. (See Box 12-2, page 296, for a discussion of another problem facing corals: coral bleaching.)

PHYLUM CHORDATA.   Chordata is a large and very important phylum. All chordates possess a **notochord,** a rodlike support structure for the animal's skeleton. This phylum can be divided into several subphyla. Two are especially important in the ocean: Tunicata and Vertebrata.

Tunicates are marine filter feeders that eat by filter-ing organisms and detritus from the water. Some are attached to the bottom, while others are planktonic forms.

Vertebrates—all animals with vertebrae—are the most highly developed life forms. Vertebrata includes seven important marine classes: Agnatha (primitive fish), Chondrichthyes (cartilaginous fish), Osteichthyes (bony fish), Amphibia (amphibians), Reptilia (reptiles), Aves (birds), and Mammalia (mammals).

Agnatha includes primitive fish such as lampreys and hagfish. They have neither articulated jaws nor scales, and many are parasites or scavengers. The lamprey attaches itself to the side of another fish, makes a hole through the body wall, and then sucks out the fish's body fluids, thus killing it.

Chondrichthyes are cartilaginous fish such as sharks (Figure 11-25a), skates (Figure 11-25b), rays, and chimaeras. Their scales do not overlap, as do the scales on bony fish. Sharks are the largest of these fish, some over 15 m (50 ft.) long. At one time sharks were fished extensively for the nutritional value of their livers. Today, sharks are commonly eaten and in some areas are considered a gourmet food.

Sharks have fascinated humans for centuries, often as an object of great fear. *Jaws* and similar movies (Figure 11-26) have presented sharks as killers, which is misleading because they are simply predators like many other animals in the ocean. The difference is that some

FIGURE 11-23 Members of the phylum Echinodermata: (a) sea star; (b) sea urchin near the Galápagos Islands; (c) close-up of the sea urchin; (d) the starfish "Crown of Thorns," which has destroyed coral reefs in some parts of the world. (Photograph (a) courtesy of P. J. Oldham, Marine Biological Laboratory; (b) and (c) courtesy of T. M. Rioux.)

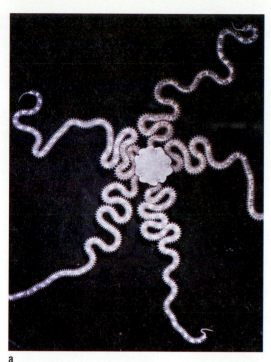

a

b

c

d

sharks will eat an occasional human if given the chance. The number of authenticated shark attacks is actually small, about 25 per year. Interestingly, elephants kill about ten times as many people as sharks do over the course of a year, but elephants do not invoke nearly the same degree of fear as sharks. Sharks have been especially successful creatures—the oldest-known shark appeared in mid-Devonian times, about 350 million years ago (Figure 11-27a). Sharks live in all the oceans, and freshwater species occur in some lakes and rivers.

Osteichthyes, or bony fish, are characterized by overlapping scales, an anterior mouth, and a bony skeleton. Bony fish are the marine animals most important to humans (Figure 11-28). About 90 million tons of fish are caught (Figure 11-29, page 272) and consumed by people each year (see Chapter 15).

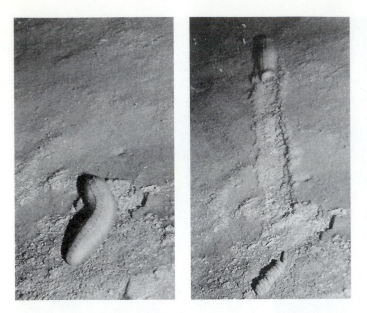

FIGURE 11-24 (a) A deep-sea holothurian, or sea cucumber, eating its way along the seafloor.

(b) Large group of holothurians on the continental slope off Virginia at 1,615 m (5,298 ft.) depth. The concentration may be due to a large piece of food. (Photograph (a) courtesy of the Institute of Oceanographic Sciences, Deacon Laboratory.)

FIGURE 11-25 (a) A white tip shark.

(b) A barn door skate caught on Georges Bank. (Photograph (a) courtesy of Dr. Laurence P. Madin, Woods Hole Oceanographic Institution.)

FIGURE 11-26 This is not what it looks like at first! Dick Edwards, a marine technician, is in the mouth of a shark, but it is the artificial shark used in the movie *Jaws.* In Mr. Edwards' mouth are several sticks of dynamite that were used to detonate the "shark" in the final scene of the movie. (Photograph courtesy of Cliff Winget.)

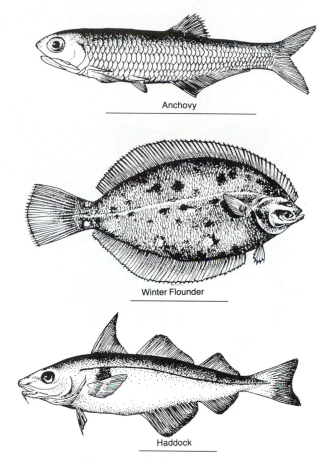

Anchovy

Winter Flounder

Haddock

FIGURE 11-28 Some commonly harvested fish. The line beneath each fish represents about 15 cm (6 in.) length.

FIGURE 11-27 (a) Scientist working on a reconstructed extinct shark at the Smithsonian Institution. The outer teeth in the jaw are real. The shark is estimated to have been about 12 m (40 ft.) long. The largest modern white shark known is about 7 m (21 ft.).
(b) Scientist dissecting a giant white shark. (Photograph (a) courtesy of Chip Clark, Smithsonian Institution; (b) courtesy of Shelley Lauzon, Woods Hole Oceanographic Institution.)

a

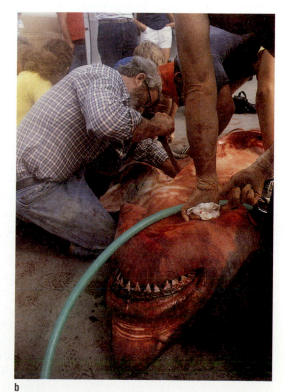

b

a

b

FIGURE 11-29   Two examples of game fish.
(a) A beautiful 53-pound striped bass that was tagged and released by the late fisherman-scientist Dr. Dennis Sabo of the Massachusetts Maritime Academy. (b) Two young fishermen holding a small bluefin tuna. (Photograph (b) courtesy of Virginia Institute of Marine Science.)

Over 25,000 fish species exist. Most are pelagic, living in all depths of the ocean. However, many species prefer to stay near the bottom, and others carry out most of their activities near the surface. **Anadromous** fish—such as salmon, striped bass, sturgeon, and smelt—are born in freshwater, spend most of their time in the ocean, and return to freshwater to spawn. **Catadromous** fish—such as some eels—do the opposite: they live in freshwater but spawn in the ocean.

Most ocean organisms are cold-blooded, which means that their internal temperature fluctuates with the surrounding temperature, in this case the temperature of the water in which they swim or float. Some marine fish such as tuna, however, are warm-blooded, meaning that they maintain a specific body temperature independent of the water around them. There appear to be two reasons why some fish are warm-blooded. First, it allows them to roam and hunt into deeper, colder parts of the ocean. Second, warm-bloodedness permits them to be more active. Tuna, for example, are among the fastest-swimming fish in the ocean, and they can feed in polar waters. Some fish, such as marlin and swordfish, can attain a speed of 120 km/hour (75 miles/hour) but keep only part of their bodies warm, such as their brains and eyes.

Amphibia are creatures that live mainly in fresh water, although some can tolerate a modest salinity content. Familiar members of this class are frogs and salamanders. Amphibians are thought to be organisms that evolved from a saltwater environment to a land environment.

Reptilia commonest in the ocean are turtles (Figures 11-30a and 11-30b) and snakes. Neither are very important to the overall economy of the ocean. However, there has been some serious concern recently about the impact of human activities on turtles, including destruction of their breeding habitats and their harvest for food or ornaments. Some species of turtles may be close to extinction.

Aves (birds) are warm-blooded animals. They generally do not live in the ocean, but many depend on the sea for food and return to land only to breed. Penguins, which cannot fly, are one group of birds that has a very close relationship with the ocean (Figure 11-31b); others are gulls, pelicans, and ospreys. Guano, the accumulated fecal waste of birds and other animals, occurs in some nesting areas and can be a valuable fertilizer deposit due to its phosphate content.

Mammalia, the most advanced class of organisms living in the sea, are also warm-blooded. Two orders with marine genera are worth our attention: Carnivora and Cetacea. Carnivores include sea otters, polar bears, seals, sea lions (Figure 11-31a), and walruses. Cetacea are dolphins, porpoises, and baleen whales.

Cetacea, or whales, are the largest living creatures on earth. The blue whale can be over 30 m (about 100 ft.) long and weigh about 150,000 kg (150 tons). Most whales, such as the sperm whale, have teeth and feed on organisms like squid or fish. A small number of whales are filter feeders whose diet consists mainly of plankton

a

a

b

b

c

c

FIGURE 11-30   (a) Adult female leatherback turtle nesting on a beach at St. Croix, Virgin Islands.
(b) Giant tortoise, a land creature, photographed on the Galápagos Islands.
(c) Green moray eel. (Photograph (a) courtesy of Dr. Tundi Agardy; (b) and (c) courtesy of T. M. Rioux, Woods Hole Oceanographic Institution.)

FIGURE 11-31   (a) Sea lion swimming off the Galápagos Islands.
(b) Penguins in Antarctica.
(c) Fluke of a humpback whale. (Photograph (a) courtesy of T. M. Rioux; (b) courtesy of Paul Dudley Hart; (c) courtesy of Dr. L. P. Madin, all from the Woods Hole Oceanographic Institution.)

such as krill—small shrimplike crustaceans (see Figure 11-19) usually no longer than 5 cm (about 2 in.).

To humans, the whale is a very special ocean creature, being both the largest animal and a fellow mammal. Whales have been immortalized in literature (*Moby Dick*, for example) and in song (especially touching are the "songs" made by the whales themselves). (See Box 11-3.) The ten principal types of whales are listed in Table 11-5. Largest is the blue whale, believed to be the largest animal that has ever lived (Figure 11-32).

Whales have been noted for a wide variety of behavior. Figure 11-33 shows a series of such natural behavior patterns (sometimes called an ethogram) observed in the humpback whale in its Hawaii breeding grounds. Although it is tempting to attribute reasons to this behavior, little is known about such actions of these magnificent animals. In the examples shown in Figure 11-33, communication (to other whales) is probably an important factor. The patterns interpreted as aggressive are supported by direct observation.

In spite of the obvious intelligence and beauty of whales, they have been hunted for over 1,000 years for their meat, oil, and fat. Some whales may become extinct because of extensive harvesting. Many people have questioned the ethics of continuing to kill whales, especially since so many alternative sources of food and oil exist. Furthermore, can the present whale population stand continued hunting without some species becoming extinct? Because many whales inhabit the open ocean, international cooperation is necessary to stop or significantly reduce their harvest. Important progress has been made in this direction in recent years (see Box 11-4).

Box 11-3
Sounds in the Ocean

Marine scientists long have recognized various sounds in the ocean, but until recently were unaware of their quantity and sometimes of their source. Advances in technology and improving relations between the United States and Russia are revealing that the ocean is a very noisy place. Sensitive new seafloor seismographs now routinely detect numerous rumblings from earthquakes, sometimes occurring in swarms, as well as volcanic activity along the mid-ocean ridges.

Many marine organisms, especially whales and porpoises, are known to be noisemakers. Whales apparently compose and transmit complex sound patterns of a musical quality. Some of these whale "songs" have been used on recordings by Judy Collins, Neil Diamond, and others. The songs of some whales—humpback whales in particular—are especially interesting. Some can last 10 minutes or longer before repeating. Within a general locality, humpbacks have similar "choruses" in their song. These differ from those in other localities.

It wasn't until the U.S. Navy released data collected from its Integrated Undersea Surveillance System (IUSS) in 1993 that scientists discovered just how much sound exists in the ocean. For years the Navy had maintained a network of underwater listening devices, primarily to detect noises from submarines. But these devices listened to the entire ocean, collecting other sounds. The release of this data to civilian scientists will have a considerable impact. Now, instead of hearing only a few whales alongside a research vessel, marine scientists can monitor sounds from a large population of animals across an ocean. The challenge will be to separate these sounds, relate them to specific species, and interpret their significance.

Previous research has shown that whales and porpoises use sound to detect food, to navigate, and perhaps to communicate with other members of their species. Some Navy scientists, who had prior access to the IUSS data, claim to be able to identify individual whales by their sound and to follow their migration across large parts of the ocean. One blue whale was reportedly followed for 43 days while it roamed the North Atlantic.

Among the early results from the IUSS data are that whale sounds can travel in the ocean for 1,600 km (1,000 miles) or more. The availability of the Navy data may become an "acoustic rosetta stone" for marine biologists, enabling them to understand how these animals communicate.

The recent Heard Island experiment (see Box 8-4, page 194) produced such a high intensity of sound that some feared it might interfere with the normal behavior of whales and other creatures. After considerable debate, it was agreed that sound levels for future parts of this experiment would be reduced. This awareness of possible harm from human-generated sound has raised other questions. Humans add other noises to the ocean, such as those from ship engines and seismic profilers (see Figure 4-2, page 79). These also might affect the ability of some organisms to communicate, navigate, or search for food.

Oceanographers will learn much about underwater sound from the IUSS system. Discoveries about geological processes and how marine organisms use sound may lie ahead. Most fascinating is the possibility that organisms are actually communicating with each other in our very, very noisy seas.

TABLE 11-5
Principal Characteristics of Large Commercial Whales

Type	Distribution		Breeding Grounds	Breeding Behavior	Usual Length (feet)	Food
Sperm whale	Worldwide	Breeding herds in tropical/temperate	Oceanic	Polygamous	Male 35–60 Female 30–38	Squid, fish
Right whale	Worldwide	Cool temperature	Coastal	Mixed breeding herds	40–60	Copepods, plankton
Bowhead whale	Arctic	Close to edge of ice	—	Mixed breeding herds	40–60	Krill
Gray whale	N. Pacific, along coasts	Large N–S migrations	Coastal	Mixed breeding herds	35–46	Benthic invertebrates
Humpback whale	Worldwide, along coasts	Large N–S migrations	Coastal	Mixed breeding herds	35–50	Krill, fish
Blue whale	Worldwide	Large N–S migrations	Oceanic	Mixed breeding herds	70–100	Krill
Fin whale	Worldwide	Large N–S migrations	Oceanic	Mixed breeding herds	58–85	Krill, other plankton, fish
Sei whale	Worldwide	Large N–S migrations	Oceanic	Mixed breeding herds	45–57	Copepods, other plankton, fish
Bryde's whale	Worldwide, temperate	Tropical/warm	Oceanic	Mixed breeding herds	40–50	Krill
Minke whale	Worldwide	N–S migrations	Oceanic	Mixed breeding herds	23–33	Krill

FIGURE 11-32 Relative size of whales. (From Dale E. Ingmanson and William J. Wallace, *Oceanography: An Introduction,* 2d ed., Belmont, Calif.: Wadsworth Publishing Company, 1979; reprinted by permission of Wadsworth Publishing Company, Inc.)

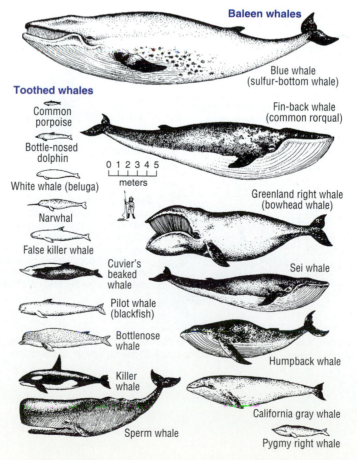

Baleen whales

Blue whale (sulfur-bottom whale)

Toothed whales

Common porpoise

Bottle-nosed dolphin

White whale (beluga)

Narwhal

False killer whale

Cuvier's beaked whale

Pilot whale (blackfish)

Bottlenose whale

Killer whale

Sperm whale

Fin-back whale (common rorqual)

0 1 2 3 4 5 meters

Greenland right whale (bowhead whale)

Sei whale

Humpback whale

California gray whale

Pygmy right whale

SLAPS

Flippering. Flippering appears as repeated slapping of either one or both of the 9 to 12-foot-long flippers of the humpback whale. Although typically seen with the animal lying in a sideways posture with one flipper being raised and lowered repeatedly, it may also involve rolling and slapping the right and left flippers alternately.

Fluke Slap. Also called "lobtailing," fluke slaps are usually seen with the whale in a normal dorsal (backside-up) posture with the flexible tail-stock, or peduncle, lifting repeatedly out of the water and slapping the water with the ventral (or underneath) side of the fluke. Sometimes, however, this may involve a head-down posture of the whale and back-and-forth action alternately slapping both ventral and dorsal surfaces of the fluke.

LEAPS

Headslap. Although considered by some researchers to be simply a "minibreach," the headslap actually appears quite distinct in form. The relatively inflexible head is raised swiftly out of the water, usually not further than where the flippers meet the body, and then brought down hard with a loud slap and visible splash.

Breach. Certainly the most spectacular and renowned of whale behaviors, the breach is a true leap whereby the whale generates enough vertical force with its powerful flukes to lift nearly three-quarters of its body length out of the water. The "true" breach typically involves a twisting motion once the body reaches its highest vertical travel such that the force of impact is taken on the dorsal surface. The twisting motion may in some cases be only partial. Breaching may occur singly or in a series. A series of as many as 30 breaches or more have been witnessed by researchers in Hawaiian waters.

AGGRESSIVE BEHAVIOR

The following behaviors have been observed in humpback whale pods with two or more "escorts" accompanying a cow-calf pair. The escorts typically attempt to displace one another in order to maintain exclusive contact with the cow. These behaviors are presented below in increasing order of intensity.

Headlunging. Thought to be a form of "threat," headlunging involves a forceful forward thrust which brings the head partially or fully above water as the whale swims towards its competitor. The mouth and throat are sometimes inflated with water which serves to exaggerate the apparent size of the threatening animal.

Peduncle Slap. A more powerful and even violent version of fluke slapping is the peduncle slap which occurs when the peduncle section strikes the water with a forceful downward diagonal motion. Individual whales have been seen to actually lift out of the water when struck with a heavy peduncle slap by a rival whale.

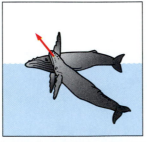

Strike. With escalating aggression, one whale may actually butt up against its competitor, sometimes with tremendous force. These strikes may be directed at different parts of the whale's body, although more often they occur as "broadside" strikes against the body and flanks.

FIGURE 11-33 Behavioral ethograms of the humpback whale, based on observations of the whale in its natural environment. (From *Humpback Whales in Hawaii: Guide for the Amateur Whale-Watcher*, University of Hawaii Sea Grant College Program, 1985.)

Box 11-4
Whaling and the International Whaling Commission

Whales have been hunted by humans for their meat and oil for over a thousand years. There is evidence of whaling as early as A.D. 800 by Norsemen and by the Basque people of Spain. In later centuries, whales were pursued by hunters from several other countries, including Japan, the United States, and by Eskimos of Canada and Alaska. In these early times, the hunting probably did not seriously impact the whale population.

About a hundred years ago, however, technology changed the character of the hunt. First, explosive harpoon guns were developed. Then ships started using engines instead of sail. Next, large factory ships were developed, permitting at-sea processing of the killed whales. Finally, aircraft and electronic devices were employed to find the whales. These methods increased the efficiency of the whale-catching process and led to the drastic reduction in numbers and near depletion of some species.

In 1946, Australia, Argentina, Great Britain, Canada, Denmark, Iceland, Japan, Norway, the former Soviet Union, the United States, and others established the International Whaling Commission (IWC). Among its objectives was reduction of the killing of certain whale species. The member nations developed regulations to control the harvest of various whale species.

In reality, however, IWC had no power beyond the goodwill of the member countries. In subsequent years, the killing of whales continued and for some species actually increased. By the end of the 1970s, however, large-scale commercial whaling slowed, and only Japan and the then–Soviet Union (now Russia) remained as whaling nations.

A 10-year moratorium on whaling began in the mid-1980s, but this did not stop the killing of some whales.

For example, countries were allowed to catch whales for scientific research. Likewise, subsistence-type whaling by some tribes (such as Eskimos in Alaska) was allowed to continue. Nevertheless, the total annual catch of whales, which had reached over 60,000 in the 1960s, was reduced to less than 10,000 in parts of the 1980s.

Unfortunately, the success in reducing the catch and increasing the numbers of whales has enticed some countries to resume whaling on the rationale that the IWC conservation procedures have worked. Iceland, Japan, and Norway feel that the whale population has stabilized, so some species now can be harvested without threat of extinction. Iceland has withdrawn from the commission. In 1993, Norway caught 157 minke whales.

On the other side, several organizations and countries, including the United States, wish to halt any form of commercial whaling. Their argument is generally based more on ethical reasons than on scientific ones. It is also argued that since numerous substitutes for whale meat and oil exist, whales need not be harvested.

Regardless of the ultimate outcome, the IWC and the goodwill of many countries has helped reduce the killing of several whale species. In the early 1990s, a similar view was extended to porpoises, many of which were accidentally killed during the harvesting of tuna. The concern was sufficient to motivate many countries, in particular the United States, to adjust their method of harvesting tuna—which frequently travel with porpoises—so that the porpoises could escape. The United States initiated laws that prevented the importation of any tuna that was harvested in ways that killed porpoises.

SUMMARY

The ocean can be divided into two major biological environments: the benthic (seafloor) and pelagic (the overlying water). Further subdivisions can be made on the basis of depth and light penetration. The upper parts of the water column where light is sufficient for photosynthesis is called the photic zone. Deeper and darker water below is the aphotic zone. Life is generally more abundant in the nearshore regions, where the food supply is more abundant.

The ocean contains a wide variety of bacteria, plants, and animals. They can be classified by their mode of locomotion or habitat: benthos, those that live on or in the bottom; nekton, those that swim; and plankton, those that float. They can also be classified by their physical characteristics into a five-kingdom taxonomy.

The most common type of plants are algae, many of which are planktonic. Attached plants are restricted to shallow water, because of their need for sufficient sunlight for photosynthesis. Over 90 percent of the photosynthesis in the ocean is done by phytoplankton.

All known animal phyla have marine representatives. Creatures, some very unique, have been found in all parts of the marine environment, and it is very likely that many more marine organisms remain to be discovered. The key to the economy of the ocean is the phytoplankton, which are the primary producers of organic food.

QUESTIONS

1. What are the main biological divisions of the ocean and how are they determined?

2. Describe important environmental factors that distinguish different marine environments, from the littoral to the deep sea.

3. How are organisms classified?

4. Describe the various modes of locomotion of marine organisms.

5. What is the difference between autotrophs and heterotrophs?

6. Discuss the distribution and varieties of marine algae.

7. How do plants in the ocean differ from those on land?

8. Which is probably the most important group of marine organisms in the ocean? Why?

9. What are some unique features of marine animals?

KEY TERMS

Agnatha
Amphibia
anadromous
aphotic zone
Asteroidea
autotroph
Aves
benthic environment
benthic life
benthos
biomass
biotope
bivalve
catadromous
cephalopod
Cetacea
Chondrichthyes
Coccolithophoridae

community
deep-sea system (environment)
diatom
dinoflagellate
gas bladder
gastropod
heterotroph
holdfast
holoplankton
Holothuroidea
indicator species
intertidal
kelp
kingdom
Kingdom Animalia
Kingdom Fungi
Kingdom Monera

Kingdom Plantae
Kingdom Protista
littoral system (environment)
macroplankton
mangrove
Mammalia
marine snow
meroplankton
microplankton
nanoplankton
nekton
neritic system (environment)
notochord
oceanic system (environment)
Osteichthyes

pelagic environment
photic zone
phytoplankton
plankton
Reptilia
species
stipe
subtidal
supratidal
taxonomy
tests
tunicate
ultraplankton
vertebrate
zooplankton

FURTHER READING

Allen, K. R. 1980. *Conservation and Management of Whales.* Seattle: University of Washington Press; London: Butterworth & Co.

Beck, W. S., Lien, K. F., and G. G. Simpson. 1991. *Life: An Introduction to Biology,* 3d ed. New York: HarperCollins. A solid textbook on all things biological.

"Biological Oceanography." *Oceanus* 35, no. 3 (Fall 1992). A special issue on recent research in marine biology.

Buchsbaum, R., M. Buchsbaum, J. Pearse, and V. Pearse. 1987. *Animals without Backbones,* 3d ed. Chicago: University of Chicago Press. A solid biological text.

Gage, J. D., and P. A. Tyler. 1991. *Deep-Sea Biology: A Natural*

History of Organisms at the Deep-Sea Floor. Port Chester, N.Y.: Cambridge University Press. A valuable book for those interested in becoming students of marine biology.

Hardy, A. C. 1965. *The Open Sea: Its Natural History.* Boston: Houghton Mifflin. A classical text on the biology of the ocean.

Jannasch, H. W., and C. O. Wirsen. 1977. "Microbial Life in the Deep Sea." *Scientific American* 236, no. 6, pp. 42–65.

Lutz, R. A. 1991–92. "The Biology of Deep-Sea Vents and Seeps." *Oceanus* 34, no. 7, pp. 75–83. A complete update on the biology of ocean vents.

Margulis, L., and K. V. Schwartz. 1984. *Five Kingdoms.* San Francisco: W. H. Freeman and Company. A good discussion of the biological classification of organisms used in this chapter.

Nelson, J. S. 1993. *Fishes of the World,* 2d ed. New York: Wiley. Comprehensive overview of fish.

Nybakken, J. W. 1993. *Marine Biology: An Ecological Approach,* 3d ed. New York: HarperCollins. A good, readable text on marine biology.

Ricketts, E. F., J. Calvin, and J. Hedgpeth. 1985. *Between Pacific Tides,* 5th ed. Stanford, Calif.: Stanford University Press. A classic text on West Coast marine organisms.

Sumich, James L. 1992. *An Introduction to the Biology of Marine Life,* 5th ed. Dubuque, Iowa: Wm. C. Brown Publishers. A good marine biology text.

Tait, R. V., and R. S. Desanto. 1972. *Elements of Marine Biology.* New York: Springer-Verlag. A good text, but out of print, so may be difficult to find.

Valiela, I. 1984. *Marine Ecological Processes.* New York: Springer-Verlag. A valuable text on marine ecology.

Webber, H. H., and H. V. Thurman. 1991. *Marine Biology,* 2d ed. New York: HarperCollins. A comprehensive marine biology text.

CHAPTER 12

The Marine Biological Environment

BIOLOGICAL OCEANOGRAPHY IS ONE OF THE MOST INTRIGUING FIELDS OF OCEANOGRAPHY. EVEN AFTER MORE THAN A CENTURY OF STUDY, MANY NEW ANIMALS AND PLANTS CONTINUE TO BE DISCOVERED IN THE OCEAN. THIS IS BEST SHOWN BY RECENT DISCOVERIES OF UNIQUE CREATURES FROM VENT AREAS ALONG THE MIDOCEAN RIDGES. INDEED, THESE FINDINGS HAVE SHOWN A WHOLE NEW TYPE OF LIFE PROCESS BASED ON *CHEMOSYNTHESIS* RATHER THAN *PHOTOSYNTHESIS.*

BIOLOGICAL OCEANOGRAPHERS ARE JUST BEGINNING TO COMPREHEND THE COMPLEXITY OF *OCEAN ECOLOGY,* WHICH IS THE RELATIONSHIP BETWEEN ORGANISMS AND THE OCEAN. THIS RELATIONSHIP HAS SEVERAL FACETS, INCLUDING POLLUTION AND FOOD. THE POLLUTION HAZARD TO THE BIOLOGICAL ENVIRONMENT IS CLEAR, BUT LITTLE IS KNOWN ABOUT ITS SPECIFIC OR LONG-TERM EFFECTS.

AS FOR NUTRITION, SOME FEEL THAT THE OCEAN HAS THE POTENTIAL TO PROVIDE A MUCH LARGER SHARE OF HUMAN FOOD SUPPLY. THIS POTENTIAL SEEMS OBVIOUS IN LIGHT OF THE FACT THAT THE SEA COVERS 71 PERCENT OF EARTH'S SURFACE *BUT YIELDS ONLY ONE-THOUSANDTH AS MUCH HUMAN SUSTENANCE AS THE LAND.* HOWEVER, NUTRIENTS IN THE OCEAN ARE GENERALLY LESS CONCENTRATED THAN ON THE LAND. THUS, SOME SCIENTISTS FEEL THAT THE

A blue angelfish feeding on the outer part of the Great Barrier Reef of Australia.

(Photograph courtesy of the Great Barrier Reef Marine Park Authority.)

OCEANS ALREADY ARE BEING FISHED CLOSE TO THEIR PRACTICAL LIMIT.

THIS CHAPTER EMPHASIZES THE SEA AS A BIOLOGICAL ENVIRONMENT, THE INTERACTIONS OF THE ORGANISMS THAT INHABIT IT, AND THE PROCESSES INVOLVED IN PRODUCING ORGANIC MATTER (FOOD) IN THE SEA.

Introduction

Much current research in marine biology concerns itself with how marine animals and plants interact with each other and with their environment. Often this involves a detailed knowledge of water chemistry and sometimes of physical oceanography and marine geology, along with an understanding of basic biological processes. Remote sensing from satellites has given biological oceanographers a worldwide picture of ocean surface characteristics.

Applied research in marine biology often emphasizes pollution and its detection, effects, and remedies. Applied research also focuses on ways to increase food production from the sea. The fields of biotechnology and genetic engineering offer exciting possibilities for using the biological aspects of the ocean, such as, for example, improving fish or shellfish production, both in the natural ocean environment and in artificial habitats such as fish hatcheries.

Global change is an important issue in biological oceanography. An important research program in this area is Global Ocean Ecosystems Dynamics (GLOBEC). GLOBEC's main premise is that fluctuations of marine populations can be better understood with improved knowledge of the ocean's physical processes, such as mixing and turbulence, and how they influence feeding, dispersal, and mortality. Thus, GLOBEC's goal is to understand ocean ecosystems fully so that scientists can predict how the population of sea organisms will vary, and how they will respond to a changing global climate.

Underlying all of this research are four questions important to the field of biological oceanography:

1. What are the plants and animals that live in the ocean?
2. How are these organisms distributed in time and space?

3. What are the factors that control their distribution?
4. How do the organisms in the ocean live and survive?

The first two questions were discussed in Chapter 11. The last two are addressed in this chapter.

Biological Oceanography: Sea Food and Ecosystems

Since ancient coastal dwellings frequently have large quantities of marine shells and other evidence that "sea food" was popular, we know that ancient people exploited food from the sea. Drawings in ancient Egyptian tombs occasionally show fish caught from the Nile River, Mediterranean Sea, or Red Sea.

Early marine scientists were mainly concerned with the depth and shape of the ocean. Some biologists, however, were interested in the distribution of life, especially at great depths in the ocean. Samples of bottom life were occasionally obtained with devices used in deep-sea soundings and from dredgings.

In 1844, Edward Forbes, considered the father of marine biology, divided the ocean into eight zones on the basis of their marine organisms. He observed that animal life existed to about 550 m (300 fathoms or 1,800 ft.), with decreasing amounts at increasing depths. Below 550 m, he thought no life existed. Forbes's observation is surprising, considering that animal life had been brought to the surface from greater depths, even in his day.

The resulting controversy about the existence of life in the depths stirred interest in the field and furthered the blossoming science of oceanography. It helped launch the Challenger expedition of 1872–76 (see Figure 1-5, page 8), which collected large quantities of data on ocean animals and plants. Many other important expeditions and scientists made significant contributions to the field of biological oceanography, although they were less publicized. These early scientists were mainly interested in collecting, describing, and studying organisms (Figure 12-1).

The hard work of many biological oceanographers has helped us understand the intimate relationships between organisms and their seawater environment. Descriptive work continues, but as the organisms become better known, the need for these studies decreases. Emphasis is shifting to studies of the seawater environment, interrelationships among organisms, marine ecosystems, and production of organic matter.

FIGURE 12-1   The Chariot, a big net designed to roll across the seafloor and collect bottom-dwelling organisms. The picture was taken in 1940. (Photograph by Herb Gardner, courtesy of Woods Hole Oceanographic Institution Archives.)

FIGURE 12-2   Plankton net being lowered for sampling. The net is towed through the water and small plants and animals are trapped and retrieved. (Photograph courtesy of Dr. Susan H. Humphris.)

Instruments of the Biological Oceanographer

Biological oceanographers are primarily concerned with the distribution and relationships of animals and plants in the ocean, so they need a way to sample the population of the sea. SCUBA is one method of in situ observation and collection. In recent years, SCUBA techniques have been used even in the open parts of the ocean, where the practice is called **blue-water diving.**

A more typical sampling method is to drag some device, usually a net, through the water. A common one is the plankton net (Figure 12-2), which is used to catch small floating plants and animals. A newer development is to use a series of nets that can be opened and closed at specific depths or times.

Do samples obtained by nets really represent the environment? Slower-moving organisms are caught easily in nets, but fast fish can avoid them. In addition, certain species tend to be caught more than others. It appears that net-type sampling devices discriminate against certain organisms, thus presenting an incomplete or unrepresentative picture of the amounts and types of organisms present.

With nets, the biological population is usually sampled in only one direction—horizontally—because the net is dragged through the water at essentially a constant speed. Lowering the net to a certain depth and then bringing it up samples the vertical distribution of life. It also is possible to open and close nets at selected depths for more specific sampling (Figure 12-3).

Specialized nets have been designed to sample just the surface layers of the ocean (Figure 12-4). Biological oceanographers want to know both the total count of organisms collected as well as the concentration of organisms per volume of water (in other words, the population density). To accomplish this, the water volume passing through the sampler must be estimated or measured, perhaps by a small current meter across the front of the net.

Bottom-living fauna (benthos) can be sampled by devices dragged along the bottom (Figure 12-5), or by grab devices similar to those used by marine geologists (see Figure 5-1, page 107). Other methods of acquiring information on bottom fauna include bottom photographs, rocks and submarine cables brought to the surface, and by diving in a submersible or using SCUBA.

Electronic devices such as echo sounders can be used to detect large individual animals or groups of them in the water (see Figure 15-17, page 384). Many fishing

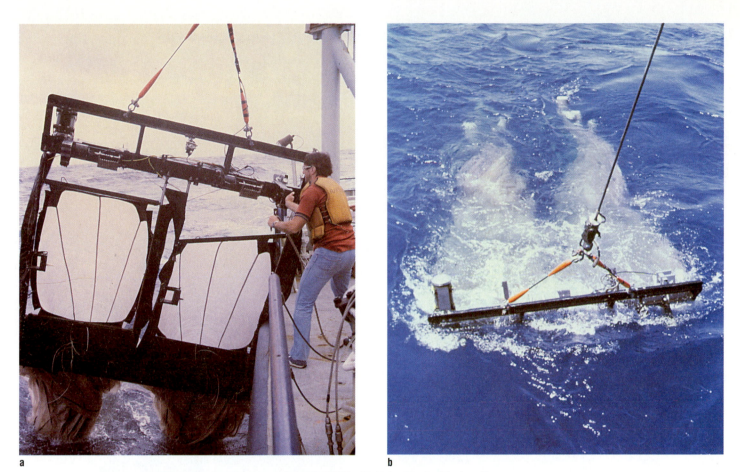

a   b

FIGURE 12-3 A more complex sampling system called MOCNESS (Multiple Opening/Closing Net-Environmental Sensing System); (a) shows sampler being launched; (b) shows the sampler in the water. Sensors in the net are linked to a shipboard computer that allow monitoring of the speed and angle of the net and salinity, temperature, depth, and light. The system contains several nets that can be opened and closed at appropriate positions to sample the organisms in the water. (Photographs courtesy of Dr. Peter Wiebe, Woods Hole Oceanographic Institution.)

FIGURE 12-4 A neuston sampler being towed. The net is kept at the water surface where it can sample surface living organisms as well as some pollutants, such as floating tar balls. (Photograph courtesy of Dr. George Grice.)

boats have such devices to locate schools of fish. Other sources of much information about migratory patterns of fish are fish-tagging programs and the accounts of people who fish.

In recent years, oceanographers have developed sophisticated instruments to measure subtle changes in various parameters in the marine environment. These data can be studied to determine how each parameter affects biological activity. Automated instruments measure temperature, salinity, light, and other characteristics of the environment. These instruments often are used with observing devices such as television and still-picture cameras. Data from such systems allow scientists to evaluate better the effect of the environment on marine life. Laboratory facilities that can simulate marine conditions are also among the important tools of the biological oceanographer.

a

c

b

FIGURE 12-5 (a) A biologic dredge. This device can be dragged along the ocean bottom, collecting bottom-dwelling (benthic) organisms and sediment in the net, which is held between the two runners. Most fine sediment passes through the net, but larger material and animals are caught. (b) Dredge being emptied. (c) Contents of a successful dredge. (Photographs courtesy of Woods Hole Oceanographic Institution.)

The Biological Community

The study of relationships among organisms and their environment is called **ecology.** An understanding of these relationships is fundamental because organisms almost never live as individual creatures, independent of others and their surroundings. Rather, they are closely influenced by physical and chemical aspects of their environment, as well as by other plants and animals with which they must interact. Such interactions, in turn, can affect the environment. Thus, one must consider animals and plants as part of a complex *system* involving physical, chemical, and biological aspects.

Such a system is commonly called an **ecosystem,** meaning *an entire range of organisms and the environment in which they interact.* Two broad categories of organisms exist: **autotrophs** (those that produce their own food) and **heterotrophs** (those that can't produce their own food and thus eat autotrophs). Prominent among the autotrophs are green plants, which produce organic matter by photosynthesis, and some bacteria that use chemical forms of energy. The amount of organic matter produced by autotrophs is called **primary production.**

Marine biologists also consider the environment according to biotopes or niches. A **biotope,** such as a reef, is an area where the habitat conditions and the living forms adapted to them are relatively uniform. Life in a biotope is not static; some animals may wander freely from one biotope to another. It is generally thought that the more rigorous the conditions, the fewer the variety of species and the greater the numbers of individuals of those species. This situation is typical of marsh environments but may not apply to the deep sea.

The relationship between some organisms can be categorized as **symbiotic,** meaning that they live together. Symbiotic relationships are of three types:

1. In **mutualism,** benefits exist for each interacting organism. For example, some species of fish "groom" one another by removing parasites and thereby get a free meal in the process.

2. In **commensalism,** only one organism benefits but no harm occurs to the other, such as when one animal lives in the protective shell of a larger and different organism. A fascinating commensal relationship is described in Box 12-1.

3. In **parasitism,** one organism benefits at the expense of another. Examples include worms that get into the internal organs of other creatures and cause damage or eventually death.

When speaking of the inhabitants of a biotope, we also must use the concept of a biological **community.** A community includes populations of different species of organisms that appear to interact with each other and with their common environment. The community may simply be the animals living together on a leaf of a plant. Or it may include a more extensive relationship, such as that between plankton and the animals that consume them, a dependency primarily based on the need for food.

Box 12-1

A Wonderful Example of Commensalism

Commensalism is a relationship between two organisms where one organism gets some benefit while the other suffers no harm. A fascinating example of commensalism was recently discovered in Antarctic waters by James McClintock of the University of Alabama at Birmingham and John Janssen of Loyola University.

As they dived, they observed an amphipod (a shrimp-like animal) carrying a sea butterfly (a type of slug or pteropod) on its back. Electron microscope pictures (Figure 1) show that the amphipod was holding the slug with its feet. One writer described it as holding the slug as one would hold a backpack.

It appears that the amphipods have a clear reason for this action. Fish that eat the amphipod find the sea butterfly very distasteful. In laboratory experiments conducted by McClintock and Janssen, they noted a clear avoidance by fish when presented with amphipods wearing their "backpack." Indeed, the fish would actually spit them out if accidently swallowed. On the other hand, amphipods without their slugs were readily eaten.

There are other examples of organisms that have creatures living with them with a similar result, such as scallops covered with sponges, which fish do not like to eat. But there seems to be no encouragement on the part of the scallop to have the sponge grow on its shell. With the amphipod and the slug, the amphipod clearly initiated the relationship and received the benefits. At

FIGURE 1 Scanning electron micrograph of an amphipod (*Hyperiella dilatata*) carrying a sea butterfly (a pteropod). Note how the rear legs of the amphipod grasp the sea butterfly. (Photograph by Phil Oshel, courtesy of James McClintock, University of Alabama at Birmingham.)

present no benefit seems evident for the slug, other than being able to get around a little quicker.

REFERENCE

McClintock, James B., and John Janssen. 1990. "Pteropod Abduction as a Chemical Defense in a Pelagic Antarctic Amphipod." *Nature* 346, no. 6283, pp. 462–64.

The food chain in the ocean is a complex system of several levels, starting with phytoplankton (the producers), then different kinds of herbivores (plant eaters), and then carnivores (meat eaters). Yet another level includes the decomposers, such as scavengers and bottom feeders.

Before examining how plants and animals live in the ocean environment, we must look at the important properties of seawater that affect the organisms dwelling in it.

The Ocean, Seawater, and Their Biological Consequences

As explained in Chapter 11, life in the ocean is considerably different than on land, because most marine organisms are constantly immersed in seawater. The properties of water afford advantages and disadvantages to organisms living in it.

TEMPERATURE. Seawater **temperature** ranges from about −2°C (28°F) in polar regions and at great depths to more than 40°C (104°F) in the Persian Gulf and other locales. Even though this range is broad, large areas of the ocean have fairly uniform temperature, thanks to the ocean's heat capacity (see Figure 8-12, page 191).

SALINITY. **Salinity** varies from nearly ‰ (parts per thousands) in some estuaries and nearshore areas to about 40‰ in areas having high evaporation rates, such as the Red Sea. Salinity is nearly constant in the surface waters of the open ocean away from river inflow, ranging narrowly between 33 and 37‰. Only in isolated areas does it get higher. The salinity of the deeper water is even more uniform, having a normal range of 34.6 to 35‰.

LIGHT. **Light** penetration varies, but hardly a trace of sunlight penetrates deeper than about 1,000 m (3,300 ft.).

PRESSURE. Ocean depths exceed 10,000 m (33,000 ft.), with **pressure** from the overlying water ranging from 1 atmosphere at the surface to over 1,000 atmospheres at great depths (1 atmosphere = 1 kg/cm^2 or 14.7 1b./in.^{2}; 1,000 atmospheres = 1,000 kg/cm^2 or 14,700 1b./in.^{2}, or over 7 tons/in.^{2}). Even though pressure does affect some biological aspects, many organisms are adapted to range vertically over considerable depth. If an organism has an internal gas bladder for buoyancy, water pressure changes with depth can present difficulty, such as the need to expel the gas when the pressure increases. Many animals, however, have adapted to this

problem. For example, some whales and fish that have gas bladders can dive from the surface to deeper than 1,000 m (3,300 ft.), a pressure change of 100 atmospheres enough to crush many objects. For marine organisms that have no air bladders, the pressure inside and outside of their body is the same, just as for organisms living on land, so no special protection against pressure is needed.

STABILITY. As is evident from the preceding paragraphs, the physical and chemical characteristics of seawater have a fair amount of **stability,** or at least are slow to change. Therefore, most marine organisms are not exposed to sudden environmental changes of temperature, moisture, pressure, turbulence, chemistry, and pollutants like their terrestrial counterparts. (Shallow coastal areas are an exception.)

SURROUNDING FLUID. Animals and plants constantly immersed in the ocean have another advantage over land varieties in that the **surrounding fluid** keeps them from *desiccation*, or drying out, from exposure to sun, wind, or drought. An exception is intertidal animals, which can have desiccation problems during low tides. Most animals of terrestrial environments have evolved impervious skins or scales to retain tissue fluids and to protect against attack from predators and parasites.

Most terrestrial plants have evolved extensive root systems to obtain water from soil. Such root systems are not needed in the ocean because marine plants are surrounded by water, which may contain their necessary nutrients.

DISSOLVING POWER. Water can dissolve nutrients and transport them, so it is essential for food production by plants. Since water can dissolve and carry more kinds of solids, other liquids, and gases than any other common liquid, it can thus store and transport the critical gases and minerals necessary for animal and plant life.

VERTICAL CURRENTS. The supply of oxygen could be a limiting factor for most life forms in the deep sea were it not for **vertical currents,** caused by the sinking of cold surface waters at high latitudes (see Figures 9-19 and 9-20, page 217). This transports oxygen-rich waters to the bottom and near-bottom, making life possible at essentially all depths. An exception can be where an extreme oxygen-minimum zone exists within the water column.

HORIZONTAL CURRENTS. The movement of seawater can be biologically very important. Motion can move nutrients from deep water to surface water, where they can be used by plants. **Horizontal currents** can also disperse waste products, eggs, larvae, and adult life forms. Water motion also may have some nonbeneficial results, such as carrying animals from their natural environment into an unfavorable one. This can happen where warm water and cold water come in contact, as in a Gulf Stream Ring (see Figure 9-16, page 214).

BUOYANCY. Seawater, because its density is generally greater than that of most marine organisms, provides **buoyancy** for many organisms in the ocean. In some instances, such as with jellyfish and other small floating animals, this eliminates the need for complex skeletal structures. Another effect of this support is to permit extremely large animals, such as whales, to develop and exist in the ocean. Most fish have a gas bladder they use to adjust their buoyancy, which controls the depth at which they float.

BUFFERING. Seawater is a **buffered solution.** This means that it acts like a chemical shock absorber, making it difficult to change its acidity/alkalinity, or pH (see Box 7-1, page 154). In general, seawater is slightly alkaline, having a pH between 7.5 and 8.4. The alkaline state of seawater is helpful for organisms that secrete calcium carbonate shells; if seawater were acidic, the carbonate could be dissolved. An additional advantage of seawater's buffering is that abundant carbon, in the form of carbon dioxide, can be present in the water without greatly changing the pH. The carbon is necessary for plants in the production of organic matter.

TRANSPARENCY. The **transparency** of water is biologically important because it allows light to penetrate seawater to a considerable depth (Figure 12-6). Because photosynthesis is light-driven, this important process is not restricted to the upper few meters of the ocean but can take place to depths of 200 m (660 ft.) or more, depending on the clarity of the water. If light penetrated thousands of meters, as it does in the atmosphere on land, photosynthesis would be active to considerable depth, expanding the biological productivity of the ocean. (See also Box 6-3, page 141.)

HEAT CAPACITY. Seawater has considerable **heat capacity,** meaning that it can store a large amount of heat. Seawater also has a high latent heat of condensation (the heat released during the change of state from gas to liquid—condensation—and the heat absorbed during the opposite change of state—evaporation). Both of these characteristics—great heat capacity and great latent heat—prevent rapid temperature changes (see also page 138). Because of this, most marine organisms do not require thermal regulation systems like some terrestrial animals, such as humans. On the other hand, the lack of such a system makes organisms vulnerable to small temperature changes.

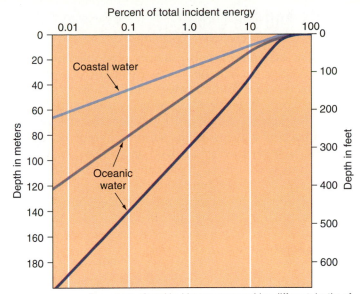

FIGURE 12-6   Percentage of total incident energy reaching different depths of the ocean for different types of water. Note that the degree of light penetration varies with water type. For some water types, only 5 percent of the light penetrates to a depth of 20 m (about 65 ft.). (After Jerlov, 1951.)

OSMOSIS. Many elements dissolved in seawater are biologically very important. The ratio of some of these elements in seawater is amazingly similar to the ratio in the body fluids of most marine organisms but less similar to the body fluids of land animals. This similarity between the external medium (the ocean) and the internal medium (the body fluids) is critical to **osmosis**—the passage of fluid from one solution into another, through a membrane (see Figure 6-10, page 146).

Imagine two solutions of different concentrations separated by a **semipermeable membrane** (like skin), which mostly isolates the fluids and prevents ions from migrating back and forth. Where a difference exists in the concentration on either side of the membrane, **osmotic pressure** builds. If the membrane is semipermeable, water will move through the membrane, from the less-concentrated solution into the more concentrated solution. The greater the difference in concentration, the greater the osmotic pressure.

Organisms must work against osmotic pressure to maintain the composition of their internal fluids. In marine organisms, the similarity of body fluids to the external medium (seawater) means that only a small osmotic pressure exists. This in turn means that marine organisms need not use as much energy as freshwater organisms to maintain their body fluids.

As discussed in Chapter 7 and later in this chapter, the influences of the ocean and its organisms are mutual. Marine organisms directly influence seawater chemistry by their life processes. Some of the biologically important physical and chemical characteristics of seawater are discussed in the following sections.

Plants and the Ocean

With one exception, plants, mainly phytoplankton, are the foundation for all other life in the ocean. (The exception is ocean-vent communities, which are discussed in a later section.) If every plant were removed from the ocean, food chains would quickly collapse, and seawater would become a lifeless saline solution except for bacteria and those organisms that feed on bacteria. Plants are the primary producers of organic material, and almost all other forms of life are dependent on them for food. The principal process of organic production—photosynthesis—is a process common to all plants, whether they float in the ocean or are anchored to rocks in the shallow sea or are rooted in terrestrial soil.

Photosynthesis

Photosynthesis is an endothermic reaction. This means that it requires an input of energy to drive it and sufficient nutrients and trace elements for the plants to grow. The needed energy is in the form of light; the nutrients include nitrogen and phosphorus compounds. Plants contain chlorophyll, a green-colored pigment that allows them to use energy from sunlight to perform photosynthesis. The general photosynthetic equation is:

$$CO_2 + H_2O + \text{solar (light) energy} \rightarrow \text{organic matter} + O_2$$

Expressed in words: solar energy, combined with carbon dioxide gas and liquid water, yields organic matter (generally expressed as some variation of CH_2O) and oxygen gas.

The reverse of this reaction, in which organic matter is consumed, is called **respiration.** This process uses oxygen and ultimately returns any nutrients involved to the water. Photosynthesis and respiration together make up the **organic cycle** in the ocean (Figure 12-7).

The organic matter produced by photosynthesis is the basic food in the ocean. It can be in several forms. A common one is the carbohydrate called *glucose,* $C_6H_{12}O_6$. As we shall see in the section on organic production, when describing or comparing the rate of photosynthesis, or production of organic matter, it is common practice to consider the amount of *carbon* that is included (or "fixed") into the organic matter.

Because light is essential for plant photosynthesis, it determines the depth to which plants can live in the ocean. As you can see from Figure 12-6, little light is present below about 200 m (660 ft.). In some areas of the ocean, significant light doesn't penetrate nearly that deep. Thus for plants to survive, they must live above this depth. Benthic plants (attached to the bottom) are therefore restricted to a small, shallow portion of the ocean. Most marine plants, therefore, are planktonic—they float.

Flotation

For plants to float is not as simple as it sounds. The density of living protoplasm is generally *greater* than that of seawater. Shells or tests are even denser. Plants have evolved several adaptations to enhance their floating ability, including special shapes of some shells that retard sinking, thin (low-density) shells, secreted oils or fats that lower the bulk density of the plant, and gas bladders. Some phytoplankton occur in colonies of long chains or ribbons, which may also help prevent sinking.

Compensation Depth

Let us sink a plant and observe what happens. Near the surface, where sunlight input is plentiful, a plant photosynthesizes (produces) plenty of food and respires (uses) little of it. If we sink the plant to a depth where insufficient light exists for any photosynthesis, it will consume

FIGURE 12-7 The organic cycle in the ocean, using a diatom as an example: (a) photosynthesis; (b) respiration.

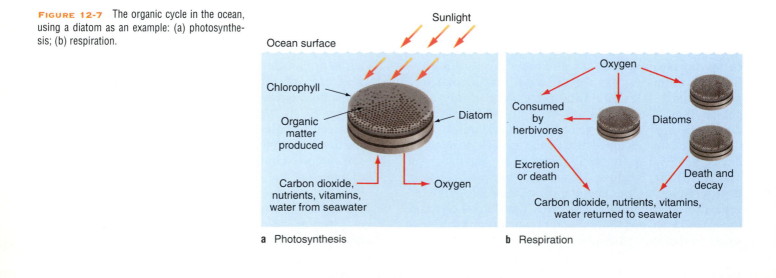

a Photosynthesis

b Respiration

its food supply (respiration), run out of food because it doesn't have sunlight to make more, and then die. Without sunlight, the plant no longer is a producer—it *consumes* more food than it *produces*.

As we move the plant back toward the surface, we reach its **compensation depth,** the depth at which enough sunlight is available to drive the plant's photosynthesis at a rate equal to its respiration rate (Figure 12-8a). Compensation depth depends on the light intensity, which is affected by factors such as solar angle, season of the year, time of day, suspended material in the water, and turbulence. With this many variables, the compensation depth frequently changes.

In coastal waters, the compensation depth is generally around 20 to 30 m (about 65 to 100 ft.). In the open ocean it can be 200 m (660 ft.) deep or more. The amount of photosynthesis is closely related to available light (Figure 12-8b). In the upper few meters of the ocean, photosynthesis can actually be inhibited, apparently due to excessive light.

Nutrient Cycles

Read the label on a container of plant food and you will see that it is a fertilizer rich in nitrogen and phosphorus. All plants need these essential elements to thrive, whether broccoli growing on the land or phytoplankton in the sea. We call these fertilizers for ocean plants *nutrients*. Their availability is necessary for plants to grow and to be healthy.

Some elements essential for plant growth, such as potassium, sulfur, and magnesium, are usually available in sufficient quantities in seawater. Other important nutrients, like nitrogen, phosphorus, silicate, and iron, are present in smaller amounts. In some instances, inadequate nutrients may limit the plant's ability to photosynthesize.

NITROGEN AND PHOSPHORUS.   The two main nutrients for plant growth in the ocean are nitrogen and phosphorus. Silica (in the form of silicate ions) is also

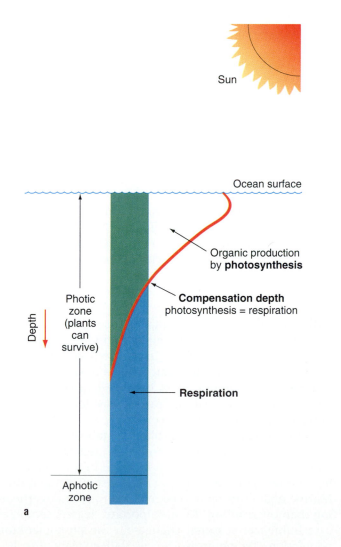

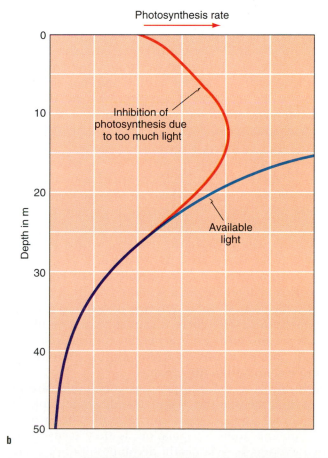

FIGURE 12-8   (a) Photosynthesis, respiration, and compensation depth. Note that production is highest near the surface, but that some inhibition exists right at the surface (see b). In the model shown, respiration is constant with depth. Compensation depth is where the photosynthesis rate and respiration rate of a plant are equal.
(b) Photosynthesis, in near-surface waters, related to available light.

needed by some plants, such as diatoms, to produce parts of their skeleton. These three nutrients are frequently supplied by rivers, but can be removed from seawater quickly by plants during periods of rapid growth.

Once the nutrients are incorporated into the plants, they return to the seawater only when the plants die (or the animals that eat them die) and decompose, or when the animals excrete the nutrients in their waste. If this occurs below the zone where photosynthesis is active, some mechanism, such as upwelling, is needed to return the nutrients to the surface waters for future use.

Earth has many cycles in which a substance is processed through a series of chemical reactions. Nutrients in the ocean are no exception. The nitrogen cycle is shown in Figure 12-9a and the phosphorus cycle in Figure 12-9b. Note that in each cycle, some of the element comes from the atmosphere and land, is processed through organism life cycles, and eventually ends up as sediment on the ocean floor.

In the ocean, nitrogen occurs in several forms: molecular nitrogen (N_2), ammonium (NH_4^+), nitrite (NO_2^-), and nitrate (NO_3^{2-}). Although molecular nitrogen is the most abundant, plants can't use it in this form. Plants require nitrate to produce **amino acids,** the building blocks for proteins. **Nitrogen-fixing bacteria** convert molecular nitrogen into nitrites, nitrates, and ammonia. Other bacteria then turn the nitrites and ammonia into nitrates for use by plants. When organisms die, **denitrifying bacteria** decompose the dead plant and animal tissue back into its component parts, including molecular nitrogen, thus completing the nitrogen cycle.

The phosphorus cycle is similar but simpler (Figure 12-9b). It also is quicker, which is important. The faster process allows phosphorus-bearing organic compounds to break down into the inorganic phosphorous compounds needed by plants, while the material still is in the photic zone. The phosphorus nutrient is thus provided right where the plants are ready to use it.

By contrast, nitrogen is frequently carried below the compensation depth before it is broken down. It then requires some vertical water movement to return it above compensation depth, where it can be reused by plants. For this reason, nitrogen is generally more of a limiting factor in photosynthesis than phosphorus, even though it is more abundant in the ocean.

Both nitrogen and phosphorous compounds are removed from the water by plants (mainly phytoplankton) as they grow. Some of these nutrients may be incorporated into the animal population as they feed on the plants. Eventually—via death, decomposition, or excretion—all of these organisms release the nutrients back into the seawater for the next stage of the cycle.

NUTRIENT DISTRIBUTION.   Note that nutrient distribution in the ocean is not uniform. Seasonal changes are evident in Figure 12-10. These seasonal changes relate to nutrient use by phytoplankton during their periods of extensive growth and reproduction, called *blooms.* During these periods of high growth, the phytoplankton remove large amounts of nutrients from the surface waters. If not replaced, this eventually leads to reduced growth (Figure 12-11). Thus the blooms themselves can eventually lead to a period of reduced development and growth of the phytoplankton population.

Geographic differences also influence the concentration and growth of phytoplankton. In the Arctic, annual production by phytoplankton is relatively low because of the ice cover, which limits the penetration of light. In the Antarctic, however, strong water circulation delivers huge quantities of nutrients to the surface waters. They are not depleted, despite the high productivity.

Nutrient distribution also shows vertical changes. Nutrient concentrations are generally low in surface waters, where they have been used by phytoplankton, and relatively high in deeper waters because of their subsequent decay and decomposition (see Figure 7-5, page 161). This difference is also a seasonal phenomenon, the contrast generally being greater during periods of rapid growth.

Sometimes phytoplankton production decreases even though nutrients are present in apparently sufficient quantities. In these instances, growth of the plants may be limited by another essential or trace element. Many elements are found in very small (trace) quantities in plants, but their significance is not fully understood.

Vitamins are also important in the photosynthesis process. Elements such as copper, zinc, and manganese help some phytoplankton growth. Silicon, as mentioned, is important to many plants, especially the diatoms, which require it for shell production. In areas where the silicon supply is low, diatoms tend to have thinner shells, apparently reflecting the scarcity of the element.

Other Factors Affecting Plants

Salinity is important to plants because it influences the osmotic pressure between the plant and its environment (see Figure 6-10, page 146). In this respect, salinity can limit the type of plant that can grow in a particular environment. This salinity effect is limited to coastal areas and estuaries where salinity can vary widely; elsewhere, salinity values are relatively constant.

Temperature can affect plant metabolism, which is the rate at which the plant respires or uses food. For example, a twofold or threefold increase in metabolic rate may occur with only a 10°C (18°F) rise in temperature. Marine organisms tend to be more sensitive to overheating than overcooling. To an important degree, temperature influences seasonal changes in the phytoplankton population. Temperature can also affect plant distribu-

FIGURE 12-9   The nitrogen cycle (a) and the phosphorus cycle (b). These processes are critical for the cycling of these nutrients in the ocean. (See text for additional information.)

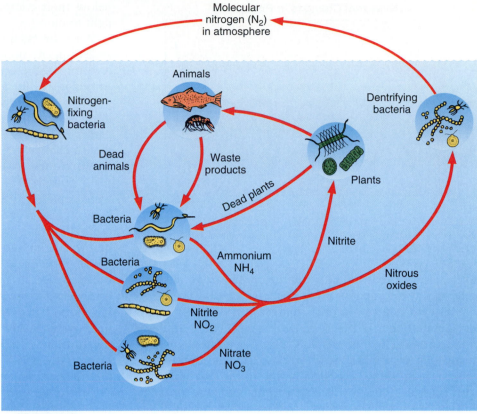

a   **Nitrogen cycle**

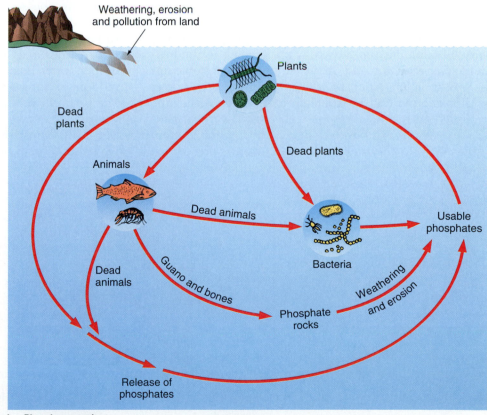

b   **Phosphorus cycle**

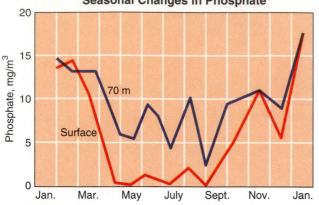

Seasonal Changes in Phosphate

FIGURE 12-10 Seasonal changes in phosphate concentration as measured in the surface waters of the English Channel and at 70 m (about 230 ft.) depth. (After Arkins, 1926.)

tion. Certain species are common to polar areas; others, to temperate or equatorial regions.

Light influences seasonal changes in the phytoplankton population. In the temperate and high latitudes, phytoplankton blooms usually occur sometime in the spring (Figure 12-11). After quickly peaking, the population rapidly decreases, usually due to the phytoplankton using up all the available nutrients. In some areas, a similar but usually smaller phytoplankton growth increase occurs in the autumn, followed by a decrease extending over the winter, in turn followed by the spring bloom.

Seasonality controls the time of rapid growth. The spring bloom may be caused by an increase in the incident solar radiation and the development of a stable upper water layer. When the thickness of the upper layer is thin enough that phytoplankton are kept within the well-lighted region, a bloom can occur. If vertical mixing by winds extends too deep, the phytoplankton, which float at the mercy of such currents, are carried

below their compensation depth and have insufficient light to bloom.

With the beginning of spring, water temperature increases in the upper layers. This causes a *stratification*, or layering, of the water, which suppresses vertical mixing. In regions of ice and high river runoff, a sharp salinity gradient, due to overlying fresher water, suppresses vertical mixing in a similar manner. This means that the plants are confined to the upper sunlit area and thus have a chance to bloom.

Nutrients in high concentration are necessary for rapid growth. The nutrients in the surface waters that were depleted by the spring and autumn blooms are replenished over the winter by mixing of the water, especially during storms. During the spring and summer months the surface waters are heated, producing a thermocline (see Figure 8-9, page 189), or temperature stratification. (Recall that the thermocline is the transition layer in which temperature changes rapidly with depth. It is found between the well-mixed surface layer and deep water.) This stratification restricts the transport of the nutrient-rich bottom waters into the surface-water layers, where they could be used in photosynthesis. Thus the thermocline can act to brake phytoplankton growth.

During autumn and winter, storms and cooling surface waters disrupt the thermocline and permit mixing of the water layers. In this manner, a nutrient supply is carried to the surface, sometimes causing an autumn bloom. This bloom is generally smaller than the one in spring because less light is available.

In some areas of the ocean, enough nutrient-rich deeper water is moved to the surface by vertical water motion to support a large phytoplankton population. The vertical water movements include upwelling, divergence, turbulence, and convection (see Chapter 9, pages 214–216). Thus, the regions of upward water motion may also be areas of high organic production.

FIGURE 12-11 Yearly changes in phytoplankton growth and environmental factors such as temperature, nutrients, light, and grazing.

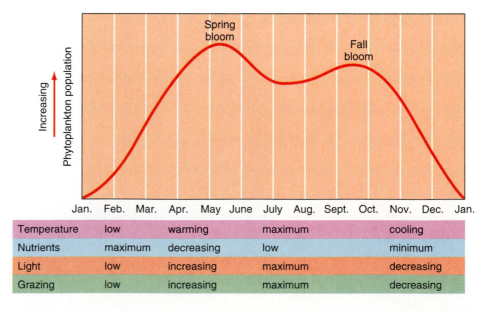

	Jan.	Feb.	Mar.	Apr.	May	June	July	Aug.	Sept.	Oct.	Nov.	Dec.	Jan.
Temperature	low		warming			maximum					cooling		
Nutrients	maximum		decreasing			low					minimum		
Light	low		increasing			maximum					decreasing		
Grazing	low		increasing			maximum					decreasing		

Upwelling can occur along a coast where the winds blow away the surface water. This allows nutrient-rich deeper water to rise to the surface (see Figures 9-17 and 9-18 pages 215 and 216). Turbulence is especially important in shallow areas having strong tidal action, such as Nova Scotia's Bay of Fundy. Convection occurs in areas where seasonal temperature changes are pronounced. In winter, the water is cooled, becoming denser, and eventually may sink, to be replaced at the surface by nutrient-rich deeper waters.

In summary, growth processes of marine plants are similar to those of land plants: both must have enough light, nutrients, and an otherwise hospitable environment. The difference between marine plants and land plants results from (a) smaller nutrient concentrations in the ocean, and (b) the fact that marine plants must float in surface waters to avoid being carried to the depth where available light is insufficient for photosynthesis. The growth of these floating plants is of paramount importance in the ocean, because all forms of animal life depend on them for food.

Animals and the Ocean

The study of all sea animals, regardless of size, is complicated by their mobility. If phytoplankton are difficult to sample, zooplankton are even more so because some have the ability to make vertical movements of tens to even hundreds of meters.

Another problem is their distribution, which occurs in three forms (Figure 12-12):

1. In *even distribution*—which is rare—each organism is approximately equidistant from its neighbor. An example is the ophiuroid, or brittle star; they sometimes live on the seafloor in such close proximity that each animal's five arms barely avoid contact with a neighbor.

2. In *random distribution*—also rare—no obvious relationship exists between any two individuals.

3. In *clumped distribution*—the most common distribution—animals or plants occur in patches or groups.

With luck, you can sample a clumped population with a plankton net. However, the probability of sampling that same population again is small.

This sampling difficulty is even more pronounced with free-swimming fish. Experiments show that two similar nets towed at the same depth and a few meters apart catch widely divergent numbers of organisms. Thus it is very difficult to obtain duplicate samples of the same group of organisms, which is necessary for most statistical work.

Adaptation to Environment

Ocean organisms must adapt to their environment. Those living in shallow water must withstand the greater temperature and salinity changes that can occur there. In very shallow water, at low tide, organisms may have to spend time out of seawater, and thus must avoid desiccation, or drying out. Shelled animals in the nearshore environment are usually streamlined and strengthened to withstand the crash of waves and being tumbled. Some burrow into the sediment for shelter. Barnacles and similar forms strongly attach themselves to the bottom or to rocks.

Despite the more arduous physical conditions in the nearshore region, this area has ecological advantages: abundant food, oxygen, and light. Growth of floating and attached plants is usually great in this environment, so organic matter (food) is abundant. This growth is sufficient to feed the benthic (bottom-dwelling) organisms.

Moving toward the deep sea, you can observe zonations of marine organisms. The reasons why animals inhabit one region of the ocean rather than another generally are not fully known. However, factors such as temperature, feeding habits, and available light are obviously important.

Temperature establishes faunal boundaries in the ocean. These latitudinal boundaries exist horizontally—there are animals typical of equatorial, temperate, or polar regions—and vertically, with depth. Some swimmers, however, move freely from one area into another. For example, the migration pattern of bluefin tuna shows that they can range over most of the ocean. Other animals are more restricted and spend their entire lives within a narrow temperature range.

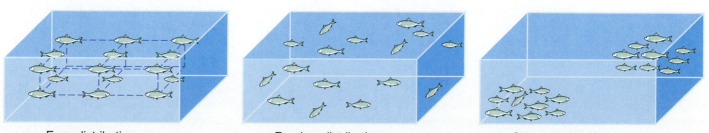

Even distribution   Random distribution   Clumped distribution

FIGURE 12-12   Types of distribution of sea animals. Most organisms tend to be clumped.

Temperature also affects the development of some animal forms. Fish tend to develop more rapidly and reach sexual maturity sooner in warmer regions. In colder areas, fish generally take longer to develop but usually grow larger than warm-water varieties.

Salinity usually is not an important factor in the deep ocean because it is nearly constant. Perhaps because they are adapted to this constancy, oceanic animals generally cannot tolerate large salinity changes. Most will die if currents carry them into an area of significantly different salinity.

Food is perhaps the most important factor in the distribution of animals in the deep ocean. It is certainly critical in the deep sea, where much of the food has to settle through the overlying waters. Bacteria also can be an important source of food in the deep sea. Nevertheless, oceanographers believe that the deep-sea food supply is very small. This probably is true in areas very far from land and in areas underlying surface waters that have low organic production. In other localities, the food supply may be adequate for the relatively small deep-sea population.

Any material, such as plants or dead animals, that sinks below about 2,000 m (6,500 ft.) depth stands a good chance of reaching the bottom because of the small number of organisms living in the water below this depth. Therefore, there is the correspondingly small probability that the material will be eaten before reaching the bottom.

Origin of Deep-Sea Fauna

Did deep-sea fauna originate where we find them today, or elsewhere? Remember that the ocean-bottom temperature has changed over geologic time. About 25 million years ago, the bottom water temperature may have been warmer by 7 or 8°C (12 to 14°F). This is based on good evidence from isotope studies of shells preserved in sediments. As the temperature cooled to its present value, some deep-sea organisms might not have survived. Animals tolerant of colder conditions could have migrated from shallower waters into the deep sea. Likewise, some of the original inhabitants of the deep sea may have evolved so that they could survive the changing conditions.

Recent discoveries of unusual life forms around deep-sea vents have added another confusing dimension for the evolution of life in the deep ocean. This is discussed later in this chapter.

Bioluminescence

You have seen that light is a very important factor in the ocean. The deep sea is essentially completely dark, if you consider only sunlight. However, many deep-sea animals and even some plants produce their own light, called **bioluminescence** (Figure 12-13). Apparently, this phenomenon occurs in all parts of the ocean. It can be observed at sea or from beaches when certain dinoflagellates (yellow-green algae) are in the water. They can give an eerie blue color to a ship's wake or to breaking waves. The light is caused by a chemical reaction, similar to the flashing of a firefly (lightning bug) on land.

There may be several possible reasons for bioluminescence. In the dark sea, it may help individuals of the same species locate each other for breeding, just as it helps terrestrial fireflies. For some deep-sea fish, light is used for feeding. These animals have evolved light-producing organs that lure other creatures close enough to be captured and eaten. Some squid use bioluminescence for protection. When attacked, they eject a cloud of glowing material that often confuses their antagonists. For many animals of the sea, however, the function of their bioluminescence is not understood.

Color

The color of many marine organisms is directly related to the presence of light in their environment. Fish in shallow water sometimes have protective coloring, dark on top and whitish underneath, to make them obscure against their usual background. For example, a predator looking at them from above would have difficulty distinguishing their dark top from a likewise dark ocean. Looking from below, a predator would have trouble telling the light-colored belly from a similar light background. Other organisms can change color or are mottled to make themselves inconspicuous in their surroundings.

Deep Scattering Layer

An interesting light-related ocean phenomenon is the **deep-scattering layer,** first observed on depth sounders as echoes. In water several hundreds or thousands of meters deep, observers noticed a broad area of sound reflection. At first the reflection was confused with the bottom, but these echoes moved, rising toward the surface at dusk and sinking at dawn. The apparent relationship of the reflection to sunlight strongly suggested a biological origin.

Nets towed in the layers sometimes collected shrimp-like animals. At other times, they caught fish or squid. Apparently the organically caused, sound-reflecting, deep-scattering layer occurs in many areas of the ocean but it is not always caused by the same type of organism. One type of scattering layer, called Alexander's Acres and observed by scientists in *Alvin*, is due to schools of small lantern fish (Figures 12-14a and 12-14b).

Why do these animals migrate up and down, sometimes as fast as 5 m (16 ft.) per minute? When these

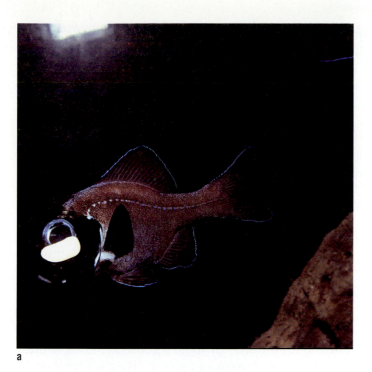

a

b

FIGURE 12-13 Two examples of bioluminescence.
(a) The flashlight fish has a luminescent organ just beneath the eye. This organ contains luminescent bacteria that emit light.

(b) A luminescent jellyfish from the Pacific Ocean. (Photograph (a) from Ken Lucas/Biological Photo Service; (b) from R. Degoursey/Visuals Unlimited.)

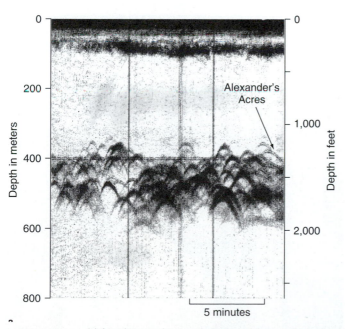

FIGURE 12-14 (a) An echo-sounder record of a type of deep-scattering layer called Alexander's Acres. The curved layers in this instance are caused by the reflection of sound (emitted by the ship's echo-sounder) from the small fish shown in (b). About 20 minutes of record are shown.

(b) The lantern fish (as observed from *Alvin*) and sampled by net that causes the scattering layer called Alexander's Acres. These fish are about 5 cm (2 in.) long. (Photograph (a) courtesy of R. H. Backus.)

organisms move, they may enter an environment having a different temperature, increased food supply, and pressure. Because the layers usually rise when the sun sets and sink when the sun rises, a reasonable explanation is that the movements are a response to light. In fact, when a bright moon is present, the layers sometimes descend slightly. However, the answer is probably not that simple. Other factors, such as internal physiological rhythms and food that also migrates up and down, may be important.

Before considering organic production in the ocean, a brief discussion of two especially interesting biological communities is appropriate.

Coral Reef Communities

Coral reefs are among the most interesting biological environments in the ocean. They occur along various coasts such as Florida and Australia (see Figure 14-21, page 353). They also exist thousands of miles from the nearest continental land mass, as is the case with South Pacific coral atolls. In either type of area, they represent a delicate balance between the forces of the ocean and the strength of the reef.

A **reef** is a seafloor biological community that forms a solid limestone (calcium carbonate) structure, built upon many generations of dead coral. The predominant organism in most communities are corals. Corals are colonies of animals whose tissues contain **coralline algae.** These algae are very important in constructing and maintaining the coral reef because they encrust and cement together pieces of coral, providing strength and forming a solid structure. This is crucial because most reefs grow slightly below sea level, where they must be strong enough to withstand the erosive power of breaking waves.

Coral reefs require specific conditions for growth. Water temperature must be at least 18°C (64°F) and generally is below 30°C (86°F). Temperatures above 30°C in some areas may lead to **coral bleaching** (see Box 12-2). Because of the need for warm water, most coral reefs are tropical. Corals cannot tolerate low salinities and can be killed by freshwater. In some instances, the reefs must grow fast enough to keep up with rising sea level or, in some regions, with the slow sinking of the islands upon which they are growing.

BOX 12-2

Coral Bleaching

Coral reefs are beautiful and important. They attract both fish and tourists. For coastal regions, they can be a buffer against large waves or storms. Loss of a coral reef is serious, and a damaged reef can take up to a hundred years to recover.

Coral reefs can be very colorful environments (see Figures 12-16 and 12-17). The main source of the color is the symbiotic algae that live within some types of coral. The algae, which are green, brown, or even gold, are called zooxanthellae (zoo-oh-ZAN-thell-i), Greek for "little yellow creatures." They live in the transparent coral polyp, producing their food by photosynthesis. The coral in turn uses waste from the algae for nourishment and to construct its limestone skeleton.

It was recently noted that, in some areas of the Caribbean, corals were looking bleached, or white (Figure 1). Examination showed that they were losing their symbiotic algae. Besides affecting their color, this eventually could lead to the coral's death by starvation.

What Causes Coral Bleaching?

If corals experience stress, they expel their algae, for reasons unknown. Once expelled, only the white lime-

FIGURE 1 Bleached brain coral, observed in 1988 at the shelf edge near La Pargura, Puerto Rico. Compare the color of the coral with healthy corals such as those shown in Figure 12-17a. (Photograph by Professor Lucy Williams, courtesy of Professor Manny Hernandez, both University of Puerto Rico, Sea Grant College Program.)

Box 12-2 continued

stone color of the coral skeleton remains visible, making it look "bleached."

Coral bleaching is not new. It has been known since 1918, but generally occurred only as an isolated event. Its incidence became more widespread in the late 1980s, and is continuing. Significant bleaching has been detected in corals off Puerto Rico, Colombia, Panama, Jamaica, the Virgin Islands, and the Florida Keys. In the early 1990s, bleaching was reported in several reefs in the south Pacific.

When bleaching occurs, it doesn't necessarily kill the coral. If the symbiotic algae return, recovery is possible. However, if they do not return, or if repeated events weaken the coral, death is inevitable. Recent bleachings already have killed nearly 30 percent of the affected corals.

Is Global Warming the Culprit?

Some scientists suspect that rising seawater temperatures, possibly due to global warming, are stressing the coral. Although much research remains to be done, it seems that stress can start when water temperature exceeds 30°C (86°F). Such temperature increases have been noted in some areas of bleaching.

Other factors such as pollution may initiate the bleaching process. However, the fact that some bleached reefs occur far from land, and therefore far from sources of pollution, supports the idea that temperature increase is the main cause. The occurrence of bleaching over different areas at about the same time also argues against a pollution cause. Other possible stresses could be human handling of the coral, salinity changes, or large amounts of sediment.

Whether the bleaching is due to a warming of seawater, and this in turn is resulting from a global warming, is a question that should be answerable within this decade.

Three types of coral reefs exist (Figure 12-15):

1. The **fringing reef** grows out from a landmass but is attached to it. An example is the reef bordering the Florida Keys.

2. The **barrier reef** is separated from the landmass by a lagoon. Exemplified by the Great Barrier Reef of Australia, this type can be very imposing (Figure 12-16).

3. The **atoll** is an oval-shaped reef surrounding a lagoon, associated with a landmass but not obviously so. Atolls commonly rise abruptly from the deep sea and some are very large. Kwajalein, for example, in the Pacific, is about 65 km (40 miles) long and 30 km (18 miles) wide.

The conditions on a reef vary from the quiet of the lagoon to the breaking waves on the outer part of the reef. These conditions are reflected in the different types of coral and animals living in each area (Figure 12-17). In areas of intense wave action, the coral must be strong enough to withstand the waves; more delicate forms can grow only in the quiet areas.

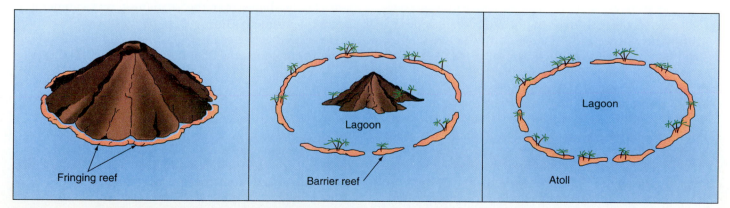

FIGURE 12-15   The three main categories of coral reefs. (See Figures 12-16 to 12-18 for real examples.)

FIGURE 12-16 Various parts of the Great Barrier Reef Complex of Australia: (a) outer shelf reefs in the Pompey Complex; (b) Lady Musgrave Island; (c) Hardy Reef; (d) Shute Island. (All photographs courtesy of Great Barrier Reef Marine Park Authority.)

Formation of Coral Atolls

The finding of numerous coral atolls in the middle of the deep South Pacific, without any visible landmass, puzzled many early scientists. How could these shallow-water animals have established an existence in water several kilometers deep? Why were the atolls ring-shaped?

During his voyage on the *Beagle* in the 1830s, Charles Darwin observed reefs and atolls. On the basis of this trip, he formulated a hypothesis to explain their origin. His idea, shown in Figure 12-18, is very simple. He suggested that a volcanic island initially provides a shallow-water base for the growth of a fringing reef. If the island slowly subsides or sea level rises, the coral continues to grow, building vertically on the bodies of its parents. Eventually, the island and the reef become separated by a lagoon, like a barrier reef. Further sinking then drowns the volcanic island, making it invisible. The result is an atoll.

A century later, Darwin's hypothesis was found to be essentially correct. The evidence came from drillings on Eniwetok and Bikini atolls. These drillings found over 1,200 m (almost 4,000 ft.) of coral underlain by volcanic rock. The coralline material originated in shallow water and clearly shows that the seamount was submerging during deposition. During part of its history, the atoll was elevated above sea level, as indicated by weathered zones of coral material obtained during the drilling.

Vent Communities

One of the most exciting marine discoveries of the twentieth century was the finding of deep-sea vents. Hot water is discharged from these vents, ranging from a few degrees above normal seawater temperature to over

FIGURE 12-17   Some forms of life in a coral reef area: (a) corals *Acropora* and *Platygyra;* (b) a soft coral, *Xenia elongata;* (c) an anemone and anemone fish living in a commensal association. (All photographs courtesy of Great Barrier Reef Marine Park Authority.)

b

c

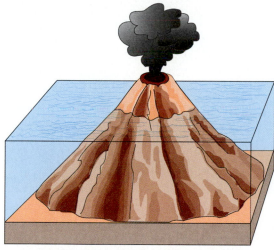

a Active volcano

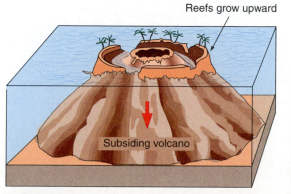

b Fringing reef

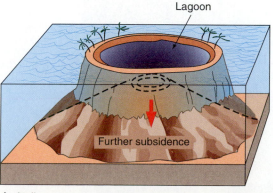

c Barrier reef

d Atoll

FIGURE 12-18   The four main stages in Charles Darwin's hypothesis for the formation of an atoll. Drilling on several atolls showed that his idea was essentially correct.

315°C (600°F). Water on the land or ocean surface, under 1 atmosphere of pressure, boils at the familiar 100°C (212°F). The 300°C or hotter water at the vents, however, cannot boil because of the high pressure that confines it.

As discussed in previous chapters, vents are associated mainly with active areas of seafloor spreading. As geologically impressive as the vents are, it turns out that their marine biology is equally fascinating. Associated with active vents are a striking collection of organisms, *most of which were previously unknown,* including a variety of worms, fish, crabs, shrimp, mussels, and clams, as well as some bizarre creatures (Figure 12-19).

Vent Dwellers and Chemosynthesis

Once these discoveries were made, an immediate question arose: What do these organisms feed on? Food is generally scarce at the depths of most deep sea vents (generally at least around 1,800 m, or 6,000 ft.), because photosynthesis is restricted to surface waters. Careful observation revealed the answer.

Water from the vents often looks milky, indicating that it contains suspended material. The suspended material was found to consist of very small particles of sulfur and other minerals. Further study showed that vent organisms were converting inorganic matter into organic carbon (food) in a manner similar to the photosynthetic process, but that bacteria—rather than plants—were the agents (Figure 12-19c). The energy source, rather than being sunlight, was hydrogen sulfide gas coming from the sulfur. This unique process can be termed **chemosynthesis,** or synthesizing food from chemicals instead of light.

Long tube worms (see Figure 12-19a) named *Riftia pachyptila* illustrate how chemosynthesis works. Blood

a

c

FIGURE 12-19 (a) Numerous unique organisms are found around active hydrothermal vents. These large tube worms are a common example. Many of these organisms feed on the chemosynthetic bacteria that use hydrogen sulfide as an energy source to convert inorganic material into organic matter that can be food for other organisms.
(b) Large clams found in the Galápagos Rift hot vent area.
(c) Scanning electron micrograph of a mussel shell found in the Galápagos Rift hot vent. The strings are stalks of bacteria; the protuberances are bacterial cells coated with manganese and iron. (Photograph (a) by Jack Donnelly, courtesy of Woods Hole Oceanographic Institution; (b) and (c) courtesy of Carl Wirsen, Woods Hole Oceanographic Institution.)

red tips (actually more like a fan or plume) on the worms extract hydrogen sulfide and oxygen from vent water. They carry these gases to an internal organ containing sulfur-digesting bacteria; these bacteria then combine the hydrogen sulfide and oxygen, oxidizing the sulfide and using the released energy to produce organic compounds (food) in a manner similar to photosynthesis in green plants.

The worms and their waste products form a food chain that can be used by other creatures. Hence the biological communities found around the vents are unique on Earth, being dependent on *geothermal* energy rather than solar energy. (See Box 12-3 for more on chemosynthesis.)

Vent Dwellers: Insight into the Past?

The answer to the question of food source led to other exciting ideas. For example, because such creatures live independent of the ocean surface, they could survive many kinds of environmental stress and might be examples of what life could be if it existed on other planets.

Box 12-3

Chemosynthesis

The finding of biological communities around deep-sea hot-water vents is an exciting scientific discovery. The creatures are startling to behold—giant tube worms, large clams, and shrimp without eyes that yet can detect light. Just as startling is the fact that many of them depend on a different food-making process than is typical in the ocean: **chemosynthesis.**

You have learned that the typical process uses light energy to drive photosynthesis to produce organic food molecules. Because of this, most biological communities require sunlight for survival. The process occurring at the vents, however, does not require light. Here, bacteria use chemical energy to produce organic food molecules.

Chemosynthesis is not unique to the ocean vents. It occurs in a few other areas, such as marsh sediments. The ocean vents, however, are the first large-scale sites of chemosynthesis discovered. The big headline from all this is that *life can develop in conditions previously thought impossible.* Perhaps some forms of life might even exist deep within Earth.

How Vent Creatures Eat

Among the most obvious of the vent organisms is the large tube worm *Riftia pachyptila*, up to 3 m (10 ft.) long, and several varieties of large clams, 25 cm (about 10 in.) long (see Figure 12-19). These animals have no gut and get their food from chemosynthetic bacteria *living in their tissues.* Other species may directly eat the bacteria or graze on dead ones.

The chemosynthetic bacteria use hydrogen sulfide (H_2S) from the vents to convert carbon dioxide into the organic food molecules that can be consumed by them and other creatures (Figure 1). The tube worms extract hydrogen sulfide from the vent water, transferring it to an organ containing bacteria. Here, the bacteria oxidize the hydrogen sulfide and turn it into food that can be used by the tube worm.

Other organisms such as the large clams and mussels maintain bacteria in their tissue. The relationship, like that in the tube worm, is a symbiotic one. The process is similar to a plant using sunlight to produce food in the photosynthetic process, but with the substitution of hydrogen sulfide for the light energy.

Much Research Lies Ahead

Vent organisms generally grow very rapidly. This may be due to the vents only being active for short periods—on the order of tens of years—which would require the creatures living there to mature quickly. When the vents become inactive, the energy source for generating food stops, and the community probably perishes.

Actually, the life history of a vent community has yet to be studied. Clearly, these vent communities are unique, being literally a self-contained and highly productive oasis on the seafloor.

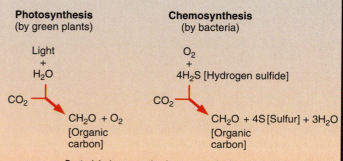

Photosynthesis (by green plants)

Light + H_2O

CO_2 → $CH_2O + O_2$ [Organic carbon]

Chemosynthesis (by bacteria)

O_2 + $4H_2S$ [Hydrogen sulfide]

CO_2 → $CH_2O + 4S$ [Sulfur] $+ 3H_2O$ [Organic carbon]

FIGURE 1 Bacterial chemosynthesis compared with plant photosynthesis. (Adapted from Jannasch and Wirsen, 1979.)

Indeed, some have argued that these communities might represent what some of the first life on Earth could have looked like, because the conditions within the vents are similar to what some postulate for the early Earth.

Apparently the animals that live in the vent communities produce numerous young, some of which succeed in floating to other vent areas. How they travel such distances is still unknown, but the organisms seem to be widely distributed. Many other questions exist. How, for example, do the organisms match their lifetimes and reproduction to the relatively short duration of a vent in a particular area, which is on the order of decades?

Initially found in the Pacific, the vents have also been found along the Mid-Atlantic Ridge (see Figure 7-10b, page 168). Some unique forms of life were found in the Atlantic, including a six-sided creature with rows of black dots (Figure 12-20). Fossils similar in appearance have been found in terrestrial rocks of about 40 million years age. Also found at the Atlantic site were new forms of eels, snails, eyeless shrimp, and worms. The Atlantic site also seems to have more freely swimming organisms than the Pacific area.

Another exotic organism, called the "vent fish," was collected from a Pacific site (Figure 12-21a). The first of these fish, caught near a vent accidentally, is a previously unknown species and may be the first vertebrate to be discovered whose existence is not ultimately tied to sunlight. Another new creature is known as the "Pompeii" worm (Figure 12-21b), which lives in a tube that is in the discharged hot water.

Have these environments and their associated fauna existed in the past? The answer appears to be yes.

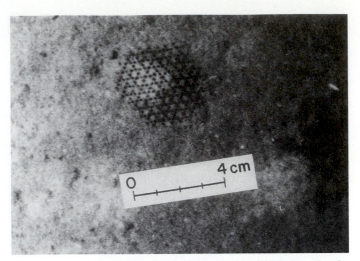

FIGURE 12-20 Photograph of a six-sided animal thought to be extinct for millions of years. This picture was taken by NOAA using a towed camera near a vent along the Mid-Atlantic Ridge. The site is about 2,880 km (1,800 mi.) east of Miami. (Photograph courtesy of Woods Hole Oceanographic Institution.)

Several mineral deposits have been known for years (on Cyprus, for example) that have similarities to vent deposits. More recently, a tube worm fossil was obtained from rocks in a sulfide ore body in Oman.

Organic Production

Organic production is the photosynthesis of organic matter by ocean plants (mainly phytoplankton). It is the

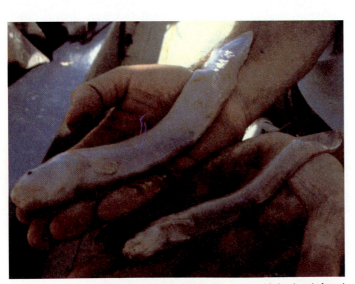

FIGURE 12-21 (a) Small vent fish (about 25 cm, or 10 in. long) found trapped in the exterior of the research submersible *Alvin* during dives near vents along the East Pacific Rise.

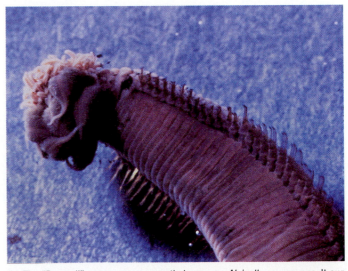

(b) The "Pompeii" worm, more correctly known as *Alvinella pompeyana*. It survives within the hot water of the vent. (Photograph (a) by Jean-Louis Michel of CNEXO, courtesy of Woods Hole Oceanographic Institution; (b) courtesy of John Porteous, Woods Hole Oceanographic Institution.)

most important biological process in the ocean; without it, marine life as we know it would be impossible. To review, the photosynthesis process combines carbon dioxide, water, and solar energy (light) to produce organic matter and oxygen. The organic matter is some form of CH_2O (the carbohydrate glucose, $C_6H_{12}O_6$, is an example). The amount of organic matter produced is related to the amount of carbon incorporated (fixed) into the organic matter formed. The unit of production generally used is *grams of carbon per square meter per day*, or $gC/m^2/day$, which represents the average growth of the plants in a general area over a day.

Organic production varies considerably with time, location, and other factors (Table 12-1). In the marine waters of the polar regions, for example, production is very high in the summer months because the sun shines almost 24 hours a day. During polar winter, the days are essentially dark and production is very low. In the tropics, where the light reaching the ocean is fairly constant, production proceeds at a relatively constant rate throughout the year.

Primary production is the rate at which organic matter is produced in a specific area (square meter) or volume (cubic meter) during a specific time (day, year, or other). **Gross production** is the total amount produced, some of which is used by the plant itself during its respiration process.

Total Organic Production

Biological oceanographers speak of organic production in terms of *carbon*. This refers to the carbohydrate food produced in photosynthesis. Areas of high production, such as Georges Bank, can produce 300 grams of carbon per square meter per year ($gC/m^2/yr$.). In other areas, organic production can be one-hundredth of this amount. Several estimates place the average gross production of the ocean at about 50 $gC/m^2/yr$. (see Table 12-1). Some areas of high productivity are shown in Figure 15-15, page 380.

One estimate is that 20 billion metric tons (20 trillion kg) of carbon are incorporated into living plant material each year in the ocean (see Table 15-7, page 381). On land the production is about 25 billion metric tons per year, or about 25 percent more. Terrestrial production, however, is more impressive when you consider that only 29 percent of Earth is exposed land, and that much land is incapable of production, being rocky or ice-covered. In the ocean, organic matter can be produced nearly everywhere within the photic zone, if there is no ice cover and there is daylight. The photic zone varies in thickness but can attain 200 m (660 ft.). On the other hand, the nutrient concentration in the ocean is considerably less than on land.

TABLE 12-1
Organic Productivity Measurements from Some Different Areas

Area	g C/m²/day	g C/m²/year
Open ocean waters	0.05–0.15	18–55[a]
Sargasso Sea	0.10–0.89	72
Continental shelf off New York	0.33 (mean)	120
Equatorial Pacific	0.50	180[a]
Upwelling areas—Coastal Chile	0.50–1.00	180–360[a]
All oceans, estimated mean	0.137	50

[a]Seasonal cycle assumed negligible; annual production computed from daily rates.
Source: Data from Ryther, 1969.

Estuaries are semi-enclosed coastal bodies of water with a source of fresh water. They are areas of high organic productivity, due in part to plentiful nutrients carried by rivers into the estuaries and commonly kept there by estuary circulation systems. Many estuaries are flanked by marshes and tidal flats, also extremely productive areas. Organic production within some estuarine areas can exceed that of the most productive land areas. Estuaries are also important as breeding areas for fish and other marine life. Unfortunately, it is this critical environment that also suffers most of the immediate effects of pollution (see Chapter 16).

Measuring Organic Production

The question of how much organic matter is produced in the ocean is more than just an interesting academic issue. In coming years, as Earth's human population continues to grow, increased demands will be placed on the food supply. New sources of food will need to be discovered, and increased use of the ocean's biological resources may become a necessity. To better exploit and manage these marine resources, scientific evaluation of what is available is certainly needed. Similarly, it is appropriate to determine the effects of marine pollution and the effects of excess fishing of certain fish species.

Organic production can be measured by several techniques, most based on the photosynthetic equation (see page 288). Methods include measuring the flow of nutrients going through the organic cycle. One widely used technique involves the radioactive isotope carbon-14 (^{14}C). The carbon atoms in carbon dioxide can be either common, nonradioactive carbon-12 (^{12}C) atoms, less common carbon-13 (^{13}C), or rare radioactive carbon-14 (^{14}C) atoms. A known amount of carbon dioxide containing ^{14}C is added to seawater that contains a known amount of carbon dioxide containing carbon-12 (^{12}C). Then some phytoplankton are added. The amount of carbon fixed by the plants into organic matter can be calculated by measuring the amount of ^{14}C absorbed into the plants at the end of the experiment. The amount of ^{14}C lost by the plant during respiration also must be accounted for.

Another way of measuring organic production is to monitor the change in the total nutrients in an area. Loss of nutrients, such as nitrogen and phosphorus, can be assumed to be caused by their being added to organic matter during photosynthesis. There are complications, however, such as additions due to recycling of dead material (see Figure 12-9).

Another estimate of productivity can be made by measuring chlorophyll, because the only place it can come from is plant cells. However, the amount of chlorophyll relative to the amount of carbon in the plants can vary. Also the quantity of chlorophyll varies with different species.

In general, all the methods of measuring organic production have some drawbacks. They can, however, give a fairly good approximation of relative organic production in different parts of the ocean.

Standing Crop

The **standing crop** is the number of organisms in a given area at a given time. Standing crop and productivity might seem to correspond, but it is not that simple. Productivity can be high, but the standing crop can be low, due to consumption (by herbivores) of the phytoplankton as they grow.

Determining the standing crop of plankton can be difficult. Net collection misses many small forms, such as coccolithophores (see Figure 11-12d, page 260), which can flow through the finest nets. Only when the plankton population is entirely of larger forms can net capture give an accurate estimate of the standing crop. A very fine filter will trap the tiny phytoplankton (nanoplankton) and give a more accurate estimate of the standing crop.

Factors Influencing Organic Production

There are two major factors that influence organic production: distance from shore and available nutrients, and the weather.

Distance from Shore and Available Nutrients. From a production standpoint, we can view the ocean in three general distances from shore: nearshore (water depth 50 m, or 165 ft., or less), intermediate (out

to 200 m, or 660 ft., depth), and offshore. Daily production in each area may be similar. Over a year, however, considerable difference can be observed (Figure 12-22).

Organic production and phytoplankton population are greatest in nearshore waters most of the time. The likely reason is that nearshore waters are richer in nutrients. Light appears to be less important, because light penetrates less in turbid nearshore waters. The turbidity results from river inflow and stronger wave action nearshore, which stirs bottom sediment. Production in intermediate waters usually exceeds that of the offshore waters, because of a more available supply of nutrients.

Weather. Clouds can affect productivity locally. With a massive cloud cover, much less light reaches the sea surface. Wind also affects productivity because choppy water presents a greater surface area to the sun, reflecting more light than a smooth surface. Temperature, through its effect on the thickness and depth of the thermocline, is also important. If the thermocline is well defined, it is very stable and resists mixing, so nutrients are not recycled from deeper waters.

Grazing

Decreases in phytoplankton population generally coincide with spring and autumn blooms. In some areas, however, an abrupt decrease in phytoplankton is often due to **grazing**—the consumption of the phytoplankton by zooplankton. Many herbivores, such as copepods,

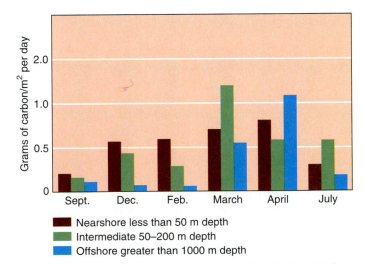

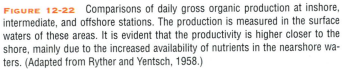

Figure 12-22 Comparisons of daily gross organic production at inshore, intermediate, and offshore stations. The production is measured in the surface waters of these areas. It is evident that the productivity is higher closer to the shore, mainly due to the increased availability of nutrients in the nearshore waters. (Adapted from Ryther and Yentsch, 1958.)

have very large appetites for phytoplankton. Most of these animals feed by filtering water to remove phytoplankton and in this manner they can rapidly reduce a local plant population.

The efficiency of zooplankton in catching phytoplankton decreases considerably as the number of available phytoplankton decreases, so some plants almost always survive. This is very important, because it provides the nucleus of a phytoplankton population with which to start increasing again when conditions become favorable.

If the important factors such as penetration of solar energy, nutrients, and quantity of zooplankton can be estimated, a model for predicting the population of an area can be developed (Figure 12-23).

Summarizing Primary Production

In summary, primary production of organic matter in the ocean is by floating plants or phytoplankton. Of these, the tiniest nanoplankton are probably the most significant, with considerably lesser amounts contributed by attached plants in shallow water and by bacteria.

The key ingredients in producing organic matter by photosynthesis are water, carbon dioxide, and sunlight. Growth rates of plants are generally controlled by the availability of nutrients and the amount of sunlight. In certain areas where the water is well mixed, either seasonally or continuously, production can be high because of replenishment of nutrients to the near-surface layers from deep waters.

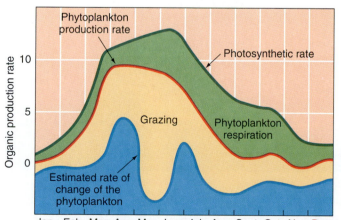

FIGURE 12-23   Estimated rates of production and consumption of carbon by plankton. This is a subtractive graph: the top curve is the photosynthetic rate. When the phytoplankton respiration rate is subtracted from the photosynthetic rate, the result is the middle curve: the phytoplankton production rate. Then, by subtracting the grazing rate, the estimated rate of change of the phytoplankton (bottom curve) is obtained. (Adapted from Riley, 1946.)

Some organic matter produced by plants is consumed by plants themselves for respiration. The remaining amount, or net production, is the main food source, either directly or indirectly, for most animals in the ocean.

The Food Cycle

Up to now we have considered ocean organisms and their responses to various factors in their environment. Another fundamental relationship in the sea is the **food cycle.** In the open ocean, primary production by plants is the initial part of a complex nutritional system that involves all marine organisms. The primary producers—plants—are consumed by herbivores (mainly zooplankton), who in turn are consumed by carnivores. Zooplankton convert the organic matter in plant tissue into animal tissue, which in turn becomes food for higher-order organisms. Humans may be a part of this system.

This pattern of each organism being eaten by the one above it is a **food chain.** A highly simplified food chain, showing the trophic (or feeding) levels and the energy gradient, is shown in Figure 12-24. In this simple trophic scheme, we assume a 10 percent efficiency in energy conversion when going from one level to the next higher level. For example, for a large fish such as a tuna to add 1 kg (2.2 lb.) of weight, it must consume 10 kg (22 lb.) of a moderately sized fish such as a bluefish. Using this model, 10,000 kg of phytoplankton are needed to produce 1 kg of tuna for human consumption. The 90 percent "loss" of energy at each level occurs because the organism uses some energy to feed, breath, move, and reproduce.

The trophic diagram helps us understand the biological efficiency of the ocean. In reality, however, the system is more complex. It is more realistic to visualize the food cycle in the ocean as a **food web** of complex interrelationships among numerous organisms. Figure 12-25 is a generalized food web diagram; Figure 12-26 is a more specific one.

Areas Having Different Food Cycles

Some ocean areas have food cycles that vary from the models shown in Figures 12-24 and 12-25. Sometimes they have fewer or different steps. An example is ocean vents, where the primary production comes from autotrophic bacteria using chemical energy in chemosynthesis rather than solar energy in photosynthesis. Although the number of trophic levels in this environment is not yet known, it appears that there are fewer than in the oceanic example in Figure 12-24.

FIGURE 12-24 A marine food chain. This simplified view shows 5 trophic or feeding levels in the ocean. The process is driven by solar energy, which phytoplankton use in the photosynthesis process. A 10 percent efficiency is assumed in going from one trophic level to the next trophic level.

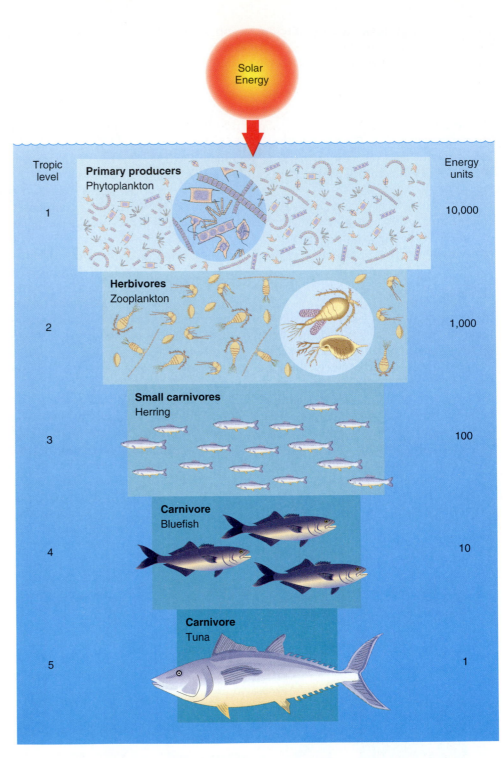

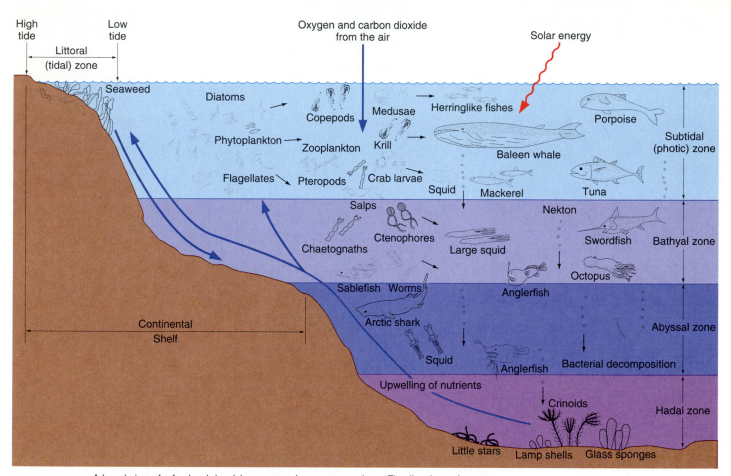

FIGURE 12-25 A broad view of a food web involving some major ocean organisms. The directions of energy flow are shown by arrows. Also shown are some divisions of the marine environment. The organisms and the dimensions of the divisions are not to scale. (Figure adapted from Beck, Lien, and Simpson, *Life: An Introduction to Biology*, 3d ed., (New York: HarperCollins, 1991).)

Coastal regions and many estuaries may also differ from the open ocean. Instead of floating plants, plants attached to the bottom may be the primary producers, if the water is shallow enough. In addition, water in these shallow areas is frequently well mixed, ensuring a continuous supply of nutrients, which can lead to higher primary production, sometimes year-round.

Clearly the marine food cycle differs from the terrestrial food cycle mainly because the primary producing plants in the ocean must be small and must float. Animals have evolved that feed on these plants, thereby concentrating the organic matter. Because plants are so numerous, it follows that their consumers must be numerous, too.

Implications of the Food Cycle

To understand completely the food cycle among the plankton, nekton, and benthos, scientists must examine the feeding habits and interrelationships of every individual species—a very large task.

The matter of efficiency also is very complex. Does a 10 percent efficiency actually exist between trophic levels? The specific efficiency varies, but each additional step in the cycle does cause a net loss of organic material. Generally, therefore, there is low efficiency in the transfer of organic matter from phytoplankton to fish.

Efficiency must be considered if we want to feed more people from the sea. The choice is either to take our food from a lower part of the food cycle (phytoplankton, algae, krill, and the like—generally not considered gourmet food) or find a way to make the transfer of organic matter more efficient. (Biological resources of the ocean are discussed further in Chapter 15.) Much research lies ahead.

FIGURE 12-26 A more detailed food web than that shown in Figure 12-25, but still considerably simplified. Arrows indicate the flow of food. This food web exists in the Antarctic. (Illustration from Harris, *Concepts in Zoology* (New York: HarperCollins, 1992).)

SUMMARY

Biological oceanography emphasizes the study of the animals and plants in the ocean and their interaction, both among themselves and with their environment. In spite of a long history of biological research on the ocean, much basic information still remains to be discovered. Dramatic evidence of this is the recent discovery of new organisms and methods of food production from vents along the ocean-ridge system.

Because of sampling limitations and the way plants and animals are distributed in the ocean (generally clumped), biological oceanographers can have considerable difficulty counting the organisms in an area or duplicating a sample. Biological research is often done un-

der simulated oceanographic conditions in a laboratory or under controlled conditions in the field.

Organisms that live in the ocean have several advantages over terrestrial ones, but they also have disadvantages. Light sufficient for photosynthesis cannot penetrate below 200 m (660 ft.) at best, so only a small portion of the seafloor can grow attached plants. Thus most ocean plants have evolved as floating organisms (phytoplankton) to remain in the surface waters of the ocean, where they can receive light for photosynthesis.

Production of organic matter by plants is a key process in the ocean because it produces food needed by all the other organisms for survival. Major ingredients for producing organic matter by photosynthesis are water, carbon dioxide, sunlight, and nutrients. In general, nutrients and sunlight are the limiting factors for growth. Certain areas of the ocean, especially where the waters are well mixed and the nutrients are recycled to the surface waters, are especially productive.

Most marine organisms are fundamentally related to each other via the food cycle. The cycle is initiated by or-

ganic matter production by phytoplankton, which are then consumed by herbivores (zooplankton), which in turn are consumed by a higher form of life, such as small fish like sardines. Ultimately these are eaten by bigger predators, such as tuna. The transfer of food to animals through the food chain is a complex process involving numerous interactions.

QUESTIONS

1. What principal methods are used to sample life in the ocean? Why is sampling so difficult?

2. What makes the ocean different from land as an environment for life?

3. How has the discovery of hydrothermal vents along ocean spreading centers changed our view of ocean biology?

4. What is a biological community?

5. Discuss how plants live and survive in the ocean.

   How do they differ from land plants? What are the factors that determine their success?

6. How do the key nutrients—nitrogen and phosphorus—get into the ocean? How are they recycled?

7. How is organic production measured? What factors influence it?

8. What are the main differences for animals and plants between the benthic and pelagic environments?

9. How does the marine food web differ from that of a land-based food web?

KEY TERMS

amino acid	coral reef	light	primary production
atoll	deep-scattering layer	mutualism	reef
autotroph	denitrifying bacteria	nitrogen	respiration
barrier reef	dissolving power	nitrogen-fixing bacteria	salinity
bioluminescence	ecology	nutrient	seasonality
biotope	ecosystem	nutrient distribution	semipermeable membrane
blue-water diving	food chain	organic cycle	stability
buffered solution	food cycle	organic production	standing crop
buoyancy	food web	osmosis	surrounding fluid
chemosynthesis	fringing reef	osmotic pressure	symbiotic
commensalism	grazing	parasitism	temperature
community	gross production	phosphorus	transparency
compensation depth	heat capacity	photosynthesis	
coral bleaching	heterotroph	pressure	

FURTHER READING

Beck, W. S, Lien, K. F. and G. G. Simpson. 1991. *Life: An Introduction to Biology,* 3d ed. New York: HarperCollins. A solid textbook on all things biological.

"Biological Oceanography." *Oceanus* 35, no. 3 (Fall 1992). A special issue on recent research in marine biology.

Brown, B. E., and J. C. Ogden. 1993. "Coral Bleaching."

Scientific American 268, no. 1, pp. 64–70. Discussion of coral bleaching and its effects, including whether such bleaching signals global warming.

Darwin, C. 1959 (reprint). *The Voyage of the Beagle.* New York: Harper and Row. The story of the famous expedition that, among other things, led to Darwin's hypothesis of the origin of coral atolls.

Gage, J. D., and P. A. Tyler. 1991. *Deep-Sea Biology: A Natural History of Organisms at the Deep-Sea Floor.* Port Chester, N.Y.: Cambridge University Press. A valuable book for anyone interested in becoming a student of marine biology.

Hardy, A. C. 1965. *The Open Sea: Its Natural History.* Boston: Houghton Mifflin. A classic early book on marine biology.

Harris, C. L. 1992. *Concepts in Zoology.* New York: HarperCollins. Covers a broad range of biology.

Jannasch, H. W., and C. O. Wirsen. 1979. "Chemosynthetic Primary Production at East Pacific Sea Floor Spreading Centers." *BioScience* 29, no. 10, pp. 592–98. A good discussion of some important biological processes occurring along the mid-ocean ridges.

Jannasch, H. W., and C. O. Wirsen. 1977. "Microbial Life in the Deep Sea." *Scientific American* 236, no. 6, pp. 42–52. Describes experiments in the deep sea to determine how the metabolism of microbial organisms is affected by low temperature and high pressure.

Libes, S. M. 1992. *An Introduction to Marine Biogeochemistry.* New York: John Wiley and Sons. A slightly advanced textbook covering marine chemistry and how it relates to some aspects of marine geology, biological oceanography, and pollution.

Lutz, R. A. 1991–92. "The Biology of Deep-Sea Vents and Seeps." *Oceanus* 34, no. 7, pp 75–83. A good update on the biology of ocean vents.

"Mid-Ocean Ridges." *Oceanus* 34, no. 7 (Winter 1991–92). A special issue devoted to the geology, geophysics and biology of ocean ridges.

Mullineaux, L. S., P. H. Wiebe, and E. T. Baker. 1991. "Hydrothermal Vent Plumes: Larval Highways in the Deep Sea." *Oceanus* 34, no. 3, pp. 64–68. Discusses the interesting question of how animals found at ocean vents get there.

Nybakken, J. W. 1993. *Marine Biology: An Ecological Approach,* 3d ed. New York: Harper & Row. A solid marine biology text.

"Reproductive Adaptations in Marine Organisms." *Oceanus* 34, no. 3 (Fall 1991). An entire issue on strategies that marine organisms use to reproduce and survive.

Ricketts, E. F., J. Calvin, and J. Hedgpeth. 1985. *Between Pacific Tides,* 5th ed. Stanford, Calif.: Stanford University Press. A classic text on West Coast marine organisms.

Szmant, A. M., and N. J. Gassman. 1991. "Caribbean Reef Corals." *Oceanus* 34, no. 3, pp. 11–18. Coral reefs, although under considerable environmental pressure, are still doing well, according to the authors.

Valiela, I. 1984. *Marine Ecological Processes.* New York: Springer-Verlag. A solid book on marine ecology.

Webber, H. H., and H. V. Thurman. 1991. *Marine Biology,* 2d ed. New York: HarperCollins. A good marine biology text.

CHAPTER 13

Climate, the Ocean, and Global Change

ONE OF THE MOST COMPELLING IS-
SUES FOR OCEANOGRAPHERS IS THE
QUESTION OF WHETHER EARTH IS
EXPERIENCING A GLOBAL WARMING.
IF SO, WHAT ROLE DOES THE OCEAN
PLAY IN THIS PHENOMENON? WE
SAW IN PREVIOUS CHAPTERS THAT
THE OCEAN AND ATMOSPHERE CON-
STANTLY INTERACT. IN THIS CHAP-
TER WE WILL CONSIDER THE MULTI-
PLE ROLES OF THE OCEAN IN
INFLUENCING WEATHER, CLIMATE,
AND OTHER POSSIBLE FUTURE
GLOBAL CHANGES.

The ocean, atmosphere, and biosphere are all
closely related.
*(Photograph courtesy of National Oceanic and Atmospheric
Administration.)*

Introduction

Weather is the short-term effect of atmospheric conditions over just a few days. *Climate,* on the other hand, extends over much longer periods—months, seasons, and years. Weather obviously is easier to predict, being most directly influenced by the present temperature, pressure, humidity, and wind conditions. Climate, however, is controlled by broader and more widespread conditions, such as the annual cycle of changing radiation from the sun, seawater temperature, and complex interactions among the ocean, land, air, ice, and the biosphere. The role of the sea in controlling Earth's climate is the subject of this chapter.

At present, we can accurately forecast the weather for only two or three days into the future. Even "five-day forecasts" become shaky toward the end of the five-day period. Predictions of climate, which attempt to look ahead over entire decades, have been even less successful. An important challenge for scientists who specialize in the oceans and meteorology is to predict future climate and perhaps even influence it.

The ability to predict weather and climate would have great benefits for agriculture, tourism, and hazard pre-vention. Long-range climatic prediction could allow for stockpiling of heating fuels or planting of certain crops for specific weather patterns in specific areas. Recent knowledge of specific events such as an El Niño, described later in this chapter, have revealed how the ocean and its interactions with the atmosphere can produce worldwide climatic phenomena.

The effects of past climate changes were discussed in previous chapters, particularly the several ice ages (glaciations) over the last million years or so. The most recent glaciation resulted in the lowering of sea level about 130 m (about 426 ft.), followed by the rise of the sea to its present level as Earth rewarmed. Droughts of the late 1960s and early 1970s, the abnormal cold periods in the United States during the late 1970s, the devastating droughts in Africa in the 1980s, the warm and dry periods of the late 1980s, and the increased incidence of very destructive hurricanes along the East Coast of the U.S. in the 1990s have all focused worldwide attention on the importance of climate.

At present, the most important technology used to study weather and its causes is the **weather satellite** (Figure 13-1). Satellite observations—now so common that their images appear daily on television—permit meteorologists to monitor weather phenomena and allow

FIGURE 13-1 Weather-satellite image with superimposed outline of North America and latitude-longitude grid. Note the strong storm system centered in New England that is spreading snow over the northeastern and middle Atlantic states. Snow cover from the wake of the storm can be seen in the Ohio River Valley and Mississippi River Valley areas. The cloud pattern in the Gulf of Mexico shows an interesting wave-like pattern. (Photograph courtesy of National Oceanic and Atmospheric Administration.)

for early warning of potential storms. We still have far to go, however, for successful long-term weather prediction.

Improved weather prediction will combine worldwide satellite images with extensive conventional measurements of temperature, cloud cover, winds, and humidity (Figure 13-2), all integrated into sophisticated models on high-speed computers. Several new scientific programs using satellites will greatly increase our understanding of the ocean and its interactions with the atmosphere. Examples of these are WOCE (World Ocean Circulation Experiment) and TOGA (Tropical Ocean and Global Atmosphere).

Most experts believe that much of our new understanding of climate will come from the oceans. Why? Because numerous fundamental links appear to exist among the atmosphere, oceans, and climate. These critical links are just beginning to be understood.

For example, the upper 3 m (about 10 ft.) of the ocean contains as much heat as the *entire* overlying atmosphere. The reason is that water has about 4 times greater specific heat per unit mass than air. Also, seawater molecules are close together, whereas air molecules are widely spaced, so the large mass of molecules in the ocean exists in a much smaller volume. As a result, the thermal capacity of the ocean (its ability to store heat) is about 1,000 times greater than that of the atmosphere. Consequently, much more energy is needed to heat water than to heat air.

In addition, the atmosphere generally moves about 10 times as fast as the ocean (comparing average atmospheric winds to average surface ocean currents). Therefore, large temperature differences can exist between the two, causing considerable interaction to occur.

Interaction Between the Ocean and the Atmosphere

In Chapter 9 we looked at how solar energy drives the atmosphere, and how the atmosphere drives oceanic circulation and waves. The atmosphere and the ocean are so closely intertwined that it is hard to avoid one when talking about the other (Figure 13-3). Both atmospheric winds and ocean currents transport heat from low-latitude (equatorial) regions, where there is more incoming radiation, to high-latitude regions, where there is less (see Figures 9-5 and 9-6, page 205). Variations in this pattern cause variations in climate.

Worldwide, climates are controlled by interaction among the ocean, atmosphere, land surfaces, and ice (either on land or sea). By studying Figure 13-4, you can see how a change in any one affects the others. Note especially the continuing interaction between the atmosphere and the ocean.

FIGURE 13-2 Large high-altitude weather balloon being released in Antarctica. Such devices collect and transmit routine weather data (temperature, wind, humidity) from the upper atmosphere. They can also be used to track pollutants and other constituents of the atmosphere. (Photograph courtesy of National Oceanic and Atmospheric Administration.)

The key to predicting climate effects is an understanding of the net effect of these components working together, especially the ocean and atmosphere. Mathematical models and long-term observations of weather and climate have led to some hypotheses about climate, which we will examine shortly.

Unequal Heating of Land and Water

Unequal heating of land and ocean surfaces, as we saw in Chapter 9, creates both atmospheric circulation and oceanic circulation (see Figures 9-5 to 9-11, pages 205–210). This unequal distribution of heat is responsible for major ocean currents and the wind system around our planet.

Because of its high heat capacity, the ocean is an effective buffer for both the seasonal and latitudinal differences in heat received on Earth. In the wintertime you

FIGURE 13-3 Waterspouts off the Bahamas provide an excellent and dramatic example of the interaction between the ocean and the atmosphere. (Photograph courtesy of National Oceanic and Atmospheric Administration.)

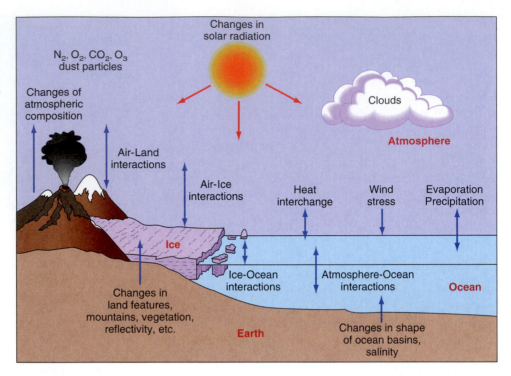

might expect the atmosphere to cool much more than it actually does. However, the atmosphere receives heat energy from the warmer ocean, which helps maintain its temperature. Likewise, in the summertime, you would expect the atmosphere to get much warmer than it does. However, it loses heat to the cooler ocean.

The net effect is that of a giant heat regulator, maintaining worldwide atmospheric temperature within remarkably narrow extremes, roughly −80°C to +50°C (−112°F to 122°F). Compare this to what temperatures could be without this regulation, roughly −170°C to +130°C (−274°F to 266°F), which is the case on the moon, where neither air nor sea exist to buffer temperatures.

The heat capacity of land materials (rock, soil) causes land to heat and cool rapidly, but the greater heat capacity of seawater causes it to heat and cool slowly (Figure 13-5). This difference is in large part responsible for the large-scale variations in atmospheric circulation. In the wintertime, for example, the ocean gives more heat energy to the air than does the land, resulting in denser air and, therefore, a high-pressure area over land. The opposite is true in the summertime, resulting in lower pressure over land.

Differences in the response of land and water to heat also affect local wind conditions. Because the land has a lower heat capacity than water, it is often warmer during the daytime in summer than the adjacent water. The air over land is heated, expands, and becomes less dense, and consequently rises. Cooler, denser air from the ocean moves toward the land to replace the rising air, producing an onshore wind or **sea breeze** (Figure 13-6).

At night or in winter, the situation reverses. The land is often cooler than the ocean, and an offshore breeze, or **land breeze,** may result. In summer, the air from over the ocean may be moist, leading to precipitation over the land, especially as the air becomes heated by the land, rises, cools, and the moisture in it condenses. In areas where these patterns are extreme and seasonal, a monsoon condition is said to occur (see Figure 9-11, page 211).

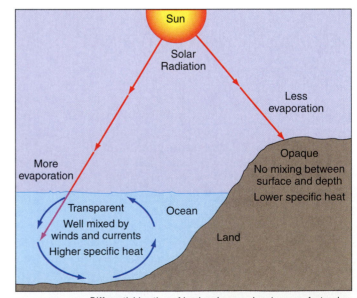

FIGURE 13-5   Differential heating of land and ocean. Land warms faster during the day and cools faster during the night than the warming and cooling of adjacent water, for the reasons shown. This difference in heating and cooling rates of land and water can cause a reversal of winds, commonly referred to as sea breezes and land breezes. (See also Figure 13-6.)

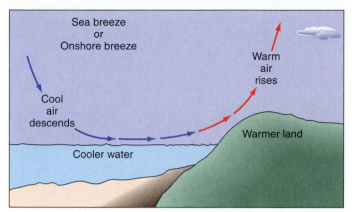

a Summer or daytime

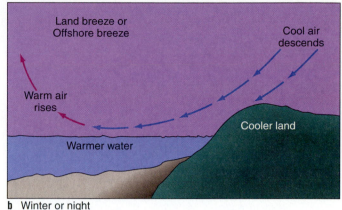

b Winter or night

FIGURE 13-6 Sea breeze and land breeze—wind patterns resulting from the relative difference in temperature between land and water. In the daytime (a), land is heated more than water (see also Figure 13-5). Land also cools more at night (b). When the air is heated, it expands, becomes less dense, and rises. Relatively cooler, denser air moves in to replace it, causing a breeze. The breeze is named for the direction from which it comes, not for where it is going. Under less-pronounced conditions, the winds change direction daily, landward during the day and seaward at night. Where these conditions are pronounced (for example, when summer conditions prevail even during the night), a monsoon pattern may last for months (See also Figure 9-11 page 211).

Summary of Oceanic-Atmospheric Interaction

The physical processes of the ocean affect the atmosphere, weather, climate, and ultimately global changes (see Figure 13-4):

- The stronger the wind, the greater the evaporation, and the stronger the oceanic currents.

- Air and water can heat or cool each other.

- Air can supply moisture to water, and water can supply moisture to air.

- The ocean, because of its high heat capacity and its density stratification, is a much more effective storehouse of solar radiant energy than air. Therefore, the ocean changes temperature much more slowly than the atmosphere.

The principal thing to remember is that Earth's atmosphere and ocean are constantly interacting, exchanging heat energy, exchanging moisture (evaporation and precipitation), and disturbing one another. No more dramatic example exists than in the great storms we call hurricanes.

Hurricanes

Hurricanes are tropical storms in which winds exceed 118 km (74 miles) per hour. They usually begin in tropical regions when ocean temperature reaches about 26.5°C (80°F). Hurricanes are called **typhoons** in the western Pacific and **cyclones** in the Indian Ocean. Whatever their name, they are an especially damaging aspect of weather.

The exact mechanism for hurricane formation is not completely understood, but they start with a low pressure area containing moist air. The moist air becomes heated because condensation releases heat, expands and becomes less dense, and rises (Figure 13-7a). As the air continues to rise, the moisture in it cools and condenses, forming clouds and thunderstorms. The column of rising air acts almost like a chimney, continually pulling additional warm moisture-laden air upward. The coriolis effect starts to rotate the rising air and clouds. At the center of most hurricanes is a well-defined eye. This is generally a calm area, where you literally can look up through the inner walls of the storm.

Energy released from the condensation of the rising, warm, moist air literally fuels the hurricane. This explains why a hurricane weakens considerably when it passes over land: its supply of warm moist air from the ocean (its energy) is cut off. The dimensions of a hurricane are impressive, averaging about 600 km (375 miles) in diameter and 12,000 m (40,000 ft.) in height. The energy released by just one hurricane can exceed that of all the electricity used in the U.S. over a 6-month interval. Considering the size and energy of a hurricane, its destructive ability is readily apparent.

In the early part of this century, numerous hurricanes hit the U.S. Gulf Coast and East Coast, with severe loss of life and damage (Table 13-1). More than 13,000 people have been killed in the United States by hurricanes since 1900. Figure 13-8 shows an apparent periodicity to hurricanes. Those in the 1940s tended to make landfall in the Florida region; in the 1950s, along the East Coast; and during the 1960s and 1970s, mainly in the Gulf of Mexico. Several especially strong hurricanes have struck

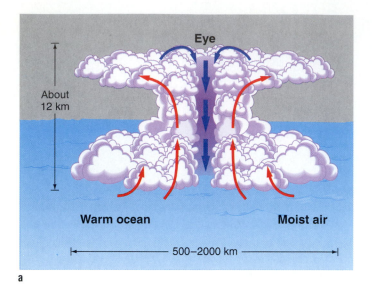

a

b

FIGURE 13-7   (a) A cross section through a typical hurricane, showing its circulation pattern. Note the rising air and outflow at the top of the storm, producing a chimneylike effect that continues to draw in more moisture-laden air.
(b) Hurricane Bonnie, near Bermuda, photographed with a special large-format camera aboard a 1992 Space Shuttle flight. This is a top view of the cross section in Figure 13-7a. Note the clouds spiraling away from the center of the storm and the well-developed eye of this hurricane. (Photograph (b) courtesy of National Aeronautics and Space Administration.)

along the East Coast, starting late in 1989 (see Box 13-1 for a possible explanation for the periodicity of hurricanes).

Hurricane monitoring has improved considerably through the use of satellites, reconnaissance aircraft flights, and radar. However, protecting people from hurricanes is still difficult because one cannot accurately predict their strength and path. In addition, no proven way exists to diffuse a hurricane.

Many areas have undergone extensive coastal development since their last hurricane, so many who live in these regions have no memory of, and little appreciation

---

TABLE 13-1
Some Exceptionally Strong U.S. Hurricanes

Year	Damage/Fatalities	Year	Damage/Fatalities
1900	6,000 killed when hurricane hit Galveston Island, Texas.	1983	Hurricane Alicia caused $676 million in damage and killed 17 in Texas.
1909	300 killed when storm flooded much of Louisiana coast.	1985	Hurricane Gloria was one of the strongest hurricanes ever and one of the most closely monitored. Fortunately, Gloria lost most of its energy before coming ashore in southern New England, and damage was relatively light.
1915	275 killed in Mississippi Delta region.		
1919	500 killed by hurricane that hit both Key West, Florida, and Corpus Christi, Texas.		
1928	1,800 killed in Lake Okeechobee, Florida.	1989	Hurricane Hugo, although one of the strongest storms to hit the United States, claimed only 26 lives in the U.S. Property damage, however, was the highest at that time, exceeding $9 billion.
1935	400 killed in the Florida Keys.		
1955	600 killed in New England.		
1957	390 killed by Hurricane Audrey, in Texas and Louisiana.	1991	Hurricane Bob caused over $1 billion in damage and the loss of 9 lives in New England.
1965	75 killed and $1.4 billion in damage from Hurricane Betsy in south Florida and Louisiana.		
1969	300 killed by Hurricane Camille, which hit Mississippi.	1992	Hurricane Andrew caused at least $20 billion in damage in southern Florida and 2 days later several billion dollars more when it hit Louisiana. Over 20 people were killed and 200,000 made homeless. Winds exceeding 225 kph (140 mph) hit both states.
1979	Hurricanes Frederic and David caused considerable damage.		
1980	Hurricane Allen caused considerable damage and loss of life in the Caribbean.		

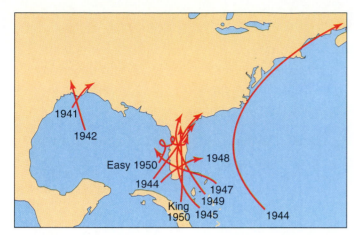

a   Major hurricanes 1941–1950

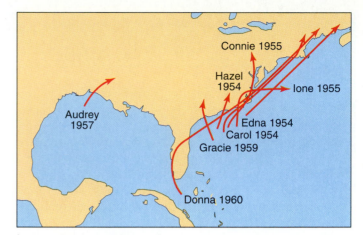

b   Major hurricanes 1951–1960

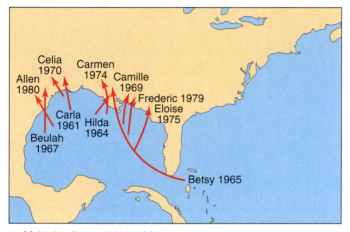

c   Major hurricanes 1961–1980

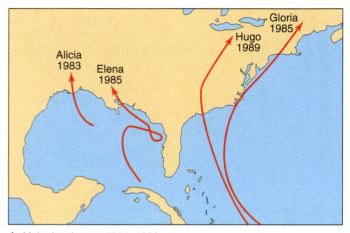

d   Major hurricanes 1981–1990

FIGURE 13-8   Paths of major hurricanes (category 3 or stronger) that have struck the continental United States: (a) 1941 to 1950; (b) 1951 to 1960; (c) 1961 to 1980; and (d) 1981 to 1990. Note the general lack of East Coast and Florida hurricanes from 1961 to 1985. (Saffir-Simpson Hurricane Scale categories are explained in Table 13-2.) (Photographs courtesy of the National Hurricane Center of the National Weather Service of NOAA.)

for, the real danger of a severe hurricane. Areas most vulnerable to hurricane damage are low-lying, heavily populated coastal regions, especially barrier islands along the Atlantic and Gulf Coasts. Many experts feel that, in some areas, even a moderate storm could produce a major economic and human catastrophe. One recent hypothesis, discussed in Box 13-1, predicts an increased incidence of hurricanes along the U.S. East Coast due to the effects of global climate change.

Hurricane Frederick in 1979 was, at the time, the most costly hurricane in U.S. history, resulting in $2.3 billion in damages. Fortunately, due to good early warning, only five deaths occurred. A decade later (September 10–22, 1989), Hurricane Hugo became the strongest storm to hit the mainland U.S. in the last 20 years, a record it held for only 3 years. Prior to reaching the mainland, Hugo's winds were recorded over the Leeward Islands at 260 km/hour (160 miles/hour). This qualified Hugo as a Category 5 storm, the greatest

ranking on the Saffir-Simpson Hurricane Scale (Table 13-2). Hugo crossed the U.S. Virgin Islands and Puerto Rico, and came ashore at Charleston, South Carolina (Figure 13-9).

At landfall, Hugo's winds were estimated as high as 220 km/hour (135 miles/hour), causing a storm surge up to 6 m (20 ft.) above normal sea level in an area where the average elevation is only a couple of meters (a few feet) above sea level. Fortunately, loss of life was limited to 49 directly related storm fatalities.

Property damage, however, was great (Figure 13-10). Estimates are that a total of $9 billion in damage occurred on the mainland United States, Puerto Rico, and the Virgin Islands. In 1991, a fast-moving hurricane called Bob caused considerable damage in Rhode Island and Cape Cod, Massachusetts. In September 1992, Hurricane Iniki pounded the Hawaiian island of Kauai, killing 4 people and badly damaging about 50 percent of the houses on the island.

Box 13-1

Hurricanes and Global Change

There is a definite periodicity to hurricanes (see Figure 13-8). For example, numerous dangerous storms reached the eastern United States during the 1950s and 1960s, but not during the 1970s and 1980s. Professor William M. Gray of Colorado State University has developed techniques for predicting hurricane severity. His data indicate that the incidence of severe storms may be considerably greater during this decade and the next.

Gray has noted that when rainfall is relatively abundant over the western desert regions of Africa, strong hurricanes develop in the Atlantic Ocean. When drought conditions prevail over Africa, fewer strong hurricanes develop. In studying rainfall records going back over a century, Gray observed a consistent pattern of wet and dry periods, each lasting a decade or longer.

The 23 years spanning 1947 to 1969 were years of high rainfall in the western desert of Africa, and thirteen hurricanes with winds exceeding 160 km/hour (100 miles/hour) hit the East Coast of the United States. On the other hand, drought prevailed during the 18 years from 1970 to 1987, and only one such storm occurred. This relation is not just a statistical one. It is based on physical processes that reflect changes in ocean circulation between the wet and dry periods. This in turn is tied in a complex manner to overall global climate.

The drought may be ending and another wet interval may be starting in the African desert. If so, the incidence of strong hurricanes reaching the U.S. East Coast may soon increase. Indeed, four very strong storms have already occurred: Gilbert, a Class 5 hurricane, struck Jamaica and Mexico in 1988; Hugo (see Figure 13-9) hit the United States in 1989; Bob reached New England in 1991; and Andrew ravaged Florida and Louisiana in 1992 (see Figure 13-11). Clearly there are limitations to such predictions, which Professor Gray updates once or twice a year. Nevertheless, they are valuable for those who plan coastal zone development.

What affect, if any, will the possible warming of our atmosphere from the greenhouse effect have on hurricane development? This is not certain, but the greater availability of heat energy could intensify storms. The coming years will certainly interest storm watchers.

TABLE 13-2

Saffir-Simpson Hurricane Scale

Category	Definition and Effects
1	*Winds 74–95 miles/hour or storm surge 4–5 feet above normal.* No significant damage to building structures. Damage primarily to unanchored mobile homes, shrubbery, and trees. Some coastal road flooding and minor pier damage.
2	*Winds 96–110 miles/hour or storm surge 6–8 feet above normal.* Some roofing material, door, and window damage to buildings. Considerable damage to vegetation, mobile homes, and piers. Coastal and low-lying escape routes flood 2–4 hours before arrival of hurricane center. Small craft in unprotected anchorages break moorings.
3	*Winds 111–130 miles/hour or storm surge 9–12 feet above normal.* Some structural damage to small residences and utility buildings. Mobile homes are destroyed. Flooding near the coast destroys smaller structures, with larger structures damaged by floating debris. Terrain lower than 5 feet above sea level may be flooded inland 8 miles or more.
4	*Winds 131–155 miles/hour or storm surge 13–18 feet above normal.* More extensive damage with some complete roof structure failure on small residences. Major erosion of beach areas. Major damage to lower floors of structures near the shore. Terrain continuously lower than 10 feet above sea level may be flooded, requiring massive evacuation of residual areas inland as far as 6 miles.
5	*Winds greater than 155 miles/hour or storm surge greater than 18 feet above normal.* Complete roof failure on many residences and industrial buildings. Some complete building failures, with small utility buildings blown over or away. Major damage to lower floors of all structures located less than 15 feet above sea level and within 1,500 feet of the shoreline. May require massive evacuation of residential areas on low ground within 5–10 miles of the shoreline.

Source: National Oceanic and Atmospheric Administration

FIGURE 13-9 Visible-spectra satellite imagery of Hurricane Hugo, on September 21, 1989. At noon that day, Hugo was upgraded to a category 3 hurricane. At 3 P.M., it was again upgraded to a strong category 3, with sustained winds of about 200 km/hour (125 miles/hour). By 6 P.M., when this image was recorded, Hugo had attained category 4 status, with sustained winds of 217 km/hour (135 miles/hour). It reached category 5 prior to coming ashore. (Photograph courtesy of the National Hurricane Center of the National Weather Service of NOAA.)

However, the most damaging U.S. storm was Hurricane Andrew, which devastated southern Florida in August, 1992. Then, regathering strength from the warm, moist air as it crossed the Gulf of Mexico, the storm struck Louisiana (Figure 13-11). Winds were reported to 320 km/hour (200 miles/hour) and a seawater elevation of 7 m (23 ft.) was measured in the Bahamas. Hurricane Andrew killed over 50 people and caused more than $20 billion in damage. Over 63,000 homes were destroyed in Florida, leaving 180,000 people homeless.

Though we tend to be more aware of the few hurricanes that occur in the United States, about 80 of these tropical storms occur around the world each year, striking not only the U.S. but also Southeast Asia, Australia, India, and East Africa. They are variously known as typhoons and cyclones. In other parts of the world, deaths from these storms are sometimes very high. A 1991 cyclone in Bangladesh caused 70,000 deaths, and one in 1970 resulted in 300,000 deaths.

With better understanding and ability to predict such severe storms, loss of life and property damage can be reduced. The extent to which prediction has improved can be seen in the information supplied by fax and the media during the passage of Hurricane Emily off the North Carolina coast in 1993 (Figure 13-12). The progress of this storm was updated literally by the minute and could be monitored by anyone who turned on a radio or TV. After several days offshore, the hurricane's center missed Cape Hatteras by about 30 km (18 miles)—a fact that was also predicted.

a

b

FIGURE 13-10 Some effects of Hurricane Hugo, 1989.
(a) Close-up of "Dos Marinas," a 29-story high-rise on Puerto Rico's Fajardo waterfront showing the gutted interior apartments and blown-in windows.
(b) Fishing fleet at McClellanville, South Carolina, deposited on dry land. The person in the picture, captain of the *Patricia Anne*, stayed on his boat during the hurricane and sailed across a two-story building at the storm's height. At this location, winds were probably near 210 km/hour (130 miles/hour) and the water rose about 5.2 m (17 ft.) above sea level. (Photograph (a) courtesy Dr. Joseph Golden, NOAA Chief Scientist's Office; (b) courtesy of the National Hurricane Center of the National Weather Service of NOAA.)

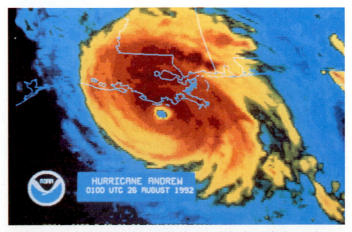

FIGURE 13-11 A false color satellite image of Hurricane Andrew, taken on August 26, 1992. The redder colors indicate more intense winds. Note the blue, lower-intensity winds associated with the eye of this storm, just off the Louisiana coast. Andrew previously passed over the Florida peninsula, where it did most of the damage. (Photograph courtesy National Oceanic and Atmospheric Administration.)

FIGURE 13-12 Actual and predicted positions and predicted track of 1993's Hurricane Emily. This is a FAX notification from the National Oceanic and Atmospheric Administration, sent while the storm was in progress. Only a few of the actual positions are shown, but the prediction of the track was very good.

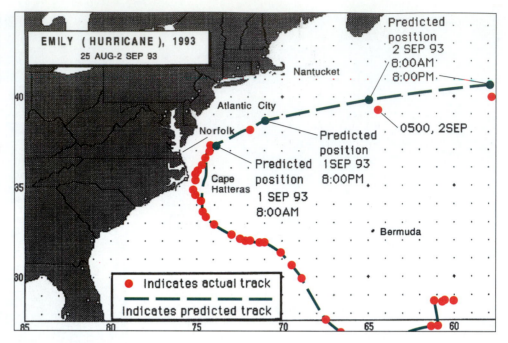

Interaction among the Air, Sea, and Climate

In what ways does the ocean affect climate? We know of several and more will undoubtedly be discovered. The following sections examine three mechanisms that significantly affect climate: carbon dioxide levels in the ocean and atmosphere, ocean surface temperature, and phytoplankton growth.

Carbon Dioxide and Climate

The Transient Tracers in the Ocean (TTO) program (see Chapter 7, page 175) found that parts of the North Atlantic were becoming less salty and colder. Noted over approximately a decade, was a salinity reduction of about 0.2‰ and a temperature reduction of 0.15°C (0.27°F). These may sound like tiny amounts, but the reductions were much greater and extended deeper into the ocean than anticipated. If such changes continue, they could eventually affect the ocean profoundly, perhaps even decreasing or modifying deep-ocean circulation. This, in turn, could significantly affect world climate.

Scientists studying this phenomenon wondered whether these changes in salinity and temperature were due to the carbon dioxide buildup in the atmosphere, which has been measured over several decades. Carbon dioxide in the atmosphere retains heat, causing the **greenhouse effect.** This effect should be most dramatic at the poles, where higher temperatures in the atmosphere would result in increased melting of ice caps and glaciers. This in turn would raise sea level and reduce ocean salinity because of the added glacial freshwater.

The process thus would be intensified, because fresher water is less dense and would reduce thermohaline circulation. This would allow less carbon dioxide to be absorbed into the ocean, keeping more of it in the atmosphere, increasing the greenhouse effect. This will certainly be a key research area in coming years.

Ocean Surface Temperature and Climate

Four major factors appear to control sea-surface temperature:

1. The amount of solar radiation absorbed in the upper layers of the ocean (this varies greatly with latitude, season, and cloud cover).

2. Horizontal and vertical movement of water. For example, upwelling brings large amounts of cold water to the surface.

3. Evaporation and cooling of surface waters (affected by the temperature of the air and wind speed).

4. Thickness of the upper (mixed) layer of the ocean.

Changes in ocean-surface temperature in the North Pacific Ocean seem to correlate with changes in air temperature in North America, so it seems reasonable that ocean surface temperatures influence climate. Studies show that sea-surface temperatures can contribute to the quality of winter weather, wind patterns, rainfall, hurricanes, and length of the season. These results are not surprising, because the ocean surface *is* the interface with the atmosphere.

Both climatic models and empirical observations indeed suggest that sea-surface temperature may be a key factor in monthly and long-term climatic patterns. In one such model, as air moves over the ocean, its temperature and humidity can change; cool, drier air will gain heat and moisture if it moves over warmer water and can affect areas downstream (by causing rain, for example).

The key question for future study is: Can ocean temperatures be used to *predict* weather and climate?

Phytoplankton and Climate

Recent research has discovered ways in which the ocean's phytoplankton can influence climate and weather. First, these mostly microscopic plants may influence the ocean's surface temperature. Phytoplankton contain the pigment chlorophyll, which they use in photosynthesis. Chlorophyll absorbs solar energy, so large amounts of phytoplankton, which can grow during a bloom, could lead to a related warming of the surface waters surrounding the phytoplankton. (Box 13-3 discusses some ways that phytoplankton can influence the greenhouse effect.)

This possibility was reported in a recent article based on satellite studies of surface water temperature and water color (caused mainly by its chlorophyll content) in the Arabian Sea. Following a period of upwelling and subsequent high biological productivity, the resulting biological heating was 4°C (7.2°F) during a one-month period. Because tropical storms tend to develop over warmer waters, the biological heating could affect the weather in the region.

Phytoplankton can also impact climate indirectly because of a sulfur compound they emit into the water. This compound, dimethyl sulfide (DMS), escapes into the atmosphere, where it oxidizes into tiny particles. These particles combine with water vapor to form droplets, then clouds, and ultimately rain. The clouds influence climate by affecting the amount of heat that can reach Earth or escape from it. It is estimated that the sulfur entering the atmosphere in this manner, about 30 million tons per year, is about half that coming from the burning of fossil fuels.

The Greenhouse Effect and Global Warming

You doubtless have heard much about the greenhouse effect. What is it? Why is it both good and bad? Why might the effect be increasing? Why can't scientists agree on its significance? Let us try to answer these questions.

What Is the Greenhouse Effect?

The **greenhouse effect** is a simple, well-understood phenomenon, illustrated in Figure 13-13. Incoming solar radiation of short wavelengths strikes Earth's surface. Some of this radiation heats the surface, and some reenters the atmosphere as longer-wavelength infrared radiation.

A portion of this infrared radiation escapes into space, but some is "trapped" in the atmosphere through absorption by the so-called **greenhouse gases.** These gases include water vapor, carbon dioxide (CO_2), ozone (O_3), methane (CH_4), nitrous oxide (NO_x), and **chlorofluorocarbons (CFCs).** The infrared radiation absorbed by these gases heats the atmosphere; the more energy they absorb, the greater the heating. Since this phenomenon is similar to what occurs in a greenhouse, it has been named the greenhouse effect. The more greenhouse gases there are in the atmosphere, the greater is the greenhouse effect, and the warmer the atmosphere becomes.

The greenhouse effect is critical to life on Earth because it controls the average temperature of the atmosphere and the surface. The problem is that the levels of gases contributing to the greenhouse effect (especially carbon dioxide) are rapidly increasing in the atmosphere, which in turn increases the greenhouse effect (Table 13-3 shows the annual rate of increase). The amount of radiation presently being trapped by greenhouses gases is about 65 percent of the energy Earth receives from the sun.

Computer models predict a gradual warming, which will vary over the planet with location and with the seasons. Because of the continuing interaction between the atmosphere and the ocean, it is possible that a changing climate could affect ocean circulation and atmospheric circulation. Whether these postulated changes will be harmful or beneficial is unclear.

Many scientists—though not all—suspect that the long-term effect of the increasing input of greenhouse gases into the atmosphere will be an increase in the temperature of Earth's surface. This could change the complex interactions among the ocean, land, ice, and atmosphere, producing climatic changes and sea-level rise. Some scientists have also suggested that this global warming and rise in sea-surface temperature could increase the number of severe hurricanes (see Box 13-1) and other ocean-spawned storms.

One obvious effect of a worldwide rise in temperature would be the melting of parts of major ice sheets, resulting in the rise of sea level and flooding of low-lying areas. Such an increase also could affect many areas now used for agriculture. It could cause coastal flooding and saltwater intrusion into coastal groundwater aquifers. Whether sea level has already risen recently due to global warming is a controversial question. (See Box 14-2, page 355; see also Chapter 2, "Ex-

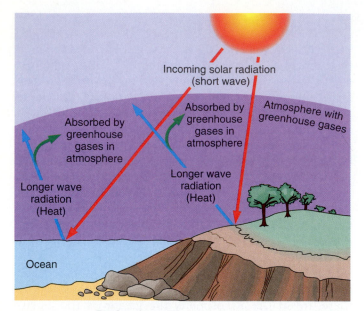

FIGURE 13-13   The Greenhouse Effect. Incoming short-wave solar radiation from the sun penetrates Earth's atmosphere and warms the surface. Greenhouse gases in the atmosphere essentially are transparent to incoming solar radiation. Some of this radiation is converted to longer-wavelength radiation (infrared or heat) when it strikes land or water. Some of this heat radiated from the surface is absorbed by greenhouse gases, thus raising atmospheric temperature.

tinctions and the Greenhouse Effect," page 31, for more background.)

Greenhouse Gases

The main greenhouse gases are water vapor, carbon dioxide, methane, nitrous oxides, ozone, and CFCs. Water vapor, the most abundant greenhouse gas, is a significant contributor to the greenhouse effect. On the other hand, water vapor condenses to form clouds, which reflect incoming solar radiation, thus *reducing* atmospheric temperature. Hence, concern about greenhouse gases has focused on those increased by human activity.

Monthly measurements of carbon dioxide at the Mauna Loa Observatory in Hawaii are accepted as a worldwide standard (Figure 13-14). Since the start of the observations in 1958, a 10 percent increase has been observed. Longer-term comparisons are even more disturbing. Analysis of the old air trapped in the air bubbles inside cores of Antarctic ice indicates a 25 percent increase in CO_2 content since the mid-1700s, the beginning of the Industrial Revolution.

The sources of methane include flatulence from domestic cattle, termites, and sheep (yes, these are significant sources) and the decomposition of organic matter in landfills and rice paddies. The sources of nitrous oxides are not completely known but may be related to use of nitrogen-enriched fertilizers, combustion of fossil fuel, burning of forests, and other activities. The increasing pollution of the atmosphere with CFCs both increases the greenhouse aspect and causes ozone depletion.

Table 13-3 compares the contribution of various greenhouse gases to the greenhouse effect, but caution must be used in interpreting these data. This information is based on currently observed concentrations in the atmosphere, which can vary. Not all the chemical and physical processes are known or fully understood. Although carbon dioxide is the dominant greenhouse gas and is increasing significantly, it is fortunately not the most efficient infrared-absorber among these gases. In other words, the effect of adding more CO_2 to the atmosphere is less than the effect of adding more of some of the other gases.

TABLE 13-3

Estimated Contribution of Various Gases to the Greenhouse Effect
(based on the currently observed increase in their concentrations in the atmosphere)

Gas	Concentration (parts per billion by volume)	Rate of Increase (percent per year)	Relative Contribution (%)
CO_2 (Carbon dioxide)	353,000	0.5	60
CH_4 (methane)	17,000	1[1]	15
N_2O (nitrous oxide)	310	0.2	5
O_3 (ozone)	10–50	0.5	8
CFC-11 (chlorofluorocarbon)	0.28	4	4
CFC-12 (chlorofluorocarbon)	0.48	4	8

[1]Recent data suggests that this rate is decreasing

Source: from Rodhe, 1990.

FIGURE 13-14 The increase in atmospheric carbon dioxide. Note that it has increased about 45 parts per million since 1958. (Data from observations by Dr. C. D. Keeling of the Scripps Institution of Oceanography and associates at the Mauna Loa Observatory in Hawaii and researchers elsewhere.) The small yearly oscillations are due to seasonal changes in the rate of photosynthesis by plants. (Adapted from Mix, Farber and King, *Biology: The Network of Life*, (New York: HarperCollins, 1992).)

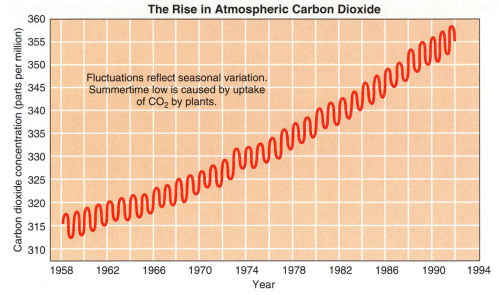

Carbon Dioxide

The release of CO_2 into the atmosphere from the burning of fossil fuels, called **anthropogenic carbon dioxide,** has increased about 0.5 percent per year since 1960. This increase is mainly due to the increased burning of fossil fuels (coal, oil, and natural gas) and the clearing of forested areas, which removes trees and other plants that otherwise would consume carbon dioxide by photosynthesis.

About 50 percent of the carbon dioxide increase seems to remain in the atmosphere. The fate of the other 50 percent is somewhat controversial. Many feel that some, if not most, may enter the ocean, but this has yet to be confirmed. Clearly, the ocean's ability to hold increasing amounts of carbon dioxide is critical in evaluations of the greenhouse effect.

There is about 50 times as much carbon dioxide in the ocean as in the atmosphere, so any change in the oceanic content of CO_2 can have a dramatic effect on the atmospheric content. The anthropogenic carbon dioxide that gets into the ocean by gas exchange at the sea surface slowly works its way into the deep sea by mixing.

The rate and exact mechanism of these important processes are not well known. Note that if the circulation rate were to change (as the previously mentioned TTO studies indicate may be happening), the amount of CO_2 being removed would also change.

The increase in carbon dioxide concentration in the atmosphere (see Figure 13-14) is not surprising when you realize that about 18 billion metric tons of carbon dioxide are added to the atmosphere each year from the burning of fossil fuel. There has been a slight reduction in the rate of CO_2 release following the worldwide decrease in fossil fuel consumption that started in the 1970s. The 18 billion tons in itself is not large when compared with the natural movement of carbon dioxide within forests and oceans. However, these movements were in a balanced steady state prior to industrialization and increased input.

Other than stopping the burning of fossil fuels, little can be done to reduce the carbon dioxide increase. The problem is further complicated by the need to distinguish which global changes are due to increasing greenhouse gases and which are due to other, unknown factors.

Is the Greenhouse Effect Increasing?

Remember that the greenhouse effect is always present. But scientific uncertainty continues over whether it is increasing, and if so, whether we can detect the increase. If we could get world climates to hold still for a few years, measurement of global warming would be easy. But the natural variability of climate makes detection difficult. Global warming, if in progress, seems to be a smaller effect than the natural climate variability. Therefore, unequivocal identification of a true warming trend is extremely difficult.

To illustrate the problem, many examples exist of large-scale temperature and climate changes. These include glacial and interglacial periods, with the last glacial period peaking about 18,000 years ago. More recent were the droughts of the 1930s and 1988, very cold winters in 1976–77 and 1981–82, and very warm years during the 1980s and into the 1990s. Among the warmest years of the past century were 1987 and 1988, and the warmest year to date is 1990. December 1989, on the other hand, was one of the coldest months on record. The 1980s was the warmest decade this century, having 6 of the 10 warmest years on record.

More recently, temperatures have moderated following this period of record highs. For example, 1992 was the coolest year since 1986 and many snowfall records were broken in the northeast in the winter of 1993–94. The reason is believed to be due to the aerosol clouds released into the atmosphere by the 1991 eruption of Mount Pinatubo (see Box 13-2).

The only way to be sure that a global warming is in progress would be to look at long-term global temperature records spanning hundreds of years, but these do not yet exist, since accurate temperature measurements have been recorded only for about 100 years and even these are misleading. Towns and cities often grew up around the site where temperatures were initially recorded. Cities are warmer than suburbs or farmland because concrete, asphalt, and buildings are better heat absorbers than trees and lawns. Thus, as a city grows, the temperature around it increases, creating a warming trend in the data. Of course, data can be adjusted for this, but the result then becomes a questionable estimate or, some might say, a guess.

Determining long-term changes in ocean temperature has also proved elusive. The problem is a technological one. In the past, ocean surface temperatures were measured by sticking a thermometer in a bucket of seawater. This technique used thermometers that were too imprecise to detect subtle temperature changes, and used a sample too small to represent a large area. Modern methods include precise electronic measurements and monitoring the temperature seawater drawn in to cool the ship's engines. Clearly, data acquired by these different methods are not comparable, so subtle changes observed in ocean temperature over time have yet to be determined.

An optimistic view of some scientists is that the increase in CO_2 and the greenhouse effect may have a completely opposite effect to what many are expecting and may even be beneficial. They suggest that the warmer air will carry more moisture and will increase snowfall in places like Antarctica. The ice sheet then would increase in volume, reducing sea level.

Still other scientists feel that the increased carbon dioxide, mixed with other atmospheric gases, will have little or no impact on the existing greenhouse effect. They note that agricultural experiments with increased CO_2 levels show increased crop production and decreased water loss by plants, and could lead to a doubling in yield per quantity of water used. If such information is correct, it suggests that the predicted increase in atmospheric CO_2 might have some positive impact.

On the other hand, if the prediction of sea-level rise and global warming comes true, the future will not be so positive. These two contrasting views highlight the difficulty of making worldwide decisions based on uncertain data. Although the evidence seems to favor the more pessimistic viewpoints, the hard fact is that *we simply do not know.* Clearly, more data and observations are needed to evaluate the carbon dioxide problem.

In a way, this is a giant experiment in which a single species (us) is determining whether our activities can change world temperature and climate. Our principal method of predicting the outcome of this experiment is to model the parameters and their possible changes by computer. The process is very complex and time-consuming, even with today's fast computer systems.

Modeling is an accepted scientific technique for determining cause and effect and making predictions. If we had enough good data, predicting the future of Earth's atmosphere and oceans would simply involve turning on a computer and manipulating the data. The difficulty here is our poor knowledge of many basic components of the problem, such as how much CO_2 the ocean can absorb, how components interact, and whether there is a "self-healing" aspect to the problem (see Box 13-3).

Several models are now in use but their results do not agree. This is not surprising, considering the many unknowns. Assuming a continuing increase in CO_2 and other gases, some models suggest that in a few decades the greenhouse effect could lead to temperature changes of 1.5 to 5°C (2.7 to 9°F) and a sea-level rise of several feet. Other models suggest a less dramatic response. A change in a single key parameter, such as cloud cover, could cause major changes in the results.

It may not be immediately obvious what a temperature difference of a few degrees implies. For perspective, the mean annual temperature difference between Boston and Washington, D.C., is 3.3°C (5.9°F). Think about the difference in the climate of these two cities and the severity of their winters. The general temperature difference between the peak of the last glacial period about 18,000 years ago and today is thought to be about 5°C (9°F). Try imagining this temperature difference where you live!

The Greenhouse Effect and Our Future

The greenhouse effect has profoundly affected past life and environmental conditions on Earth, so the concern about it is valid. Unfortunately, the implications of an increasing effect may take decades to resolve and the problem may grow worse before a cure can even start taking effect, if indeed a cure proves to be necessary or even possible.

Any solution will require international agreement, considerable funding for research, major economic

Box 13-2

The Mount Pinatubo Eruption

As our understanding of climate increases, it has become evident that volcanic eruptions can cause major changes in Earth's climate. A good example was the April 1982 eruption of El Chichón in Mexico. The exceptionally sulfurous cloud from this eruption blocked incoming solar radiation, which may have lowered worldwide air temperature by a few tenths of a degree during 1982 and 1983. If that seems insignificant, remember that it takes only a small change of temperature to disturb oceanic and atmospheric circulation.

The June 1991 eruption of Mount Pinatubo in the Philippines has had an even greater effect (Figure 1). By some estimates, it may have been the largest volcanic eruption of the twentieth century. Observations from the *Nimbus* 7 satellite indicate that about 20 million tons of sulfur dioxide gas was emitted high into the atmosphere during Pinatubo's eruption, over twice the quantity of that from the El Chichón eruption. Within ten days, a cloud extended from Indonesia to central portions of Africa, a distance of over 10,000 km (over 6,200 miles).

In the atmosphere, sulfur dioxide gas combines with moisture to become tiny droplets of sulfuric acid, which can block incoming solar radiation for up to 3 years. It was anticipated that these droplets would decrease worldwide temperature in a manner similar to El Chichón, but this turned out to be a conservative estimate. By 1993, data showed that temperatures worldwide may have been lowered by 0.4°C (0.7°F). The effect may actually have been greater, because an El Niño was in progress that would normally raise temperatures by about 0.2°C (0.36°F). By early 1994 the cooling due to the Pinatubo eruption started to decrease.

Another effect of the droplets in the atmosphere is that they produce a surface for chemical reactions that caused reductions in the ozone layer. According to NOAA scientists David Hoffman and Susan Solomon, the El Chichón eruption might have helped in the destruction of 15 percent of the stratospheric ozone in the middle latitudes of the Northern Hemisphere. A similar reduction may have occurred following the Pinatubo eruption. Such losses, although temporary, are startling compared to the 5 percent loss per decade that has been measured recently.

Volcanic eruptions like Mount Pinatubo also confuse our efforts to determine whether greenhouse warming is in progress. The lowering of temperature by a few tenths of a degree will counterbalance any greenhouse warming which, according to many experts, has been only about 0.5°C (0.9°F) during *all* of this century. The effects from Mount Pinatubo could last 3 years until all the droplets settle back to Earth. The effect of this eruption will certainly confuse the evaluation of any near-term global temperature increase.

REFERENCE
Kerr, Richard A. 1991. "Huge Eruption May Cool the Globe." *Science* 252 (June), p. 1780.

FIGURE 1 A June 1991 view of the north side of the Pinatubo crater, with a small explosion in progress. The crater is approximately 2 km (1.2 miles) in diameter. (Photograph by R. Batalon, U.S. Air Force, distributed by National Geophysical Data Center, NOAA.)

Box 13-3
An Experiment with Phytoplankton

The biology of the ocean, in particular the number of floating plants or phytoplankton, can significantly affect one greenhouse gas—carbon dioxide. Because phytoplankton use CO_2 in the photosynthesis process, large increases in their numbers could lead to more CO_2 being withdrawn from the atmosphere into the ocean. A way to achieve this could be by artificially "fertilizing" parts of the ocean to promote biological growth.

A variation of this idea was posed by John Martin of Moss Landing Marine Laboratory in California. He suggested fertilizing the ocean by adding small amounts of iron—an element whose absence may prevent plant growth—to promote the growth of phytoplankton. Increased phytoplankton growth would tie up more carbon dioxide. The subsequent sinking of dead organisms would remove large amounts of carbon dioxide from the atmosphere, reducing the impact of global warming.

Martin's idea came from observations that atmospheric carbon dioxide decreased during the most recent ice age, possibly due to increased phytoplankton growth in the Antarctic region. The increased growth may have been due to changes in wind patterns that carried in the needed amounts of iron to the ocean. Thus, his suggestion to fertilize areas of the ocean to achieve similar results: reduction of carbon dioxide. This idea was challenged by some scientists who questioned whether the increased phytoplankton growth would actually move CO_2 to the sea bottom.

A 1993 experiment in the equatorial Pacific may be supporting Martin's fertilizer idea. When oceanographers added a small amount of iron to a 20-square-mile area, the plant biomass doubled in 3 days. Additional experiments will be needed to determine if fertilizing the ocean is indeed a useful mechanism for removing CO_2 from the ocean. Sadly, Dr. Martin died just before the experiment started, but his fascinating idea will be tested further by others.

adjustments, and strong political decisions that will confront powerful opposing interests. The prospect of all this, coupled with the uncertainty of whether we are experiencing an increased greenhouse effect, promises neither a prompt nor an easy solution.

It is important to remember that humans did not start the greenhouse effect. It started in the early history of our planet, and without it (in other words, without clouds or greenhouse gases in the atmosphere), life as we know it would be impossible. Earth's surface temperature would average about −18°C (0°F) rather than its present +15°C (+59°F). What humans have done is to add large amounts of greenhouse gases, including some new ones, to the atmosphere, *possibly* increasing the greenhouse effect. Interestingly, studies of air trapped in old glacial ice have revealed similar fluctuations in CO_2 content *prior to the establishment of human civilization.* Clearly, other factors are involved beside human activities.

El Niño

One of the more dramatic effects of the ocean on climate, and therefore on the environment, is the phenomenon called **El Niño** (ell NEEN-yo). In simple terms, an El Niño is an unusually warm current that periodically flows southward along the Peruvian coast, causing severe climatic effects. This is far more important than it sounds, because it has worldwide implications. Under normal conditions, upwelling along the Peruvian coast brings to the surface deep, cold, nutrient-enriched waters that nurture phytoplankton upon which a small fish, the anchovy, feeds. The harvest of this fish can be immense; at times it has been about 20 percent of the total fish catch of the ocean.

Sometimes, however, the pattern changes, and warm water flows farther south than usual along the Peruvian coast, eventually reaching the Chilean coast. This warm water acts almost like a blanket on the ocean and prevents the cold, nutrient-enriched water from upwelling and reaching the surface. This is an El Niño. The result is a significant reduction in planktonic growth, which directly limits the growth of anchovies as well as the sea birds that feed on them. The effects of El Niño can be dramatic, reducing the fish harvest from this area by 50 percent or more.

Spanish for "The Child," El Niño is an appropriate name for a major event that generally occurs around Christmas. El Niños have been documented back to 1726 and doubtless occurred before then. They generally develop once or twice a decade and last about 12 to 18 months. In recent years, however, both the incidence and duration of El Niños have increased.

What Causes El Niño?

El Niño has been studied for decades to understand how the wind and the sea produce this phenomenon and to predict its occurrence. A breakthrough occurred in the early 1970s when Klaus Wyrtki, an oceanographer from the University of Hawaii, developed a model that appeared to explain El Niño. It had been noted that easterly trade winds built up prior to the development of an El Niño. (Remember that winds are indicated by the direction they come *from*, whereas ocean currents are indicated by the direction they go *toward*.)

Wyrtki suggested that these winds could increase the amount of warm water carried westward, "piling it up" toward Australia, but when the winds weakened, the water would slowly "slosh" back (to the east) toward the western coast of South America. This would result in a buildup of warm water all along the west coast of the Americas—an El Niño. The unusually warm water produced climate variations, not only locally but worldwide, lasting up to a year. This change in wind pattern and subsequent water movement is called the **Southern Oscillation.** The Southern Oscillation also involves a

significant pressure change over the Pacific Ocean (Figure 13-15).

The first successful prediction of an El Niño from meteorological data was made in 1975. Oceanographers traveled to the area to test the prediction and indeed observed a small El Niño. This encouraged future prediction attempts, through monitoring winds and water buildup. For a while it was believed that an El Niño was the only kind of major natural disaster that could be predicted months to a year in the future.

The ability to predict an El Niño requires much data from broad areas of the Pacific, an operation especially well suited to satellite observations (see Box 13-4). By the early 1990s, several El Niños had been successfully predicted. However, in 1991, one appeared that had been largely unanticipated. Further, it was expected to end by mid-1992, but—in another surprise to the predictors—it didn't. In fact, it regained strength and continued until late 1993. Another El Niño is anticipated for 1995–1996. These recent surprises have caused some reevaluation of models used to predict El Niños.

Recent research suggests that relatively small changes in sea-surface temperature can lead to the changes in surface winds that start the El Niño–Southern Oscillation (ENSO) process. But it still is unclear which comes first: in an El Niño, does the atmosphere drive the ocean, or

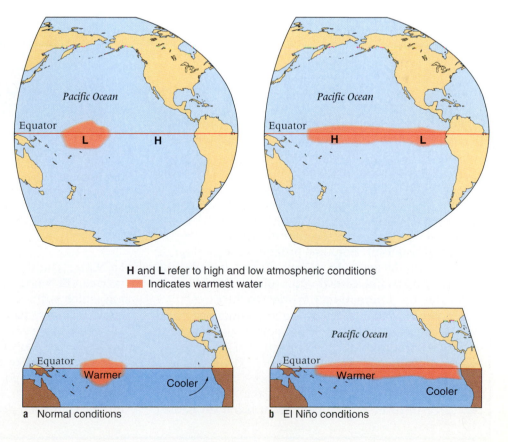

FIGURE 13-15 The thermal structure and atmospheric patterns over the equatorial Pacific Ocean: (a) normal conditions, in which the surface water of the eastern Pacific is relatively cool; (b) conditions during an El Niño and a Southern Oscillation. Note the change in the atmospheric patterns and the movement of the warm water across the Pacific.

Box 13-4
Predicting El Niño with Satellites

The ability to predict an El Niño condition is a major challenge to oceanographers, meteorologists, and planners of national economies. Being able to predict accurately a wet or dry period can be useful for deciding what crops to plant and when, or how much oil, gas, coal, or fuelwood should be stockpiled for heating.

The strongest El Niño of this century occurred during the winter of 1982–83, causing climate-related problems worldwide. Rains producing 300 times more water than normal prevailed along the west coast of South America, while Australia had its worst drought in over 200 years (see Figure 13-16). In Peru and Ecuador, flood losses attributed to the El Niño exceeded $1 billion. In the United States, strong coastal storms damaged the Pacific Coast, and heavy snowfall in the Rocky Mountains led to spring flooding.

Satellite observations of sea surface temperature (SST) can be used to better understand and sometimes to predict El Niños. Figure 1 shows 3 sets of worldwide SST observations produced by Richard Legeckis of NOAA. The upper chart shows a typical SST configuration, in this case from winter 1984. The warmest water (indicated by a "1") occurs in the western equatorial Pacific, and a tongue of relatively cold water ("2") extends westward from the coast of South America. Coastal upwelling is obvious (colder water rising from the depths).

The middle chart shows the SST pattern associated with the 1982–83 El Niño (data are for January 20, 1983). It is obvious that the coastal upwelling and the tongue of cold water off South America are gone and the waters of the eastern equatorial Pacific have a higher SST (see also Figure 13-15).

The lower chart, showing the difference between the SST in 1983 and 1984, clearly illustrates the cessation of the upwelling and the fact that temperatures were warmer in the eastern equatorial region ("3") and cooler ("4") in the western equatorial region.

Several major international programs are in progress to better understand El Niños. These include the Tropical Ocean and Global Atmosphere (TOGA) program and the international World Ocean Circulation Experiment (WOCE). Such studies, combined with satellite data and better understanding of global change conditions, will lead to better prediction of El Niños and other major climatic perturbations.

As knowledge of the El Niño process improved, an interesting observation was made: the phenomenon can occur in reverse. This new event has been named

FIGURE 1 Sea Surface Temperature (SST) as determined by satellite measurements. Blues indicate cool waters (0° to 12°C, or 32° to 54°F), greens show intermediate temperatures (13° to 24°C, or 56° to 76°F), and yellow-red-magenta indicates warm temperatures (25 to 30°C, or 78° to 87°F). (See the text for a description of these charts.) (Maps produced by Richard Legeckis of the Environmental Satellite Data and Information Service of NOAA; figure courtesy of NASA.)

La Niña ("The Girl Child"). Observations indicate that, between El Niños, periods of up to two years exist during which surface waters in the equatorial Pacific become relatively cool. Unfortunately, like its "brother," La Niña also appears to have serious environmental impact. The first La Niña observed, in the spring of 1988, may have caused or abetted the U.S. droughts, flooding in parts of Asia, and very cold temperatures in Canada and Alaska. Much remains to be learned about this brother-and-sister act.

does the ocean drive the atmosphere? Interestingly, the tropical Pacific is the only water body on Earth wide enough to allow such a coupling to the atmosphere; here the ocean width is approximately equal to the wavelength of typical fluctuations in the atmospheric jet stream. Thus, it may be that variations in water temperature from one side of the ocean to the other can affect the overlying atmosphere.

Effects of El Niño

During the 1982–83 El Niño, thought to be the most extreme of the twentieth century, sea level was lowered in the western Pacific when the water started to "slosh" back toward South America and flooding from higher sea level and storms occurred along the coast of South America.

The effect of this large-scale change in ocean circulation and temperature was far-reaching. It caused major drought in Australia, Sri Lanka, southern India, Indonesia, and parts of Africa. Combined with other environmental and ecological impacts, this El Niño affected several hundred million people around the world (Figure 13-16).

This El Niño was also responsible for the disappearance of over 17 million seabirds on just one tropical atoll. The entire adult bird population of Christmas Island in the mid-Pacific either perished or fled the island. The probable reason is that the El Niño reduced upwelling in the area, sharply curtailing phytoplankton growth and consequently the population of squid and fish on which the birds feed.

An ancillary industry that directly depends on the anchovy volume is the harvesting of seabird waste, called guano, which is used for fertilizer. The more anchovies there are, the more the sea birds eat, and therefore the more guano they produce, so more fertilizer can be harvested and sold.

How can El Niño have such far-reaching climatic effects? The warmer surface waters in the equatorial Pacific appear to change the pattern of the atmospheric jet stream in the Northern Hemisphere. This creates storms in some areas and mild weather or droughts in other regions (see Figure 13-16).

Interestingly, the development of the 1982–83 El Niño may have been aided by the eruption of the El Chichón volcano in Mexico. The eruption created an especially large cloud of particles in the atmosphere, and the particles may have reflected enough heat to affect the temperature in the underlying surface waters of the Pacific, thus triggering the El Niño–Southern Oscillation process. The 1991 eruption of Mount Pinatubo in the Philippines also affected worldwide climate (see Box 13-2).

The Ozone Problem

You probably have heard much about the "ozone problem." It is like a bad dream that will not go away. Before we explore this problem, let us examine Earth's ozone layer.

FIGURE 13-16 Some effects and climatic changes due to the 1982–83 El Niño, the most damaging in this century.

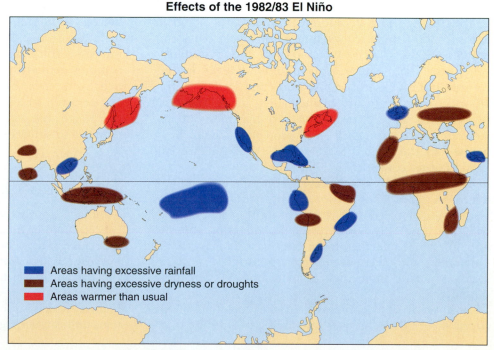

Ozone (O_3) is a variety of oxygen molecule in which three oxygen atoms are linked together, instead of the usual two (O_2). Ozone occurs throughout the atmosphere, but it is concentrated in a layer 20 to 40 km (12.5 to 25 miles) above Earth's surface, in the stratosphere. This is the **ozone layer.** It is critical to the survival of all life on Earth because ozone absorbs dangerous ultraviolet radiation from the sun.

A decreased amount of ozone in the stratosphere means that more ultraviolet radiation can reach Earth's surface, with potentially serious biological consequences. Human health hazards associated with increased ultraviolet radiation include skin cancer, eye cataracts, and damage to the human immune system. Ultraviolet radiation may also affect single-celled plants (plankton) living in the ocean, and in doing so may impact the entire food chain.

Ozone forms naturally in the atmosphere when ultraviolet light breaks the oxygen molecule (O_2) into single oxygen atoms (O). Most of these reform into the two-atom molecule of "normal" oxygen, O_2. Some, however, combine to form the three-atom ozone molecule, O_3. In this natural and continuous process of breaking up and reforming, the amount of ozone remains essentially constant. However, external factors can interfere with this natural equilibrium. These external factors include increased presence of certain chemicals in the stratosphere.

The chemicals believed to be causing the ozone depletion are the **chlorofluorocarbons (CFCs),** halons, methylchloroform, and carbon tetrachloride. These chemicals were developed over 50 years ago, when no one realized they might cause environmental problems. CFCs are the coolants in refrigerators and air conditioners (including those in automobiles), the propellants in aerosol sprays, and are used for cleaning electronic components. They can be released into the atmosphere every time a refrigerator or air conditioner leaks or is repaired. Halons, methylchloroform, and carbon tetrachloride are used in fire-extinguishing foam. They can be released during testing of fire-fighting systems. Aerosol sprays using CFCs as propellants have been banned in the United States since 1978, but are still used in other countries.

These chemicals slowly migrate into the atmosphere, a process taking up to 10 years. As they rise to altitudes where ultraviolet light is much more concentrated, they are broken down by the light, releasing chlorine and fluorine molecules from the *chloro-fluoro*-carbons. One chlorine molecule can break up hundreds of thousands of ozone molecules, and fluorine is even more destructive.

How were CFCs implicated in the depletion of the ozone layer? One clue was the finding of chlorine dioxide in the stratosphere. This chemical is a by-product of the chlorine and bromine processes that break down ozone. Even if all release of these chemicals were stopped immediately, enough are already in the atmosphere to continue destruction of ozone molecules well into the twenty-first century.

The Ozone Hole

A decrease in ozone content in the stratosphere was first detected over Antarctica (the famous **ozone hole**). Later, ozone was found to be reduced worldwide. As you have read, the causes for depletion seem to be understood, the implications are very considerable, and the solution requires international cooperation, which has begun on a limited scale. **Ozone depletion** is a major problem that could exceed global warming in significance.

Although some scientists had warned of ozone depletion as early as the 1970s, the "hole" was not discovered until 1985, when a significant decrease in ozone was measured over Antarctica. The situation had deteriorated further by 1987, when as much as 60 percent of the ozone was discovered to be missing. The Antarctic discovery was followed by later findings that ozone was decreasing all around the globe, in particular at higher latitudes (the Arctic), especially during the winter. A decrease of 4.7 percent was noted in March 1988 over a latitude of 42° N (about that of Boston, Rome, and Beijing).

The ozone-depletion process depends on the amount of solar ultraviolet radiation available, so it cycles with the seasons. Ozone breakdown accelerates during the winter, when solar radiation becomes minimal. Ozone depletion slows during springtime when increased solar radiation, specifically ultraviolet light, resumes generating ozone to "fill in" the hole. Thus, over Antarctica, ozone is reduced during local winter and restored, at least partially, during local summer. At the North Pole, the same happens during the local winter and summer.

Ozone near the Ground

As noted, ozone exists throughout the atmosphere and is concentrated in the critical ozone layer, where it shields life on Earth by absorbing incoming ultraviolet radiation. However, when ozone is near the ground, it can be a serious health hazard. The interaction of pollutants with oxygen in the air and sunlight can form an ozone "smog," which is an irritant to all animal respiratory systems and can damage plants and trees. Thus, the benefit or hazard from ozone depends on where it is and on its concentration.

Solving the Ozone Problem: A Start

In 1987 an international meeting on the ozone problem held in Montreal, Canada, produced the Montreal Protocol, which now has been signed by about 100 countries. The Protocol called for a 50 percent reduction in CFC production by 1999, and a complete ban of all ultraviolet-sensitive halogens by 2000. The full implications of such a reduction are unknown, but many fear that it may have little effect on the short-term reduction of ozone.

Nevertheless, the Montreal Protocol is an important international step toward correcting the very dangerous trend of stratospheric ozone depletion. Although the United States had pushed for a greater reduction of CFC production, the Protocol set an important precedent for resolving international environmental problems.

Developed countries such as the United States and Japan produce and use a large portion of the total CFC chemicals, so their signing of the Protocol was very important to our atmosphere's welfare. The 12 nations of the European Community have been especially aggressive in calling for greater reductions, if not total elimination, of CFCs and halons. However, these chemicals also are used in developing countries, and some, such as Brazil, India, and China, hesitated to sign the Protocol because of economic concerns. If they refuse to sign and they expand their CFC industries, the reductions accepted by signers of the Protocol could be negated.

It is generally acknowledged that more reductions must be realized before a significant impact can be made on the ozone problem. Little impact will be felt if only a few countries phase out CFCs. Global action is needed.

More recent analyses and modeling of the ozone problem clearly indicate that it can be solved only by complete elimination of CFCs and fluorine-based products, and if necessary by replacement with safer chemicals. CFCs already in the atmosphere may remain active for more than 100 years. Thus, even with a complete phaseout of CFCs, chlorine levels in the atmosphere will increase for a while before they start decreasing. The Montreal Protocol allows for periodic review and a possible increase in controls in future years.

Past Climates

Earth's climate certainly was different in the geologic past, when the geometry of the continents and ocean

basins was very different. Over 200 million years ago, the single continent called Pangaea extended almost from pole to pole and was surrounded by a major undivided ocean (see Figure 3-4a, page 46). Little is known about oceanic climate and circulation at that time because the evidence is gone—most marine sediments of that age have already been recycled into Earth's mantle by subduction associated with seafloor spreading.

The land climate 200 million years ago, however, was distinguished by ice sheets at the South Pole (located in southern Africa at that time), occasionally extending over what is now India, South Africa, Australia, Antarctica, and South America. The mid-latitude regions at that time had some evaporitic conditions, producing salt deposits and deserts. As Pangaea eventually split apart, different ocean basins and patterns of ocean currents developed, and climates evolved (Figure 13-17).

By 65 million years ago, climate had changed considerably. Conditions then are thought to have been warmer than today, with no polar ice caps. Antarctica remained attached to Australia or South America until 35 million years ago. Consequently, the West Wind Drift or Circum-Antarctic Current could not be established until a clear, around-the-world passageway existed. Once this occurred, it probably led to the present ocean circulation system, in which cold, dense waters from the Antarctic form the bottom waters common to much of the world's oceans.

One of the more dramatic examples of ocean-climate interaction occurred over the past million years when the amount of ice on Earth increased significantly and global temperatures probably dropped as much as 10°C (18°F). During this ice age, many periods of glacial advance and retreat occurred.

When sea level was lower, more land was exposed and land "bridges" existed between some continents. These bridges may have been important in the migration of land species from one continent to another. During the most recent glacial advance, from about 30,000 to 10,000 years ago, a land bridge existed between Asia and North America (between present-day Russia and Alaska) over the Bering Strait (Figure 13-18). During that time, horses migrated from America to Asia and humans and other animals traveled in the opposite direction.

Numerous hypotheses have been proposed to explain the start and stop of the ice advance. Among the most popular are those involving variations in Earth's orbit, changes in the energy received from the sun, and large inputs of volcanic dust or carbon dioxide into the atmosphere. Perhaps the true explanation involves some combination of these.

FIGURE 13-17 Oceanic circulation patterns approximately 45 to 50 million years ago, based on the study of sediment cores. Compare with the present patterns shown in Figure 9-10. (Image courtesy of Dr. Ted Moore.)

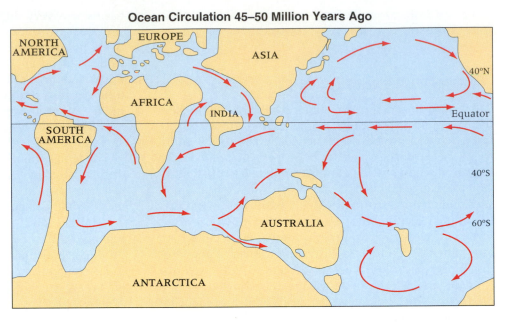

Ocean Circulation 45–50 Million Years Ago

The Milankovitch Hypothesis

One hypothesis to explain climatic changes involving variations in Earth's orbit has been favored by several scientists, and some persuasive arguments to support it have been published. This idea, commonly called the **Milankovitch hypothesis** after its originator, has three basic components:

1. Slow changes in how much Earth's axis tilts

2. Slow changes in the precession of Earth's equinox, which is a change in the season of the year when Earth is closest to the sun

3. Slow changes in the shape (eccentricity) of Earth's orbit

According to the hypothesis, these three components (Figure 13-19) can combine to produce conditions that influence Earth's climate. One likely result is climate change from glacial to interglacial conditions. Following is a closer look at these three factors.

Earth's **axial tilt,** which causes the seasons, varies between about 24.5° and 22.1°; the larger the angle, the more extreme the seasons. The change in angle occurs quite slowly, with the axis shifting from one extreme to the other over 40,000 years. At present,

FIGURE 13-18 Areas covered by ice sheets (white) during the last glacial maximum, which occurred about 18,000 years ago. The lowered sea level, at that time, exposed additional land (green areas) and formed some land bridges, between previously separated continents. Note, for example, the connection between Asia and Alaska and thus the rest of North and South America. (From C. L. Harris, *Concepts in Zoology* (New York: HarperCollins, 1992).)

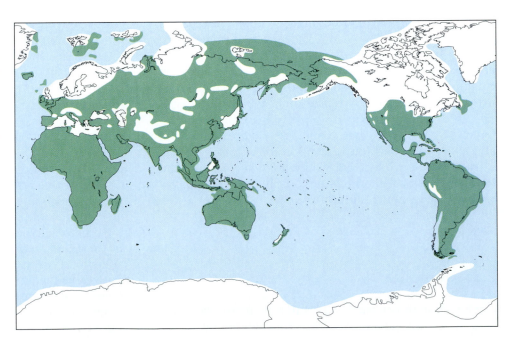

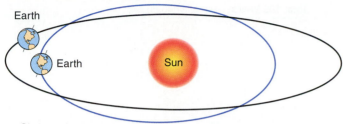

a Changes in Earth's orbit 100,000 year cycle

FIGURE 13-19 Three variations in Earth's motion relative to the sun influence Earth's climate: (a) change in Earth's orbit; (b) the precession or wobble of Earth's axis; (c) change in the tilt of Earth's axis. All distances are considerably exaggerated.

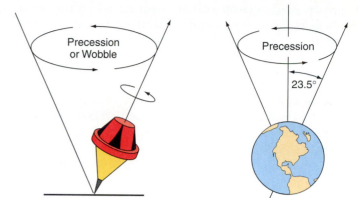

b Precession of wobble of Earth's axis 26,000 year cycle

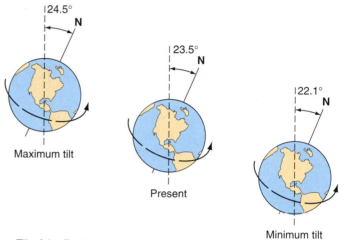

c Tilt of the Earth's axis 40,000 year cycle

Earth's axial tilt of 23.5° places us roughly halfway between the extremes.

Precession is the very slow wobble of Earth's axis as it spins. This is like a spinning top that is running down and beginning to wobble (see Figure 13-19b). Precession has a 26,000-year cycle.

Earth's orbit slowly changes over a 100,000-year cycle, elongating into a longer ellipse, then shortening into a more circular one. The more circular orbit brings Earth closer to the sun, probably making winters milder. Therefore, this change in the eccentricity of orbit should influence climate.

We are uncertain to what degree each of these motions affects climate, but they probably all do to some extent. It is the *combination* of these cycles that is important.

Evidence in the Sea

Some marine shells indicate temperature changes, and these have been studied to establish a general climatic pattern for the past 700,000 years. These data show several distinct climatic cycles that are very close to those predicted by the Milankovitch hypothesis.

It seems logical that times of extreme seasons would favor glacial growth, but the opposite is true. Seasonal extremes bring considerable glacial melting during the summer, but little buildup of the glaciers during winters that are cold and dry. However, during moderate conditions, less melting occurs in the summer, and rainfall/snowfall is greater in the winter, causing glacial growth.

Other factors contributing to climate change are possible. For example, deep-sea sediments from the last 2 million years are relatively rich in volcanic ash when compared with sediments from the previous 20 million years. Most of the ash results from large-scale volcanic activity on land. A recent examination of sediment cores obtained by drilling in the western Pacific showed an interesting and intriguing pattern. About the time the Pleistocene glaciation started, a ten-fold increase in volcanic ash is observed in the cores. Whether volcanic activity affects glacial growth (because ash in the atmosphere reduces the amount of sunlight reaching Earth's surface) or whether it is somehow caused by the glacial periods is still unclear, but some interesting possibilities are appearing.

CLIMATE TRENDS. From the end of the last glacial advance (its maximum occurred about 18,000 years ago) until about 6,000 years ago, Earth's climate has been warming (Figure 13-20, bottom). An example of this is that semitropical plants grew in Minnesota during this period. Since then, Earth has been in a cooler period. An especially cool period, sometimes called the "little ice age," occurred from A.D. 1430 to 1850 (Figure 13-20, center). During these times there was a modest advance of the glaciers and the overall weather was considerably colder than at present. Following this, until about 1940, temperatures warmed somewhat. Then a period of lower temperatures followed. Now we appear to be in a warming phase, perhaps even warmer than anticipated, due to the greenhouse effect.

The coming years will see more integration of oceanographic, geologic, and meteorological data, which should assure a better understanding of climate and could make possible better long-term climate prediction. A thorough understanding of global change may take a little longer.

Ice Cores and the Great Ocean Conveyor

To better understand past climate, scientists need better *resolution.* In other words, they must determine what has happened on the finer scale of centuries or decades, rather than the present coarse scale of thousands of years. Such an improvement in resolution recently became possible through study of ice cores taken from the Greenland ice sheet. The ice sheet has been drilled to 3 km (1.8 miles) depth, reaching deep layers that were deposited as snow over 150,000 years ago.

One of the properties studied in these ice cores is the ratio of oxygen isotopes. This ratio depends on the atmospheric temperature at the time the snow fell. Surprisingly, the oxygen isotope ratios reveal occasional large and rapid changes in atmospheric temperature. Changes as extreme as 7°C (12.6°F) over a 50-year period have been detected. Such rapid changes cannot be explained by the Milankovitch hypothesis.

A very interesting model proposed by Wallace Broecker of the Lamont-Doherty Earth Observatory may offer an explanation. The model conceptualizes an entire ocean circulation pattern in what is now called the **Great Ocean Conveyor.** It is a vast looping current that connects the Atlantic, Indian, and Pacific Oceans. The concept is simple, and follows on the earlier work of Henry Stommel (see Figure 9-19, page 217).

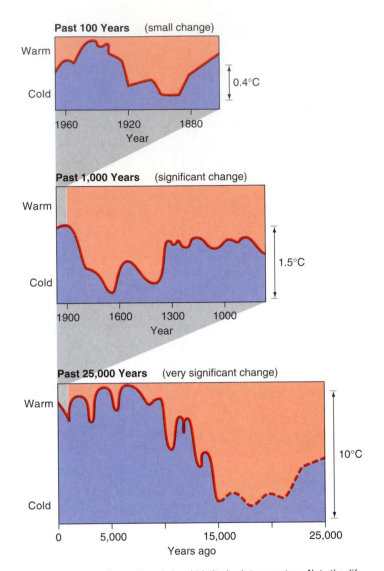

FIGURE 13-20 General trends in mid-latitude air temperature. Note the different scales to the right of each diagram, showing temperature range for each time period. (Adapted from the *Physical Basis of Climate and Climate Modelling*, GARP Publication Series No. 16 (Geneva: World Meteorological Organization).)

The conveyor starts with the warm, northward-flowing waters of the Gulf Stream. As they cool, releasing heat into the atmosphere, they grow denser and sink (Figure 13-21). Heat released by the Gulf Stream makes parts of Europe about 6°C (10.8°F) warmer than they otherwise would be. Upon sinking, these waters flow south, occupying much of deep ocean, eventually resurfacing and becoming rewarmed in the North Pacific Ocean and the Indian Ocean. Surface currents carry the water back through the Pacific, Indian, South Atlantic, and then into the North Atlantic, completing the circuit. The entire trip takes about 1,000 years and the amount of water involved is immense—more than 100 times the flow of the Amazon River.

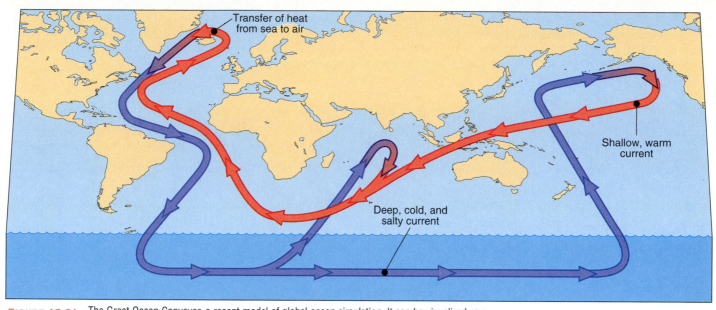

FIGURE 13-21   The Great Ocean Conveyer, a recent model of global ocean circulation. It can be visualized as a huge conveyor belt slowly moving through the Atlantic, Pacific, and Indian Oceans. Warm surface water is cooled in the North Atlantic, sinks, and slowly works it way to the Indian and Pacific Oceans, where it is rewarmed. The water then travels back on the surface to the Atlantic, completing the cycle. Interruptions in the cycle may have important implications for climate.

This conveyor model, if correct, has several implications for present-day climate and for future climate change. If for some reason the conveyor were to slow, or even to stop, it would reduce the warming of Europe and could produce the sudden, surprising temperature changes detected in the Greenland ice cores. Such a change in the Great Ocean Conveyer might also change the ocean's ability to transport and store carbon dioxide. In the coming years, advanced computers will permit detailed testing of this model and show us more about how the ocean and climate are linked.

SUMMARY

Recent studies indicate a close relationship between the ocean and climate. Many atmospheric-oceanic interactions, such as circulation, are known and clearly influence climate and weather. Other interactions, such as those involving CO_2 and the greenhouse effect, are not as well known, but their effects could be considerable.

As scientists learn more about climate, oceans, and the atmosphere, better prediction of weather and climate may become possible. The impact of weather or climate often is not appreciated. Hurricanes, for example, have killed more than 13,000 Americans since 1900 and hundreds of thousands worldwide. A cold winter instead of a warm one might mean severe agricultural problems and a difference of over $10 billion in fuel costs.

Many exciting international programs are studying ocean and climate interactions. An especially important one is World Ocean Circulation Experiment (WOCE).

A significant change in weather patterns was caused by the 1982–83 El Niño. Studies of this phenomenon by programs such as Tropical Ocean and Global Atmosphere (TOGA) and by weather satellites should add much to our understanding of tropical oceanography and climate in general.

A major environmental problem is the decrease of ozone in the upper layers of Earth's atmosphere. The decrease, which can cause serious health problems, is due to certain chemicals being released into the atmosphere. A recent meeting in Montreal produced an international agreement to reduce or eliminate the discharge of some of these dangerous chemicals. This agreement does not completely solve the problem, but it is a major step toward a solution.

The buildup of carbon dioxide and other greenhouse gases in the atmosphere is a challenging environmental problem. This buildup may be causing a worldwide global warming; if so, it could melt glaciers and raise sea levels. Among the unanswered questions are: What is the fate of the carbon dioxide and the other gases that enter the atmosphere? How much enters and remains in the ocean?

Climate has varied over geologic time, having been influenced by past positions of the continents and more recently by the ice ages. Various hypotheses have been proposed to explain ice ages, but the one involving variations of Earth's orbit (the Milankovitch hypothesis) is presently in favor.

Recent studies of ice cores from Greenland have revealed rapid, large temperature changes in the past. These changes appear to be closely tied to changes in ocean circulation.

QUESTIONS

1. Why should scientists look to the ocean to improve our understanding of climate?
2. What are the effects of the unequal heating of the land surface and the ocean?
3. How might sea-surface temperature affect climate?
4. How might phytoplankton affect climate?
5. What are the possible implications shown by the recent results from the Transient Tracers in the Ocean (TTO) program?
6. Discuss how the ozone layer is being damaged, and the options for fixing the problem.
7. Describe the greenhouse effect and its possible causes and effects.
8. What are the major issues and questions concerning the future impacts of the greenhouse effect?
9. What is an El Niño, and how is it formed?
10. What is the Milankovitch hypothesis, and how does it work?

KEY TERMS

anthropogenic carbon dioxide
axial tilt
chlorofluorocarbons (CFCs)
cyclone
El Niño

global warming
Great Ocean Conveyor
greenhouse effect
greenhouse gas
hurricane
land breeze

La Niña
Milankovitch hypothesis
ozone
ozone depletion
ozone hole
ozone layer

precession
sea breeze
Southern Oscillation
typhoon
weather satellite

FURTHER READING

Berner, R. A., and A. C. Lasaga. 1989 "Modeling the Geochemical Carbon Cycle." *Scientific American* 260, no. 3, pp. 74–81. Discussion of processes that may have caused past intervals of global warming by the greenhouse effect.

Broecker, Wallace S. 1992 "Global Warming on Trial." *Natural History,* April, pp 6–14. A discussion of the evidence on whether Earth is warming or not.

Broecker, W. S., and G. H. Denton. 1990. "What Drives Glacial Cycles." *Scientific American* 262, no. 1, pp. 49–56. Proposes that massive changes in the ocean-atmosphere system combined with cyclic changes in Earth's orbit are the key to the beginning and end of glacial periods.

Canby, T. Y. 1984. "El Niño's Ill Wind." *National Geographic* 165, no. 2, pp. 144–83. Typically well-illustrated and well-written article.

Emery, K. O., and D. A. Aubrey. 1991. *Sea Levels, Land Levels and Tide Gauges.* New York: Springer Verlag. A little complex, but provides extensive information about sea level and its recent change around the world.

Funk, B. 1980. "Hurricane." *National Geographic* 158, no. 3, pp. 346–79. Well-written and well-illustrated article.

Houghton, Richard A., and G. M. Woodwell. 1989. "Global Climatic Change." *Scientific American* 260, no. 4, pp. 36–44. Evidence that human-produced carbon dioxide and methane already have caused some changes to our climate.

Jones, P. D., and T.M.L. Wigley. 1990. "Global Warming Trends." *Scientific American* 263, no. 2, pp. 84–91. Analysis of past climatic records indicate that a slight warming has occurred; the future trend, however, is uncertain.

The Oceans and Global Warming. Oceanus 32, no. 2 (1989). A special issue on this subject.

Ramage, C. S. 1986. "El Niño." *Scientific American* 254, no. 6, pp. 7683. Good article on the subject.

Revelle, R. 1983. "The Oceans and the Carbon Dioxide Problem." *Oceanus* 26, no. 2, pp. 3–9. Roger Revelle was one of the first to call attention to the possible carbon dioxide problem.

Rhode, H. 1990. "A Comparison of the Contribution of Various Gases to the Greenhouse Effect." *Science* 248, pp. 1217–19. Discusses the data presented in Table 13-3.

Sathyendranath, S., A. Gouveia, S. Shetye, P. Ravindran, and T. Platt. 1991. "Biological Control of Surface Temperature in the Arabian Sea." *Nature* 349, pp. 54–56. The data for the satellite study of temperature and water color mentioned in the section on phytoplankton and climate.

Schneider, S. H. 1987. "Climate Modeling." *Scientific American* 256, no. 5, pp. 72–80. Discusses how computer models of the Earth's climate yield clues to its future and past climates.

White, R. W. 1990. "The Great Climate Debate." *Scientific American* 263, no. 1, pp. 36–43. A fascinating discussion of the prospect of global warming and what should be done—when we are not sure what may actually happen.

CHAPTER 14

Coastal and Estuarine Environments

To many people, coastal and estuarine (river mouth) environments are the most important part of the ocean. Of the 32 largest cities in the world, 22 are situated on estuaries (New York, San Francisco, and Rio de Janeiro, for instance). About two-thirds of the world's population live near the seacoast. In the United States, if the Great Lakes are included, about 75 percent of the population lives within 80 km (50 miles) of a coast, and these numbers are increasing. Coastal and estuarine environments form less than 1 percent of the world ocean area, but unfortunately, they are the part of the ocean most vulnerable to receiving human pollution.

The coastal zone along the Torrey Pines, in Southern California. In this tectonically active area, uplift has formed steep cliffs along the beach.

(Photograph courtesy of Dr. David G. Aubrey, Woods Hole Oceanographic Institution.)

Introduction

The coastal and estuarine region is the boundary of Earth's three major environments: land, ocean, and atmosphere (Figure 14-1). An **estuary** is a body of water, partially enclosed by land, that generally connects the ocean to a river, an area where saltwater and freshwater mix. The precise land-ocean boundary is the **shoreline,** which constantly changes position with the tides, deposition, and erosion.

Generally, the **coastal zone** is that part of the ocean affected by the land, and that part of the land affected by the ocean. It is a complex region of numerous biological, chemical, physical, geologic, and meteorological interactions. There is no uniform definition of the coastal zone. Seaward, it may include the continental shelf. Landward, it includes estuaries, marshes, seacliffs, the coastal plains, and other similar environments. Thus it varies greatly in width, depending on the nature of the land-sea contact. Some people have even suggested that the entire "exclusive economic zone" (EEZ) should be considered as part of the coastal zone—in other words, out to 200 nautical miles (370 km) from the shoreline (see Chapter 19).

Figure 14-2 is helpful in understanding coastal zone terminology. A **shore zone** straddles the shoreline and includes the beach and surf zone. The **nearshore region** is that area seaward of the shoreline. The worldwide shoreline has a total length of about 400,000 km (250,000 miles), or ten times the distance around the world! The shoreline is very dynamic, but most of its features are temporary because their position is affected by the height of sea level. Sea level, in turn, is influenced by tides, wind direction, and the strength and height of breaking waves, especially during large storms. Thus, the shoreline is constantly changing and may migrate over a large area in a very short time. In addition, sea level has risen and fallen considerably over the last 35,000 years because of widespread growth and melting of glaciers (see Figure 2-5, page 31). During this time, the rise and fall of the ocean has caused the shoreline to migrate out and back over most of the continental shelf.

The extent of the coastal zone is considerable. Inland waters, such as semi-enclosed bays, estuaries, and lagoons, exceed over 100,000 km² (over 38,000 miles²) in the United States alone. The region immediately seaward of the shoreline, the continental shelf (see Chapter 4), has a worldwide area of about 50 million km² (about 19 million miles²).

The importance of the coastal region to the other areas of the ocean cannot be overstated. A large percentage of important marine organisms either begin their lives in estuaries and marshes or spend a major portion of their lives there. Even more inhabit the continental shelf, and over 90 percent of worldwide fishing resources are within the coastal zone.

Rivers enter the ocean through the coastal zone, carrying beneficial nutrients and harmful pollutants from coastal industries and waste from municipal dumping. The world's rivers carry about 2.8×10^9 tons (2.8 billion tons) of dissolved material into the ocean each year. As immense as this number is, it is only about one twenty-millionth of that already in the ocean. Much of this material eventually reaches the deeper ocean by processes involving waves, currents, and tides.

Circulation in the nearshore zone eventually exchanges the water in this region with the more offshore areas. In this manner nutrients, pollutants, and other material carried by the rivers can reach the more offshore parts of the ocean. Coastal currents, driven by rivers, winds, and tides, keep most coastal waters well-mixed, resulting in adequate nutrients for plant growth throughout the vertical water column from surface to bottom.

FIGURE 14-1 General features and landforms of the coastal zone.

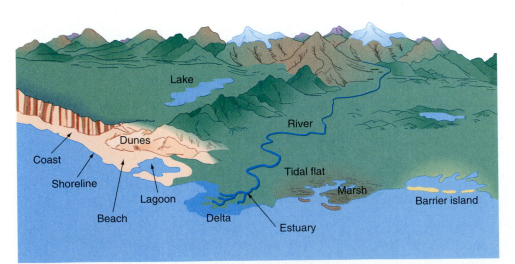

FIGURE 14-2 Some specific features of the coastal zone. Berms are flat portions of a beach formed by wave action. The two shown were caused by previous storms. Dunes are mounds of sediment moved by the wind. If colonized by plants, they become stable, and thus protect inshore areas from storm erosion. (See also Figure 14-3.)

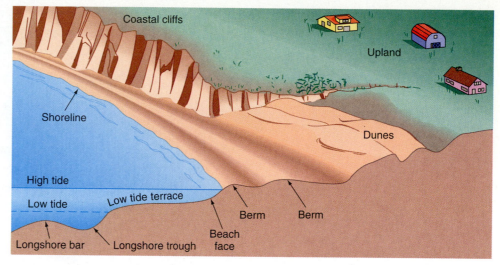

In the United States and many other countries, management of many activities in the coastal zone is divided among federal, state, and local agencies. Many have overlapping and conflicting responsibilities. More often, however, regulations are not enforced, especially with regard to pollution. Local governments often control land use and waste disposal, whereas state agencies may control water, pollution discharge, highways, and ownership of state lands. This fragmented control can lead to complexities and incorrect decisions concerning the coastal zone. Before considering some of these problems, let us look at the components of the coastal zone and its oceanographic processes.

Beaches

Beaches are the unconsolidated sediment (mainly sand or gravel) that covers most parts of the shore (Figure 14-2). Beaches are influenced continually by waves. They generally are somewhat permanent features of the coast, but can be rapidly changed by storm waves. Change also results from daily and monthly tides and the seasons. Human activity strongly affects beaches, and beach erosion from both natural and human causes has prompted efforts to preserve them (Figure 14-3). In many instances, these efforts have been unsuccessful (see Figures 14-9 and 14-14).

Figure 14-4 illustrates how waves erode the shore. If the wave crests are not parallel to the shore as they approach the coast, and they usually are not, a bending of wave direction, called **refraction,** occurs when they reach shallow water (see in Chapter 10, page 227). Refraction causes the wave crests to turn toward shallow areas, so the crests tend to align themselves parallel to the bottom contours. The wave energy *converges* on projecting points such as offshore bars and sea cliffs and *di-*

verges in open bay areas. Erosion by waves is therefore greater at points of convergence. Sediment usually moves toward, and accumulates in, the quieter divergent areas. If this process continues undisturbed for many years, a straight shoreline results.

Longshore Current and Rip Current

After waves break, water moves forward toward the beach. The beach, however, is an essentially impermeable barrier. If waves approach the beach at an angle (not parallel to the beach), the water, after the waves break, will move seaward but also along the beach. The next wave pushes the water back up the beach, and the net result is a zigzag movement of water along the beach. This forms an "along-shore current," known as a **longshore current** (see Figure 10-8, page 228). The longshore current transports sediment parallel to the beach.

FIGURE 14-3 A method for preserving a beach. The planting of grasses will eventually stabilize a beach or dune area and make it less vulnerable to erosion. (Photograph courtesy of Dr. David G. Aubrey, Woods Hole Oceanographic Institution.)

FIGURE 14-4 Sediment movement in the coastal region due to the convergence and divergence of wave energy. The convergence and divergence is caused by the refraction of waves as they approach shallow water. (See also Figures 10-7 and 10-8, pages 227 and 228.)

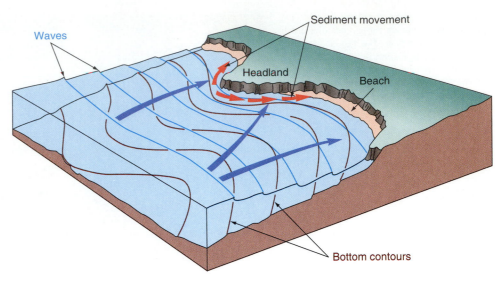

Water is continuously supplied to the longshore current by incoming waves, so eventually some of this water must return seaward. When this happens, a **rip current** is produced (Figure 14-5). Its position depends on the bottom topography and the height and period of the waves. Rip currents can carry sediment seaward, only to be moved landward again with the next approaching waves.

Beach Erosion

Waves approaching a coast keep the bottom sediment in a transient state: eroding in one place, depositing in another, ultimately smoothing the coastline, unless people interfere. If sediment is not returned to the beach by incoming waves, the beach is gradually eroded. This can result in damage to homes or other structures if they are built too close to the shoreline (Figure 14-6).

Some beaches lose their sediment because they are situated near submarine canyons (Figure 14-7). Sand carried by longshore currents can be intercepted by these canyons and eventually be carried into the deep sea by sandfalls, slumping, or turbidity currents.

Longshore currents cause a net along-shore movement of sediment, which eventually straightens a shoreline through repeated erosion and deposition. In some areas, this erosion can be considerable. Sand being moved by longshore currents also can be trapped by structures built along the coast, and is usually deposited on the upstream side of these structures (Figure 14-8).

Waterfront property owners often construct jetties or groins to interfere with the longshore current, thus preventing erosion along their property. (A **groin** is a collection of large rocks placed at right angles to the beach and extending seaward.) To some degree these structures work, but they generally cause erosion downstream, commonly leading to construction of more groins or jetties (Figure 14-8).

FIGURE 14-5 A series of rip currents along a southern California beach. They are distinguished by light-colored, sediment-laden water beyond the breaking waves. Land is in the upper right-hand corner near the airplane wing. (See Figure 10-9, page 229, for a description of the process that forms rip currents.) Photograph courtesy of Dr. D. L. Inman, Scripps Institution of Oceanography.)

FIGURE 14-6 A Southern California house built too close to the shoreline and in danger of being undercut by waves.

FIGURE 14-7 This diagram shows how the presence of a nearshore submarine canyon can cause the loss of sediment from a beach. The sequence of events is from left to right (from 1 to 4).

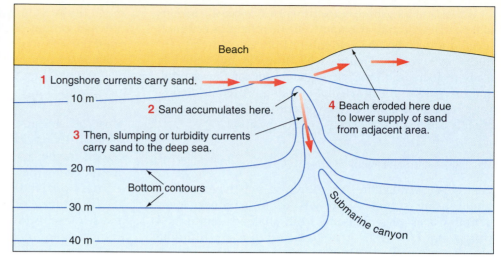

Beach

1 Longshore currents carry sand.

10 m

2 Sand accumulates here.

4 Beach eroded here due to lower supply of sand from adjacent area.

3 Then, slumping or turbidity currents carry sand to the deep sea.

20 m

Bottom contours

30 m

Submarine canyon

40 m

FIGURE 14-8 (a) Methods for modifying a beach by coastal structures. Sand moving along the coast can be trapped, causing deposition in one area and erosion in another. For example, the beaches are much narrower on the sheltered side of the jetty because the sand they would have received was intercepted by the jetty. Two examples are shown in (b) and (c).
(b) Coastal erosion along a New England shore. The direction of wave and sediment motion is from the top to the bottom of the figure, the same as shown in (a). In this manner, sediment builds on the downstream side of the prevailing current direction.
(c) A similar situation of coastal erosion along the California coast near Ventura. In this example, the direction of sediment motion is from left to right. (Photograph (b) courtesy of Jack Silver; (c) courtesy of Dr. David G. Aubrey, Woods Hole Oceanographic Institution.)

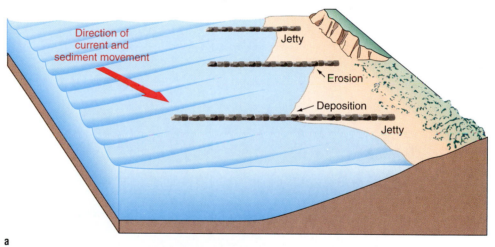

Direction of current and sediment movement

Jetty

Erosion

Deposition

Jetty

a

b

c

Waves and longshore currents will erode coastal regions and beaches, despite attempts to thwart them. A common technique is to build a **seawall** in front of a beach to reduce wave erosion (Figure 14-9). However, this generally leads to increased erosion, undercutting of the seawall, and even loss of the beach and dwellings. Attempts to prevent coastal erosion are rarely successful, and when they are it is usually at the expense of nearby areas.

Seasonal Sand Movement

Figure 14-10 illustrates the different net movement of sand during summer and winter. In summer, long-period waves pick up sand from shallow depths and carry it onto the beach. The backrush of water from the wave carries the sand back seaward. But the backrush has much less energy than the incoming wave, so the sand is not carried back to its original seafloor location; thus the net movement of sand is toward the beach.

In winter, there are usually more offshore storms that produce higher waves with short periods. This keeps the sand suspended and prevents it from settling. Therefore, much of the sand washed off the beach by the backrush of water cannot settle until it reaches deeper water, beyond the action of subsequent waves. This situation generally results in the loss of beach sand during the winter.

Glaciers, Sea Level, and Slowly Migrating Beaches

Most beach sand comes from shallow parts of the seafloor. Its original source, however, is the land. It is carried to the ocean by rivers, wind, glaciers, and cliff erosion along the coast. One might think that the supply of sand from land to sea is constant, but it is not. During glaciated times, when more water is tied up in glaciers and sea level is lowered, erosion of shallow offshore areas increases, and the beaches migrate seaward (Figure 14-11). During such times of lowered sea level, rivers carry their sediment to the outer parts of the continental shelf or upper continental slope. When sea level rises once again, the existing beaches are eroded and sediment is deposited in shallow offshore areas. A rising sea level also floods or drowns existing river valleys, forming extensive estuaries.

FIGURE 14-9 How building a seawall to protect an area from erosion can actually increase erosion.

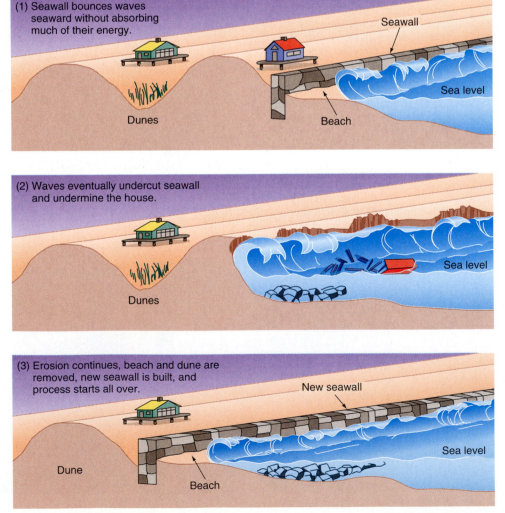

(1) Seawall bounces waves seaward without absorbing much of their energy.

(2) Waves eventually undercut seawall and undermine the house.

(3) Erosion continues, beach and dune are removed, new seawall is built, and process starts all over.

a Winter Beach Profile

Net sand movement is offshore and coarse-grained sediments remain. Beach profile is low and flat.

b Summer Beach Profile

Net sand movement is onshore and there is a buildup of the beach. The sands cover up the coarse-grained material that dominated the winter beach.

Cliff

Beach

Cliff

Beach

FIGURE 14-10 Typical seasonal changes along a beach: (a) common winter situation on a beach at La Jolla, California; (b) common summer situation at the same beach. About 0.6 m (2 ft.) of sand has been added. A similar amount was removed during the winter by storm waves. (Photographs courtesy of Dr. David G. Aubrey, Woods Hole Oceanographic Institution.)

Today, in most areas of the world, relatively small quantities of sediment are being supplied to the ocean, because the recent rise in sea level has caused most rivers to deposit their sediment directly into their estuaries rather than into the ocean. Since the end of the glacial maximum about 18,000 years ago, the melting glaciers and rising sea level have caused beaches to migrate across the entire continental shelf to their present position.

Barrier Beaches and Barrier Islands

Barrier beaches are long, narrow ridges of sediment that are above high tide and run parallel to the main coast (Figure 14-12). When separated from the mainland, barrier beaches are called **barrier islands.** These features form some of our finest and most popular beaches: Fire Island in New York, Atlantic City in New Jersey, Ocean City in Maryland, Cape Hatteras in North Carolina (Figure 14-13), Miami Beach in Florida, and Padre Island in Texas.

Because of their position, barrier beaches protect the mainland from offshore storms. Unfortunately, their location has also made them prime sites for development, and barrier islands are among the most urbanized regions of the United States. The urbanization continues, although at a reduced rate. In some regions, such as Miami Beach, large sums are being spent—usually unsuccessfully—to restore the beaches to their original condition (Figure 14-14).

Movement of Barrier Beaches and Islands

Barrier beaches and barrier islands, like almost everything else in the coastal zone, are frequently in some state of movement. Indeed, the process of formation of

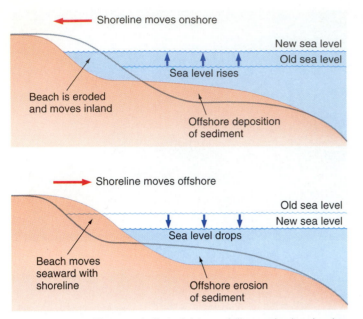

FIGURE 14-11   The general effect of rising or falling sea level on beaches and nearshore sedimentation.

FIGURE 14-13   A barrier island chain that forms Cape Hatteras and Cape Lookout, North Carolina. The barrier islands enclose Pamlico Sound. Note the suspended sediment being carried from the rivers (estuaries) on the left into Pamlico Sound and then to the Atlantic Ocean through the breaks in the barrier islands. The many small white areas to the right are clouds. (Image courtesy of National Aeronautics and Space Administration.)

barrier beaches and barrier islands (Figure 14-15) clearly shows how dynamic the coastal region is. In one area sediment is being eroded, while in other areas it is being deposited. When the source of sediment or the amount of sediment changes, a corresponding change occurs in the size of the beach or other coastal features.

The direction of movement of a barrier beach or island can be determined by, and will follow, the rise or fall of sea level, such as due to glacial periods (see Figure 14-11) or to changes in the supply of sediment to the coastal zone. Box 14-1 describes a coastal area that has gone through some recent changes.

Estuaries, Lagoons, Fjords, Marshes, and Mangrove Swamps

Estuaries, lagoons, and marshes are common to many coastal areas (Figure 14-16). Within an estuary, the seawater is usually diluted by freshwater, reducing its salin-

ity below the average 35‰ of the open sea. In a small number of estuaries, the evaporation rate exceeds the river inflow, thus causing relatively high salinity.

Estuaries are biological, chemical, physical, and geological transition areas and may be the most complex part of the ocean. Estuarine water can undergo very large changes in temperature and salinity due to the mixing of fresh river water and seawater, often with marked biological consequences. Human activity has compounded further the complex natural processes within estuaries. Important changes can occur during just a few hours in estuaries, whereas comparable changes in the open ocean can require from a few years to thousands of years.

Estuaries reveal the significance of the most recent glacial period to the world environment: most present-day estuaries owe their existence to the sea-level rise of

FIGURE 14-12   Some different types of barrier beaches: barrier island, bay barrier, and barrier spit. (See also figures in Box 14-1.)

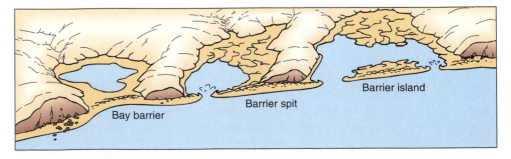

FIGURE 14-14 Before (left) and after pictures show the effect of an intensive sand addition program along the Florida coast. In most instances, however, much of the new sand will eventually be removed by subsequent storms. (Photograph courtesy of U.S. Army Corps of Engineers.)

about 130 m (426 ft.) that occurred during the past 18,000 years. Because of this sea-level rise, most rivers of the world now flow into estuaries instead of directly into the ocean. The return of a severe ice age could eliminate the world's estuaries in a few thousand years.

Another type of estuary is a **lagoon,** a broad, shallow arm of the sea partially restricted from the ocean by barrier beaches or islands (see Figure 14-13). If a lagoon's opening to the ocean becomes closed, the lagoon can eventually become a marsh or saline lake.

FIGURE 14-15 The formation of a barrier spit, lagoon, and barrier beach by longshore transport of sediment. The process is a continuing one that will also be influenced by sea-level changes.

a Coastal estuary and cliffs

b Sediment transport by longshore current builds a barrier spit

c Continued transport of sediment along coast extends barrier spit almost across mouth of estuary forming a lagoon

d Spit is split probably by a storm forming a barrier island

Box 14-1

Cycles of Coastal Erosion

Coastal erosion is a natural phenomenon. Incoming waves and currents continually adjust the shape of the shoreline and the coastal region. While erosion is active in one place, deposition is in progress elsewhere. Some of the changes are predictable, short-term, and cyclic, such as the buildup of beaches in the summer and their erosion during winter (see Figure 14-10).

Some cycles are longer and are just beginning to be understood. A cycle of barrier beach erosion with a period of about 150 years was described in a 1988 report by coastal geologist Dr. Graham Giese. By combining old charts with modern observations, he noted a pattern in the **breaching** (opening due to storms) of barrier beaches and formation of new tidal inlets on Cape Cod, Massachusetts.

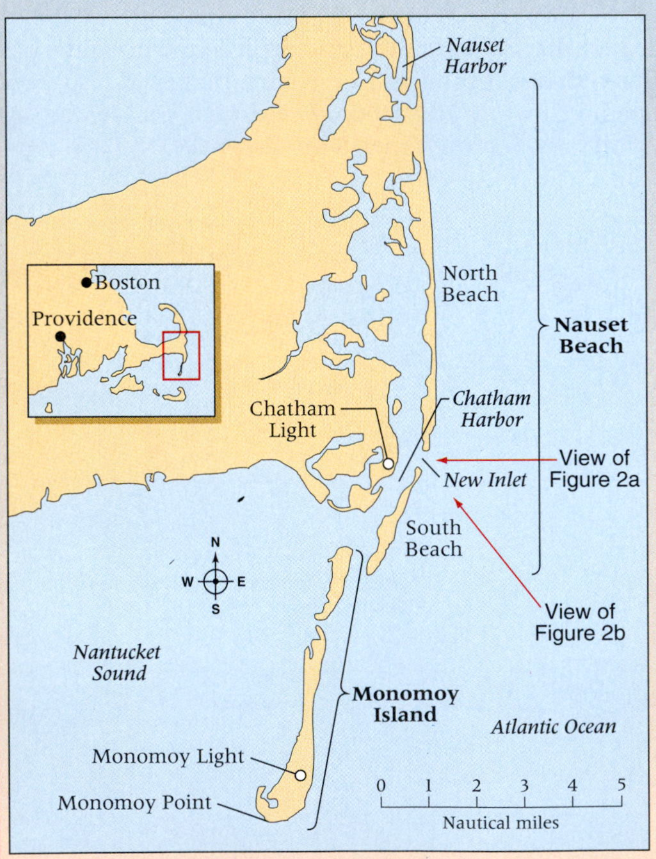

The area's outer coastline is a series of barrier beaches (Monomoy Island and Nauset Beach). It is exposed on the east to the Atlantic Ocean (Figure 1), as are other barrier beaches in southern New England. As these barrier beaches are changed by waves and currents, the protected areas behind them also change and adjust.

During a strong storm in January 1987, the Nauset Beach area was breached (Figure 2). Most residents

FIGURE 1 General location chart showing the Nauset and Monomoy Barrier Beach System. Note the new inlet created in 1987. The location of Figures 2a and 2b is shown. (Adapted from Giese, 1990.)

FIGURE 2 Nauset Barrier Beach: (a) the 1987 breach, one day after it occurred, viewed from the Atlantic Ocean westward; (b) the same area about two and a half months later, looking more northward. (Photograph (a) by Kelsey-Kennard Photographers, Chatham, Mass.; (b) by Richard Miller; both courtesy of the Coastal Research Center of the Woods Hole Oceanographic Institution.)

Box 14-1 continued

were not surprised, because the event had been anticipated. Changes along the Nauset Barrier Beach System have been observed for 200 years, and a similar break had occurred in 1846 in almost the same location. In 1871, many residents of the area moved their homes, correctly anticipating a break in a nearby location.

The impact of the 1987 breach was considerable: seven houses were destroyed and at least twenty damaged, with property loss exceeding $5 million. Some owners have placed rock revetments along their coastal property to prevent further damage. Chatham Harbor was dredged to remove the sand carried in by the storm.

In 1978, Giese reconstructed the history of the erosion and deposition over 20-year increments from 1770 to 1970. He then predicted changes to be expected during 10-year increments from 1975 to 2005 (Figure 3). By anticipating such changes, better development regulations may be made. The word *may* is appropriate: despite such knowledge of potential changes, people love the sea, and tremendous pressure exists to build as close to the shoreline as possible.

REFERENCES

Giese, S. Graham. 1988. "Cyclic Behavior of the Tidal Inlet at Nauset Beach, Chatham, Massachusetts." In *Lecture Notes on Coastal and Estuarine Studies,* vol. 29, pp. 269–83. New York: Springer-Verlag.

———. 1990, "The Story Behind the New Tidal Inlet at Chatham." *Nor'easter* 2, no 1, pp. 29–33.

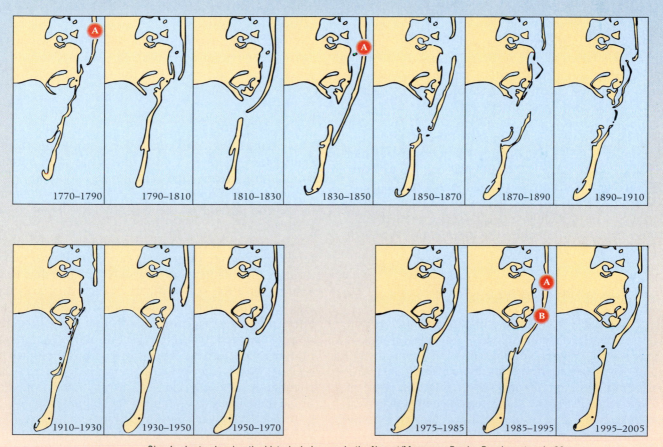

FIGURE 3 Giese's charts showing the historical changes in the Nauset/Monomoy Barrier Beach system in 20-year increments during the two centuries from 1770 to 1970, plus his predictions for 1975 to 2005 in ten-year increments. **A** marks the approximate inlet location in 1740 and 1846 and Giese's predicted location for 1985 to 1995. **B** is the actual location of the new inlet formed in January 1987.

FIGURE 14-16 A complex coastal area (Cape Cod Bay and Barnstable Harbor, Massachusetts) in autumn. Reddish areas in the foreground are marshes, crossed by numerous tidal channels. Green upland areas are somewhat developed. (Photograph courtesy of Woods Hole Oceanographic Institution.)

A third type of estuary is the **fjord** (fee-YORD), common in glaciated areas such as Scandinavia, Alaska, and Canada. (An example is the Strait of Juan de Fuca between Washington State and Vancouver Island.) A fjord generally has a deep basin, frequently carved by a glacier, behind a shallow ridge or sill (usually a glacial deposit) at its seaward end. Although fjords generally have large amounts of river inflow, they are poorly mixed due to their shallow sill and limited tidal action. In some fjords, the deep water may eventually become stagnant, which can occur when the oxygen in the deeper water is consumed by respiration and decay processes and the sill prevents its replenishment by the inflow of oxygen-rich water from the ocean.

Where Estuaries and Lagoons Occur

Estuaries generally occur where continental shelves and coastal regions have remained stable during the recent rise in sea level. Where a coastal region is tectonically active, there tend to be few estuaries. This explains why estuaries are more abundant on the stable East Coast of the United States relative to the tectonically active West Coast. However, some estuaries have formed on the West Coast through a combination of subsidence and faulting of the coastal area; San Francisco Bay is an example. Most estuaries are drowned river valleys or fjords. Most lagoons are formed by the buildup of offshore bars, and generally occur where the shelves and coastal regions are wide and smooth and have low relief.

Well-known estuaries along the U.S. coast are Chesapeake Bay (Maryland–Virginia), the Hudson River in New York, and Delaware Bay. One of the best examples of a lagoon is Laguna Madre along the south Texas coast. Here a long barrier island, Padre Island, isolates much of the Texas coast from the Gulf of Mexico.

Short-Lived Estuaries

Like beaches, estuaries and lagoons are temporary features when considered on a time scale of hundreds or thousands of years. Over time, both may disappear, either by the buildup of marshes (which trap sediment) or by marine erosion of the coastline (see Figures 14-11 and 14-15).

Marsh buildup, and thus sediment accumulation, tends to obliterate estuaries because freshwater marshes form in the inland areas and saltwater marshes or bars tend to close the seaward part of the estuary. Estuaries and fjords that are cut into solid rock are less likely to be affected by these processes. Rivers that carry large quantities of sediment may fill in their estuaries. Large quantities of sediment can also be carried into an estuary from offshore areas. If an estuary is an important navigation channel, like Chesapeake Bay, it may need almost continual dredging to keep it open for shipping.

If sea level were to lower rapidly, most estuaries would disappear, with their rivers eroding their own deposits. If sea level remains constant or slowly rises (due, as many experts predict, to global warming, discussed in a later section), estuarine destruction would continue by marsh buildup and sediment accumulation. Only a rapid *rise* in sea level would maintain or rejuvenate present estuaries. From this it follows that estuaries were not common in the geologic past, except during periods of rising sea level or lowering of the land.

Water Mixing in Estuaries

Most estuaries have a two-way water movement. Freshwater from the river flows seaward along the surface (because it is less dense than saltwater), while the denser seawater flows landward along the bottom. The degree to which they mix is determined by wind, tidal range, estuary shape, and the relative flow rates and volumes of river water and seawater. Three general styles of mixing occur: salt-wedge, partially or poorly mixed, and well-mixed (Figure 14-17):

- A **salt-wedge estuary** develops where mixing is slight, which occurs where a large, rapidly flowing river enters an area where the tidal range is small. Here the saltwater and freshwater remain almost com-

FIGURE 14-17 *Salinity-depth profiles across three types of estuaries: a salt-wedge estuary, a well-mixed estuary, and a poorly mixed or partially mixed fjord-type estuary. Also shown is a typical salinity-depth profile across a fjord. Arrows indicate direction of mixing.*

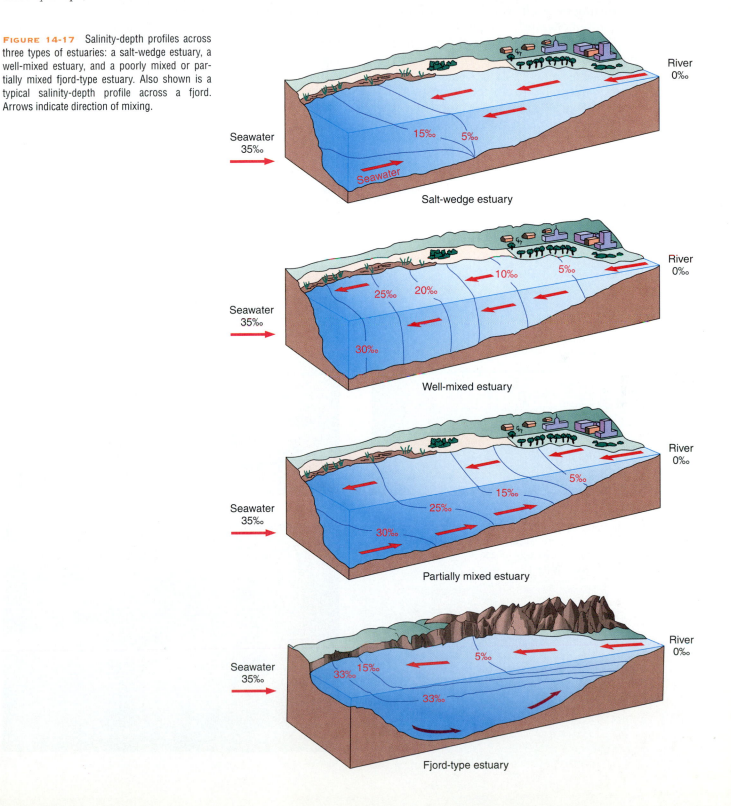

pletely separated. Examples are the mouths of the Hudson River in New York, the Columbia River in Washington–Oregon, and the Mississippi River.

- A **partially** or **poorly mixed estuary** will occur when the estuary is deep, tidal ranges are moderate, and river flow is high. Puget Sound, San Francisco Bay, and many fjords are of this type.

- **Well-mixed estuaries** result when the river flow and tidal range are both moderate. Examples include Chesapeake Bay and Delaware Bay. A fourth type of flow, called a **reverse estuary,** can result when there is no river flow. Some lagoons fall into this category, where a reverse flow of water moves from the ocean back into the estuary. This happens when water evaporates from the lagoon surface and is replaced by ocean water.

Estuarine Life

Many estuaries or coastal areas are bordered by a narrow strip of vegetation or wetland area containing either salt marshes or mangrove swamps (Figure 14-18). In tem-

perate climates, **salt marshes** occur. In tropical-to-semi-tropical areas between 30°N and 30°S latitude (the equatorial region), **mangrove trees** occur (see Figure 14-21). Some mangroves are rooted in water and produce offshoots that grow further out into the water. Their root systems can be very effective in trapping sediment. Mangrove distribution is generally limited to tropical and subtropical areas because their seedlings cannot survive colder temperatures.

Salt marshes generally occur in intertidal areas, along the banks of tidal rivers, or behind barrier beaches (see Figure 4-16). These extremely productive areas provide a source of food that supports a large marine population of fish, birds, shellfish, and plants (Figure 14-19). Marshes also protect the land behind them from storms and high seas. The organic production of marshes is helped by nutrients supplied by rivers and from the ocean carried in by tidal flow.

Marsh organic production may attain 5 to 10 tons of organic matter per acre per year, compared to 1 ton for a wheat field or under 0.5 ton for the open ocean or a desert. This great production of organic matter results in

a

b

c

FIGURE 14-18   Two common coastal environments: (a) coastal marsh; (b) and (c) mangrove swamp. (Photograph (a) courtesy of Coastal Research Center of the Woods Hole Oceanographic Institution; (b) and (c) courtesy of University of Puerto Rico Sea Grant College Program.)

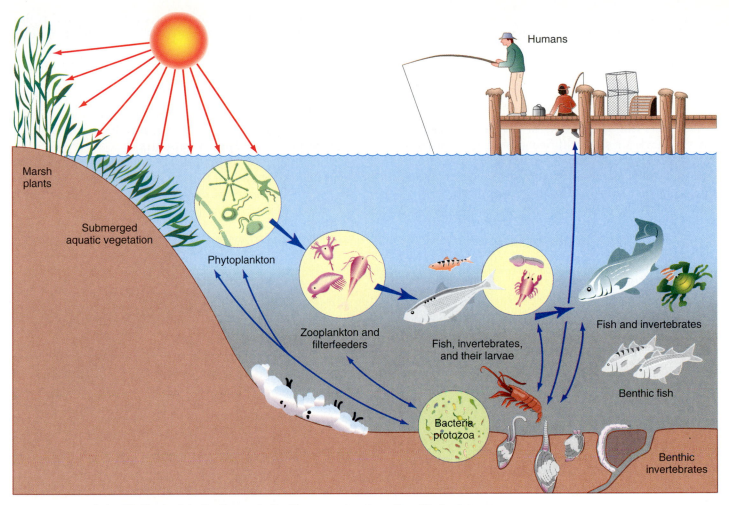

Humans

Marsh plants

Submerged aquatic vegetation

Phytoplankton

Zooplankton and filterfeeders

Fish, invertebrates, and their larvae

Bacteria protozoa

Fish and invertebrates

Benthic fish

Benthic invertebrates

FIGURE 14-19 A simplified food web for the Chesapeake Bay. The arrows show the motion of food and energy through the system. A coastal area like this supports a very rich variety of life. (Adapted from Kaufman and Franz, *Biosphere 2000: Protecting Our Global Environment* (New York: HarperCollins, 1993).)

food that can be consumed by the numerous organisms that inhabit the area.

Most marshes have a small number of species, but these usually are quite abundant. Common in most marshes is a plant zonation related to exposure to the sea (Table 14-1). The plant *Zostera* is mostly covered by water. Proceeding toward higher ground and shallower water, the plant sequence is usually *Spartina, Salicornia,* and *Distichlis* (see Figure 11-9, page 257).

A sediment core from a marsh can show how the marsh is forming. For example, if *Spartina* is found to be growing above decaying *Salicornia*, it indicates a relative rise in sea level in the marsh area, because *Spartina* grows closer to sea level than *Salicornia*.

The accumulation of large quantities of plants such as *Salicornia* and *Spartina* in the marsh can form a deposit called **peat.** Because peat forms at or near sea level, it can also indicate the past position of sea level. Peat deposits have been left on the shallow portions of

the continental shelf because of the recent rise in sea level. In fact, dating of these peat deposits by the carbon-14 method has given clear evidence that in general worldwide sea level has been rising during the last 18,000 years.

TABLE 14-1

Typical Plant Zonation in a Saltwater Marsh

Plant (Genus)	Common Name	Exposure
Zostera	Eel Grass	Always submerged
Spartina	Cord Grass	Submerged twice daily
Salicornia	Spike Grass	Submerged once daily
Distichlis	Salt Hay	Submerged a few times a month

Deltas

A **delta** results from the accumulation of sediment at the mouth of a river. The name comes from the resemblance of the Nile River's triangular sediment accumulation to the Greek letter delta (Δ) (Figure 14-20).

All rivers transport sediment that is deposited where the river enters a slower-flowing estuary, ocean, or lake. In the ocean, this sediment may be removed from the river mouth by longshore currents, ocean waves, and tidal currents. However, many large rivers carry more sediment than these marine forces can remove. In this instance, sediment accumulates at the river mouth, forming a delta.

As the delta builds, the river usually forms a **distributary channel** like the Rosetta Branch of the Nile Delta shown in Figure 14-20a. Coarse-grained sediment is deposited along distributaries, forming the channel. Finer-grained sediment washes into the area between channels and becomes **floodplain deposits.**

Eventually, the distributary builds itself so far seaward that the river flow can no longer maintain it. When this happens, the river changes course and forms a newer,

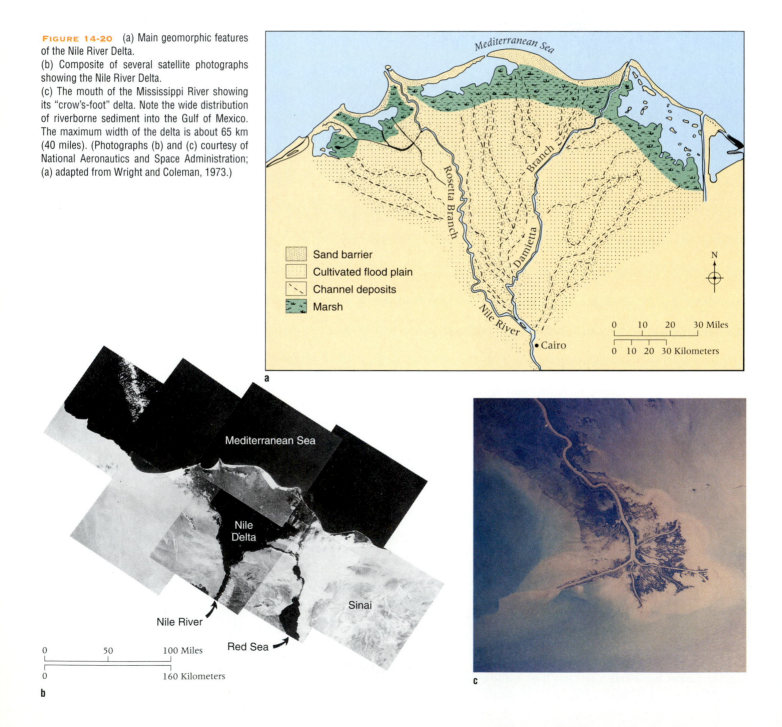

FIGURE 14-20   (a) Main geomorphic features of the Nile River Delta.
(b) Composite of several satellite photographs showing the Nile River Delta.
(c) The mouth of the Mississippi River showing its "crow's-foot" delta. Note the wide distribution of riverborne sediment into the Gulf of Mexico. The maximum width of the delta is about 65 km (40 miles). (Photographs (b) and (c) courtesy of National Aeronautics and Space Administration; (a) adapted from Wright and Coleman, 1973.)

shorter distributary. In this manner the delta continuously builds seaward. As a distributary is abandoned, the reworking of waves and currents carries the fine-grained deposits seaward, leaving a coarse-grained deposit at the delta's edge.

Deltas are generally very flat and extremely fertile because the sediment deposited by the river is rich in nutrients. Consequently, large settlements have been built on deltas and many early civilizations began in such areas. Unfortunately, their low elevation and relief makes deltas extremely vulnerable to flooding.

The volume of sediment carried by some rivers is immense. The Mississippi River, for example, carries up to 300 million tons per year, and the Nile carried as much as 140 million tons before the Aswan Dam was built. Both rivers empty into fairly quiet water, allowing huge deltas to form. The present Nile Delta covers an area of over 100,000 km^2 (38,000 miles^2), larger than the state of Indiana.

A Quick Tour Around the U.S. Coast

This brief tour illustrates the coastal and estuarine features examined thus far. As you read this section, refer to Figure 14-21, which shows the location of some common coastal features around the world.

The U.S. East Coast is a passive coast—a tectonically inactive one. Sea level has been rising slowly along most of it, so estuarine environments prevail and barrier beaches are common.

Much of the coastline north of New York City has been glaciated recently (geologically speaking), and many beaches there are composed of glacial debris. Parts of Maine and much of Canada have rocky beaches with limited sand. South of New York, coastal rocks are poorly consolidated and provide a good source of sand to the many barrier beaches and other coastal features in the area (see Figure 14-13, for example). Coral reefs occur in the southernmost parts of the United States, in the Florida Keys.

The Gulf Coast is dominated by the Mississippi River Delta and the great volume of sediment that the river contributes to the Gulf of Mexico (see Figure 14-20c). The area is slowly subsiding and coastal erosion is extensive, with Louisiana losing over a million acres in this century alone. In recent years, the area has been hit by several hurricanes, causing considerable coastal damage. Further to the west, the Texas coast is bounded by extremely long barrier islands that shelter lagoons. One of them, Padre Island, extends over 160 km (100 miles).

Unlike the East Coast, the West Coast is a tectonically active area, where volcanism and earthquakes are common. In general, the coast is rising there relative to sea level. Numerous coastal cliffs and other elevated features reflect the instability of the region (see opening figure to this chapter). To the north, the Alaskan coast is dominated by glacial features, including fjords.

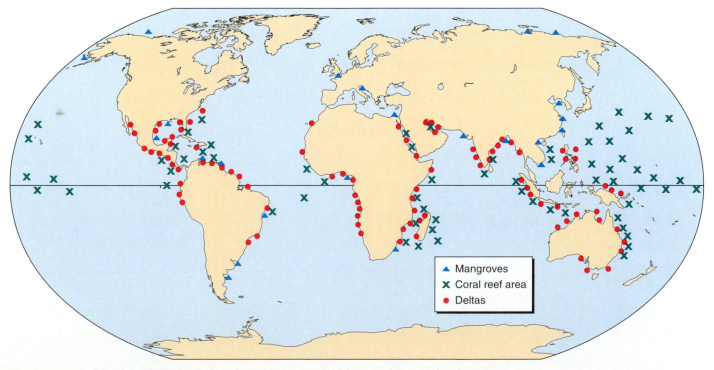

FIGURE 14-21   General worldwide distribution of some common coastal features: deltas, coral reefs, and mangroves.

The West Coast is exposed to high-energy waves from large offshore storms. These can cause considerable beach erosion during winter (see Figure 14-10). In some areas, nearshore submarine canyons intercept sand transported by the longshore current, causing some coastal erosion. The Hawaiian Islands have similar conditions, but with dramatic examples of recent volcanic activity, and the darker sands that are produced from volcanic rock.

Coastal Problems

Erosion is a major problem of the coastal zone. If Alaska is excluded, about 42 percent of the U.S. shoreline is eroding seriously (Table 14-2). The problem is most critical in the North Atlantic, where population is densest and up to 85 percent of the shoreline is privately owned. The estimated average rate of shoreline erosion is about 0.4 m per year (about 1.3 ft.), greatest along the Gulf Coast at about 1.8 m (about 6 ft.). Locally, rates can be even greater, such as 2.3 m (7.5 ft.) per year along some parts of Cape Hatteras and Louisiana.

Important factors in coastal erosion are exposure to waves and currents (especially during storms), sediment supply, coast shape, tidal range and intensity, coastal climate, and changing sea level. At least some coastal erosion is related to the present slow rise in world sea level, due to glacial melting from global climate change, local tectonic movement, and subsidence of some coastal areas (see Box 14-2).

Storms are especially destructive to beaches. In 1983, coastal storms cost California an estimated $500 million in damage. In 1989, Hurricane Hugo, one of the most expensive disasters in U.S. history, did over $9 billion damage, much of it in South Carolina's coastal zone. Hurricane Andrew, in September 1992, caused over $20 billion in damage, much of it in coastal Florida. Similar large hurricanes have caused considerable damage in other countries.

It is dismaying to realize, however, that human activities, including construction, probably have affected coastal areas more profoundly than have storms in recent years (see Figures 14-8 and 14-9).

Global Change

Increasing greenhouse gases in the atmosphere may cause higher temperatures worldwide in the coming decades, which in turn could slowly melt the polar ice caps, causing a worldwide rise in sea level. Predictions of such a rise vary from tens of centimeters to several meters over the next 30 to 50 years (see Box 14-2). Any significant rise may affect the coastal zone.

A sea level rise will inundate low-lying coastal marshes, beaches, barrier islands, deltas, and parts of estuaries. With a rise of as little as a meter (3.3 ft.), 30 percent to 70 percent of present U.S. wetlands would be underwater. A smaller rise still could cause considerable change to beach areas, causing severe erosion. Most beaches and wetlands would migrate inland, keeping pace with the rising sea. In the process, however, they would damage any development in their way (see Figure 14-6).

TABLE 14-2

Erosion of U.S. Shoreline

Region	Total Shoreline Length		Significant Erosion		Percentage of Shoreline Being Seriously Eroded
	Kilometers	Miles	Kilometers	Miles	
North Atlantic	13,870	8,620	12,035	7,480	87%
South Atlantic—Gulf	23,524	14,620	4,537	2,820	19%
Lower Mississippi	3,121	1,940	2,542	1,580	81%
Texas Gulf	4,022	2,500	579	360	14%
Great Lakes	5,921	3,680	2,059	1,280	34%
California	2,912	1,810	2,494	1,550	86%
North Pacific	4,570	2,840	418	260	9%
Alaska	76,106	47,300	8,206	5,100	11%
Hawaii	1,496	930	177	110	12%
Total for nation	135,542	84,240	33,047	20,500	24%

Source: National Shoreline Study of the U.S. Army Corps of Engineers.

Box 14-2

Is Sea Level Rising?

It is evident that sea level has risen and fallen many times over geological time. One "recent" cycle (18,000 years ago and earlier) resulted in flooding of vast areas of Earth's land surface. Likewise, periods of lowered sea level exposed the present continental shelves and upper parts of the continental slope.

The most recent cycle of sea-level rise and fall is associated with the glacial advance and retreat that occurred at the end of the Pleistocene Epoch (see Figure 2-5, page 31). About 18,000 years ago, sea level was 130 m (426 ft.) below its present position. The water was tied up in glaciers, which then extended southward as far as New York City. Since that time, world climate has warmed, the glaciers have melted somewhat, and a portion of their water has returned to the ocean, raising its level.

Accurate determination of sea-level change over a short period is difficult. The sea surface continually fluctuates with tides, wind, and waves. On a longer time scale, land itself may rise and lower. Areas once depressed by the weight of glaciers now are rising slowly, a process called *glacial rebound*. An example is the Great Lakes and parts of Canada. In such regions, the *net effect* is that sea level is "falling." In other words, because the land is rising faster than sea level is rising, sea level drops relatively, so the water appears to recede seaward.

Likewise, coastal areas having extensive construction and development may experience local subsidence, making the sea level rise relative to the land surface. Extensive pumping of underground oil or water can also contribute to local subsidence, as has happened in Long Beach, California, and along the Louisiana coast. Another problem with determining the rise or fall of sea level is that geological processes such as faulting and seafloor spreading also cause the land to rise or fall, affecting relative sea level.

So, while sea level in one area is falling, it is rising in another area, and determination is relative to the status of the adjacent land. A trend may be determinable locally, but the simultaneous trend worldwide is elusive and often confusing in direction. And there is another problem: tide gauges—devices that measure the height of sea level—have only recently been placed in many critical regions of the world.

Sea level rise could be caused by global warming, as a result not only of the melting of glaciers and snow fields, but also of the expansion of seawater as it is warmed. The past 100 years witnessed a sea-level rise of only about 0.10 to 0.15 m (about 4 to 6 in.). This modest rise, according to some experts, is due to the warming climatic conditions associated with the increasing greenhouse effect. Predictions and articles concerning sea-level rise due to global warming appeared in the late 1980s, suggesting that very dangerous times lie ahead for those living in the coastal environment.

The estimates were for a rise of up to 2 m (about 6.5 ft.) by the year 2050. Even more extreme suggestions were for complete melting of the ice caps of Greenland and Antarctica, leading to a catastrophic rise of about 60 m (about 190 ft.). However, more detailed investigations indicate that such dramatic increases are highly unlikely. More recent estimates are closer to a third of a meter (about a foot) over the same time interval.

Nevertheless, in extremely low-lying areas like Bangladesh, with its huge population living in simple housing on a vast delta, such a rise could cause serious coastal flooding. Bear in mind that small rises or falls may seem slow in our lifetime, but they can have significant effects over geologic time, or when large storms hit.

Remember, however, that some scientists question whether the increase in greenhouse gases will have such effects, or if the increase will be as great as predicted. Attempting to make an accurate prediction from sparse data and complicating factors (like volcanic eruptions; see Box 13-2, page 325) is a great challenge to oceanographers and meteorologists. Even if sea level were to remain stable, coastal erosion problems will continue.

Another global change problem, the depletion of atmospheric ozone (see page 329) may indirectly affect our use of the coastal zone. Recall that the ozone layer shields us from ultraviolet radiation, so with a decrease in ozone, the ultraviolet radiation reaching our beaches increases, and so does the risk of skin cancer. This concern has already reduced the appeal of sunbathing, a favorite activity of tourists who visit the beaches of the coastal zone.

Beaches

As you have seen, erosion is especially pronounced on beaches and coastal cliffs. Change is slower where the coast is rocky or where beach sediment is mainly gravel, because more energy is needed to erode rock or move gravel. Especially dramatic effects can occur along gently sloping sandy beaches, where the sand is moved easily by waves, currents, and storms.

People affect coastal and shoreline areas in three basic ways:

1. Recovery of land by dredging, filling, and/or damming of rivers

2. Construction of jetties, seawalls, or other coastal structures (see Figures 14-8 and 14-9)

3. Development or destruction of coastal dune areas.

Coastal dunes are a protective barrier to inland erosion. When they are removed for development or are broken up, increased erosion can occur (Figure 14-22). Dunes are easily damaged by vehicular traffic (such as trucks and dune buggies), footpaths, and roadways. Any of these activities disrupt the fragile plant structure that anchors the dune, and once a gap exists in the dune pattern, the dunes are quickly eroded by storms.

Estuaries and Marshes

Estuaries, salt marshes, and mangrove swamps are unique and extremely fragile environments that deserve special care. Marsh and estuary areas worldwide have been filled and developed with homes, factories, farms, or harbor facilities (Figures 14-23 and 1-1, page 3). In the United States, it has been estimated that the contiguous 48 states (excluding Hawaii and Alaska) have lost half their wetland areas, including those around rivers, since the Pilgrims arrived in 1620. The United States originally had about 215 million acres of wetlands, which has been gradually reduced to 100 million acres. The annual loss is about 458,000 acres, half the area of

Rhode Island. In some localities, as cities grow, the marshes are drained and built over and developed (Figure 14-24). In recent years public action against such development has slowed the destruction of these valuable environments.

Marshes and estuaries contain clams, oysters, scallops, and fish that may be important to the local economy. They may also have other important values. For example, marshes can remove some pollutants from the water, particularly nitrogen and some metals.

Human Activity in the Coastal Zone

The following types of coastal structures often used in combination, are commonly built to prevent erosion:

1. **Groins** or **jetties** are structures built out into the water, perpendicular to the beach, to interrupt the longshore movement of water and sand.

2. **Seawalls** are rigid structures built parallel to beaches to withstand and reduce the impact of incoming waves.

3. **Breakwaters** are massive structures built parallel to the shore but 100–300 m (330–1,000 ft.) seaward to absorb wave energy ("break the wave") before it reaches shore. Breakwaters produce a quiet zone behind them for anchorage.

4. **Revetments** (ri-VET-ment) are layers of strong, nonerodible material placed on the shore to prevent erosion.

5. **Artificial fill** is material added (often yearly) to replenish the beach, an alternative to preventing erosion.

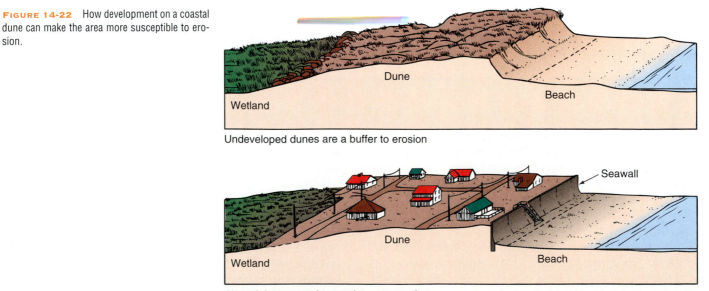

FIGURE 14-22 How development on a coastal dune can make the area more susceptible to erosion.

Undeveloped dunes are a buffer to erosion

Altered dune area—less resistant to erosion

FIGURE 14-23 Some examples of industrial development in coastal areas.
(a) On Cape Cod, Massachusetts, this marsh and estuarine area has been developed into a marina and residential area.
(b) The Ebro Delta in Spain has been developed for mariculture and rice growing.
(c) Venice, Italy, a city literally under sea level. (Photographs courtesy of Dr. Ivan Valiela.)

FIGURE 14-24 The development and filling in of Boston's Back Bay area from 1814 to 1976. In 1814, the main part of Boston was to the northeast. By 1836, areas were filled to support railroad tracks over Roxbury Flats. Today almost all the marsh and shore area are gone. (From Valiela and Vince, 1976.)

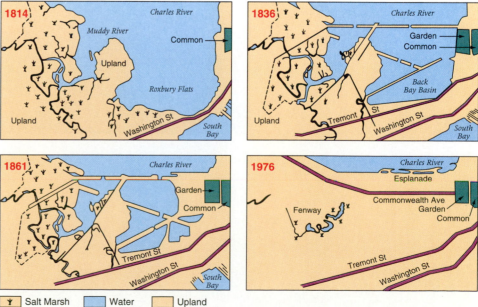

The problem with coastal structures is that they can interfere with water and sediment circulation in the nearshore zone, increasing deposition in some locations and erosion elsewhere. These structures can interfere with longshore transportation of sediment and prevent replenishment of beaches, generally resulting in erosion on the downstream side of the structure and accretion on the upstream side (see Figure 14-8).

The effect can be compounded by a series of structures that develops because the owner of property downstream is "forced" to subdue the now-increased erosion of his or her beach. It is difficult to visualize the volume of sand that may move along a beach, but many ocean beaches have 0.5 to 1 million m³ (17.5 to 35 million ft.³) of sand transported across them in an average year.

Some innovative structures have been used to diminish and dampen the effect of waves that cause coastal erosion. For example, there are floating breakwaters that adjust their orientation with changing currents or winds. Such devices could shelter even large vessels, although none that size have yet been built.

One especially innovative idea came from a group of scientists at Scripps Institution of Oceanography, led by Professor John Isaacs. His concept was simple: ocean waves would lose much of their energy if they met rows of spherical floats attached to the seafloor but floating close to the surface (Figure 14-25). The floats behave somewhat like upside-down pendulums that have a back-and-forth movement faster than that of the incoming waves. Thus, the floats move in the opposite direction to the waves and dissipate much of their energy. As Isaacs said, "The system beats the waves to death."

Preliminary tests showed dissipation of wave energy up to 50 percent at a small fraction of the cost of conventional breakwaters. Other advantages are that the system can be portable and apparently does not interfere much with the movement of sediment.

Uses of the Coastal Zone

Many uses of the coastal zone are possible. Unfortunately, one use often impacts on another. For example, consider these conflicting uses:

- Developing offshore or nearshore oil terminals versus protecting the shore zone for recreation and preserving its natural biological resources from pollution

- Developing waterfront homes or larger buildings that destroy, damage, or change the beachfront area versus preserving this area for its aesthetic value

Some believe that a multiple-use approach to the coastal zone can be achieved. If so, it will require considerable knowledgeable management, skill, and compromises among the varied interests.

Coastal Zone Facts

To better understand the demands on the coastal zone, consider the following facts. The specific figures are for the United States, but many aspects apply to other coastal areas in the world.

FIGURE 14-25 Part of a model of a tethered-float breakwater anchored in San Diego Bay. The tethered floating balls dissipate the energy of wind-generated or boat-generated waves. (See text for a description of how these floating balls "beat the waves to death.") Official U.S. Navy photograph.

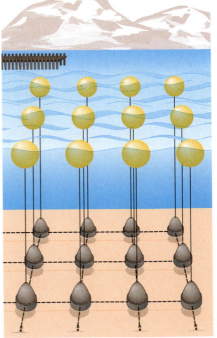

1. The U.S. coastal area is only 11 percent of our total land area (excluding Alaska), yet about 31 percent of U.S. gross national product (GNP) originates along our coasts.

2. Approximately 75 percent of the U.S. population lives within 80 km (50 miles) of the coastal region, if the Great Lakes are included. The population density in coastal regions is about 4 times that of inland areas.

3. The 48 contiguous states have about 96,000 km (about 60,000 miles) of shoreline, but only about one-third presently has recreational potential. Only about 6.4 percent of that is publicly owned; the rest is private. Over 100 million U.S. residents participate annually in ocean-related activities.

4. Seven of the largest U.S. cities are on the seacoast or on the shores of the Great Lakes. Over 33 percent of the population lives in coastal counties, which contain only about 15 percent of total U.S. land area. Many of the world's largest cities are in coastal areas (Tokyo, New York, London, Shanghai, and Osaka, for example).

5. Shoreline property has continually increased in value, usually at more than 10 percent a year. In many areas, shorefront property prices favor availability only to the wealthy. Clearly, since the demand is great and the supply is limited, land prices will not decrease.

6. About 40 percent of U.S. manufacturing takes place in coastal counties. Coastal facilities handle about 350 million tons of foreign trade and 630 million tons of domestic cargo.

7. Since 1970, about 50 percent of U.S. construction has occurred in coastal counties.

8. Numerous power plants, including nuclear facilities, and oil refineries are on or near the coast.

9. Coastal developments have already destroyed or built over a large percentage (about 50 percent) of our coastal habitat areas.

10. A very large percentage of commercial fish spend a major portion of their life cycle in the coastal zone. Some East Coast shellfish, such as clams, oysters, and mussels, spend their entire life cycle in the more restricted waters of estuaries, bays, and marshes. Worldwide, the coastal regions provide about 90 percent of the fishing harvest.

11. Daily waste disposal from municipal and industrial activity is about 30 billion gallons. Other solid wastes are about 100 million tons per year.

12. Coastal waters receive the major portion of material from land, including sewage, radioactive and thermal discharge, and dredging, mining, and construction wastes.

13. At most times, one-third of U.S. shellfishing beds are closed to commercial or recreational harvest due to pollution problems.

Some authorities feel that the destruction of the coastal zone in many parts of the world has proceeded so far that it can never be returned to its original state. A more positive approach is to minimize pollution and destruction and to make choices having long-term benefits for as many people as possible. This may require trade-offs, such as having a nuclear power plant in one place on the coast and an undeveloped coastal park in another.

In some instances, decisions concerning the coastal zone have been postponed for the next generation to deal with, or plainly have been wrong choices. Among the reasons are conflicting jurisdiction and lack of coordination among government agencies, lack of clear development plans, lack of knowledge and data about the resources and marine interactions of the coastal zone, lack of funds to manage the coastal zone, and conflict between economic uses and ecological uses. Local governments sometimes choose to see only the local view and assume the national government will consider the broader picture (see Box 14-3).

Box 14-3
The U.S. Approach to Managing the Coastal Zone

The United States has taken several legislative steps to protect its coastal zone. The first occurred in 1972, when Congress passed the National Coastal Zone Management Act, promoting a strong role for individual states in managing their coastal zones. The Office of Coastal Zone Management provides federal money to states that develop a comprehensive plan for using their coastal zones. At present 29 states participate, covering about 95 percent of the U.S. coastline.

To start a coastal zone management program, a state defines its coastline, generally extending 5.5 km (about 3 miles) seaward from its coast, including intertidal areas, beaches, dunes, and wetlands. The state designates areas of concern, such as oil and gas, envi-

continued

Box 14-3 continued

ronmental problems, and recreation potential. It then develops policies to guide use of its coastal zone. Public hearings are held, the program is approved by the state's governor, and a state agency is created to administer the program.

Three other important U.S. coastal zone programs have been developed:

- The Marine Protection, Research, and Sanctuaries Act permits the federal government to designate national marine sanctuaries from the high-tide line to the edge of the continental shelf. An example is the site where the *U.S.S. Monitor* sank off Cape Hatteras, North Carolina, during the Civil War (see Figure 17-5, page 424). Long-term management is planned for each sanctuary area. Numerous sanctuaries have been added or are proposed (Figure 1).
- The National Estuarine Research Reserve System, a similar program, covers estuaries, protecting several hundred thousand acres of estuarine waters and

marshes. It establishes field stations and supports environmental research programs.
- The Fishery Conservation and Management Act gives the United States a procedure for conserving and managing its coastal fish resources. Fishery Management Councils develop management plans for their geographical areas.

Some legislation may not be beneficial for the coastal zone. An example is the Federal Flood Plain Insurance Program, which compensates individuals whose homes are damaged or destroyed by coastal flooding or storms. Some feel that the availability of such insurance encourages extensive development on barrier beaches because of the knowledge that damage will be compensated. However, the Coastal Barrier Improvement Act, a more recent piece of legislation, restricts the use of federal funds that can be used to insure development on designated barrier islands. At present, close to 788,000 acres are covered by this act.

The National Marine Sanctuary Program and the National Estuarine Research Reserve System

FIGURE 1 Areas designated or proposed as *National Marine Sanctuaries* or *National Estuarine Research Reserves* in the United States. These programs were established to provide long-term protection and management for special ecological, historical, recreational, and aesthetic areas. (Illustration by Jayne Doucette, Woods Hole Oceanographic Institution.)

SUMMARY

The coastal zone, which is situated at the boundary of sea and land, is the most important portion of the ocean to many people. In the United States about 75 percent of the population lives within a short drive of the ocean or one of the Great Lakes.

The coastal zone—broadly defined as that part of the ocean affected by land and that part of land affected by the ocean—is an area of many complex biological, geological, physical, and chemical interactions. Many marine features of the coastal zone are temporary, being controlled by sea level, wave height, and tide conditions. The recent rise in sea level has caused or modified many coastal zone features, such as estuaries, deltas, and barrier beaches.

Although attractive in a recreational sense, the coastal zone is also the site of considerable urbanization. Thus, many often-conflicting uses coexist in this environment. Development, both industrial and residential, has damaged or removed large portions of the land and parts of the coastal zone from public use. In recent years, control by individual states and the federal government has resulted in improved management plans for coastal regions in the United States.

QUESTIONS

1. Describe the key features of the coastal zone.

2. Why are beaches generally an area of considerable change?

3. How do waves and beaches affect one another?

4. Why is coastal erosion such a worldwide problem?

5. Why do the physical properties of seawater such as temperature and salinity vary more in coastal waters than in the deep sea?

6. What would happen to coastal features such as estuaries, marshes, and beaches if the sea level were to rise?

7. What are the different types of estuaries and what is their cause?

8. What are the important factors causing coastal zone and beach erosion and what methods are used to prevent or reduce this erosion?

9. Why are estuaries such a unique environment?

KEY TERMS

artificial fill	distributary channel	mangrove tree	salt marsh
barrier beach	estuary	nearshore region	salt-wedge estuary
barrier island	fjord	partially mixed estuary	seawall
beach	floodplain deposit	peat	shoreline
breaching	groin	refraction	shore zone
breakwater	jetty	reverse estuary	well-mixed estuary
coastal zone	lagoon	revetment	
delta	longshore current	rip current	

FURTHER READING

Aubrey, D. G. 1980–81. "Our Dynamic Coastline." *Oceanus* 23, no. 4, p. 413. Readable article on the processes that influence the coastline.

Bascom, W. 1980. *Waves and Beaches: The Dynamics off the Ocean Surface*. Garden City, N.Y: Anchor Press/Doubleday. Popular and readable pocket book on coastal processes.

"Coastal Science and Policy." *Oceanus* 36, nos. 1–2 (1993). A two-issue series devoted to various coastal-zone problems.

Emery, K. O., and D. A. Aubrey. 1991. *Sea Levels, Land Levels and Tide Gauges*. New York: Springer Verlag. A little complex, but provides extensive information about sea level and its recent change around the world.

Flanagan, R. 1993. "Beaches on the Brink." *Earth* 2, no. 6, pp. 24–33. A very readable discussion of the effects of rising sea level on beaches, and what, if anything, should be done.

Fox, W.T. 1991. *At the Sea's Edge: An Introduction to Coastal Oceanography for the Amateur Naturalist*. New York: Prentice-Hall. A good and uncomplicated introduction to coastal marine science.

Horton, Tom, and W. M. Eichbaum. 1991. *Turning the Tide: Saving the Chesapeake Bay*. Washington, D.C.: Island Press. A discussion of some of the environmental problems of one of the most important estuaries in the United States.

Kaufman, D. G., and C. M. Franz. 1993. *Biosphere 2000: Protecting Our Global Environment*. New York: HarperCollins. A good, readable discussion of our many environmental problems.

Kaufman, W., and O. Pilkey. 1979. *The Beaches are Moving*. Garden City, N.Y: Anchor Press/Doubleday. A classic book that called attention to the fragile aspect of beaches.

Shepard, F. P., and H. R. Wanless. 1971. *Our Changing Coastline*. New York: McGraw-Hill. Although over 20 years old, this book gives an excellent description of the U.S. coastline and how it has been formed and modified.

Teal, J., and M. Teal. 1969. *Life and Death of a Salt Marsh*. Boston: Little, Brown. A very pleasant description of the history of a salt marsh.

CHAPTER 15

Resources of the Ocean

The preceding chapters emphasized the scientific aspects of oceanography, at times revealing how little we know about the ocean. Our knowledge of potential ocean resources is similarly scant; some were barely known only a decade ago. It also is probable that some resources have yet to be discovered.

Today, we face urgent questions of how and what biological, mineral, and energy resources of the ocean can be used without environmental damage, and at what rate. Unfortunately, these questions often are addressed without benefit of adequate scientific study.

To be expected in the coming years are new resource development methods involving aquaculture and innovative energy sources from the ocean. Their economics, however, are uncertain.

The Law of the Sea Treaty is an important related development (see also further discussion in Chapter 19), allowing each country to claim and control marine resources in its Exclusive Economic Zone, some 200 nautical miles out from shore. Most coastal nations now exercise some kind of control, encouraging exploitation of their marine resources.

Sophisticated deep-water, semi-submersible, free-floating drilling rig. These types of vessels can work in water depths of about 600 m (2,000 ft.) and drill to depths of 7,600 m (25,000 ft.) below the seafloor.

(Photograph courtesy of National Oceanic and Atmospheric Administration.)

Introduction

This chapter looks at ocean resources. We consider them in three categories: renewable, nonrenewable, and physical.

- **Renewable resources** are continually replaced by the reproduction and growth of plants and animals, usually on a yearly or seasonal basis. The ocean's renewable resource is its *biological productivity.*

- **Nonrenewable resources** are present in fixed amounts, or are replaced at geologically slow rates, and so cannot be replenished during a useful time interval. The ocean's nonrenewable resources include most mineral deposits, including oil, gas, and coal.

- **Physical resources** are the ways in which we use the physical aspects of the ocean, such as for transportation and recreation.

The renewable biological resources of the ocean—such as fish, clams, and lobsters—have been harvested by people from before recorded history, judging from archeological excavations. Understand that the term *renewable* does not imply that a biological resource can *never* be depleted. A renewable resource can in fact be depleted by overharvesting, by poor management, or through environmental damage. Harvesting of nonrenewable mineral resources is a more modern development. (An exception is the ancient practice of obtaining salt from evaporated seawater.)

Any use of the ocean must consider the potential environmental damage. Interestingly, resource mining in the marine environment is sometimes *less* harmful than the corresponding operation on land. With proper care,

the resources described in this chapter may be recoverable with minimal environmental effect. As is true in most human undertakings, it is the poorly planned and badly researched activities that cause environmental harm.

In this chapter we will look at the three categories of ocean resources, beginning with mineral resources. We will consider them by where they occur: the continental margin (including the coastal zone) and the deep sea (Figure 15-1).

Mineral Resources of the Continental Margin

Recently, marine scientists, industry, and governments have displayed interest in the ocean's mineral wealth (Table 15-1). Value estimates are staggering—billions and even trillions of dollars. For some resources these estimates are ridiculously high; for others they may prove conservative. It is important to note that many industrialized countries like the United States are *deficient* in most of the potential marine resources listed in Table 15-1.

Based on present use, the most valuable marine mineral resources are hydrocarbons—oil and gas—which are concentrated within the sediments and rocks of the continental shelf and probably in the outer parts of the continental margin.

A major difficulty in assessing marine mineral resources is that they are underwater, and therefore less accessible than similar resources on land. Conventional techniques of exploration and evaluation cannot be

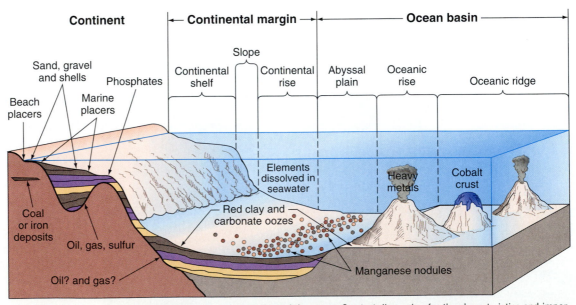

FIGURE 15-1 Potential mineral resources of the ocean. See text discussion for the characteristics and importance of the various potential resources shown.

TABLE 15-1
Possible Marine Mineral Resources and Their Sources

Mineral Resource	Principal Sources
Boron, bromine, calcium, magnesium, potassium, sodium, sulfur, and uranium	Seawater
Sand, shells and gravel, phosphorite, glauconite, and heavy minerals (magnetite, rutile, zircon, cassiterite, chromate, monazite, gold)	Sediment (continental shelf and slope)
Oil, gas, and sulfur	Subsurface (continental shelf, slope, and rise)
Copper, lead, silver, and zinc	Heavy-metal muds and nodules
Zinc, iron, copper, silver, molybdenum, chromium, gold, and platinum	Thermal vents, and ocean ridges
Nickel, cobalt, and maganese	Nodules and crusts

used. Thus, exploitation of some marine resources must await technological advance. Some resources may never be developed because of transportation cost, or the challenge of preventing pollution, or because similar land resources are cheaper.

Sediment Deposits

Most mineral resources on the continental margin are related to three important factors:

1. The large amount of sediment present

2. The reworking effect of waves and currents, combined with sea-level changes due to glaciations

3. Development of the continental margin throughout geologic time, including the effects of seafloor spreading.

Most continental margins have a large thickness of sediment on their continental shelves. Much less sediment exists on the continental slope, but if a continental rise is present, it has a very thick sediment mass. These accumulations are due to the large amount of material eroded from land and carried to the ocean by rivers, wind, glaciers, and coastal erosion.

In some areas, this sediment mass has been accumulating for up to 200 million years and approaches 20 km (about 12.5 miles) in thickness. This thickness, combined with organic material from plants and animals in the water or from land, eventually can create oil and gas deposits.

Sea level has risen and fallen several times over the past million years in response to extensive periods of glaciation and subsequent melting. The most recent glaciation reached its maximum about 18,000 years ago (see Figure 2-5, page 31). During that time, sea level dropped as much as 130 m (426 ft.) as water was removed from the ocean to form large glaciers. As the glaciers melted, the water returned to the ocean and sea level slowly rose.

The shoreline would migrate seaward as the water level fell, then return landward as the glaciers melted. The ocean is now covering the previously exposed portion of the continental shelf. As the shoreline migrated, so did the beaches, marshes, and other nearshore features. Where the shoreline movement was fast enough, the beach was left behind, leaving an ancient submerged beach or ridge.

Breaking waves and nearshore currents occur along all shorelines (see Figures 10-8 and 10-9, pages 228 and 229) and can remove fine particles from the sediment and carry them seaward, leaving behind a coarser-grained sand or beach deposit. If the continental shelf is narrow and has a modest slope, reworking by waves and currents is more intense. This results in beach sands that are well-sorted, meaning that the sediment particles are mostly of the same size (Figure 15-2). This process of a rising and falling sea level helps form two types of mineral accumulations: placers and deposits of sand and gravel.

Placers

If sea level pauses at a particular level for a significant period, the lighter-weight particles can be carried away by wave and current activity. This leaves behind a deposit of particles enriched in heavier minerals, called a **placer** (PLASS-er). Minerals typically occurring in placers include gold, zircon, rutile (a source of titanium), cassiterite (a source of tin), and platinum.

Several of these minerals are rare and important, with substantial value: at this writing, platinum and gold were valued at slightly less than $400 per ounce. Marine placers are mined in several offshore areas, such as Alaska for gold. Marine placer mining should become more important as less-expensive land mines become depleted.

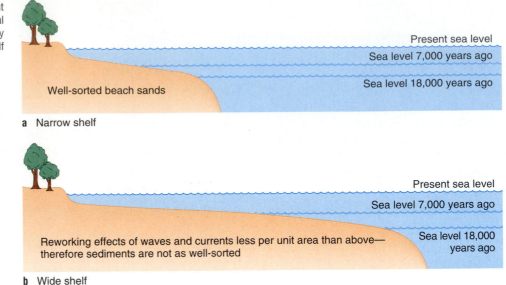

FIGURE 15-2 The slowly rising sea of recent millennia reworks the sediment on the continental shelf. Note the different effect of reworking by waves and currents on a narrow continental shelf (a) and a wide shelf (b).

Reworking of the sediment is especially effective and common on beaches, so they are often the site of placers of economic value. Ancient beaches, formed during a lower or higher stand of the sea, can also be enriched in heavy minerals. Because many beaches are better-used as recreational areas, they are rarely mined. It is more acceptable to mine ancient offshore beaches, especially for their sand and gravel, which is used in the building industry. The more common minerals found in placers are listed in Table 15-2.

TABLE 15-2

Mineral Deposits on the Continental Margin Seafloor

Mineral Resource	Use	Possible Mineral Marine Areas[1]	Value[2]
Marine placers			
Gold	Jewelry, electronics	Alaska, Oregon, California, Philippines, Australia	about $400/oz.
Platinum	Jewelry, industry	Alaska	about $400/oz.
Magnetite	Iron ore	Black Sea, Russia, Japan, Philippines	$6–11/ton
Ilmenite	Source of titanium	Baltic, Russia, Australia	$25–35/ton
Zircon	Source of zirconium	Black Sea, Baltic, Australia	$45/ton
Rutile	Source of titanium	Australia, Russia.	$100/ton
Cassiterite	Source of tin	Malaysia, Thailand, Indonesia, Australia, England, Russia	
Monazite	Source of rare-earth elements	Australia, United States	$170/ton
Chromite	Source of chromium	Australia	$25/ton
Sand and gravel	Construction	Most continental shelves	
Calcium carbonate (aragonite)	Construction cement, agriculture	Bahamas, Iceland, southeastern United States	
Barite	Drilling mud, glass, paint	Alaska	
Diamonds	Jewelry, industry	Southwest Africa	
Phosphorite	Fertilizer	United States, Japan, Australia, Spain, South America, South Africa, India, Mexico, New Zealand	$6–12/ton
Glauconite	Source of potassium fertilizer		
Potash	Source of potassium	England, Alaska	

[1]Not necessarily including all areas, because exploration has been minimal or nonexistent in many localities.

[2]Data are from various sources and may not reflect current economic value.

Source: Ross, 1980, updated.

Sand and Gravel

After oil and gas, sand and gravel are the second most valuable mineral resources presently recovered from the ocean. One might think these mundane materials have no great value, but consider that a concrete basement measuring 9 by 12 m (about 30 by 40 ft.) requires 80 tons of sand and gravel aggregate. One km (0.6 mile) of a four-lane highway can use 40,000 to 60,000 tons of mineral aggregate. Large amounts of sand are also used to replenish beaches that are being eroded (see Figure 14-14 page 345).

The extensive occurrence of sand and gravel on almost all shelves, their ease of mining, and their importance to the construction industry have made this a most valuable resource. As much as 3 billion tons will be mined worldwide each year from land and marine sources by the year 2000.

The value of sand and gravel will undoubtedly increase as present land resources are consumed or become covered by extensive building in coastal areas. Transportation is a major cost in sand and gravel mining This makes the vast deposits in the marine environment, which are frequently close to industrial centers, more appealing. Although little marine sand and gravel is presently mined in the United States, 14 percent of the United Kingdom's supply comes from the sea.

Offshore sand and gravel mining can cause environmental problems, including beach erosion, since some offshore sand may be involved in the seasonal beach cycle. Mining can also disrupt ecosystems, because dredging (Figure 15-3) can introduce large quantities of suspended material into the water, with damaging environmental effects. The mining process can also destroy or disturb the physical characteristics of bottom habitats, affecting benthic organisms and other creatures who use or feed on the bottom. Yet marine mining is sometimes less damaging to the environment than the mining of land quarries near residential areas.

Phosphates

In some nearshore areas, phosphorite rocks and phosphorite-rich sands are exposed on the seafloor. These can be used for fertilizer. Marine phosphorite mining is active or is being considered in the offshore regions of New Zealand, Mexico, and the southeastern U.S.

Most phosphorite deposits, however, have not been mined due to water depth, difficult recovery, transportation cost, environmental concerns, and more-

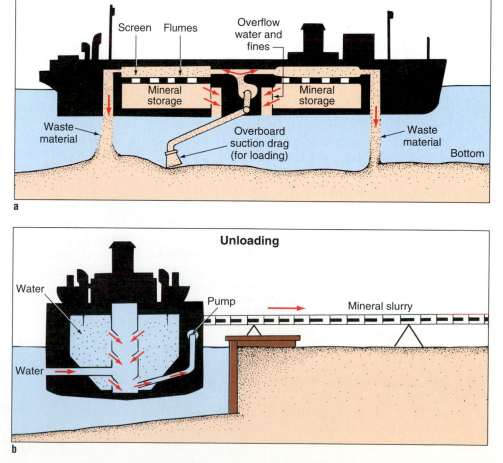

FIGURE 15-3 A typical seafloor dredging operation: (a) cross section of a dredge used in mining sand and gravel; (b) the unloading procedure. (Adapted from Construction Aggregates Corporation, Chicago.)

accessible land sources. Marine phosphorite may become more commonly mined if demand increases and the cost of land-harvested fertilizers increases. Marine phosphates could supply world demand for several hundred years.

Phosphorite deposits occur where upwelling brings to the surface cold, nutrient-rich waters containing dissolved phosphate. As these waters are warmed and their pH increases, phosphate is precipitated. Rich accumulations occur where the phosphate is not diluted by land-derived sediments. Especially favorable areas for phosphate deposition are topographic highs, such as seamounts or hills.

Other Minerals

Other minerals mined from the continental margin include calcium carbonate and barite. A pure form of calcium carbonate, called aragonite, has been mined from nearshore deposits in the Bahamas. It can be a source of lime for agriculture or used for cement.

Barite is mined off the Alaskan coast and is used by the petroleum industry as a major ingredient in drilling mud, which is used to lubricate the turning drill pipe, to flush debris from the drill hole, and to maintain pressure on the drilled rock. Another mineral that could be mined is glauconite, a potassium-rich clay mineral that is a potential potassium source for fertilizers.

In the past, diamonds were mined off the coast of South Africa. However, the operation was hampered by stormy sea conditions and small diamond concentrations, and was not profitable.

Seawater as a Resource

Freshwater is in such short supply in many localities, especially tropical islands and desert areas, that a bottle of drinking water can cost more than the same amount of gasoline. Freshwater can be obtained from seawater by salt removal, or **desalination** (see pages 454–455), so seawater itself can be considered as a marine resource. Some countries, such as Saudi Arabia, are very dependent on freshwater produced by desalination.

One difficulty with desalination is that most processes yield a hot, salty residue. If released into a restricted coastal environment, it can damage the nearshore ecology by increasing the salinity of the water. Because of environmental and cost problems, some alternatives have been suggested.

Perhaps the most imaginative is to tow polar icebergs to areas where freshwater is needed. (Their salinity is only about 5‰ or less.) Another idea is to condense water vapor from the atmosphere in the tropics by cooling it with cold ocean water, somewhat like a "sweating" cold drink glass. (Both ideas are discussed further in Chapter 18.)

Seawater contains nearly all of the chemical elements. Currently, however, only three substances are being "mined" from seawater for economic use—common salt (NaCl), bromine, and magnesium. (Common products from magnesium are milk of magnesia, Kaopectate, and Pepto Bismol; bromine is used as an antiknock compound in gasoline.) Salt is obtained by evaporating the seawater, and during evaporation some useful compounds of potassium, magnesium, carbonate, and bromine form and are extracted. About 50 percent of the world's magnesium comes from seawater. Other elements could be recovered with the appropriate technology and economic potential.

The ocean's volume is immense: $1,350 \times 10^6$ km³ or 318×10^6 miles³. Earth's vast salty reservoir contains a virtually inexhaustible supply of many elements. As explained earlier, seawater's average salinity is 35‰. In other words, 3.5 percent of seawater is not water, but various elements in solution. A cubic mile of seawater weighs about 4.7×10^9 tons. Because 3.5 percent of it is dissolved elements, each cubic mile contains about 165×10^6 tons of dissolved elements, an enormous quantity.

Worldwide, the ocean contains over 5×10^9 tons of uranium and copper, 500×10^6 tons of silver, and perhaps 10×10^6 tons of gold. For gold alone, this works out to about 4 pounds for every person on earth, a value of $23,000 per person at $400 per ounce.

These estimates are interesting, but are not very realistic when you consider the cost of extracting these elements. For example, the average concentration of gold ranges from 4×10^{-7} to 6×10^{-7} mg per liter, or about 50 pounds in a cubic mile of water. Even at today's high price, this is only about 0.001¢ per ton of seawater. However, other elements like bromine and magnesium, with values of only $0.02 and $1.00 per ton of seawater, can be extracted more easily and thus can be profitably recovered.

Oil, Gas, and Sulfur Deposits

Oil and gas are the most valuable marine resources. The yearly value obtained from the marine environment is about $110 billion for oil and $23 billion for gas (numbers can vary due to fluctuating prices), several times higher than the present value of marine biological resources.

PETROLEUM. Petroleum (an inclusive term for oil and gas) originates from the organic remains of the animals and plants that once lived in the sea, in rivers, or in lakes. About 95 percent of the world's oil occurs in sediments that were originally deposited in a marine environment. After death, the plant and animal remains settled to the bottom and were buried under sediments. The actual conversion process whereby this organic material changes into petroleum is extremely complex and not completely understood.

Some facts are known, however. If the organic material decays or is oxidized, it will not form petroleum. If the basin where the material settles has limited oxygen, the organic material can be preserved and accumulate. It can also be preserved if the sedimentation rate is sufficient to bury the organic material before it can become oxidized.

Once the organic material reaches the bottom, conversion starts when bacteria or other organisms digest the organic material and redeposit it as waste material. As the sediments are buried, further chemical change occurs, influenced largely by the heat and pressures associated with burial. The time needed for oil or gas to form is unknown. However, they rarely occur in rocks less than 2 or 3 million years old. The conditions necessary for an oil or gas deposit are summarized in Table 15-3.

PETROLEUM EXPLORATION. When exploring for oil or gas deposits, geologists seek a thick sequence of marine sediments that includes occasional porous layers. These have great potential and are called **reservoir beds** (Figure 15-4). Ancient coral reefs also have good reservoir potential, and many have hydrocarbon deposits. Identifying these reservoir beds is a first step. Drilling is always necessary to determine whether oil or gas are actually present.

Offshore drilling is conducted from a drilling platform, which can be stationary, semifixed (movable after drilling), or can float free relative to the seafloor. Floating platforms (see the opening figure to this chap-

---

TABLE 15-3
Essentials for Favorable Oil and Gas Prospects

1. A sufficient source of the proper organic matter
2. Favorable conditions for preserving organic matter, such as rapid burial or a reducing environment
3. An adequate blanket of sediments to produce the necessary temperatures and pressures for conversion of organic matter to fluid petroleum
4. Favorable conditions for movement of oil and gas from the source rocks to porous and permeable reservoir rocks
5. Presence of petroleum accumulation traps, either structural or stratigraphic (see Figure 15-4)
6. Adequate cover or cap rocks to prevent loss of oil and gas
7. Proper timing of these essentials for accumulation
8. A postaccumulation history favorable for preservation

Source: Adapted from Hedberg and others, 1979.

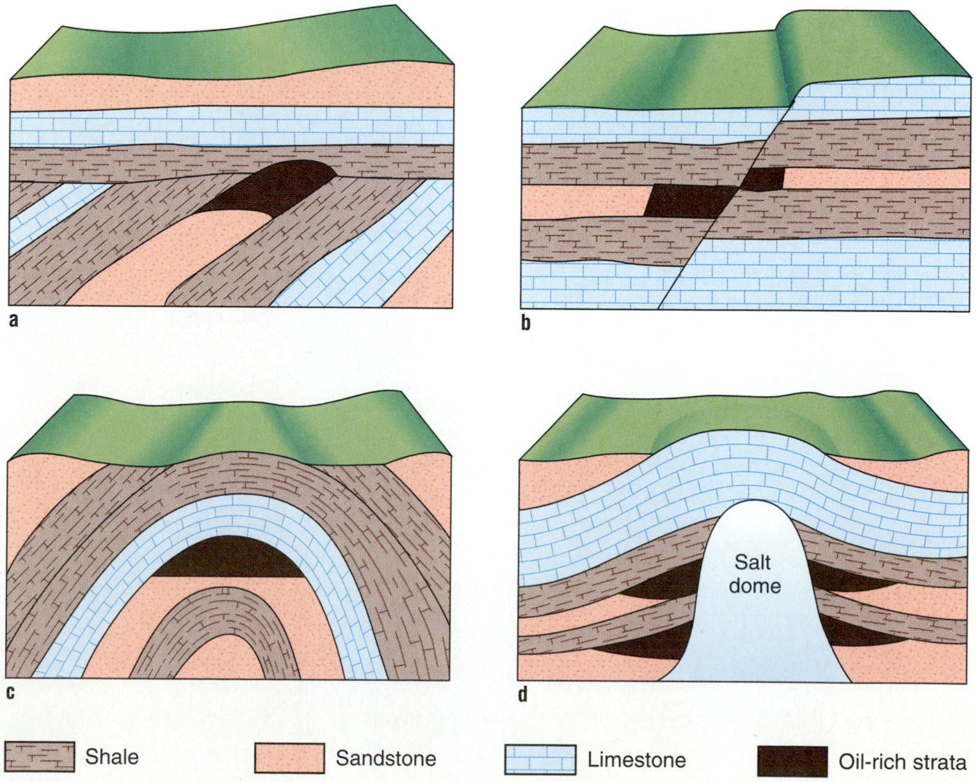

---

ter) are suited for deeper water because their position over the drill site is maintained by reference to subsurface buoys (see Figure 5-7a, page 113). A limitation of ocean drilling is the greater cost compared to drilling on land. A well drilled in 200 m (660 ft.) of water can easily cost up to $5 million.

MARINE PETROLEUM POTENTIAL. The oil and gas potential of the marine environment could be very large compared to the remaining petroleum resource on land,

FIGURE 15-4 Different types of oil and gas traps (a "trap" is a barrier that prevents upward movement of oil or gas, thus causing them to accumulate): (a) stratigraphic (rock strata) trap; (b) structural trap (a fault, in this case); (c) anticline (rocks bent upward in center); (d) salt dome, or diapir (see text). Traps of these types occur both on the continents and on the continental margin (see Figure 15-1). Note that the petroleum-bearing rocks are all shown as sandstone, because this is the most common situation. Other rock types can also contain hydrocarbons. Shale is shown as a typical rock that prevents upward migration of the oil or gas. Other rocks can also block migration.

Shale   Sandstone   Limestone   Oil-rich strata

since the continents are mostly explored and promising areas already have been drilled, but ocean exploration has so far been restricted to shallow water.

Exploration has already occurred off the coast of most countries having marine boundaries. Especially favorable for oil or gas are deltas, where thick sediments extend from land into the offshore area, and other localities having thick sedimentary sequences. Several favorable areas off the U.S. coast have yet to be fully explored (Figure 15-5). A new type of potential marine resource, called *gas hydrates,* is discussed in Chapter 18, pages 453–454.

A major future petroleum source may be the continental rise. Where present, it presents a thick wedge of sediment at the base of the continental slope. The distribution of continental rises (see Figure 4-12, page 88) is strongly controlled by seafloor spreading (see next section). Where oceanic plates are thrust beneath continents, rises do not form because the sediment is removed. Continental rises and other marine sedimentary basins cover about 19 million km² (7.3 million miles²), equal to about 33 percent of the total area of the continental shelf and slope.

DIAPIRS. Salt domes, or **diapirs,** occasionally occur beneath the sediments of the continental shelf or continental rise (Figures 15-4 and 15-6). The salt was deposited when the seawater in the basin evaporated. Later, after burial, the lighter salt may flow up and through the overlying heavier sediments. When this happens, it generally moves upward as a long plug, forming a domelike structure. Hence it is called a salt dome, or diapir (DIE-uh-peer, a Greek word meaning "to pierce"). Diapirs are extremely common in the Gulf of Mexico and Persian Gulf and have formed traps for much of the oil recovered from these areas.

SULFUR. As the salt forming the diapir nears the surface, some of it dissolves, and a **cap rock** residue of less-soluble anhydrite remains. Anhydrite can react with petroleum and bacteria to form hydrogen sulfide gas, water, and the mineral calcite (calcium carbonate). If oxygen then interacts with the hydrogen sulfide, pure sulfur can become concentrated within the cap rock.

The sulfur can be recovered fairly simply by the Frasch Process, in which superheated water and air are

FIGURE 15-5 Regions having good hydrocarbon potential in the U.S. Exclusive Economic Zone (up to 200 nautical miles offshore). (Adopted from B. A. McGregor and M. Lockwood, *Mapping and Research in the Exclusive Economic Zone,* (Reston, Va.: U.S. Geological Survey Publication, 1985).)

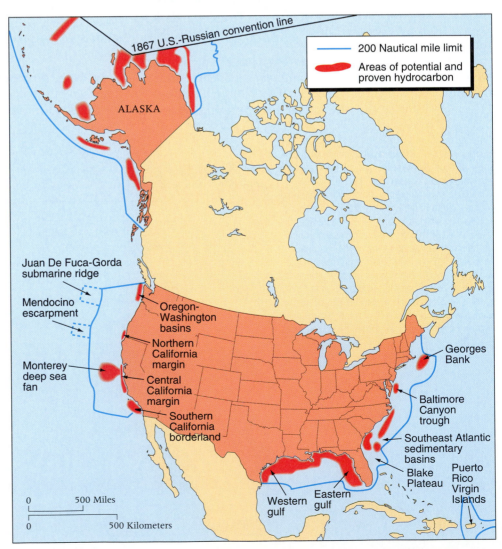

FIGURE 15-6 Seismic reflection profile showing several salt domes, or diapirs, off the coast of Angola in western Africa (compare to Figure 15-4d). The left scale is reflection time; 1 second = about 1 km (0.6 mile) depth. The top scale is the time aboard the moving ship as the record was made; an hour essentially = 15 km (8 nautical miles) of travel. (Photograph courtesy of K. O. Emery.)

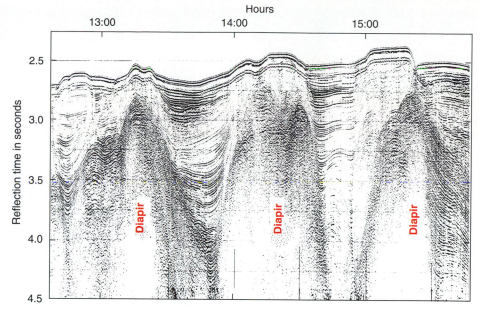

piped down the drill hole. This melts the sulfur, and the liquid sulfur and water are pumped to the surface. Salt domes off Louisiana in the Gulf of Mexico and in the eastern Mediterranean produce about 5 percent of the world's sulfur supply.

Mineral Resources of the Deep Sea

The deep seafloor includes the main ocean basins, trenches, seamounts, and oceanic ridges, comprising about 80 percent of the ocean floor and covering about 287 million km² (about 110 million miles²), or about four times the area of the continental margin.

The resource potential of the deep sea is much less known than that of the continental margin, simply because it is less accessible. However, potential deep-sea resources have captured the public's attention more than the mundane sand, gravel, oil, and gas deposits of the continental shelf. For many years, manganese nodules were the most attractive potential resource. More recently, the ocean-ridge vents and cobalt crusts on seamounts have become more attractive.

It is important to note that *no deposits have yet been commercially mined from the deep sea*. The reasons are apparent: the lack of technology for deep-sea mining; the danger of working in deep water; transportation cost; and cheaper availability of the same minerals on land. Nevertheless, economic potential of deep-sea resources continues to be an exciting prospect for the future.

Seafloor Spreading and Mineral Resources

To quickly review the seafloor spreading theory: new seafloor forms along the mid-ocean ridges and is con-

veyed slowly away from the ridge in opposite directions, forming a continually spreading ocean floor. Eventually, after millions of years, the seafloor is either subducted at the edge of a continent or becomes accreted onto the continent (Figure 15-7; see also Chapter 3).

Vents along the mid-ocean ridge emit hot, mineral-rich water (Figure 15-8). This can produce mineralization, concentrating numerous metals: copper, zinc, iron, silver, gold, nickel, vanadium, lead, chromium, cobalt, manganese, and others. A dramatic example is heavy-metal deposits in the Red Sea and in ocean-ridge vents (described in later sections). Ancient mid-ocean vents have been found in some land areas. One, near Thunder Bay, Ontario, formed the largest gold deposit ever found in North America.

We have only rudimentary knowledge of how these minerals form. Most mineralization occurs at the ridge axis, but some are found in other areas of the seafloor, because, with time, mineral-rich material spreads away from the ridge crest across the ocean floor (Figure 15-7b).

As spreading seafloor approaches a subduction zone, mineral enrichment also may occur. This is caused by movement of deeply occurring hydrothermal fluids (hot water enriched in various minerals) into the overlying rocks and sediments. This process is poorly understood, but prospecting has increased on land where ancient subduction occurred.

In some places the ocean crust and continents have moved as a unit, without developing a subduction zone. A good example is along the eastern margins of North and South America and the western margin of Africa. Consequently the continental margins in this area, especially the rises, have been available for sediment accumulation for the past 200 million years. These areas have very thick sediment sequences and therefore potential

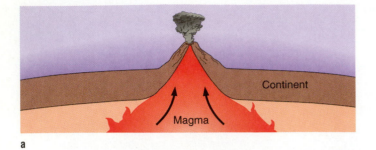

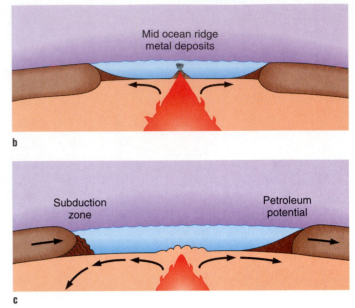

FIGURE 15-7 How seafloor spreading can influence where some mineral deposits occur. During the first phase (a), volcanic material (magma) intrudes a large continental mass along areas of weakness. The continental mass eventually breaks apart, forming two continents (b). During this phase and the next, metal deposits can form along the ridge axis by volcanic activity and from hydrothermal fluids emitted at ocean vents. New ocean floor forms at the ridge axis and sediment from continental erosion is deposited along the continental margins. Sometimes the initially small ocean basin evaporates, leaving thick sequences of salt and other mineral deposits. In the last phase (c), two things can happen. On the left, the denser seafloor moves beneath the continent, forming a subduction zone. The sediments lying atop the seafloor either are subducted with it or become accreted (stuck) to it.

This is occurring now on the west coast of South America and Central America and on the margins around much of the western Pacific Ocean. On the right, the continent remains coupled to the seafloor, and sediment accumulation on the continental margin continues. This is occurring now on the margins around the Atlantic Ocean and the east coast of Africa. These thick sediment accumulations may contain petroleum deposits.

for oil and gas accumulation. In these ways, seafloor spreading has created favorable environments for marine mineral deposits.

Metalliferous Muds

In addition to mineral-enriched *seafloor,* an interesting discovery has been the finding of *sediments* enriched in certain metals. These sediments occur especially in areas of past or present seafloor spreading. First discovered in the early 1960s along the East Pacific Rise, they were later found in the Red Sea.

The sediments are of three types: (1) *localized areas* of enriched sediment on some mid-ocean ridges associated with vent areas; (2) *dispersant deposits,* enriched mainly in iron and manganese, occurring on or near some seafloor spreading areas; and (3) *sulfide deposits* such as those in the Red Sea's brine pools. The consensus is that all these deposits, although different, are precipitated from hydrothermal solutions.

Some enriched sediments are localized to vent areas, such as those along the East Pacific Rise (Figure 15-8) and the Juan de Fuca Ridge off Oregon, or to depressions where dense hydrothermal water can accumulate (Table 15-4). They also can occur uniformly over a large area if the hydrothermal water is dispersed over the bottom. Although only a small portion of the oceanic spreading centers have been surveyed in detail, these mineral-enriched sediments already have been found in several areas (Figure 15-9).

Drill cores have penetrated deposits buried far from present spreading centers. This suggests that the mineral-enrichment mechanism has been active in the past as well as the present, which may explain the formation of some mineral deposits on land (see Box 15-1).

It was not until the late 1970s that the hydrothermal activity at spreading centers was actually observed, in the Galápagos Islands region. Here, waters up to 315°C (600°F) were seen forming mineral deposits and supporting a unique fauna (see Figures 7-10, 7-11, and 12-19 to 12-21, pages 168–170 and 300–302).

Despite optimistic reports in commercial media and the popular science press, the economic potential of these deposits is unknown (see Box 15-1). However, there is no doubting their scientific value.

Red Sea Metalliferous Muds

One of the most interesting mineral discoveries in the deep seafloor occurred in 1948 in the central part of the Red Sea. Seafloor spreading is active here, with Africa and the Saudi Arabian Peninsula slowly moving away from each other (see Figures 3-15 to 3-18, pages 58–61).

When scientists on the Swedish research vessel *Albatross* were performing a routine *hydrocast* (collecting water samples and measuring their salinity and temperature), they noted slightly anomalous temperatures and salinities at the bottom of the Red Sea's central ridge but, attributing it to instrument error, they assigned little significance to these observations.

a

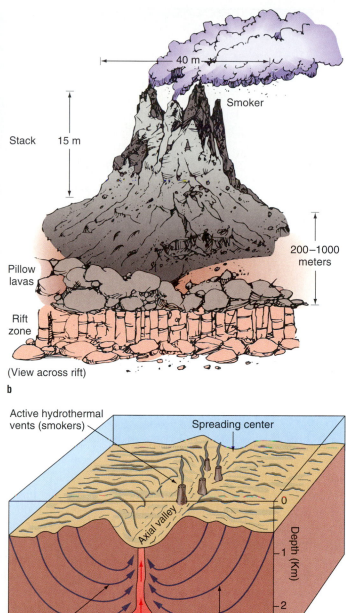

b

c

FIGURE 15-8 (a) Erupting undersea vent (a so-called black smoker, or metallic sulfide smoker) on the East Pacific Rise. Sediment in the discharging water is often enriched with sulfur, zinc, copper, and other elements. These can accumulate in sediment near the vent. The vent itself forms a stack that is rich in minerals.
(b) Typical structure of a black smoker as ascertained by numerous dives and photographs.
(c) A typical distribution of black smokers around a spreading center. Arrows indicate direction of movement of hydrothermal fluids. (Image (a) courtesy of Dr. Robert D. Ballard and Mr. Dudley Foster; (b) courtesy of David B. Duane; (c) slightly modified from David B. Duane.)

However, the British R.R.S. *Discovery* and ships from Germany and the Woods Hole Oceanographic Institution explored the area further. This led to the discovery in the early 1960s of three pools of hot, saline water at about 2,000 m (6,600 ft.) depth on the bottom of the Red Sea. Salinity attained 257‰ and water temperatures were as high as 56°C (133°F).

This salinity is not comparable to the "normal" 35‰ of the ocean. The ratio among major elements in the Red Sea is different from ratios found in "normal" seawater. Some elements, such as iron, are enriched thousands of times, whereas others are less abundant than in normal seawater.

TABLE 15-4

Metal Concentration from Selected Marine Vent Areas

Area	% Zinc	% Copper	Silver (PPM)	% Iron	% Lead
Galápagos	0.1	5	10	44	0.07
Juan de Fuca	0.6–5.4	0.3	3–290	2–50	0.06–0.25
East Pacific Rise at 21°N	0.1–42	0.1–1.3	1.6–241	0.6–26	0.04–0.8

Note: Values are based on a small number of samples.

Source: Adapted from Broadus, 1987.

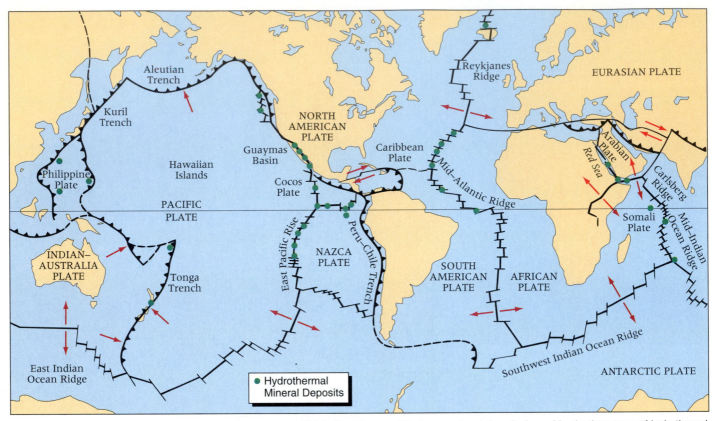

FIGURE 15-9 Distribution of some of the known mineral deposits formed by the three types of hydrothermal activity described in the text. Note that essentially all are associated with plate boundaries and that most are associated with spreading centers.

Not until 1966 were the economic implications of the area realized. At that time, Woods Hole Oceanographic Institution launched a major expedition to the area. The region was mapped and numerous sediment cores were collected, penetrating 10 m (about 33 ft.) below the sediment-water interface (Figure 15-10).

Based on a few chemical analyses of the sediment underlying one of these pools (the Atlantis II Deep), researchers estimated their value to exceed $2 billion. However, this excludes the cost of raising the sediments from the bottom and refining and marketing the minerals.

About 2 decades ago, the governments of Saudi Arabia and Sudan agreed to mine these deposits. Mining has not yet occurred, however, largely due to political differences between the two countries.

The Red Sea deposits are interesting for more than their high content of copper, zinc, and silver. Oceanographers are actually "seeing" a mineral deposit forming on the seafloor. Measurements made over a few months showed that water temperatures were increasing, a clear indication that this is a dynamic process. If deposits are now actively forming on the Red Sea floor, in this area of seafloor spreading, such deposits may also be forming in similar parts of the ocean.

Manganese Nodules

Manganese nodules are sometimes called iron-manganese or ferromanganese deposits because they also contain substantial amounts of iron. First discovered during the Challenger expedition (1872–76), these nodules commonly occur as spheres, 1 to 20 cm (about 0.4 to 8 in.) in diameter. They also occur as coatings on rock and other objects or as long slabs called *manganese pavements* (Figure 15-11).

The principal economic interest in the nodules is the manganese, used in steel making. Accessory elements in the nodules—copper, nickel, and cobalt—are also valuable. The nodules form with amazing slowness—adding about 1 mm every *million* years—and are common to all the oceans, especially where sedimentation rates are very slow (otherwise they could become buried).

According to one estimate, more than 25 percent of the seafloor is covered with nodules, with over 1.5 trillion (1.5×10^{12}) tons in the Pacific Ocean alone. Analyses of nodules reveal slight differences in elemental content from ocean to ocean. However, elemental content is generally similar over large areas of any one ocean, and even greater similarities may exist within certain physiographic provinces of a particular ocean.

Box 15-1

Mining a Deep Sea Vent

The potential for mining a deep sea vent area is intriguing—full of adventure and dollar signs. The realities, however, are less attractive.

Most active vent areas have rugged relief, which makes mining difficult. The water is deep and consequently its pressure is very great. The vented water is hot and very saline. Mining surely would be opposed by environmental interests because of the real potential damage to unique life forms. Clearly, vent mining presents formidable technological challenges.

Another problem is ownership. Although some vent areas lie in the Exclusive Economic Zone of a country, such as near Ecuador, Mexico, and the U.S. West Coast, most are outside of any country's EEZ. At present, mining of active vent areas seems improbable.

The scenario, however, is very different for ancient vent areas that are now *above* sea level. Especially attractive are the island of Cyprus in the Mediterranean, and the country of Oman in the Persian Gulf (see Figure 3-20, page 63).

Much is known about Cyprus, where several ore bodies have deposits similar to those of modern ocean vents. One, the Skouriotissa area, was mined as early as 300 B.C. by the Greeks, and later by the Romans. Mining resumed in the 1920s, but stopped in 1974 because of warfare in the region. Sulfur and copper are the main materials obtained from the mine. In fact, the name *Cyprus* comes from *cuprum,* the Latin word for copper (chemical symbol Cu). The Skouriotissa ore body is at least 670 m (2,200 ft.) long, 213 m (700 ft.) wide, and 150 m (500 ft.) thick (Figure 1).

As we learn more about the process of seafloor spreading and how it affects the ocean and land, expect more ancient vent areas to be discovered and perhaps mined.

FIGURE 1   The Skouriotissa mine site on Cyprus. The pit area is about 60 m (200 ft.) deep. Its individual benches, created by mining, are about 2.5 m (8–10 ft.) high. This mine is essentially a cross section of an ocean vent, showing many different types of mineralization. (Photograph courtesy of Dr. Deborah Sue Kelley, Woods Hole Oceanographic Institution.)

Most favorable for nodule mining is an area of red clay and siliceous ooze in the North Pacific. These deposits lie north of the equator in a broad band between 6°30'N and 20°N, extending from 110°W to 180°W, covering about 11.5 million km² (3.4 million nautical miles²). The area is attractive because of greater metal content (Table 15-5) and a comparatively smooth seafloor to make mining easy.

Nodules of the Atlantic and Indian Oceans contain smaller quantities of copper, nickel, and cobalt, below the amounts needed for economic recovery. The potential value of manganese nodules has excited many countries, increasing the intensity of international discussion of seafloor ownership (see Chapter 19).

TOP-SECRET MANGANESE?   Two decades ago, several groups moved rapidly toward mining manganese nodules. One appeared to be the Summa Corporation, owned by the late billionaire Howard Hughes. Summa built a 36,000-ton experimental ship called the *Hughes Glomar Explorer* (Figure 15-12). Press releases told how the ship would pluck nodules from the bottom by air suction and store them in a barge beneath the ship. In 1975 it was learned that the venture was a cover for a

FIGURE 15-10   Portion of a core, rich in heavy metals, collected from the Red Sea. The length of the core is about 1 m (3 ft.).

TABLE 15-5
Average Partial Composition of Manganese Nodules from the Oceans (Weight Percent)

Mineral	Pacific[1]	Indian[1]	Atlantic[1]	Favorable North Pacific Area[2]	
				Red clay	Siliceous oozes
Manganese	19.78	15.10	15.78	17.43	22.36
Iron	11.96	14.74	20.78	11.45	8.15
Nickel	0.63	0.46	0.33	0.76	1.16
Copper	0.39	0.29	0.12	0.50	1.02
Cobalt	0.34	0.23	0.32	0.28	0.25

[1] Source: Cronan, 1992.
[2] Source: Horn et al., 1972.

Central Intelligence Agency (CIA) attempt to raise a sunken Soviet submarine from the deep sea. The cover story was very successful, and it wasn't until 1992 that the CIA acknowledged the attempt, stating that they recovered only one-third of the submarine, with little of intelligence value.

This operation demonstrated that the technology exists to recover large objects from any portion of the seafloor. It also showed the rest of the world—especially less-developed countries—that an apparently peaceful operation in the deep sea can have military implications.

This point was not missed by delegates attending a Law of the Sea Conference in Geneva, Switzerland. Ironically, the submarine story broke as this meeting was in progress. Even more ironic: the *Glomar Explorer* was used later in an actual manganese nodule mining operation.

The Ocean Minerals Company, a consortium of several companies, chartered *Glomar Explorer* to test a prototype mining device in the Pacific (Figure 15-13). As the device moves along the bottom, it rakes nodules onto a conveyer belt, where they are washed, crushed, and transported to the surface.

Several critical problems exist with a deep-sea manganese nodule program, including economic feasibility,

a

b

FIGURE 15-11   (a) Manganese nodules and slabs from the Blake Plateau off the southeastern United States.
(b) Manganese nodules coming off a conveyor belt after experimental recovery from the seafloor. (Photograph (a) courtesy of Frank Manheim, U.S. Geological Survey.)

FIGURE 15-12 The *Hughes Glomar Explorer*, for many years thought to be an experimental mining ship for deep-sea manganese nodules. In the 1980s the vessel was actually used in a prototype mining operation in Pacific depths of about 5,000 m (16,400 ft.). (Photograph courtesy of Summa Corporation.)

environmental impact, and legal implications. A main issue is ownership; there are numerous articles in the Law of the Sea Treaty concerning ownership and mining of manganese nodules. A key point of the treaty is that ownership of a site would be decided by an international authority, and any mining revenues would be shared among Earth's nations. (Further details concerning the Law of the Sea Treaty are covered in Chapter 19.)

The fundamental problem, though, is the poor economics of deep-sea mining. In some instances composite materials such as plastics or graphites have replaced metals. If this trend continues, the demand for some metals may actually decrease. Alternatively, some land resources of metals are being depleted, and others are in politically unstable countries. This causes most experts to doubt that marine mining will occur unless the main elements in nodules are competitive in price to those in present land deposits. This might not occur until at least the twenty-first century—if ever.

Cobalt Crusts

A new potential resource, called **cobalt crusts,** may be exploited sooner than manganese nodules. Cobalt crusts occur as encrustations on the rocks in seamounts, ocean ridges, or other elevated seafloor areas. They occur in shallower water, about 450 to 1,800 m (about 1,500 to 6,000 ft.) deep. The crusts are only 2.5 to 5.0 cm (1 to 2 in.) thick (Figure 15-14).

These deposits are important because they sometimes contain more than 1 percent cobalt (an exceptionally high value when compared to other sources), as well as manganese and platinum. Cobalt is used in alloys, magnets, and cancer radiotherapy; platinum is valued in alloys and in catalytic converters. Cobalt crusts are widely distributed, and research on them is just beginning.

Deep-Sea Muds and Oozes

Much of the deep-ocean floor is covered by fine-grained muds that have been deposited very slowly, only a few centimeters every thousand years. The commonest type is called *brown clay,* or sometimes *red clay* (see Figure 5-15, page 121). It accumulates at a few millimeters per thousand years.

Areas of brown clay deposition are immense, covering about 100 million km² (38 million miles²). Assuming an average thickness of 300 m (1,000 ft.), the volume of brown clay deposits in the ocean is about 30 million km³ (about 7.1 million miles³).

Chemical analyses of these sediments show they may contain up to 9 percent aluminum and 6 percent iron, plus lesser amounts of copper, nickel, cobalt, and titanium. Such contents, multiplied by the immense volume, could yield enough copper and aluminum to last over a million years at our present rate of use—*if* the brown clays can be mined economically. At present, they cannot.

Among the problems to overcome are recovery from depths of 6,000 m (more than 20,000 ft.) or greater, and transportation and refining of the fine-grained material. A potential advantage of this vast deposit, however, is that it lies unconsolidated (loose) on the seafloor with no overlying rocks—ideal for dredging.

Oozes are muds that contain larger amounts (usually exceeding 30 percent) of shells, or *tests,* of dead organisms (see Figures 5-16 and 5-18, pages 122–124). Oozes are of two main types, calcareous or siliceous, depending on whether they are composed mainly of calcium carbonate or siliceous shells. Some calcareous oozes can attain 95 percent pure calcium carbonate and are a possible limestone source for cement. As with brown clays, the volume of this material is awesome. If ever mined, it could supply limestone for several million years. Siliceous oozes could be mined for silica, which is used as insulation and as a soil conditioner. This type of deposit also covers extensive seafloor areas, although less than calcareous ooze (again, see Figure 5-15, page 121).

At present, oozes and muds must be considered only as *potential* resources because of the technological difficulties in mining them. Even so, their vast extent suggests that they could become a very important resource in the near future. One especially impressive feature of these potential resources is their rate of accumulation. Although amazingly slow, it occurs over such a large area that the effective rate of accumulation of several elements in the sediment is faster than their rate of consumption or use on land.

Mineral Potential of the U.S. Exclusive Economic Zone

The United States, like many other coastal countries, claims a 200 nautical mile (370 km) Exclusive Economic

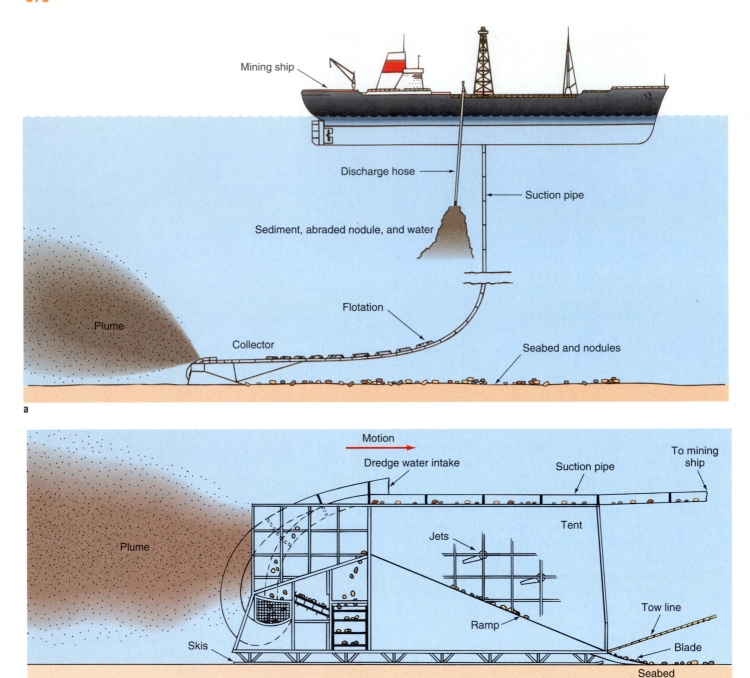

FIGURE 15-13 A prototype mining device for manganese nodules, developed by Ocean Minerals Company: (a) towed by a surface ship; (b) side view. The device, only about one-tenth the size of an operational system, is almost 14 m (46 ft.) long and 9 m (29 ft.) wide. (Adapted from J. E. Flipse, 1980.)

Zone (EEZ). This simply means that our exclusive right to economic use (building, fishing, mining, and the like) extends to 200 nautical miles offshore, a distance roughly twice the width of the Florida peninsula. The EEZ concept, a product of the Law of the Sea Conference, is almost universally accepted. Because the United States has such a long coastline, our EEZ is the world's largest (see Figure 19-5, page 472). The United States now has more marine territory than land territory.

One might expect that this "economic annexation" of extensive territory would trigger a large-scale attempt to exploit its resources. Economics have dictated otherwise, however. The value of many minerals is presently depressed, and U.S. budget problems have restricted offshore research.

On the other hand, many minerals in the EEZ have strategic importance for the United States, and at present some are obtainable only from foreign sources. Despite

a b

FIGURE 15-14   (a) A piece of cobalt crust collected from the Mid-Pacific; (b) enlarged view. Actual crust (upper dark area) is about 5 cm (2 in.) thick; the underlying rock is a phosphatized limestone. (Both photographs courtesy of F. T. Manheim, U.S. Geological Survey.)

this, the U.S. mineral program is slow-paced, with no major exploitation targets in sight. Emphasis has focused on *exploration*—finding sites that have mineral potential.

In some respects, the United States is in an enviable situation, having within its waters numerous potential mineral deposits. However, as previously stated, these are only *potential* resources, because the cost of their recovery, refining, and marketing may not be competitive with conventional resources on land.

Assessing the value of a marine mineral resource is very difficult for many reasons, including the problem of defining the dimensions of an underwater deposit, calculating the cost and time needed for development, the impact of weather (in itself unpredictable), environmental concerns, and so on.

ESTIMATED U.S. MARINE RESOURCES. Estimated sand and gravel resources are great, exceeding several hundred years of likely consumption (Table 15-6). Heavy minerals also are potentially promising—for example, titanium-rich minerals, zircon, garnet, and monazite are found on the mid-Atlantic shelf. (Titanium and zirconium are important alloying metals, garnet is an abrasive, and monazite is a source of thorium.)

On the Pacific shelf, heavy minerals occur extensively in old beaches, buried river channels, and gravels. Titanite and zircon are found off Oregon. Alaska has offshore gold deposits, some currently being mined.

Phosphate deposits off the southeastern United States cover at least 125,000 km^2 (about 48,200 miles^2), containing several billion tons. Other phosphate deposits occur off California and the Pacific Islands, often associated with cobalt-rich crusts. These are increasingly attractive because environmental and economic problems have reduced land phosphate mining in most of the United States.

Some manganese nodules occur in the U.S. EEZ but are of modest grade. Nodules on the Blake Plateau, off the southeastern United States (see Figure 15-11a) have small amounts of platinum, which has attracted some interest.

Most known metalliferous mud deposits occur outside U.S. waters, although rich samples have been obtained from the Juan de Fuca and Gorda Ridge off Washington and Oregon. Active volcanic islands off the Aleutians and Marianas and isolated seafloor volcanoes in the Pacific are possible sites for metalliferous deposits. Many favorable oil and gas areas in U.S. waters have yet to be fully explored (see Figure 15-5).

TABLE 15-6

Estimated Sand and Gravel Resources Within the U.S. Exclusive Economic Zone

Province	Volume (meters^3)
Atlantic	
Maine–Long Island	340 billion
New Jersey–South Carolina	190 billion
South Carolina–Florida	220 billion
Gulf of Mexico	269 billion
Caribbean	
Virgin Islands	>48 million
Puerto Rico	170 million
Pacific	
Southern California	30 billion
Northern California–Washington	Insufficient data
Alaska	>160 billion
Hawaii	19 billion

Source: Modified from Williams, 1975.

Biological Resources

The Protein Crisis

The study of marine biological resources is a very important part of oceanographic research. Many believe that the ocean can—in fact *must*—help solve the world's food problems, which will only increase. However, it is naive to think that the ocean provides an easy solution. Even if the marine biological harvest were doubled, it would not begin to meet the global deficiency of animal protein.

Nearly *2 billion* people suffer from a diet deficient in animal protein, a problem compounded in some less-developed countries because their population is growing faster than their ability to produce food. A daily supplement of 10 to 20 grams (0.3 to 0.6 ounce) of animal protein is considered a sufficient minimum. This is equivalent to 3.6 to 7.2 kg (about 8 to 16 lbs.) of protein per year.

Some have suggested that large amounts of protein could be obtained by harvesting organisms low in the food chain, such as plankton (see Figures 12-24 to 12-26, pages 306–308). Though they comprise the largest group of ocean organisms, plankton are microscopic, so obtaining even modest quantities requires filtering an immense quantity of water (as much as 1 million pounds of seawater to obtain 1 pound of plankton). This makes plankton a dubious resource—not to mention the problem of their taste.

For a usable source of protein, we must look to animals such as fish. Fish contribute about 6 percent of the world's total protein supply for human consumption. The proportion becomes more significant, about 24 percent, if we consider the amount of *fish meal* used to feed other animals. Worldwide, an estimated 15 million people (about 1 person in 360) are involved in the fishing industry. The total yearly catch from the sea is worth about $45 billion.

Phytoplankton and Fish

The geography of the ocean's biological resources, like those of the land, is ultimately related to production of organic matter by plants. As discussed in Chapters 11 and 12, ocean plants are primarily floaters, or phytoplankton. Because they need light to photosynthesize, they live in the upper layers of the ocean. Photosynthesis also requires nutrients, the supply of which is controlled largely by water circulation. Nutrients are especially abundant in areas of upwelling, which brings deep, nutrient-rich waters to the surface.

Phytoplankton production of organic material can be defined in four geographic regions (Figure 15-15). From low productivity to high productivity, they are:

1. Deep-ocean areas far from land, where the productivity is very low

2. Arctic and equatorial waters, which are moderately productive due to mixing by currents and winds

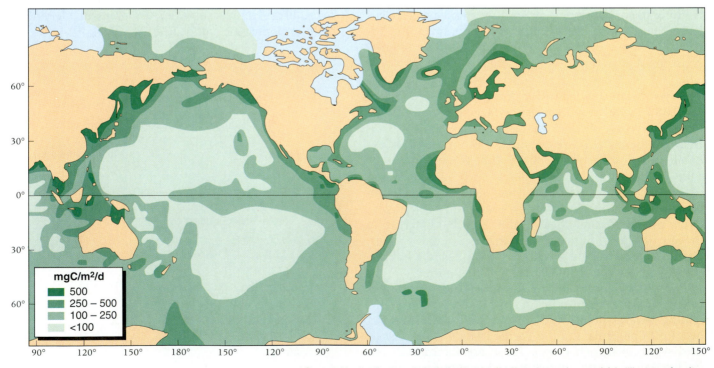

FIGURE 15-15 Geographic distribution of phytoplankton production of organic material (milligrams of carbon per square meter per day). (Adapted from *Atlas of the Living Resources of the Sea*, Food and Agriculture Organization (FAO) of the United Nations, 1972.)

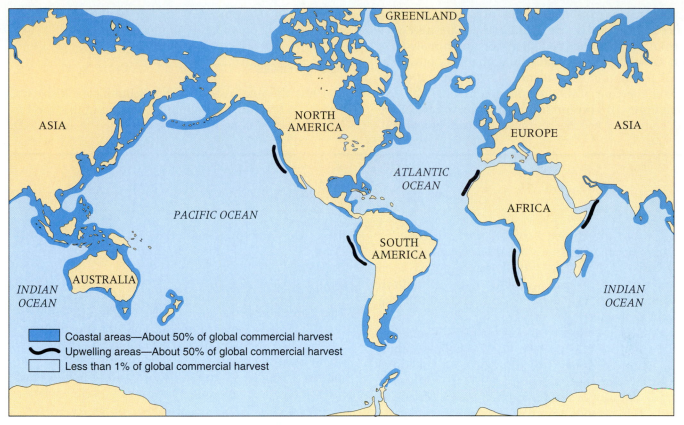

FIGURE 15-16 Major fishing areas of the world. (From *Patterns and Perspectives in Environmental Science*, report for the National Science Foundation, 1972.)

3. The shallow waters of most continental shelves, which have high production because of high nutrient availability and sunlight for photosynthesis

4. Upwelling areas such as those off Peru, California, and western Africa, where organic production is extremely high.

The world's major fishing areas follow this pattern of high phytoplankton production, especially in the nearshore parts of the ocean (Figure 15-16). This is logical because fish feed on organic material produced by phytoplankton, either directly or indirectly through eating zooplankton or other organisms that eat zooplankton.

By combining the first two regions (deep ocean and Arctic/equatorial), we can divide the ocean into three provinces, based on organic production, as shown in Table 15-7.

Note that, although the *open-ocean province* includes about 90 percent of the ocean by area, it produces less than 1 percent of the fish caught and has little potential for increased production. The *coastal-zone province* and *upwelling-zone province* occupy respectively only 9.9 and 0.1 percent of the ocean, but each produces about half of the world's fish catch.

TABLE 15-7

Organic Material Production and Estimated Fish Production

Province	Percentage of Ocean Area	Mean Productivity of Organic Material (G Carbon/m²/Year)	Total Primary Production (Tons Organic Carbon)	Number of Feeding Levels	Organic Material Loss (Efficiency of Conversion, %)
Open ocean plus Arctic	90	50	16.3 billion	5	10
Coastal zone (including offshore areas of high production)	9.9	100	3.6 billion	3	15
Upwelling zones	0.1	300	0.1 billion	1.5	20

Source: Adapted from Ryther, 1969.

A main reason for these differences is that in the open-ocean province, more feeding levels are necessary to go from microscopic phytoplankton to animals large enough to be used by humans. In going from one feeding level to another (see Figure 12-24, page 306), as much as 90 percent of the organic material may be lost. The **efficiency of conversion** is the growth rate of the animal compared to the amount of food consumed.

Efficiency of conversion is greater in the coastal zone and upwelling provinces than in the open ocean.

Fishing Methods

In 1991, the total world catch (including shellfish) was 96,900,000 metric tons. About 90 percent of this was finfish (the U.S. fishing industry is described in Box 15-2).

Box 15-2
The U.S. Fishing Industry

Fishing within the United States is not as lucrative a business as one might expect, but it has increased in recent years. In 1992 the United States ranked sixth among the countries of the world in total weight of fish caught, behind China, Japan, the former Soviet Union, Peru, and Chile (see Figure 15-21).

The total U.S. commercial harvest of fish, shellfish, and other marine life in 1992* was 9.6 billion pounds (4.4 million metric tons). This was worth $3.7 billion to the fishermen, or an average of $0.57 per kilogram ($0.26 per pound). Finfish accounted for 85 percent of the landings in pounds, but were only about 55 percent of the total value.

The most abundant commercially caught fish in the U.S. is pollock, followed by menhaden and then salmon, flounder, crab, cod, and shrimp. Salmon, although third in quantity, had the highest dollar value. Shrimp is second in dollar value.

Alaska was the leading state in weight of fish landed, followed by Louisiana, Virginia, California, and Mississippi (Figure 1). Alaska also was the clear leader in catch value, followed by Louisiana, Massachusetts, Texas, and Maine. Dutch Harbor in Unalaska, an Alaskan fishing port, led in quantity and value. New Bedford, Massachusetts was second in value.

U.S. per capita consumption of fishery products was about 14.8 pounds per person in 1992. This number is surprisingly low, considering the nutritional value of fish. In contrast, the consumption of beef, pork, lamb, veal, and poultry in the United States was about 200 pounds per person. Some fish is consumed indirectly because of its use in production of other foods. Fish is used as feed for chickens, turkeys, and other organisms, and for fish oil and fertilizer.

U.S. recreational fishermen in 1992 caught an estimated 285.5 million fish, weighing about 233 million pounds. The number of fishermen is estimated at 17 million, and they made about 52 million trips. Most of fish catches and trips occurred on the Atlantic and Gulf Coasts.

Why U.S. Fisheries Are Lagging

There are several reasons why the U.S. fishing industry has not done as well as it might. For one, the market is smaller: fish products are not as popular as other forms of animal protein, such as beef, pork, and chicken. The market is smaller in another sense, too—the choice of fish among U.S. consumers shows a preference for more expensive marine varieties, such as salmon, shrimp, and lobster.

The United States does not mind importing fish. For example, in 1992, we imported edible fish valued at $5.7 billion, while our exports were $3.5 billion.

Some laws and customs have restricted the catch of U.S. fishermen. Although many U.S. fishing boats are inferior to those of other countries, the law requires that any fishing boat unloading its catch in the United States must be *built* in the United States. This may help the boat-building industry, but it hinders the fishing industry.

Some regulations are necessary to prevent overfishing of particular species. This restricts the catch, but with good reason. The main reason for the relatively small U.S. fish catch is the dwindling supply of many species in U.S. waters. Overfishing, too many fishing boats, pollution of habitat, and natural variability in fish populations all contribute to this problem. Overfishing has become an especially important problem on Georges Bank, off northeastern U.S. and Canada. In 1994, both countries have severely reduced the numbers of cod, haddock, and halibut that can be harvested from this region. The problem also exists elsewhere; for example, in 1994 offshore salmon fishing was banned on the West Coast from Washington to Point Arena, north of San Francisco. It has also become evident that unless fishing nations are willing to control and better regulate the fish catch in their waters, their fish harvest may also decrease.

*Note: fishery statistics are always somewhat out of date because they take a few years to compile.

Box 15-2 continued

U.S. Commercial Landings by Region

Alaska
5.6 billion lb (58%)

Great Lakes
31.0 million lb (<1%)

New England
647.1 million lb (7%)

Mid-Atlantic
260.8 million lb (3%)

Chesapeake
687.6 million lb (7%)

Pacific Coast & Hawaii
708.7 million lb (7%)

Hawaii

South Atlantic
237.7 million lb (2%)

Gulf
1.4 billion lb (15%)

FIGURE 1 U.S. commercial fish landings, in pounds (and percent). (From Fisheries of the United States, U.S. Department of Commerce, National Oceanic and Atmospheric Administration, 1992.)

One positive sign for the industry is a modest increase in the percentage of U.S. catch. At one time, foreign fishermen took about twice the fish from U.S. coastal waters that American fishermen caught, and did so legally. But, with recent legislation and the U.S. declaration of an Exclusive Economic Zone of 200 nautical miles, other nations must now get permission to fish in U.S. waters.

Many marine scientists feel that this number is near the maximum sustainable yield from the sea. Others, perhaps too optimistically, think yields of 3 to even 30 times this number are possible. The world's fish catch (excluding shellfish) had risen dramatically over the last few decades but has leveled off in recent years. It was about 20 million metric tons in 1950, 40 million in 1960, 70 million by 1970, and attained 80 million in 1986.

Further increases in the yield from the sea, even if possible, will require innovative techniques. It will be necessary for people who fish to change from "hunters," which they essentially are now, to "herders." In other words, it will be necessary for them to be able to *control the movement* of fish. Innovative attempts to influence fish movement include increasing the biological productivity to create a rich feeding area, thus concentrating them for mass fishing (see Chapter 18, and Box 18-2, page 443).

Today, electronic devices (Figure 15-17) permit modern fishermen to *find* schools of fish, but not to *control* them. Countries that fish far from their home ports, like Russia, use modern **factory ships** to process their catch at sea (Figure 15-18). However, these fish still are caught by conventional techniques, using nets from small ships (Figures 15-19 and 15-20).

By-Catch

Another impediment to improving the fish harvest often occurs because large numbers of unwanted species are caught during the harvest of a desired species. These unwanted species are often edible and very tasty, but they are difficult to market because most consumers are unaware of their quality. Thus, these unwanted species, called **by-catch** or incidental catch, are returned—generally dead—to the marine environment. Harvesting one pound of shrimp from the Gulf of Mexico can create a by-catch of as much as 15 pounds. The net effect of the by-catch problem is the possible loss of a resource, the unnecessary capture and death of animals, and the removal of an organism from the food web with possible negative effects. Conversely, the reduction of by-catch can help increase our harvest from the ocean.

A well-publicized example of the by-catch problem is the capture and death of dolphins and porpoises during

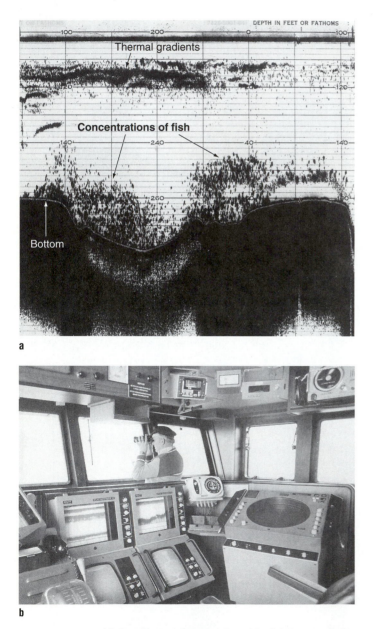

FIGURE 15-17 (a) An echo-sounding record used by fishermen to locate concentrations of fish. A large school of fish is visible above the bottom of the seafloor.
(b) An array of electronic equipment on a modern fishing boat. (Photograph (a) courtesy of Raytheon Marine Company; (b) courtesy of Krupp Atlas Elektronik.)

FIGURE 15-18 Russian stern trawler (left) about to offload its catch onto a large factory ship. (Photograph courtesy of Brenda Figuerido, National Marine Fisheries Service of NOAA.)

a

c

b

FIGURE 15-19 Commercial fishing operations: (a) preparing to remove the fish caught in a seine net (see also Figures 15-20a and 15-20b); (b) emptying a trawl net (see also Figure 15-20c); (c) some large fluke caught in the trawl net; (d) unloading of fish catch at the pier. (Photograph (a) courtesy of the National Oceanic and Atmospheric Administration; (b) and (c) courtesy of Dr. Susan Peterson; (d) courtesy of Brenda Figuerido, National Marine Fisheries Service of NOAA.)

d

the tuna harvest. These animals tend to swim above schools of tuna and can be caught when a purse seine is used to catch the underlying tuna (see Figure 15-20a and b). Upward of 100,000 dolphins and porpoises were being killed unnecessarily each year by this process. In the U.S., public pressure and eventually legislation led to rules that have considerably reduced this mortality, and these rules are now being adopted worldwide.

Natural Controls of the Fish Catch

Successful fishing is frequently influenced by factors beyond the control of fishermen or politicians. In particular, changing oceanographic or atmospheric variables can dramatically influence fishing success.

Perhaps the best example occurs off Peru, where one of the world's largest fisheries existed for many years.

FIGURE 15-20 Some methods of catching fish: (a) two-boats-and-mothership method of purse seining; (b) one-boat purse seine operation; (c) otter trawl gear; (d) pair trawl; (e) dredging operation; (f) lobster pot trawl; (g) longline gear; (h) gill nets. (From Ross, 1980.)

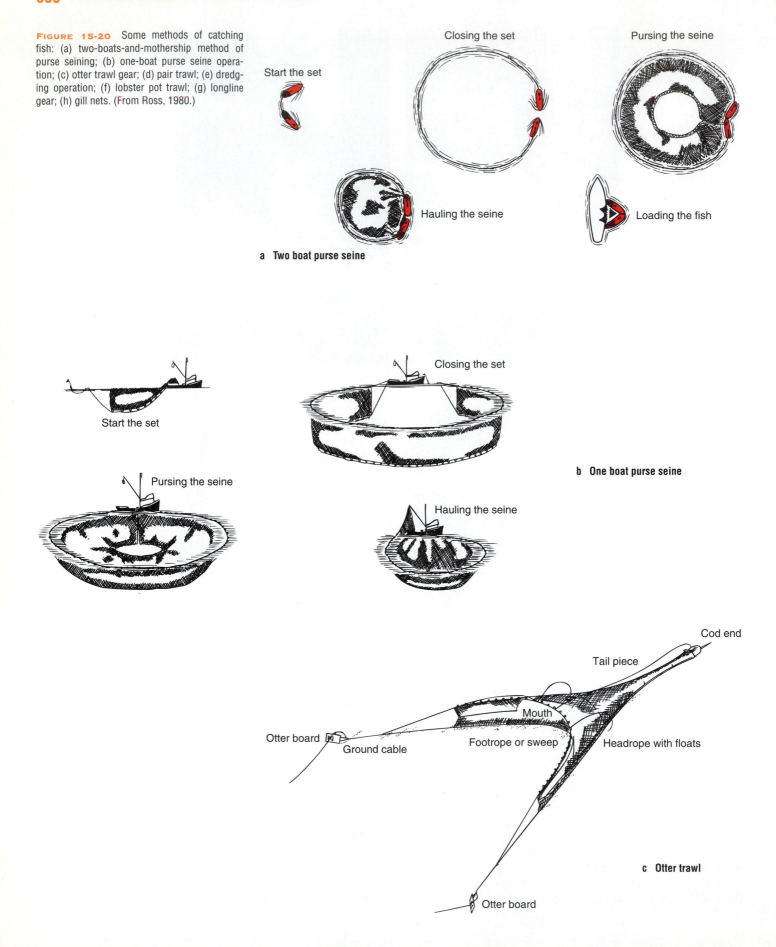

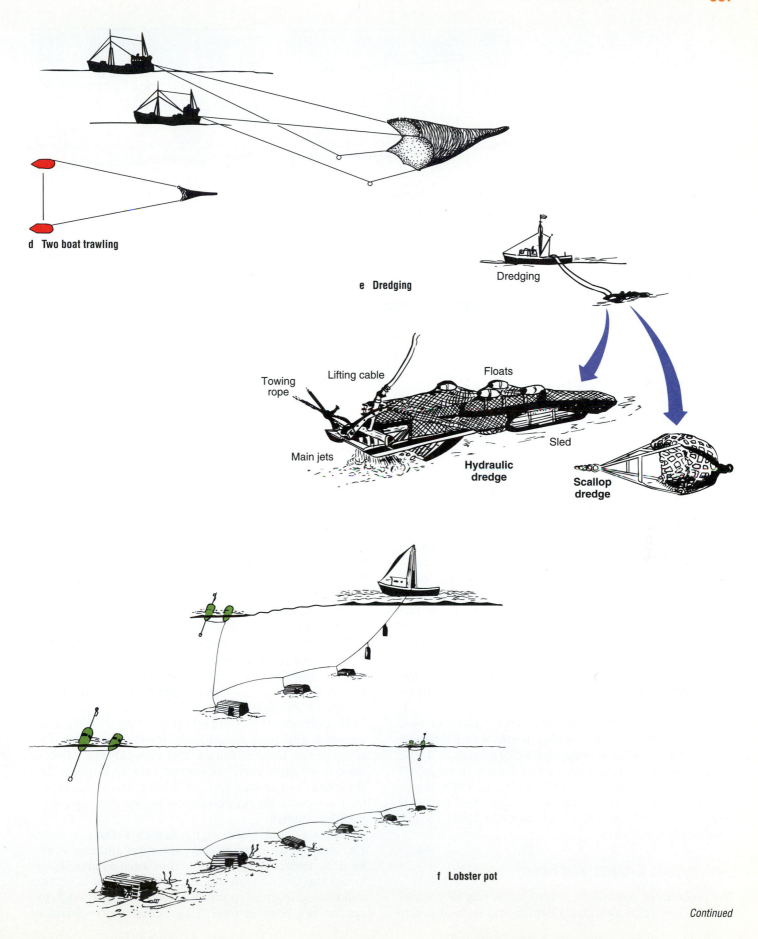

d **Two boat trawling**

e **Dredging**

Dredging

Towing
rope

Lifting cable

Floats

Main jets

**Hydraulic
dredge**

Sled

**Scallop
dredge**

f **Lobster pot**

Continued

FIGURE 15-20 *(continued)*

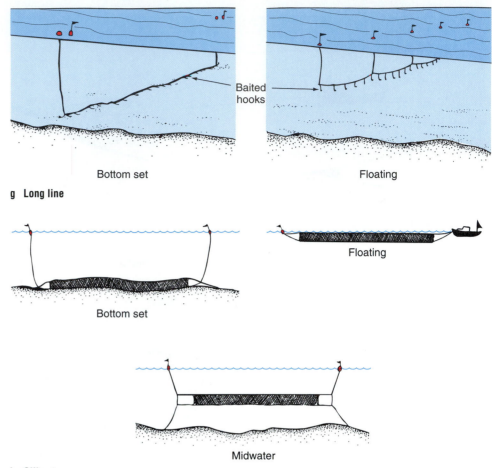

Baited hooks

Bottom set

Floating

g Long line

Bottom set

Floating

Midwater

h Gill nets

This is an area of intense nutrient upwelling, mainly due to prevailing southerly winds. Occasionally, however, atmospheric conditions change, causing more northerly winds. This results in the El Niño condition (see Figure 13-15, page 327), which causes the warming of surface waters and a slackening of the upwelling. With less upwelling and reduced nutrient availability, the nutrients can be used up quickly. Organic production by phytoplankton diminishes and the associated marine and bird life that depend on it disappear or die.

In 1971 Peru had an annual anchovy catch of over 10 million metric tons, making it the world's fishing leader. An El Niño condition that occurred in 1972, combined with overfishing, reduced their catch to 4.5 million tons and then 1.8 million tons in 1973. In recent years, however, Peru's catch recovered, and it is now the fourth-largest fish harvester (the United States is sixth; Figure 15-21).

Legal Problems of International Fishing

The problem of controlling fishing areas can be complicated. A few years ago, Great Britain and Iceland nearly came to war over Iceland's claim to a 50-nautical-mile fishing limit off its coast. The limit was imposed, in Iceland's words, to protect its stocks of herring, cod, and haddock.

In 1952, Peru, Ecuador, and Chile limited fishing within 200 nautical miles of their coasts to people of their own nations and to properly licensed foreign vessels. This led to numerous confrontations between these nations and the United States, which refused to recognize such claims for many years. However, things changed in the mid-1970s.

In 1976 the U.S. Congress passed the Fishery Conservation and Management Act, effective in 1977. This act gave the United States a 200-nautical-mile offshore zone in which it exercises control over its fisheries. The zone does not restrict foreign fishing, but requires foreign vessels to obtain permission before they can enter the EEZ and fish.

The act established eight Regional Fishery Management Councils that prepare the rules and regulations for U.S. residents who fish within each region. It has forced the United States to take a position on allocation and management of its offshore resources. Although the act has not been in force long enough to evaluate its

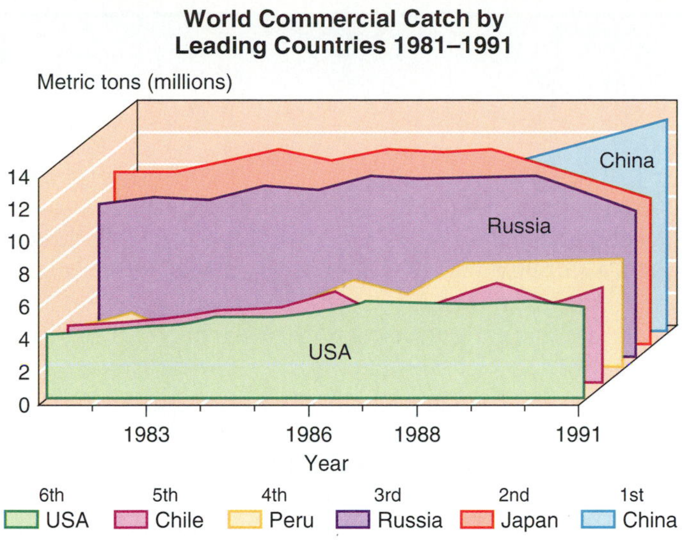

World Commercial Catch by Leading Countries 1981–1991

Metric tons (millions)

6th	5th	4th	3rd	2nd	1st
USA	Chile	Peru	Russia	Japan	China

FIGURE 15-21 Fishing catch of the six main fishing nations for the decade 1981–91. (From Fisheries of the United States, National Oceanic and Atmospheric Administration, 1992.)

goals fully, the U.S. fishing catch per unit effort generally has been decreasing.

Worldwide, the Law of the Sea Treaty could prove beneficial for managing fish stocks. Over 95 percent of the living resources of the sea occur in the various EEZs around the world. Because a coastal country controls all resources within its EEZ, it follows that each country should be concerned about controlling overfishing in its waters.

Several articles in the treaty allow coastal countries to limit the fish caught by foreign countries in their waters. Other articles encourage the formation of joint fishing ventures, in which a less-developed country could join with a developed nation to exploit its marine resources efficiently.

Regardless of the amount of fish caught, or the legal restraints on fishing, it remains doubtful that food from the ocean could ever be sufficient to feed the world's population.

Other Problems

In many countries, social and religious customs exist against fish consumption. For example, Jews and Muslims have religious rules against eating shellfish. Some countries also have problems of storage and transportation that limit fish consumption. In some nations, however, fish consumption is increasing due to public awareness of certain health benefits from eating fish (for example, a reduced incidence of heart attack; see also Chapter 18, page 440). In recent years there has been an increased popularity in eating raw fish (see Box 15-3).

Another problem facing the fishing industry is pollution. Many fish and other biological resources—such as clams, mussels, and some crabs—breed in estuaries, which are especially vulnerable to pollution (see Box 15-4). Even marine mammals are not exempt from the effects of pollution. For example, Beluga whales from

---

Box 15-3

Sashimi and Sushi

Fish is a major food and protein source in many Asian and European countries. It is less popular in the United States and Canada, but its consumption is slowly increasing. The ready availability of fresh fish is a big factor. This, combined with its health benefits, is attracting more consumers.

One very popular style of fish consumption comes from Japan. *Sashimi,* which is raw, thinly sliced fish, is available in Japanese restaurants. The most popular form of sashimi is probably the red flesh from the yellowfin tuna. *Sushi* is a piece of the fish or shellfish served on a small portion of rice and/or seaweed.

In Japan, the fresh fish used for sashimi and sushi are auctioned daily. Properly preserved fish can bring $100 a pound. Some of these fish are caught as far away as the U.S. East Coast and are flown in individual containers or "coffins" to Japan.

In recent years, sashimi and sushi have become very popular in America. Most U.S. cities have at least one Japanese restaurant with a sushi bar where individual pieces of raw fish, crustaceans like shrimp, fish eggs, or even eel may be purchased. A properly run sushi bar prides itself on the quality of its raw fish.

Buying Fresh

Fish (raw or cooked) is best when eaten fresh and properly treated, which means keeping it chilled and moist. One can determine the quality of fish in several ways:

- Fresh fish have clear eyes that are shiny and bulging. Older fish have depressed and cloudy eyes.
- Fresh fish have a firm texture.
- Fresh fish do not have a "fishy" odor.
- Fresh fish have reddish gills, rather than brown.

Raw fish obviously isn't going to be appreciated by everyone, but regardless of how the fish is eaten, its health benefits are impressive (see Table 18-1, page 441).

Box 15-4

Pollution and Fisheries

Many scientists feel that the world's fish resources are being harvested near or at their limit. Strong management is needed to rehabilitate overfished species and to maintain healthy stocks. Whether or not the world's fishing harvest is near its potential, the resource is vulnerable to human activities, such as pollution and destruction of coastal habitats. To maintain and perhaps increase world fishing, the coastal environment demands more attention and control.

The open ocean is in relatively good health. However, the same cannot be said for coastal areas, especially near densely populated regions. As discussed in earlier chapters, the coastal and estuarine areas are critical to the growth and development of many fish and shellfish. The types of pollution in the coastal and estuarine region are varied and include organic waste, pesticides, heavy metals, and synthetic compounds.

These chemicals can reduce the survival of young fish and other species, which ultimately reduces the number of adults that can breed and be harvested. Another problem is that certain chemicals—such as heavy metals, PCB, and DDT—can enter the food chain and render organisms unsafe to eat (see Figure 16-7, page 404).

Eutrophication is a condition in which excess nutrients concentrate in water that has limited or restricted circulation. These excess nutrients can increase the growth of algae and other plants. Their death and decay may deplete oxygen in the water, in turn causing mass mortality of other living organisms. Modest increases in nutrients, however, can benefit larger organisms by increasing production of their food, algae and plankton.

Domestic sewage, which contains nutrients, can contribute to eutrophication. It also can introduce into the environment pathogenic (disease-causing) organisms that often contaminate shellfish. The problem is compounded when shellfish are eaten raw. Cooking will kill many of the disease-causing organisms.

Another problem for fisheries is simply debris, including plastics, lost fishing nets (which continue to catch fish after they are lost or discarded), and general litter. Fortunately people today are much more aware of how debris affects biological resources and many coastal communities have beach or coastal cleanup programs.

The actual effect of pollution on fisheries is hard to assess, but most feel it is significant and increasing. Like fish, pollution problems do not respect national boundaries, so regional or international action is often required.

the Saint Lawrence Seaway between Canada and New England are literally classified as toxic waste because of their high tissue concentrations of polychlorinated biphenyls (PCBs). Every one of these white whales taken from the Seaway in the last few decades was found to have tumors, presumably caused by pollution.

Finally, and perhaps most important, some marine species may have been so overfished that they risk extinction. The data, frequently challenged by the fishing industry, show considerable reduction of many heavily fished species. The issue is whether the reduction is due to natural causes and cycles, or to overfishing, pollution, reduction of habitat space, or other unknown factors.

Aquaculture: Farming Seafood

Agriculture is farming on land. **Aquaculture** is farming in the sea. To a large degree they are the same—controlling or manipulating environmental conditions to promote concentrated, rapid, high-volume growth of desired plants or animals for harvest. Farming the ocean is believed to have begun in China about four thousand years ago (see also Box 18-1, page 440).

At present, aquaculture is restricted to nearshore bays and estuaries. Predictably, the target animals are clams, oysters, shrimp, lobsters, and some fish. The long-range intent of aquaculturalists is to extend this technique to other animals and plants and to the open ocean (see additional discussion in Chapter 18). Aquaculture in many countries is an important method for obtaining additional biological resources from the ocean.

Oyster farming is a potentially important industry. Baby oysters, called **spat,** float and move with ocean currents until they encounter any shell on the bottom to which they attach themselves. Oyster farmers place shells on the bottom in hospitable environments, such as bays and inlets. After the spat attach to the shells, they grow until large enough for harvesting.

Another method is to suspend shells on wires or ropes (Figure 15-22), creating a large *vertical* area for spat development. This "three-dimensional farming" can elevate baby oysters and similar creatures above their chief predator, the bottom-dwelling starfish. This method also places the oysters where more food is available. Bottom-living oysters can feed only on plankton that settles to the bottom, whereas oysters living on suspended shells can feed on the much larger quantity of plankton that float by in the currents.

In Spain, a similar technique is used to grow mussels in several estuaries (Figure 15-22). The animal protein per unit area produced with this technique is greater than anywhere else in the world. These methods have been used in Europe and Japan for decades, but are relatively new to the United States.

Lobster farming works well, especially in heated water. Lobsters grown in warm seawater in the Massachusetts Lobster Hatchery reached sexual maturity in 2 years, compared to 8 years in their natural environment. This raises the possibility of using the warm water released from nuclear or conventional power plants for lobster farming. (The cooling water used in a nuclear plant is *not* radioactive, and is released only to remove waste heat from the plant.)

Plants, especially forms of algae, are harvested directly from the sea, but they can also be part of an aquaculture effort. Several forms of algae (seaweed) are eaten in many countries and are a good vitamin and mineral source. Several alga species also provide some industrial and food products. Examples are **algin,** a food additive from brown algae, and **carrageenan** and **agar,** from red algae. Carrageenan is a food stabilizer used in products such as ice cream, salad dressing, and cosmetics. Agar is used as a growth medium for bacteria as well as in some food products. Marine plants also can be a source of new drugs (see Chapter 18, which considers innovative uses of the ocean).

Worldwide, aquaculture is fairly common. Examples include shrimp from Ecuador, fish from Israel, and oysters (and their pearls) from Japan. In the United States, aquaculture is a minor industry at present, supplying only about 2 percent of our total fish product (including most of our trout and catfish).

If we farm the sea with the same vigor applied to land farming, tremendous rewards may await us. An advantage of aquaculture is that it can be carried out in the nearshore areas of countries that most need the protein, such as some of the poorer countries of Africa. In Ecuador, shrimp farming is the second-largest industry, although the shrimp are mainly exported. Israel produces about 130 million pounds of fish yearly, mainly from inland freshwater ponds. Aquaculture is not without problems: storms can damage the artificial habitat and land-derived pollution or disease can kill organisms.

An unusual form of "sea mining" may become possible—using marine organisms to extract elements from seawater. Some marine animals concentrate trace elements like vanadium from seawater into their skeletons or tissues by factors of over 10,000. Cultivating these animals could provide a source of these elements.

a

b

FIGURE 15-22   In Spain, a very successful aquaculture program is employed in some estuaries: (a) ropes are hung from rafts and used for settlement and growth of mussels, (b) one of the "mussel ropes" being removed for harvest of the attached mussels. Note other rafts in the background. (Photograph courtesy of Dr. Donald Anderson, Woods Hole Oceanographic Institution.)

Physical Resources of the Ocean

The ocean's physical resources are sometimes less obvious than its mineral and biological resources, but they are important. For example, the ocean embodies a vast amount of energy. Tidal energy particularly interests many scientists and engineers, and various schemes have been devised to drive electrical turbines from the dependable rise and fall of the tides. Also, temperature differences of seawater and waves could be used to generate power. To extract this energy will require some unique technology, which is discussed in detail in Chapter 18.

Shipping can be considered another physical resource, for the world ocean is Earth's largest continuous highway. Shipping materials over this highway is the largest human use of the ocean—in fact, over 90 percent of the world's trade is carried by ships. Mechanized cargo handling, more efficient ship design, and expanded freighter size have increased the efficiency of ocean transport. A major ocean-transported product is oil from the Middle East to consumers in Japan, Europe, and North America. This transoceanic "pipeline" is usually safe and reliable, using large supertankers (Figure 15-23). However, serious environmental damage can result when oil spills from these ships (Chapter 16 discusses some of these environmental problems).

At this writing, in the mid-1990s, over 76,000 merchant ships are operating (counting only those over 100 tons in size). These vessels have an aggregate cargo capacity exceeding 400 million tons. Many are registered in nations different from the nation of their owner. The reasons vary, but most common are the easier or more lenient regulations within the countries of registry. The leading countries of registry, based on gross tonnage, in decreasing order, are Liberia, Panama, Japan, Russia, Greece, the United States, China, Taiwan, and Cyprus.

FIGURE 15-23   The *Texaco Italia*, a large supertanker used to transport oil.

The ocean does present some unique engineering challenges to its use. A problem afflicting all metals placed into seawater is **corrosion.** Pure water (distilled) conducts electricity poorly, but seawater conducts electricity very well because of its dissolved salts. When a metal object is placed in seawater, an electrical current is generated, similar to that in an automobile battery. The current removes particles from the metal, slowly dissolving or corroding it. The dollar cost of corrosion is estimated in the tens of billions per year as it affects ships, drilling platforms, and essentially anything metallic that comes in contact with seawater.

Another major physical resource of the ocean is simply its use as a recreational area. Sailing, fishing, skin and SCUBA diving, surfing, and swimming are enjoyable uses of the ocean as a physical resource. In many areas, however, industrial development and recreation compete for the nearshore area.

Future uses of the ocean are limited only by people's imagination, cost, and the law. Some exploitation already may have been excessive, however, such as the harvest of certain fish species.

SUMMARY

Ocean resources can be categorized as renewable (biological), nonrenewable (mineral), or physical. Scarce mineral resources in the industrialized nations, especially the United States, can be exploited from the ocean.

Important mineral resources presently obtained from the ocean are freshwater from desalination techniques, elements such as bromine and magnesium, sand and gravel, mineral placer deposits, oil, sulfur, and gas. Hydrocarbons have the greatest dollar value of any ocean resource.

Oil and gas production from the marine environment will likely increase. Potential marine mineral resources include phosphorite, manganese nodules, cobalt crusts on seamounts, heavy-metal muds, and minerals associated with seafloor spreading and vent activities at the ocean ridges. The mining of these potential resources is uncertain due to a poor world's mineral market.

The establishment of Exclusive Economic Zones (EEZs) that extend 200 nautical miles (370 km) out to sea from coastal countries will affect resource development. The EEZs allow government control of significant portions of the seafloor and its resources. The United States has not only a very large EEZ, but one with significant mineral potential.

Biological resources supply a major source of animal protein in several areas of the world. However, many feel that this exploitation may be near or even beyond its sustainable capacity. Exploitation of presently unused organisms, or the use of aquaculture, could increase the yield from the ocean. Other marine resources include energy from the sea and the use of the ocean for transportation and recreation.

QUESTIONS

1. What are the three categories of ocean resources?

2. List the principal mineral resources of the ocean. Describe their potential and the general geography of their occurrence.

3. What are placers? Where are they found? What key minerals do they sometimes contain?

4. What role does seafloor spreading play in the formation of mineral deposits? What role in the location of oil and gas deposits?

5. Why do the rocks under the continental shelf offer good potential for oil and gas?

6. What important minerals occur in the deep ocean? What is their economic potential?

7. What are the potential resources (mineral, biological, and physical) of the EEZ?

8. Where are the important areas of biological ocean resources? Why does this distribution occur?

9. What role should the declaration of an EEZ play in a country's development of its marine resources?

10. Describe the physical resources of the ocean.

KEY TERMS

agar	cobalt crust	factory ship	physical resource
algin	corrosion	lobster farming	placer
aquaculture	desalination	manganese nodule	province
by-catch	diapir	nonrenewable resource	renewable resource
cap rock	efficiency of conversion	ooze	reservoir bed
carrageenan	eutrophication	oyster farming	spat

FURTHER READING

Borgese, E. M. 1980. *Seafarm: The Story of Aquaculture.* New York: Abrams. A nicely illustrated history of the business of aquaculture.

Broadus, J. M. 1987. "Seabed Minerals." *Science* 235, pp. 853–60. A technical but understandable review of marine minerals.

Brown, E. E. 1983. *World Fish Farming: Cultivation and Economics,* 2d ed. New York: AVI Publishing Company. A discussion of aquaculture around the world.

Center for Marine Conservation. 1993. *Fish for the Future,* Washington D.C. A citizen's guide to the rules and regulations for the management of U.S. marine fisheries.

Champ, M. A., W. P. Dillon, and D. G. Howell 1984–85. "Non-Living EEZ Resources: Minerals, Oil, and Gas." *Oceanus* 27, no. 4, pp. 28–34. A general discussion of nonrenewable marine resources.

Cronan, D. S. 1992. *Marine Minerals in Exclusive Economic Zones.*

London: Chapman and Hall. A concise review of the occurrence of marine minerals in the EEZs of the world.

Hammer, W. M. 1984. "Krill: Untapped Bounty from the Sea." *National Geographic* 165, no. 5, pp. 626–43. All you want to know about krill, a potential major resource from the ocean.

Heath, G. R. 1982. "Manganese Nodules: Unanswered Questions." *Oceanus* 25, no. 3, pp. 37–41. A good discussion about the origin of manganese nodules.

Hedberg, H. D., J. D. Moody, and R. M. Hedberg. 1979. "Petroleum Prospects of the Deep Offshore." *American Association of Petroleum Geologists* 63, no. 3, pp. 286–300. The source of the material in Table 15-3.

Horn, M. H., and R. N. Gibson. 1988. "Intertidal Fisheries." *Scientific American* 258, no. 1, pp. 64–70. A good readable account of intertidal fisheries.

McGoodwin, J. R. 1990. *Crisis in the World's Fisheries: People, Problems and Policies.* Stanford, Calif.: Stanford University Press.

Manheim F. T. 1986. "Marine Cobalt Resources." *Science* 232, pp. 600–608. An important paper on cobalt crusts.

Marine Technology Society Journal 19, no. 4 (1985). A special issue devoted to marine minerals.

Marine Technology Society Journal 25, no. 1 (1991). A special issue on the use of living marine resources for food.

National Research Council. 1992. *Dolphins and the Tuna Industry.* Washington, D.C.: National Academy Press. A key document concerning the by-catch of dolphins and porpoises during the fishing for tuna.

Ross, D. A. 1980. *Opportunities and Uses of the Ocean.* New York: Springer Verlag. A general description of the many uses of the ocean.

Sissenwine, M. P., and A. A. Rosenberg. 1993. "U.S. Fisheries." *Oceanus* 36, no. 2, pp. 48–54. A readable article of the status, long-term potential yields, and stock management ideas of U.S. fisheries.

U.S. Department of Commerce, National Oceanic and Atmospheric Administration Fisheries of the United States. 1993. *1992, Current Fishery Statistics, ms. 9200,* Washington, D.C.: U.S. Government Printing Office. The source of much of the fishery statistics in this chapter.

CHAPTER 16

Marine Pollution

Locally and nationally, environmental incidents assault us daily in the media. The environment has become a major concern for nearly all nations, with special focus on issues of global change, such as the increase in greenhouse gases and reduction of ozone in the upper atmosphere. In the marine environment, pollution is an important and sometimes controversial subject.

When carefully studied, some pollutants and sources turn out to be relatively harmless, whereas many others can have a critical effect. This chapter discusses many of the types of marine pollution and their effects.

A marine bird partially covered with oil; if not treated, it will soon die.

(Photograph courtesy of Dr. David G. Aubrey, Woods Hole Oceanographic Institution.)

Introduction

What is marine pollution? A United Nations report defined it as: . . . "the introduction by [humans], directly or indirectly, of substances or energy into the marine environment (including estuaries) resulting in such deleterious effects as harm to living resources; hazards to human health; hindrance to marine activities, including fishing; impairment of quality for use of seawater; and reduction of amenities."* Note that this definition excludes *nonhuman pollution*, such as natural oil leaks, volcanic eruptions, and radioactivity.

Marine pollution is nothing new. It has occurred since civilization began. Recently, however, pollution from human activities has become so global in scale that it appears to be seriously affecting some aspects of life on our planet. Human activities clearly have contributed to the increase in certain elements in the environment, including carbon dioxide, arsenic, mercury, and lead. Human input often exceeds the natural input.

Some pollutants, notably carbon dioxide, circulate globally through the atmosphere and enter the ocean via the air-sea interface. As discussed in Chapter 12, carbon dioxide can become a critical contaminant at certain levels, and it can dramatically impact climate and sea level.

Recent studies have revealed new environmental problems, including a surprising incidence of cancer in some fish and marine mammals. In natural unpolluted waters, fish cancers are rare or nonexistent. The cause of fish cancers is uncertain, but in most cases potential carcinogens are present in the water, and some researchers suggest they may be the cause.

It does not necessarily follow that eating cancerous fish is harmful to humans. However, many experts recommend limiting consumption of fish from contaminated waters, and children and women of child-bearing age should not eat any. Fortunately, most fish consumed are from saltwater, where this problem is less frequent than in freshwater.

On the positive side, there is progress in controlling some pollutants. Major pollutants such as DDT and PCB (both discussed later in this chapter) are now used less in many countries. Oil pollution has diminished worldwide, although major spills still occur.

Marine Pollution

The preceding chapters showed that Earth, by surface area, is more of a water planet than a land planet. It should also be evident that the ocean strongly influences our environment, climate, and future. The ocean's waters, which cover about 71 percent of Earth, connect the shores of over 100 nations, and a major portion of the world's population lives near the water.

The ocean could accommodate some of the future need for energy, food, and minerals (see Chapters 15 and 18). It is also the main world thoroughfare for commerce and, unfortunately, the principal disposal area for many industrial and domestic wastes. These uses sometimes create environmental problems in marine areas, especially the nearshore zone, and can increase the potential for pollution damage to the ocean.

It should be emphasized that humans are not the only creatures dependent on the ocean. The oceans abound with lifeforms, most of which are intimately (or ultimately) dependent on each other. This marine life has evolved over hundreds of millions of years and usually cannot adjust to rapid environmental change. The decrease or elimination of one variety of life within an ecological level can seriously affect higher or lower orders of life. Likewise, increased use and pollution of estuaries and coastal areas may damage the extensive breeding areas for plants and animals on which other forms of life depend (see opening figure to this chapter and Figure 16-1).

Marine pollution is usually most severe in coastal areas or semi-enclosed water bodies, like the Mediterranean Sea. This is due to the dense human population and industrial development, compounded by restricted mixing of coastal waters with cleaner offshore waters. Within coastal areas, sewage discharge often presents the most direct danger to humans, especially if it is near beaches or harvestable shellfish beds (Figure 16-2). For these reasons, most pollution studies have focused on the coastal region. By contrast, the deep sea is *relatively* unpolluted.

The marine environment is an important resource for any coastal country. Over 75 percent of the U.S. population lives within 80 km (50 miles) of a coastal area— Atlantic, Pacific, Gulf of Mexico, or Great Lakes. Over 120 million Americans are estimated to participate each year in ocean-related activities, and to spend nearly $20 billion in doing so. Fishing, swimming, and boating, which in the past have been taken so casually, can be affected by the pollution of our nearshore waters.

Most marine scientists agree that pollution is one of the ocean's most serious potential problems. This problem often is complicated by a lack of research data: What natural pollutants did the ocean contain before human input? What pollution levels can the oceans tolerate before some environmental damage occurs? Unfortunately, the vast marine environment does not lend itself to experimentation. We cannot simply wait the necessary years to see the results from our present actions. However, the United States and other countries are committing increasing money for research on and monitoring of ocean pollution (Box 16-1).

*United Nations, Economic and Social Council, 51st Session, *The Sea: Prevention and Control of Marine Pollution*, 1971.

a

b

FIGURE 16-1   Two examples of pollution effects: (a) elephant seal trapped in discarded fishing line; (b) sewer outfall in La Palmas, Canary Islands; a hotel is downstream from the outfall. (Photograph (a) by John Domont, courtesy of the Center for Marine Conservation; photograph (b) courtesy of Dr. Ivan Valiela.)

The Great Lakes

The Great Lakes of Canada and the United States are among the world's largest freshwater resources. In decreasing order of size, the Great Lakes are Superior, Michigan, Huron, Ontario, and Erie. Over 37 million people live in the region, depending on water from the lakes for drinking, fishing, agriculture, industry, recreation, and shipping. Oceanographers, who usually concern themselves with the saltwater marine environment, also study these great freshwater lakes because of their importance and interest.

Pollution in the 1960s and 1970s caused extensive fish kills and made some water unfit for drinking. Fortunately, increased environmental controls, such as reduction of phosphate input, have improved the condition of the Great Lakes and allowed their continued use. However, the problems of these lakes are far from solved. Many rivers and harbors have impaired water quality and contaminated sediments.

Pollution: Defining, Detecting, and Measuring

A key difficulty in combating marine pollution is defining exactly what a pollutant is. Can a pollutant in one situation actually be a nonproblem or even beneficial in another?

Marine pollutants are as varied as our technological achievements. The previously mentioned United Nations report that defined marine pollution categorizes them as follows:

1. Wastes from domestic sewage, industry, and agriculture

2. Ship-borne pollutants, deliberately or operationally discharged

3. Interference with the marine environment from the exploration for and exploitation of marine minerals, including oil and gas

FIGURE 16-2   A nearshore algal bloom. Such blooms are caused by rapid and intense growth of algae, due to a large input of nutrients, often from sewage discharge. These blooms may not be toxic, but are unsightly and foul-smelling (Photograph courtesy of Dr. Don Anderson.)

Box 16-1
Monitoring Coastal Environmental Quality

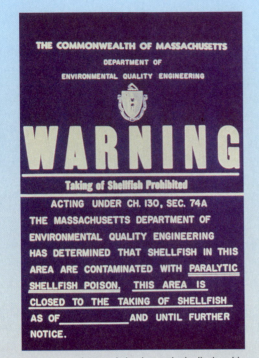

FIGURE 1 A type of sign increasingly displayed in coastal areas. (Photograph courtesy of Dr. Don Anderson.)

You commonly hear of beaches or shellfishing areas being closed due to pollution (Figure 1). Oil spills or accidental waste discharges seem to occur frequently. Media stories are generally not scientific, however, and accurate statistical assessment of coastal pollution is required to evaluate the effect of human activities on coastal and estuarine areas.

To achieve this, the National Oceanic and Atmospheric Administration (NOAA) established its National Status and Trends Program (NS&T) in 1984. Under this program, chemical analyses are made from a network of almost 300 coastal sites around the United States. This systematic study is important for accurate pollution assessment and for developing and implementing pollution remedies.

Coastal environmental monitoring is a big business in the United States, with over $130 million spent yearly. However, most funding is directed toward testing pollution sources *prior* to their release into the environment. The NS&T Program is more concerned with what happens *after* release into the environment.

The NS&T Program focuses on two major categories of chemicals, *trace metals* and *organic compounds*. The main trace metals measured are cadmium, chromium, copper, lead, mercury, silver, and zinc. These metals generally come from industrial activity, although natural concentrations do occur.

Organic compounds measured include DDT, chlordane, PCBs, and PAHs. DDT and chlordane are pesticides. DDT was banned in the U.S. in 1970 and chlordane use ceased in 1988. PCBs have been used for many purposes, such as in electrical power transmission, and were banned in the U.S. in 1976. However, all of these toxic compounds are still used in some countries, and residues remain in the environment of nations where they were once used. PAHs (polycyclic aromatic hydrocarbons) are natural compounds that occur in fossil fuels (coal and oil) and are produced when organic matter—including wood—is burned.

The significance of trace metals and the four organic compound groups is that, in sufficient concentration, each can be toxic to marine life. Under certain conditions, they also can be toxic to humans. The NS&T Program measures the concentration of these elements and compounds in sediments and in selected organisms. In particular, mussels and oysters are sampled, for they are sensitive indicators of pollution change. Their internal concentration of contaminants changes in response to change in the local environment. This fact, combined with their immobility on the seafloor, makes them very effective "pollution meters." An international program called Mussel Watch monitors pollution in this way.

Part of the NS&T study is to determine whether a contaminant has biological effects. One approach is to examine organisms such as fish at various sampling sites. Although only 36 of 5,600 fish examined so far exhibit tumors, their *geographical occurrence* may be significant. Fish discovered to have tumors include 14 winter flounder from Boston Harbor in Massachusetts and five sole from Elliot Bay in Washington State. Both areas have high concentrations of contaminants. Some fish with tumors, however, have been found where little evidence of chemical contamination exists.

In some locations, seafloor sediment cores were studied to detect concentration changes in certain chemicals over time. Recent decreases in some pollutants were observed, especially those such as DDT, which has been banned in the United States.

Early NS&T results present both good news and bad news. The bad news is that significant levels of contaminants were found, mainly in urbanized estuaries such

4. Radioactive waste disposal from peaceful uses of nuclear energy

5. Military uses of the ocean (for example, the loss of weapons at sea)

Obviously, controlling this plethora of pollutants and sources demands elaborate technology, engineering, controls, laws, and lots of money.

Immense volumes of waste material are introduced into the ocean yearly from factories, power plants, rivers, the atmosphere, and shipping. Some of these materials will degrade (decompose or break down into less harmful substances) in the ocean, but the rate at which they degrade is generally unknown. Likewise, the effect of many of these materials on the biology of the ocean is unknown. (And, if the waste material were not put into the ocean, it would end up on land, conceivably causing even greater damage.)

Our limited awareness of ocean problems is understandable. For many centuries, the ocean was seen as essentially incapable of being harmed or overfished. This view was reasonable at that time, considering human knowledge and technology. We have since found, however, that rivers and lakes (including the Great Lakes)

are polluted easily by human activity, especially now that Earth's human population exceeds 5.3 billion.

One paradox with some types of pollutants is that they can also enter the ocean through the normal processes of weathering and erosion of rock and soil, volcanic activity, and even by processes associated with seafloor spreading. The problem is to determine whether individual ocean pollutants result from human activity or from natural causes. Because few data are available, especially prior to the Industrial Revolution, it sometimes is difficult to gauge the human impact (see Table 16-1).

Human Impacts on the Marine Environment

Humans can affect the marine environment in many ways. One is by the introduction of plant or animal species into areas where they did not previously exist, often with very damaging effect (see Box 16-2). A recent marine case is the **zebra mussel,** a pest in Europe for over 200 years. In the mid-1980s it was unintentionally introduced into the Great Lakes, probably from an oceangoing ship. Since then, its rapid reproduction and adaptability have enabled its dramatic spread. It causes major problems: clogging municipal and industrial

TABLE 16-1

Pollutants Introduced into the Ocean by Natural Processes and by Human Activity

Ocean Pollutant	Natural Process or Input	Human Input
Heavy metals	River runoff (erosion of rocks), volcanic activity at subduction zones, decay of organic matter	Industrial and municipal discharges
Hydrocarbons	Natural oil and gas seeps (see Figure 16-15), river runoff, bacteria and other organisms in water, volcanoes	Shipping and drilling activities, runoff, atmospheric input
Nutrients	River runoff, reworking of bottom sediments, upwelling, and biological activity	Industrial and municipal discharge, agricultural effluents
Radioactive elements	River runoff (rocks containing radioactive elements), volcanic activity, atmospheric interactions	Nuclear power plants; nuclear weapons testing; industrial, medical, and municipal discharge
Particulate matter	River runoff, biological activity, mixing of bottom sediments, atmosphere, turbidity, currents	Fishing activity, mining, drilling, municipal and industrial discharge

Box 16-2
Marine Hitchhikers

Organisms may normally migrate from one area to another. This migration usually proceeds slowly, allowing time for the ecological balance to be maintained. Unfortunately, humans have introduced a mechanism by which marine organisms can be rapidly transported around the world and introduced into areas where they may cause ecological havoc.

The mechanism is the ballast water carried by transport ships. Most ships, if they do not have a full load of cargo, will take on large quantities of harbor water as ballast. This allows the ship to ride better as it travels to its next port. Once at that port, the ballast water is discharged, usually close to shore or even in the harbor. Traveling along in that ballast water can be a multitude of marine species, usually in their planktonic stage. They have hitchhiked to a new home, often with disastrous results for the existing biological community.

Marine biologists James T. Carlton of Williams College and Jonathan B. Geller of the University of North Carolina in Wilmington examined ballast water from 159 ships. Their ballast water originated in Japan and was discharged in Coos Bay, Oregon. In that water they identified 367 species, representing all marine phyla and most trophic levels. Some species came from environments with oceanographic characteristics similar to Coos Bay, so they had a chance of surviving in their new home. These survivors have the potential to disrupt an established biological community that has slowly evolved over thousands of years.

The eventual ecological damage from such hitchhiking (perhaps invasion is a better term) is hard to predict. Several recent examples show that the damage can be considerable. The most recent instance is the zebra mussel invasion, which probably began in the Caspian Sea and then spread into the Baltic Sea and to England. The creatures showed up in the Great Lakes in 1988. The mussel is expected to cause problems along the Atlantic coast, as it has already reached the Hudson River in New York. Damage from the zebra mussel can already be measured in the tens of millions of dollars.

Marine hitchhiking can proceed in any direction, wherever ships travel. In 1982, the comb jellyfish was introduced from U.S. waters into the Black Sea, and quickly reproduced to dominate parts of that environment, eventually destroying an anchovy fishery.

Reference

Carlton, J. T., and J. B. Geller. 1993. "Ecological Roulette: The Global Transport of Nonindigenous Marine Organisms." *Science* 261, pp. 78–82.

water-intake pipes, littering beaches (the shells are quite sharp), and growing on almost anything in the water, including boats (Figure 16-3).

Zebra mussels are removed by scraping and treatment with chemicals, which, however, can cause other environmental difficulties. The problem is especially bad in Lake Erie and Lake Ontario and promises to spread to other areas. The large-scale floods that inundated the midwestern parts of the United States in 1993 have spread the zebra mussels far south of Chicago and into the Mississippi River. Considerable research is underway to learn more about the zebra mussel and how to control or eliminate it.

Acid rain is an especially widespread pollutant. This general term includes acid snow, acid fog, and even acidic dry particles in the atmosphere. Its origin is still debated by some, but most agree that it occurs when sulfates (which can come from fossil-fueled power plants) and nitrates (often from automobile emission) combine with water to form acidic compounds, which then fall to Earth (Figure 16-4).

These compounds can reduce the pH of lakes and soils enough to severely affect or even kill organisms that live there. Acid rain can also react with limestone or marble buildings and statues, causing corrosion. The effect of acid rain on forests and lakes—sometimes thousands of miles from the source of the sulfates and nitrates—can be devastating.

The dilemma is that reducing the amounts of sulfates and nitrates emitted will cost literally billions of dollars and conceivably tens of thousands of jobs. This cost is too high for some. Fortunately, the ocean, because of its carbonate content and buffering ability, can neutralize enormous amounts of acidic material with little change in its pH (see Box 7-1, page 154). Thus, acid rainfall over the ocean is not a problem. The ocean, however, cannot neutralize the acid rain that falls on the continents.

Pollution is frequently an emotional subject because it involves people's health, livelihoods, tax dollars, and the environment we all share. Ideally, all forms of pollution would be eliminated. However, some pollutants simply cannot; some risks just cannot be avoided. In some

a

b

c

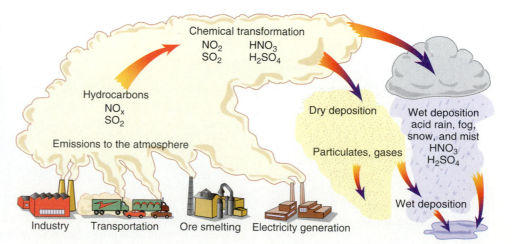

d

FIGURE 16-3 (a) Zebra mussels (*Dreissena polymorpha*) are small mollusks, less than 7.5 cm (3 in.) long. The light and dark bands on their elongated shells give them a zebralike appearance. They use strong threadlike material to attach themselves to hard surfaces.
(b) Zebra mussels growing on a mollusk shell.
(c) Thousands of Zebra mussels attached to a car recently fished out of Lake Erie.
(d) Zebra mussel shells litter a beach. (Photograph (a) courtesy of C. Czarnechi, Michigan Sea Grant College Program; (b) and (d) courtesy of Ohio Sea Grant; (c) courtesy of the Ontario Ministry of Natural Resources.)

FIGURE 16-4 How acid precipitation forms. The sources of sulfur and nitrogen oxides in the atmosphere include fossil-fuel power-generating plants, motor vehicles (cars, trucks, buses), and factories. In *wet deposition,* sulfur and nitrogen oxides react with atmospheric water vapor to form sulfuric and nitric acids, which reach Earth's surface as rain, mist, snow, and fog. In *dry deposition,* the sulfur and nitrogen combine with water in a lake, pond, or stream to form the acids. (Illustration from Kaufman and Franz, *Biosphere 2000: Protecting Our Global Environment* (New York: HarperCollins, 1993).)

Chemical transformation
NO_2 HNO_3
SO_2 H_2SO_4

Hydrocarbons
NO_x
SO_2

Emissions to the atmosphere

Dry deposition

Particulates, gases

Wet deposition
acid rain, fog,
snow, and mist
HNO_3
H_2SO_4

Wet deposition

Industry   Transportation   Ore smelting   Electricity generation instances, society must be willing either to pay the price to maintain a quality environment or to accept the effect of more pollution.

An example is the vast daily accumulation of garbage by large cities such as New York, where land disposal sites are filling rapidly. New land sites or disposal at sea are possibilities, but neither is an ideal solution. Additional land sites involve expensive land acquisition, transportation, and environmental problems. Innovations such as using garbage, dredged material, or other unwanted material to build new offshore islands for airports or recreation are clever alternatives. But this is expensive and many oppose the idea. Other alternatives should always be considered, including recycling, reduction, and reuse of waste.

Pathways to the Sea

Individual pollutants have several pathways to the sea, including direct input by rivers, runoff, sewer outfalls, dumping (deliberate or accidental), and from the atmosphere.

By the atmospheric route, up to 40,000 tons of industrial lead may reach the oceans yearly by rainfall and atmospheric fallout, whereas over 250,000 tons may enter via rivers and outfalls. However, the rain and atmospheric inputs are spread over the entire ocean, while the river and outfall inputs are mainly coastal. In this manner, the atmosphere can be the principal source of offshore pollution. Atmospheric input of radioactive materials from nuclear bomb testing is another example.

Many elements enter the environment through combustion of fossil fuels (coal, gas, oil, and wood). For example, most gasoline used to contain lead to help it burn smoother. Lead entered the atmosphere via automobile exhaust, and then was deposited in the ocean. In fact, more lead used to enter the ocean from the atmosphere than was carried in by rivers. Lead-free gaso-

line is now standard, reducing this atmospheric form of heavy-metal pollution. Other dangerous substances, such as sulfur and mercury, can also enter the ocean in this manner.

Most oceanographers feel that domestic sewage, agricultural waste, and industrial waste are the most serious marine pollutants today, especially in the nearshore area. However, pollution from oil spills often receives more attention because it is usually associated with a dramatic event like a shipwreck or because energy is such an emotional issue. Sludge and sewage are relatively unattractive and do not make for exciting television news.

The movement of all substances—both harmful and safe—to the ocean and through it is controlled by a very complex series of chemical, physical, and biological processes (Figure 16-5). Once an element or compound enters the ocean, its behavior and ultimate fate are determined by these processes.

Processes vary with type of area (estuary, river, bay, or open ocean), geographic region (tropical, polar), or general ocean properties (salinity, oxygen content, circulation, and so on). Thus it is very difficult to predict a pollutant's potential harm or ultimate fate as it passes through the marine environment. Clearly, understanding these factors and how they interact is a critical area of pollution research.

Controlling Marine Pollution: Options

Six principal options exist for addressing most pollutants:

1. Ideally, stop producing the pollutant. This option ranges from "doable" (banning DDT and CFCs) to virtually impossible (stop using coal, oil, and gas).

2. Develop methods for recycling the potential pollutant without environmental damage. An example is recycling aluminum cans.

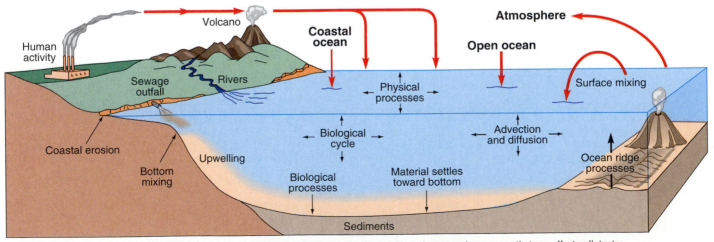

FIGURE 16-5 A simplified model of the pathways and processes that can affect pollutants.

3. Develop safe methods of disposal without environmental damage. A frequently mentioned example is to use old cement building blocks or tires to form marine habitats in coastal regions.

4. Dispose of the pollutant on the ground or below it (burial). Landfills and disposal areas are common examples of this option.

5. Dispose of the pollutant in the ocean. River runoff and ocean outfalls are typical methods.

6. Dispose of the pollutant in the atmosphere, by burning it or releasing it (if a gas). An example is the discharge of sulfates and nitrates from fossil fuel plants.

The first three are obviously the best options. In most instances, however, they are not viable, or can only be partially adopted. This leaves the latter three options, which can have harmful environmental effects. If one of the last three is to be used, the challenge is to find the option with the least damaging effect.

Domestic, Industrial, and Agricultural Pollution

Domestic, industrial, and agricultural pollution is an extremely broad category that includes human sewage, chemicals, organic material, and pesticides. Each can reach the ocean by different pathways, including rivers, runoff, and outfalls (see Figure 16-1b).

Many chemicals once thought to be safe now are known to be dangerous, sometimes even **carcinogenic** (cancer-inducing). It is estimated that over 60 million tons of hazardous waste are produced each year in the United States. This is about 227 kg (500 lbs.) for each citizen, and only a small portion of it receives adequate disposal. In spite of various laws, little effective action has resulted. The following are some of the problems that can result from such pollution:

1. *Oversupply of plant nutrients:* As discussed in Chapters 7 and 12, certain nutrients are critical for plant growth. But if they are oversupplied, plankton growth can accelerate, causing eutrophication or red tides (discussed further in a later section). Eutrophication results from a rapid growth and eventual oxidation of organic matter that consumes all dissolved oxygen in the water. It can be lethal to life in a closed or restricted body of water, as in some estuaries and lakes.

2. *Oxygen-consuming products:* Most organic waste, including human waste, oxidizes when it reaches the marine environment. However, if mixing and oxygen supply are

insufficient, all the dissolved oxygen could be consumed, killing most marine animals.

3. *Toxic chemicals and minerals:* Many industrial and agricultural wastes are extremely poisonous to marine life. In some cases, what appears to be a relatively innocuous waste may have unknown long-term effects. Examples include the pesticides DDT, Aldrin, and Dieldrin, which are described later in this chapter.

4. *Sediments:* U.S. rivers discharge nearly 500 million tons of soil and sediment per year (about 1.3 million tons per day) into their estuaries and nearshore areas (Figure 16-6). An input of large amounts of soil and sediment can harm bottom-living organisms such as oysters or clams. Sometimes dredging is required to keep waterways open. The disposal of the dredged materials is yet another pollution problem.

For several of these pollutants, the harm results from their depletion of *dissolved oxygen* in restricted or poorly mixed bodies of water. However, the effect on the ocean's total oxygen content and the implications for humans has sometimes been exaggerated. For example, it has been suggested that oxygen decrease or elimination in the ocean would drastically affect atmospheric oxygen content. This is not correct, because the production and consumption of oxygen in the ocean is essentially steady-state, and the net exchange with the atmosphere is fairly small.

It has been estimated that if all *marine* photosynthesis were to stop, the atmospheric concentration of oxygen would drop about 10 percent in 1 million years. Such a

FIGURE 16-6 Sediment plume (left) from the Mississippi River as it enters the Gulf of Mexico. The river here carries about 5 tons of suspended sediment per second into the Gulf. (Photograph courtesy of Dr. Nenad Iricanin, Florida Institute of Technology.)

loss, although undesirable, probably could be tolerated by most nonmarine species of life. Life in the ocean, however, would be devastated.

Pesticides

Among the more dangerous pesticides found in the marine environment are **DDT** (dichlorodiphenyltrichloroethane), **Aldrin,** and **Dieldrin,** all chlorine-based poisons. Most of the publicity has focused on DDT, and its use is restricted in many areas of the world. Unfortunately, DDT is a stable compound that tends to concentrate in organisms to much greater quantities than normally exist in water. This process, known as **biological magnification** (Figure 16-7), can be very effective.

In an exceptional example, oysters exposed continuously to waters containing only 0.1 part per billion (ppb) of DDT were able to concentrate up to 7.0 parts per million (ppm) of it in their tissue, a *70,000-fold increase.* DDT's effect on organisms varies. Its presence in some marine birds, such as pelicans, causes very thin eggshells, making reproduction difficult if not impossible (heavy birds incubating their eggs literally crush them).

Some phytoplankton significantly reduce photosynthesis when DDT is present in concentrations of about 10 ppb, yet other plankton suffer no adverse effects.

Once the DDT reaches the ocean, it initially remains in the upper mixed layers. Eventually it is carried to deeper layers by incorporation into organic matter or by sedimentation. These processes are not completely understood.

When DDT was phased out in the U.S. in 1973, the pesticides Aldrin and Dieldrin were used as replacements. However, they are even more toxic than DDT. Fortunately, in the United States, the Environmental Protection Agency (EPA) has restricted their use. This action, challenged by various industries, has been moderately successful, but large amounts of these pesticides still remain in the environment. Both are more soluble in water than DDT, although they are carried to the ocean mainly by the atmosphere. Like DDT, Aldrin and Dieldrin become concentrated in animals higher in the food chain at levels exceeding normal concentration occurring in the water by factors of thousands or more (see Figure 16-7).

The control of pollution by such pesticides depends almost entirely on stopping their use. Prohibition is diffi-

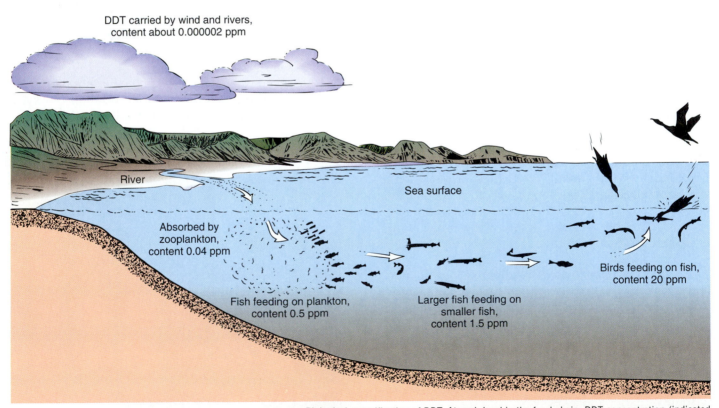

DDT carried by wind and rivers, content about 0.000002 ppm

River

Sea surface

Absorbed by zooplankton, content 0.04 ppm

Fish feeding on plankton, content 0.5 ppm

Larger fish feeding on smaller fish, content 1.5 ppm

Birds feeding on fish, content 20 ppm

FIGURE 16-7 Biological magnification of DDT. At each level in the food chain, DDT concentration (indicated in parts per million, or ppm) increases and is passed to the next level. Animals high in the food chain, like humans and birds, can thus receive pesticide concentrations millions of times greater than generally found in the environment.

cult, however, because they are used for a reason: they are badly needed (and still used) in many countries to control disease-carrying insects (for example, the mosquitoes that carry malaria). Thus a dilemma exists, because frequently no alternative pesticide exists.

Synthetic Organic Compounds

Synthetic organic compounds comprise a wide and unfortunately increasing variety of chemical pollutants, including materials initially thought to be quite harmless, such as plastics, fibers, polymers, solvents, and fertilizers. Many of these materials reach the ocean via river and atmospheric discharge or through shipping accidents. Once in the ocean, their pathways, their effect on marine animals and plants, and their fate are rarely known. Among the more dangerous are the *chlorinated* and *halogenated hydrocarbons*, including **chlorofluorocarbons (CFCs).** These are common in flame retardants, fire extinguishers, air conditioners, solvents, and aerosol propellants (until the latter was recently banned in the United States). These compounds cause liver damage in humans and are sometimes carcinogenic. (The effect of CFCs on Earth's ozone layer was described in Chapter 13.)

One dangerous chemical in this category is **polychlorinated biphenyl (PCB),** used until recently in making plastics, electrical insulation, fire retardants, and in heat exchangers. Toxic effects of PCBs were noted in the 1930s, but not until 1979 did it become illegal to manufacture them. Over 600,000 tons of this pollutant have entered the ocean, and an estimated 50 to 80 percent of it may be in the North Atlantic. PCBs have been shown to reduce the growth rate of some phytoplankton even when present in amounts as small as 10 ppb (parts per billion). The reduced growth rate in turn could affect the phytoplankton species composition and influence the entire food chain.

PCBs are fairly common in the environment and appear to show biological magnification (Table 16-2). Studies of PCB pollution in the Great Lakes reveal that some salmon and trout have PCB contents 100,000 to 1,000,000 times that of the surrounding water. These concentrations can, in part, be passed on to the organisms that eat such fish. As yet, no confirmed human deaths have been reported due to eating marine organisms that contain high PCB content. However, there have been instances where PCB content within certain organisms (fish and shellfish) has been found to exceed official U.S. government standards.

The effect of PCBs on humans varies: nausea and vomiting, abdominal pain, and jaundice. Long-term exposure can weaken the ability to recover from other diseases and can affect reproduction, extending even to the possibility of birth defects.

TABLE 16-2
Biological Magnification: PCB Concentration from Organisms, Sediments, and Water of the Atlantic Ocean

Source	Average PCB Concentration (parts per billion)
Sea mammals	3,000
Sea birds	1,200
Mixed plankton	200
Finfish from upper ocean waters	50
Finfish from midwaters depths	10
Bottom-living invertebrates	1
Deep-sea sediments	1
Seawater	0.001

Data: From Harvey, 1974

Excess Nutrients

Nearshore waters may receive large amounts of industrial and human waste that contain nitrogen and phosphorus nutrients. If this creates an oversupply of these nutrients, or if the waste is not sufficiently circulated, the excess can cause extremely heavy growth of algae and other phytoplankton. This growth colors the water, makes boating and swimming unappealing, and can clog water intakes and filters (Figure 16-8). If the process continues, decay of this extraordinary number of organisms can deplete available oxygen in the water and result in large fish kills.

One such growth, called a **red tide,** has become a major problem. Red tides used to occur along Florida's Gulf Coast about once every 16 years, but now they occur almost every year. Red tides are caused by the rapid growth of dinoflagellates, normally present in very small numbers but whose population can increase by factors of a million or more under certain conditions (Figure 16-9). The phytoplankton themselves generally are not poisonous to the fish, mussels, and clams that feed on them. However, toxins produced within these animals sometimes can kill fish as well as poison humans who eat organisms that have fed on the dinoflagellates. The incidence of red tides may be increasing due to the increased supply of nutrients from pollution.

Heavy-Metal Pollution

Among the more dangerous pollutants are the so-called heavy metals, introduced into the marine environment by waste and sewage. These include mercury, cadmium, silver, nickel, and lead. Some heavy metals are present in

FIGURE 16-8 In Venice, Italy, most homes discharge sewage directly into coastal canals. Note the large algal growth nourished by such a discharge. (Photograph courtesy of Dr. Ivan Valiela.)

the factory were being concentrated in shellfish eaten by the townspeople. By 1969, at least 110 people had acquired Minamata disease and 45 had died. Several unborn children were also found to have symptoms.

Two unanswered questions are: What heavy metal intake can humans tolerate, assuming an otherwise normal diet? And what heavy metal input do marine organisms need in their diet? Several, especially copper and zinc, appear to be necessary in some amount for life processes. As with DDT, however, heavy-metal content can be amplified on its way up the food chain (see Figure 16-7).

Lower members of the food chain, such as plankton, can concentrate heavy metals. When the plankton are eaten by herbivores or carnivores higher in the food chain, they pass on their heavy-metal content. High in the chain are predators such as swordfish and tuna, which may have concentrations great enough to raise concern about human consumption. The problem is complex because it is unclear how much of these metals these fish may have contained "normally," prior to our awareness of the problem and prior to our polluting the marine environment with these metals. For example, the mercury content of museum specimens of tuna, some caught almost 100 years ago, is similar to the mercury content in recently caught fish.

It is difficult to evaluate the human input of heavy metals because some enter the ocean at similar or greater rates from natural activity. For example, uncontrolled human input of mercury to the ocean is 4,000 to 5,000 tons per year. Natural weathering of rocks supplies sediment to the ocean containing 5,000 tons of mercury per year, most carried by rivers. However, some of the highest mercury values in the deep sea occur near the ocean ridges (probably associated with the ocean vents), suggesting still another source (for example, see Figure 7-10, page 168).

This example highlights our paucity of data on the pathways elements take into the ocean. Relatively few samples have been analyzed for mercury, and its ocean-wide distribution is unknown. Studies made along the Atlantic coast of Europe show concentrations as low as 16 parts per billion; concentrations along oceanic ridges can attain 400 ppb; and even higher values occur near industrial areas. The important questions are: How much is being introduced? Is it being consumed and retained by organisms that we eat?

Although mercury values in seawater are generally very low, biological magnification does occur. *Mercenaria mercenaria*, the common quahog (hard-shelled clam) of estuaries, may contain 0.1 to 0.4 ppm of mercury. This value is below—but near—the U.S. limit of 0.5 ppm for safe human consumption. (Limits like these are frequently determined by laboratory tests on animals, and

extremely small concentrations, on the order of 1 ppb. In certain situations, marine plants and animals concentrate these metals without apparent harm to themselves. If humans consume these organisms, however, health problems can result. Heavy-metal effects are often difficult to identify because they can be disguised by other pollutants. A classic example is "Minamata disease."

This disease takes its name from Minamata, a town on the west coast of Kyushu Island, Japan, where a factory produced chemicals such as mercury compounds. As early as 1953, some sort of convulsive disease was killing animals, including dogs and cats. The symptoms also occurred among the fishermen of the town, who were found to have extensive nervous-system damage. Investigators received little help from the factory, which would not reveal the chemical agents they were using or what products were being introduced into the sea. Eventually, it was found that mercury compounds discharged from

FIGURE 16-9   (a) A dramatic picture of a red tide in a Japanese bay.

(b) A red tide off the Florida coast.

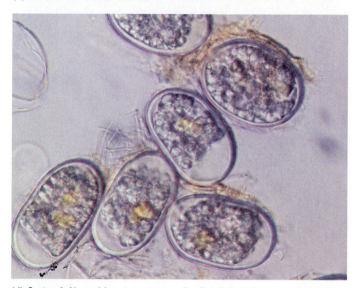

(c) Fish kill from a red tide off Florida.

(d) Cysts of *Alexandrium tamarensa*, a dinoflagellate that can cause red tides. The organism is about 50 microns long. (All photographs courtesy of Dr. Don M. Anderson; (b) and (c) taken by the Florida Department of Natural Resources.)

their actual applicability to humans is always questionable.) Organisms with this amount of mercury are supposedly safe to eat, but the question remains: How much of such organisms can a person safely eat?

Ocean Dumping

Ocean dumping is often the most preferable of three alternatives, the others being placement of the waste on land or into the atmosphere. The ocean may be the final choice for economic reasons: land, especially coastal land, is generally considered more valuable than the ocean. The atmosphere option is limited to certain pollutants and geography (so toxics don't blow over heavily populated areas).

Another argument for dumping waste at sea is that it may quickly oxidize and decompose some substances, perhaps even benefiting the marine environment. Some studies show, however, that microbial decay of organic matter in the deeper ocean is anything but quick, proceeding 10 to 100 times slower than at similar temperatures on land. This surprising discovery came from a unique and unplanned experiment.

The research submersible *Alvin* once sank in about 1,540 m (5,052 ft.) of water, fortunately with no loss of human life. Aboard the submersible were sandwiches and fruit. The submersible was recovered about 10 months later, with the food surprisingly well preserved. The implication is that decay may be slower in the deep sea than anticipated, so using it as a dumping site for organic waste may need rethinking since the waste may remain there for very long periods. Further experiments are certainly appropriate, although better planned than the *Alvin's*!

Disposal of waste at sea is a major form of pollution. Many countries routinely dump waste in shallow waters off their coasts, assuming that the ocean can easily absorb these ingredients (Figure 16-10). In late 1990, however, most major nations agreed to a global ban on industrial waste dumping at sea. The ban phases out marine dumping by 1995 and is binding in over 60 countries, including the United States, Japan, Britain, Germany, Russia, France, and most other industrialized nations. Individual countries will monitor ships and prosecute violators. If successful, this ban should eliminate about 10 percent of marine pollution.

Plastic Problems

Recent expeditions have found surprisingly large amounts of plastic and polystyrene particles floating on the ocean surface and stranded on islands. Concentrations up to 12,000 particles per km^2 (0.39 miles^2) have been detected in the Sargasso Sea off the U.S. East Coast (Figure 16-11). Such large concentrations probably reflect increased production of plastics on land and subsequent dumping at sea. The particles can serve as an area of attachment for small plants and animals, but bigger pieces may prove hazardous to larger marine animals through entanglement or ingestion (see Figure 16-1a). The simple plastic rings that hold together a six-pack of beverage cans may survive *over 400 years* in the ocean, and continue to trap or choke animals during that period.

Recently, volunteers started a yearly scouring of beaches around the world to remove litter. Coordinated by the Center for Marine Conservation in Washington, this effort involves over 160,000 volunteers and covers more than 8,000 km (5,000 miles) of beach in 33 countries. The types of litter collected are documented, and various forms of plastic usually are about 60 percent of the debris collected.

International rules and standards for plastic disposal in the ocean (called MARPOL Annex V) were adopted in 1988 by the United States and over 30 other countries. These rules prohibit the dumping of plastic and other garbage in the ocean and require coastal facilities for handling plastic waste. Although this certainly will reduce future plastic input into the ocean, it will still take several hundred years for the plastic already there to decompose!

Ocean Incineration

Ocean incineration is an innovative waste-disposal method, one especially appealing for chlorinated hydro-

FIGURE 16-10 Acid-iron waste disposal at sea off New York City. (Photograph courtesy of Mark Dennett, Woods Hole Oceanographic Institution.)

FIGURE 16-11 Plastic pellets collected during several plankton hauls off the eastern United States. (Photograph courtesy of Dr. R. Jude Wilber.)

carbons. Burning converts them into carbon dioxide, hydrochloric acid, and water, substances easily absorbed by the ocean and less harmful than the chlorinated hydrocarbons. The burning process is relatively complete, destroying over 99.99 percent of the cargo.

Recent years have seen several proposals for ocean incineration off the U.S. coast, in the Gulf of Mexico and off New Jersey. The Gulf of Mexico proposal was to burn nearly 80 million gallons of pollutants, including PCBs, aboard *Vulcanus II*, a vessel especially designed for the purpose (Figure 16-12). Incineration was to occur over 300 km (180 miles) from any Texas city and over an area of 7,770 km² (3,000 miles²) where water depths exceed 900 m (3,000 ft).

At hearings concerning the proposed burning, the public was strongly opposed. Concerns included marine life, potential effects on tourism, the possibility that materials would be carried back to land via the atmosphere, and the lack of knowledge about the process and its effects on the environment. Transportation of the wastes, especially over land to the ship, was also a potential problem. A similar public concern occurred following a proposal to burn 700,000 gallons of PCB-laden liquid waste about 200 km (120 miles) off New Jersey. The EPA has decided not to issue incineration permits until specific rules are written and accepted.

How such rules can be adequately developed without test burnings and other studies is unclear. The issue of ocean incineration (which is done by some European countries) is a clear example of how emotions can color pollution issues. To some, ocean incineration appears the most modern and safest way of disposing of some wastes; to others, it represents a clear danger. But land disposal does not seem a reasonable solution, and to ignore the waste is becoming impossible because its volume is steadily growing.

FIGURE 16-12 *Vulcanus II*, a vessel especially equipped to burn waste at sea. Material is burned at over 1,100°C (over 2,000°F) at 25 tons per hour. Tests indicate that this may be an efficient method for disposal of some toxic substances. (Photograph courtesy of U.S. Environmental Protection Agency.)

Oil Pollution

Oil is a major source of ocean pollution. More than 5,000 spills are estimated to occur yearly in the waters of the United States alone; on a worldwide scale the numbers are even worse. Over 2 billion tons of crude oil are used worldwide each year, and about half is carried by tanker (Figure 16-13).

Of some 6,000 tankers operating worldwide, about 360 (6 percent) are involved in collisions or groundings each year. Estimates of the total hydrocarbons released into the ocean vary, but a detailed 1975 study by the U.S. National Academy of Sciences concluded that about

FIGURE 16-13 Major international marine shipping routes for petroleum. Arrow size reflects petroleum volume being transported. You can see the importance of the Persian Gulf to worldwide oil supply. (Adapted from British Petroleum, Ltd.)

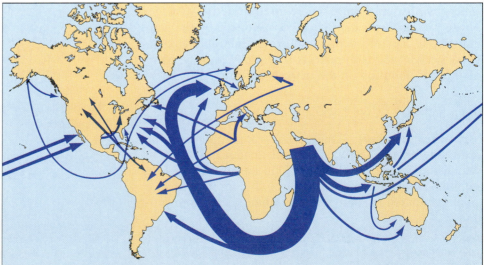

FIGURE 16-14 Sources of petroleum pollution in the oceans. (Adapted from Parricelli and Keith, 1974.)

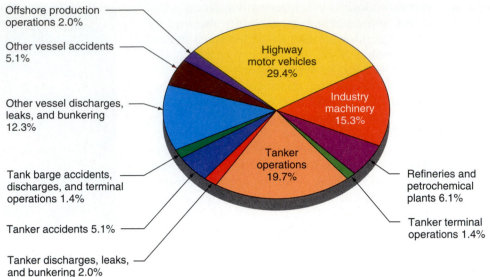

Offshore production operations 2.0%

Other vessel accidents 5.1%

Other vessel discharges, leaks, and bunkering 12.3%

Tank barge accidents, discharges, and terminal operations 1.4%

Tanker accidents 5.1%

Tanker discharges, leaks, and bunkering 2.0%

Highway motor vehicles 29.4%

Industry machinery 15.3%

Tanker operations 19.7%

Refineries and petrochemical plants 6.1%

Tanker terminal operations 1.4%

6 million tons (about 34.5 million barrels, or 1.9 billion gallons)* entered the environment. This loss came from several sources, but transportation was the major polluter (Figure 16-14).

A 1985 evaluation by the National Academy of Sciences concluded that petroleum input into the marine environment from all sources ranged from 1.7 to 8.8 million metric tons per year, with the best estimate being 3.2 million metric tons (Table 16-3). Although the 1985 number is half the 1975 value, it does not necessarily represent a decline in input. It may just reflect a better estimate of individual sources.

Discharge at sea can occur because of leaking or accidents, as well as from the illegal practice of cleaning oil tanks at sea by flushing them with seawater. Oil and gas spills are especially notable when a large ship goes aground or when a drilling platform has an accident. These events take place over a few days, or even months, and receive high visibility in the media. Such spills obviously can cause severe environmental damage, but so can small accidents that go unnoticed by the general public. Despite many studies, considerable controversy persists about the effects (especially long-term) of oil and gas pollution on the marine environment.

Major Spills

TORREY CANYON. One notable catastrophe was the grounding of the *Torrey Canyon* off England in 1967, in which the tanker lost over 700,000 barrels (around 30 million gallons) of crude oil. A large ship for its time, it is small compared to some modern tankers. Efforts to con-

---

*A ton of petroleum can contain 6.5 to 7.5 barrels of oil, depending on the oil's density. A barrel of oil contains 42 gallons, so a ton of petroleum contains between 273 and 315 gallons of oil.

TABLE 16-3
Input of Petroleum into the Marine Environment

Source	Probable Range[a]	Best Estimate
Natural sources		
Marine seepage	20–2,000	200
Sediment erosion	5–500	50
Offshore production	40–60	50
Transportation		
Tanker operations	400–1,500	700
Drydocking	20–50	30
Marine terminals	10–30	20
Bilge and fuel oils	200–600	300
Tanker accidents	300–400	400
Nontanker accidents	20–40	20
(Total transportation)	(950–2,620)	(1,470)
Atmospheric deposition	50–500	300
Wastewaters, runoff, and ocean dumping		
Municipal wastes	400–1,500	700
Refineries	60–600	100
Nonrefining industrial wastes	100–300	200
Urban runoff	10–200	120
River runoff	10–500	40
Ocean dumping	5–20	20
(Total Wastes and Runoff)	(585–3,120)	(1,180)
Total	1,700–8,800	3,200

[a]The input rates are expressed in thousands of metric tons per annum. The lowest figure is the sum of individual minimum input estimates, and the highest figure is the sum of individual maximums.

Source: From *Oil in the Sea: Inputs, Fates, and Effects.* Washington, D.C: National Academy of Sciences, 1985.

trol the spreading oil generally were ineffective and much of it reached the shore, causing severe environmental effects.

Ironically, it was estimated that 90 percent of the marine animal deaths were caused by the *detergent* used to clean up the oil. Detergents cause the oil to form smaller drops, which are more easily dispersed into the marine environment and are more detrimental to marine life. In addition, detergents themselves are a pollutant. A lesson from this accident is that better methods are needed to contain oil movement once a spill occurs.

SANTA BARBARA.   A major U.S. oil spill occurred off Santa Barbara, California, in 1969. A high-pressure offshore oil well blew out, introducing nearly 17,000 barrels (700,000 gallons) of oil into the environment. Much of it drifted ashore, coating beaches and sea birds that dived unknowingly into the oil-covered water. Oil leaked into the area for over 300 days, causing several million dollars' damage.

Due to inadequate data, the environmental effects of this spill have been widely debated. The quantity and quality of marine life in this area before the spill were not well known. The offshore Santa Barbara area is especially vulnerable to oil pollution because natural petroleum seepage in the area may attain 50 to 75 barrels per day (about 2,100–3,150 gallons or 8,000–12,000 liters). This natural rate could equal the spill in only 200–300 days. Natural oil and gas seeps are common in some areas (Figure 16-15) and contribute a natural and modest amount of oil in the environment (see Table 16-3). The presence of natural seeps does not make the

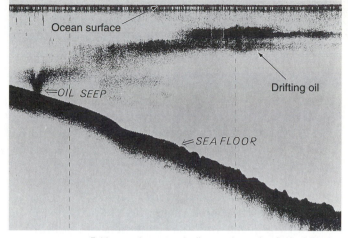

FIGURE 16-15   Evidence of a natural oil seep from the Gulf of Mexico. (Photograph courtesy of Dr. Richard A. Geyer, College of Geosciences, Texas A&M University.)

Santa Barbara spill all right, but it does contribute to the difficulty of evaluating its impact.

ARGO MERCHANT.   In 1976, the Liberian-registered *Argo Merchant* went aground off Nantucket, Massachusetts, spilling about 181,000 barrels (7.6 million gallons) of oil (Figure 16-16). At the time, this was the largest spill to occur in U.S. waters. But the dubious honor for the biggest tanker accident in the world belongs to the *Amoco Cadiz*, which broke up off the Brittany coast of France in 1978. Much of the vessel's 68 million gallons (257 million liters) of oil reached the nearby coast, damaging it and nearshore shellfisheries.

a

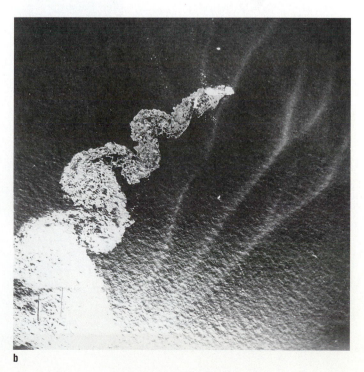

b

FIGURE 16-16   (a) The *Argo Merchant* as it starts to sink. (b) Oil is clearly visible from the *Argo Merchant* from an elevation of 1,676 m (about 5,500 ft.). (Photograph (a) courtesy of Dr. John Farrington; (b) courtesy of National Aeronautics and Space Administration.)

IXTOC I. The record for the largest overall oil spill was held for about 11 years by the Mexican *Ixtoc I* offshore platform blowout in the southern Gulf of Mexico, Bay of Campeche area. The well blew out on June 3, 1979, and continued spilling oil and discharging gas until March 23, 1980 (Figure 16-17). As the disaster proceeded, Mexico and the United States debated into whose waters some of the oil reached. The issue, of course, was responsibility for the accident, cleanup, and damage compensation. The total oil released may never be known, but as much as 30,000 barrels of oil per day was believed to be escaping. One estimate is that 500,000 tons was released over a nine-month period. The Mexicans attempted several techniques to stop the flow: they plugged the hole with steel balls, drilled an adjacent relief well, and finally capped it successfully with a steel sombrero-shaped device.

Scientists studying the spill found damage assessment difficult. Clearly it affected the entire marine ecosystem, but the ocean, as one scientist said, showed a definite ability to clean itself up. The potential harm of this spill may have been mitigated by favorable winds, the well's offshore location, and much of the oil being burned off at the surface.

PERSIAN GULF. A significant challenge to the record for world's largest oil spill occurred in 1983 in the Persian Gulf. In February a storm damaged a drilling platform in the Nowruz field at the northern end of the Persian Gulf. This occurred during the severe warfare between Iran and Iraq. Later that year, missiles added to the damage. Up to nine wells were discharging an estimated 5,000 to 7,000 barrels per day into the Gulf. Experts predicted the worst spill ever, and numerous fish kills were observed. However, Iran successfully capped the wells. Unfortunately, during the Persian Gulf War of 1990–1991, all oil spill and burn records were shattered—hopefully never to be so challenged again (Box 16-3).

EXXON VALDEZ. The worst U.S. marine oil spill occurred in 1989, when the *Exxon Valdez* ran aground in Prince William Sound, Alaska, spilling about 11 million gallons of crude oil into a nearly pristine environment. Fortunately, another 42 million gallons were safely removed from the grounded vessel.

The nearly new *Exxon Valdez* was 300 m (987 ft.) long, with a 211,000 ton capacity. Because the vessel had sophisticated navigation equipment to avoid problems like

FIGURE 16-17 The location and oil movement from the *Ixtoc I* blowout. The photograph was made during a NOAA cruise to the region. Note the fire-fighting equipment spraying water on the burning oil erupting from the blowout. (Photograph courtesy of National Oceanic and Atmospheric Administration.)

Box 16-3

The Persian Gulf Oil Spill and Fires

One appalling result of the 1990–91 Iraqi invasion of Kuwait and the subsequent Persian Gulf War was the discharge of large amounts of oil into the Persian Gulf. An international group of scientists estimated the spill at between 6 and 8 million barrels (252 to 336 million gallons). If the estimate is correct, this spill tops the previous estimated record of 4.2 million barrels from the 1979 *Ixtoc* incident in the Gulf of Mexico. The long-term impact to the Persian Gulf, which already had some oil pollution, cannot be known for years but surely will be substantial.

The Persian Gulf is a very shallow sea (median depth of 35 m, or 115 ft.) of very high salinity (about 40‰) due to the greater evaporation rate in this hot, dry area. Many organisms living in this already stressed region are near their limit of environmental tolerance. The Gulf oil spill badly affected a shrimp fishery that had produced about 10,000 tons per year, numerous species of fish, turtles, porpoises, and birds, and several coral reef areas.

One species especially vulnerable to oil pollution in this area is the dugong, a slow-moving mammal similar to the manatee (sea cow) of Florida and the Caribbean. The dugong was thought to have been eliminated by previous oil spills in the Gulf. Following the 1991 spill, however, over 50 dead animals washed up on various beaches. The drifting oil also damaged several desalination plants, a major source of freshwater in this desert region.

Another problem was the burning oil wells of Kuwait. At the peak of this catastrophe, more than 600

FIGURE 1   Some of the burning oil wells set afire during the Iraqi invasion of Kuwait in 1990–91. The conflagration released large amounts of sulfur and other atmospheric pollution, much of which fell into the Persian Gulf and Indian Ocean. (Photograph courtesy of National Oceanic and Atmospheric Administration.)

burning wells released about 40,000 tons of sulfur dioxide and half a million tons of carbon dioxide into the atmosphere every day (see Figure 1). These numbers are greater than the combined daily industrial discharge of France, Britain, and Germany.

Fortunately, the smoke did not rise high enough to be carried out of the general area and have a global impact. Locally, however, the smoke plume extended over the Arabian peninsula to Oman, comparable to a cloud extending from New York City to Florida. The last burning well was extinguished early in 1992.

---

running aground, negligence was suspected, and some have called it the largest drunk-driving accident on record. The spill and its consequences were front-page news for several years. Oil from the spill eventually reached islands in the Aleutian Chain, over 1,100 km (700 mi.) from where the ship went aground.

The *Exxon Valdez* oil spill was unique in ways beyond the size of the disaster. First, the cold water of Prince William Sound slowed the natural decay processes that break down oil. Second, the frigid and foggy weather prevented a continuous cleaning effort, and delays were frequent. Finally, the many gravel beaches in the area made it easy for the oil to soak into them and then slowly leak back out (Figures 16-18c and 16-18d). Public concern was great, in part because of the variety of animals that live in Prince William Sound, including

sea otters, whales, sea lions, seals, and numerous resident and migrant birds. The spill also threatened several species of salmon, including many that are spawned in local hatcheries. Fishing is a big business in the region, with a catch of about $120 million anticipated in the year following the spill.

The spill killed over 30,000 sea birds and migratory birds, about 900 sea otters, and over 100 bald eagles. Ironically, the cleanup and attempt to save animals was a boon to the local economy. Workers were paid up to $2,000 per week, and those who owned ships could earn even more. At one time, 10,000 people worked on the cleanup program, making it one of the better studied and managed U.S. oil spills.

Controversy persists as to the degree of recuperation of the environment, and several lawsuits are pending

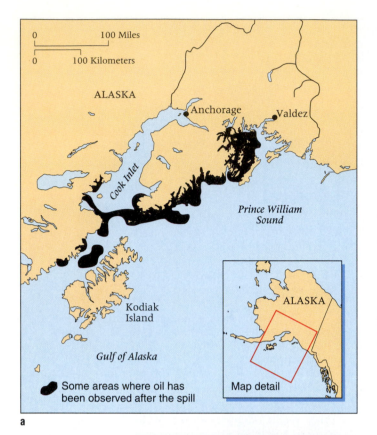

a

b

c

d

FIGURE 16-18   (a) The location and main areas where oil was observed after the *Exxon Valdez* oil spill.
(b) *Exxon Valdez* (center) with a vessel that is helping to remove some oil still in the ship. Note the long boom in the water that is restraining some of the spill.
(c) Some oil from the spill on a local beach.
(d) Workers cleaning a beach with steam. (Photographs (b) and (c) courtesy of Dr. David G. Aubrey.)

between Exxon Corporation and the State of Alaska and other parties. Few question the initial ecological impact of the spill. However, some suggest that cleanup techniques—such as hot-water pressure-hosing of beaches (Figure 16-18d)—may have introduced additional problems. Some innovative methods for oil cleanup were successful (see Figure 18-1, page 439) and the prospect for recovery of Prince William Sound is fairly good. However, the long-term effects on the breeding of some species is unknown. The *Exxon Valdez* accident also

shows how hard it is to clean up a big spill—even when the proper gear and organization are in place.

KHARG-5. In December of 1989 an Iranian tanker, *Kharg-5*, exploded in the Atlantic off Morocco, releasing over 20 million gallons of oil. This spill, barely noticed on this side of the Atlantic, was about twice the size of the *Exxon Valdez*.

FLORIDA. Probably the best-studied oil spill occurred in 1969 when the oil barge *Florida* ran aground in Buzzards Bay, Massachusetts, off West Falmouth. The grounding discharged about 180,000 gallons of Number 2 fuel oil (standard heating oil) into the coastal waters. Onshore winds and tides carried much of it into very productive marshlands, killing many organisms.

The area had been studied by scientists of the Woods Hole Oceanographic Institution before the spill occurred. Thus they had a unique opportunity to document the damage and found that the oil retained its toxicity for several years. In some areas the oil penetrated 60 cm (about 2 ft.) into the sediment. Shellfish were especially affected. Total fishing and shellfish losses eventually exceeded $1 million.

The area appeared "normal" to casual visitors a few weeks after the disaster, but careful study showed that the area had not returned to its original condition even a *decade* later. Indeed, two decades later, some slight traces of oil were still detectable below a few centimeters' depth in the sediment; otherwise the area appeared to have recovered. Note that the Falmouth spill was a relatively small one; what distinguished it was the detailed studies that showed how long-lasting oil pollution effects can be.

Acute and Chronic Effects of Oil Spills

Oil spill effects can be acute or chronic. *Acute effects* result from a single spill or discharge, which can kill numerous organisms, but—depending on the area and oceanographic conditions—the fauna and flora usually recover, eventually. About 3 to 4 percent of the total annual oil input to the environment is estimated to result from acute spills. *Chronic effects* occur when the oil pollution happens continuously or repeatedly without sufficient time for the area to recover. Chronic effects last longer than acute effects.

Are Platforms Bad Polluters?

Oil pollution is an emotional issue, so hard facts are often ignored. Here are some to consider: according to the 1985 National Academy of Sciences report, the oil entering the marine environment from offshore drilling (50,000 tons per year; see Table 16-3) is four times less than from natural oil seeps (200,000 tons per year) and

29 times less than that from accidents involving tanker transportation (1,470,000 tons per year). Therefore, offshore drilling is *relatively* safe. Despite these facts, most public concern centers on offshore drilling rather than on oil pollution from transportation by tankers.

A more realistic approach is to determine the environmentally best way to obtain oil. From 1964 to 1971, there were 16 major spills from 10,234 producing wells. The oil released in this eight-year period was about 46,000 tons. This number is less than twice the oil spilled in the single *Argo Merchant* incident, and only a small fraction of that spilled by the *Amoco Cadiz*. An offshore drilling program with buried pipelines (a relatively safe way of transporting oil) and onshore refineries should be several times safer ecologically than bringing oil in by tankers. Oil clearly can be dangerous to the environment, but some aspects are not as threatening as might be perceived from media reports. The best alternative, of course, would be to reduce oil use.

How Oil Behaves in the Marine Environment

Oil is a very complex mixture of thousands of different organic compounds. When it enters the marine environment, several things happen. Some of the compounds dissolve into the water; some of the lighter and more volatile compounds evaporate into the atmosphere; some sink; some decay by biological weathering; some get into the food chain if eaten by organisms; and some may enter the bottom sediments (Figure 16-19). These processes are affected by the wind and waves (which aid in dispersal and mixing), by temperature (with colder temperature the oil lumps and degrades less), and by the original composition of the oil.

Oil is very toxic to most marine life. It can kill organisms that ingest it. It can injure or kill organisms whose gills or body surfaces become contaminated with it. The effects of oil can persist for many years, especially if it enters the bottom sediments and is then slowly released when the sediment is disturbed, as by a storm. After a spill, an area needs considerable time to return to its natural state. The process resembles a forest fire in the initial destruction and the time it takes for normal conditions to return.

Even small amounts of oil can affect biological processes, such as feeding and reproduction. Oil spills are potentially more damaging in shallow water, because this is a more biologically fertile area and has greater potential for oil to reach and enter the bottom sediment. Of course, shallow water is where tankers go aground. There is some good news, however: recently the quantity of oil spilled in U.S. waters has decreased. This may be due to the Oil Pollution Act of 1990, which increased the liability of tanker owners and thus encouraged increased safety precautions.

Figure 16-19 Some pathways that an oil spill can take in the marine environment. (From Ross, 1980.)

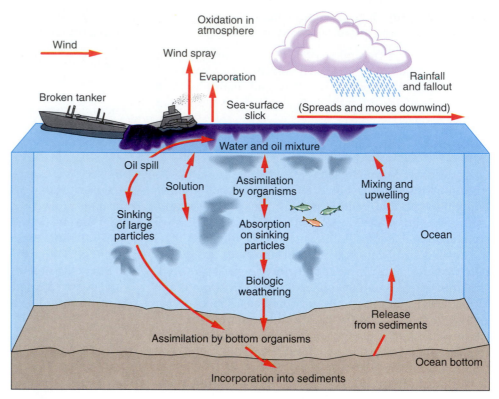

Pollution from Mineral-Resource Exploration/Exploitation

In shallower waters, important mineral resources exist in the unconsolidated sediments—sand and gravel, heavy minerals, and phosphates. Deeper waters offer manganese nodules, polymetallic sulfides, and cobalt-rich seamount crusts (see Chapter 15). In Great Britain and Japan, sand and gravel are already being mined, as are tin deposits in Indonesia.

There can be environmental problems with resource exploitation. For example, dredging offshore sands can cause beach erosion by interfering with the normal beach cycle (see Figure 14-10, page 343). Dredging affects bottom-dwelling creatures, both by the dredging itself and the suspended matter introduced into the environment. Most coastal states regulate nearshore dredging, but this may change as land sources of sand and gravel become exhausted (Figure 16-20).

Mining of manganese nodules and ocean vent minerals presents interesting oceanographic and pollution problems for the future. Some techniques for mining

Figure 16-20 Potential environmentally damaging effects from sand dredging or similar bottom mining operation. (From Ross, 1980.)

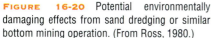
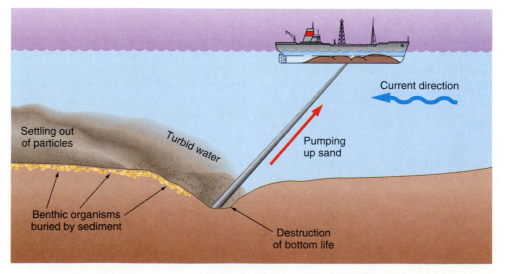

nodules bring colder water of different salinity to the surface, plus bottom mud, which generally has a greater nutrient content than surface waters. If the deep water spreads out at the surface, these nutrients could benefit phytoplankton growth. Its suspended matter content, however, could reduce light penetration and therefore reduce the water depth available for photosynthesis. Suspended material could take months to settle back to the bottom, and its extended presence could alter the water chemistry, affecting organisms in the water or on the bottom.

The effect on the bottom environment is much more difficult to ascertain. Deep-water ecology is poorly understood, but mining would certainly disrupt it. Falling suspended matter from sediments carried to the surface could bury benthic organisms, although the organism density in manganese nodule areas is low (some oceanographers consider the deep sea to be a biological desert). Thus, nodule mining may have a relatively small impact on the overall biological system of the ocean. Nevertheless so little is known about this part of the ocean that mining could have unanticipated results. On the other hand, land mining of copper, nickel, cobalt, and manganese is very destructive and could be several times more environmentally damaging than a marine operation would be.

Radioactive Waste

On land, dramatic incidents of radioactive pollution include the 1986 explosion of the nuclear power plant at Chernobyl in the Ukraine. Detectable radioactive material from this event has been found in the Black Sea. In general, however, radioactive material from nuclear explosions does not appear to have created marine environmental problems. In fact, some of the radioactive isotopes from nuclear bomb fallout have been very valuable for oceanographic research (see Chapter 7, page 172).

It has generally been thought that marine radioactive pollution would remain small because of international agreements. Atmospheric testing of nuclear weapons has been banned, and nuclear weapons are prohibited on parts of the seafloor by treaty. Following the dissolution of the Soviet Union in the early 1990s, however, came the distressing news that the Soviets had disposed of several nuclear reactors from their submarines by putting them into the ocean. In addition, several Soviet nuclear submarines have sunk recently. A private group, headed by a former Soviet admiral, has even proposed raising one that sank off Norway in 1989. The scientific community at this time is divided as to the environmental damage posed by these submarines.

International rules exist against deliberate ocean dumping of nuclear plant radioactive waste. Even so, some still suggest using the ocean as a "final" repository for nuclear waste (one technique is discussed in Chapter 18). In late 1993, Russia was caught discharging 237,000 gallons of low-level nuclear waste in the Sea of Japan, which led to an international meeting where 37 countries voted to ban permanently the dumping of nuclear waste at sea. However, Britain, Russia, France, and China—all important nuclear powers—abstained from supporting the ban. The danger from nuclear dumping is probably not so much in polluting the entire ocean but rather in introducing a large amount into a restricted area where it cannot be dispersed or diluted.

Thermal Pollution

Power plants, whether fossil fuel or nuclear, need "cooling water." Each year, a plant may take in millions of gallons of water, run it through piping to absorb waste heat, and return it to the original source—a river, estuary, lake, or the coastal ocean. The water is often returned to the environment without cooling and can be 10 to 15°C (18 to 27°F) warmer than the water into which it is released. This **thermal pollution** can be harmful or even fatal to many marine organisms.

Thermal pollution can disturb the metabolism of organisms and decrease oxygen solubility in the water. Collateral environmental damage can occur when marine life, especially plankton or small fish, are drawn into the cooling system and killed by passing through the pumps or other mechanisms.

There can be a beneficial side, however. The heated effluent could be used in an aquaculture system to increase growth of oysters, lobsters, or other organisms of economic interest. The heated effluent, in northern areas, also may prevent ice from forming, which might benefit ducks, migrating birds, and some fish.

SUMMARY

Pollution is an emotional and controversial subject sometimes presented in the media without adequate scientific basis. Some aspects are exaggerated, whereas others are understated. The inputs of some pollutants into the marine environment are very large; others are minor compared to the natural input. Many different kinds of pollution occur in the ocean, especially in the nearshore zone. In most instances, data concerning pollution input or prepollution conditions are nonexistent. It is probable that some forms of marine pollution may become worse before controls are established. Reduction of much pollution requires international cooperation and large-scale financing.

Pollution has three main effects in the marine environment:

1. It can directly destroy organisms in the polluted area.

2. It can alter the physical and chemical properties of the environment, favoring or excluding specific organisms.

3. It can introduce substances that are relatively harmless to simpler lifeforms but dangerous to those higher up the food chain, due to biological magnification.

Pollutants can take numerous pathways to the ocean, including direct input via rivers, sewer outfalls, runoff, and ocean dumping, and from the atmosphere. Most pollutants come from human activity, but some result from natural processes.

Most critical is the large quantity of pollutants that enter the coastal zone and estuaries, often exceeding the capacity of these waters to cleanse themselves. Even with proper treatment and management, large amounts of potentially damaging materials can remain in the bottom sediments of the coastal zone and estuaries.

Oil and gas production, including drilling and transportation, are highly visible polluters. More rigid rules concerning the transportation of oil could reduce this toxic and widespread pollution.

Most marine pollution problems are nearshore. The open ocean receives pollutants too, but its great capacity (due to its immense volume) makes it a less critical problem. One reasonable solution to pollution is to recycle more.

QUESTIONS

1. Discuss the types of pollutants in the ocean and how they get there.

2. What is the probable source of acid rain? Why are its effects so restricted to land and freshwater lakes and relatively negligible in the ocean?

3. Discuss the harmful effects of marine pollution. Try to cite examples where the benefit of producing the pollutant is outweighed by its harmful effect.

4. What options exist for stopping or reducing marine pollution?

5. What are some of the effects of oil being introduced into the marine environment? Discuss the pathways that an oil spill might take in the ocean.

6. What have we learned from oil spills like that from the *Exxon Valdez*?

KEY TERMS

acid rain

Aldrin

biological magnification

carcinogenic

chlorofluorocarbons (CFCs)

DDT

Dieldrin

Exxon Valdez

marine pollution

ocean incineration

polychlorinated biphenyl (PCB)

red tide

synthetic organic compound

thermal pollution

zebra mussel

FURTHER READING

Anderson, D. M. 1994. "Red Tides." *Scientific American* 271, no. 2, pp. 62–68. An excellent summary of our state of knowledge about red tides.

Brown, B. E., and J. C. Ogden. 1993. "Coral Bleaching." *Scientific American* 268, no. 1, pp. 64–70. Discusses coral bleaching and its effects, including whether such bleaching signals global warming.

Buesseler, Ken O. 1987. "Chernobyl: Oceanographic Studies in the Black Sea." *Oceanus* 30, no. 3, pp. 23–30. Discusses the finding of radioactive material from the Chernobyl disaster in the Black Sea.

Farrington J. W., J. M. Capuzzo, T. Leschine, and M. A. Champ. 1982–83. "Ocean Dumping." *Oceanus* 25, no. 4, pp. 39–50. A readable discussion about ocean dumping.

Goldberg, E. D. 1981. "The Oceans as Waste Space: The Argument." *Oceanus* 24, no. 1, pp. 2–9. Kamlet, K. S. 1981.

"The Oceans as Waste Space: The Rebuttal." *Oceanus* 24, no. 1, pp. 9–17. An interesting debate about the pros and cons of ocean dumping.

Horton, Tom, and W. M. Eichbaum. 1991. *Turning the Tide: Saving the Chesapeake Bay.* Washington D.C.: Island Press. Discusses some environmental problems of one of the most important estuaries in the United States.

Kaufman, D. G., and C. M. Franz. 1993. *Biosphere 2000: Protecting Our Global Environment.* New York: HarperCollins. A good, readable discussion of our many environmental problems.

Kitsos, Tom R., and Joan M. Bondareff. 1990. "Congress and Waste Disposal at Sea." *Oceanus* 33, no. 2, pp 23–28. Discusses the history of laws concerning ocean dumping and what will happen after dumping is outlawed.

Libes, S. M. 1992. *An Introduction to Marine Biogeochemistry.* New York: John Wiley and Sons. A slightly advanced textbook covering marine chemistry and how it relates to some aspects of marine geology, biological oceanography, and pollution.

MacInnis, J., ed. 1993. *Saving the Oceans.* Toronto: Key Porter Books. A beautifully illustrated book that focuses on some of the ocean's environmental problems.

McDowell, Judith E. 1990. "Effects of Wastes on the Ocean." *Oceanus* 33, no. 2, pp. 39–44. A discussion of how waste disposal in the coastal region degrades the water and affects fishing and mariculture.

Ocean Disposal Reconsidered. Oceanus 33, no. 2 (Summer 1990). A special issue devoted to various aspects of marine waste disposal.

Oil in the Sea: Inputs, Fates, and Effects. Washington, D.C.: National Academy of Sciences, 1985. An update of the 1975 National Academy of Sciences study on marine oil pollution.

Petroleum in the Marine Environment. Washington, D.C.: National Academy of Sciences, 1975. Important early study about marine oil pollution.

Teal, J. M. 1993. "A Local Oil Spill Revisited." *Oceanus* 36, no. 2, pp. 65–70. The story of the *Florida* oil spill after twenty years.

United Nations, Economic and Social Council, 51st Session. 1971. *The Sea: Prevention and Control of Marine Pollution.* Defines various types of marine pollution.

United Nations, Environmental Program. 1982. *The Health of the Oceans.* UNEP Regional Seas Reports and Studies, no. 16.

Wastes in Marine Environments. A report of the U.S. Congress Office of Technology Assessment, 1987.

Young, John. *Sustaining Earth.* Cambridge, Mass.: Harvard University Press. An interesting book describing the beginnings of the environmental movement.

CHAPTER 17

Marine Archeology

IN RECENT YEARS, NEW TECH-
NIQUES AND TECHNOLOGIES HAVE
BEEN DEVELOPED FOR EXPLORING
THE SEAFLOOR: REMOTELY OPER-
ATED VEHICLES (ROVS), SOPHISTI-
CATED CAMERAS, AND VERY ACCU-
RATE NAVIGATION SYSTEMS. THESE
INSTRUMENTS HAVE EXPANDED THE
SCIENCE OF MARINE ARCHEOLOGY
AND IN SOME INSTANCES HAVE LED
TO THE RECOVERY OF MARINE
TREASURE.

A diver prepares to map the location of a group
of stone anchors found in the lower section of
the Ulu Burun, discovered off the coast of
Turkey in the Mediterranean Sea. The vessel is
about 3,400 years old.

*(Photograph by Donald A. Frey, courtesy of Institute of
Nautical Archeology.)*

Introduction

The possibility of finding a sunken shipwreck or buried treasure has always intrigued people, whether scientists, treasure hunters, or the simply curious. Discoveries by marine archeologists, amateur divers, and professional treasure hunters are often in the news. One of the most fascinating of recent times has been the finding of the R.M.S. *Titanic,* an "unsinkable" British luxury liner that sank on its maiden voyage in 1912 after colliding with an iceberg, about 550 km (340 miles) off Newfoundland. People sought the wreck for decades with no success, until 1985. The story of the *Titanic* will be told later in this chapter.

Treasures such as gold coins, statues, and other ancient artifacts have been collected from the sea for centuries. Sometimes people walking the beaches accidentally discover items that have washed ashore. Other finds are quite deliberate, based on research into probable locations and using systematic diving. Expeditions today rely on detailed studies of where and how wrecks occurred. Finding a wreck, however, requires a great deal of patience, money, and luck.

In the past, areas that could be explored were limited by diving capabilities. With modern technology, discovery potential has increased: magnetometers are used to detect iron and large areas are explored visually with cameras, side-scan sonar, and remote television. Thus, the number of discoveries has increased without the time-consuming, depth-limiting problem of keeping people underwater. Several companies exist through which investors can fund a treasure expedition, although at great financial risk. We will examine the search for the S.S. *Central America* as an example.

Competing Interests

These same technologies have expanded the vistas and techniques of the scientific discipline of **marine archeology.** An inevitable conflict has ensued between those interested in scientific exploration and those favoring salvage exploration and exploitation for treasure. The basic conflict between archeologists and salvagers centers on *time* and *money.* The archeologist wants to preserve and study historical information from a site, which requires slow, systematic investigation. The salvager seeks maximum economic value in minimum time.

Thus the professional marine archeologist is usually opposed to treasure hunting, and with cause: treasure hunters usually damage or destroy a site before proper documentation, including photography, can be done. Important artifacts may be sold to the highest bidder rather than becoming available for study or museum dis-

play. Marine archeologists believe in "preserve and learn," while treasure hunters believe in "finders, keepers," creating a complex and emotional controversy.

The discovery of shipwrecks and recovery of artifacts also has a legal side. Insurance companies may claim their share of recovered property many years after the event. Most U.S. coastal states have laws that require sharing of recovered treasure by the finder and the state. In some instances, however, no laws protect a site, so it is fair game for anyone to explore or exploit.

Sport divers comprise one of the largest groups interested in marine wrecks. Fortunately, their interest lies in recreation and site preservation. As so often happens, however, many divers who each take a "little sample" from a marine archeological site collectively damage it. In some areas, such as around the U.S.S. *Monitor* site, marine parks have been established that simultaneously protect them while giving the diving public limited access.

The Discipline of Marine Archeology

The professional marine archeologist is interested in how people lived in ancient times, and underwater archeology demands the same skill and effort as does similar work on land. Today's marine archeologist may still use hearsay and old sea tales to help find prospective sites, but more often depends on high-tech methods and gear.

Newer technologies like sophisticated submersibles and single-person diving devices (Figure 17-1) may soon become standard tools of the marine archeologist. Conventional diving using SCUBA can constrain marine archeologists, since extended work below depths of 45 m (about 150 ft.) is limited and dangerous. Submersibles expand the depth that can be explored and the time available on the bottom. Newer submersibles that allow divers to exit for short periods may be used for deeper work (see Figure 5-9, page 115).

Study of a marine site typically starts by establishing a grid system on the seafloor for reference and mapping (Figure 17-2). Items collected are photographed while still in place on the seafloor. Small objects can be excavated by "vacuum cleaners" that suck up bottom sediment. The sediment then is screened to detect artifacts. Using the same method, large items are uncovered but may be left in place. Some items are transported to the surface for study and display (Figure 17-3).

A major problem with underwater artifacts is preserving them. In most instances, wood has already been destroyed or made very fragile by the drilling action of marine mollusks and worms and by bacterial action (see Figure 17-10c). Surviving wood that becomes exposed to air quickly dries out and decays. Other materials, especially some metals, are damaged by chemical reactions.

a

b

FIGURE 17-1 (a) Famous diver Sylvia Earle prepares to descend in a diving suit named WASP (for its resemblance to the insect). This device allows a diver to work down to a depth of 600 m (1968 ft.). It is tethered to a surface ship. (b) *Deep Rover,* a one-person submersible capable of working to depths of 1,000 m (3,300 ft.). Such submersibles offer considerable visibility. Note the two arms for collecting or sampling. (Both photographs courtesy of Dr. Sylvia Earle, Deep Ocean Engineering.)

a

b

c

FIGURE 17-2 Institute of Nautical Archeology divers at an eleventh-century site off Turkey: (a) expedition barge anchored above the site; (b) divers work on the site (note the grid that marks the area); (c) divers carry small sections of the hull to the surface. (Photographs courtesy of Institute of Nautical Archeology, Texas A&M University.)

FIGURE 17-3   (a) Ancient amphorae and other artifacts found on the ocean floor off Sicily, in the Mediterranean, photographed by the ROV *Jason*.

(b) An amphora is removed from an elevator shortly after its recovery from the ocean floor. The elevator was designed for this purpose. (Photograph (a) copyrighted by Quest Group, Ltd.; (b) copyrighted by Joseph H. Bailey, National Geographic Society, picture taken by Joseph E. Bailey; both courtesy of Woods Hole Oceanographic Institution.)

Seawater is a good electrolyte,* so severe rusting may occur (see Figure 17-12b). Fortunately, gold is only slightly chemically active in an electrolyte solution, so gold coins and jewelry often survive marine burial very well.

A Capsule History of Marine Archeology

A world leader in underwater archeology is Dr. George Bass, president of the Institute of Nautical Archeology at Texas A&M University. In a 1980 paper, Professor Bass divided marine archeology into five historical stages:

1. *Approximately 1800–1948.* Marine discoveries were made mostly by fishermen and sponge divers. Some beautiful sculptures were found and crude attempts at excavation were made.

2. *1948–60.* The pioneering stage of marine archeology, made possible by the invention of **SCUBA (Self-Contained Underwater Breathing Apparatus)** by Jacques Cousteau and Emile Gagnan. With SCUBA, divers (generally not archeologists) could freely explore the shallow seafloor. Some techniques for artifact recovery and mapping procedures were developed during this stage.

---

*An *electrolyte* is a fluid in which electrical current can flow easily, which also moves ions around. In seawater, this ion flow causes most metals to rust quickly.

3. *1960–70.* Trained archeologists began working on the seafloor. During this decade, several detailed marine excavations were performed (see Figure 17-2) and sophisticated mapping and searching techniques using sonar and metal detectors were developed.

4. *1970–76.* Emphasis shifted to ships and shipping. Marine archeology gained stature as a discipline and several journals were inaugurated. A number of institutes were established, including the Institute for Maritime Archeology at the University of St. Andrews in Scotland, the Center for Maritime Studies at the University of Haifa in Israel, and the Institute of Nautical Archeology, now affiliated with Texas A&M University.

5. *1976–80.* Marine archeology clearly had become an academic discipline, with advanced degrees available.

In his paper, Professor Bass correctly anticipated that the next stage would belong to a new generation of students using new technology. He also advocated protection of marine shipwrecks from people who were just hunting for treasure.

It seems appropriate to add a sixth stage, which extends into the present time. This stage is dominated by sophisticated technology, such as *Jason Jr.* (described shortly), that allows all portions of the seafloor to be explored by archeologists and other scientists. This new period, unfortunately, also has seen an increase in the number of treasure hunters using advanced technology, ready buyers for their harvest, and greater media coverage.

Let us now look at some recent finds that are important and exciting in marine archeology.

Important Discoveries

U.S.S. *Monitor*

An important underwater archeological find was the U.S.S. *Monitor*. During the U.S. Civil War, this "iron-clad" ship, heavily armored with steel plates, was built by the Union side and launched in 1862. It fought only one battle, with the Confederate iron-clad *Merrimac*. While under tow, it sank on New Year's Eve, 1862 (Figure 17-4).

Monitor was discovered in 1973 by scientists aboard the Duke University *R/V Eastward* in waters about 64 m (210 ft.) deep off Cape Hatteras, North Carolina (Figure 17-5). Dives and underwater photographs revealed the ship's remains to be very fragile and thin due to severe rusting; raising the vessel probably would destroy it. Thus, the site has been designated a national marine sanctuary area, and underwater operations and study are strictly controlled. More recent studies with ROVs and sophisticated cameras are determining the extent of corrosion of the vessel, surveying the debris field around

FIGURE 17-4 Drawing of rescue operations aboard the U.S.S. *Monitor* on December 31, 1862. The vessel sank during a storm off the North Carolina coast while being towed by the U.S.S. *Rhode Island* (background). (Photograph courtesy of Naval History Division, Department of the Navy.)

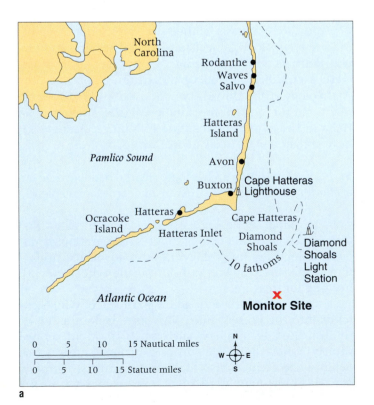

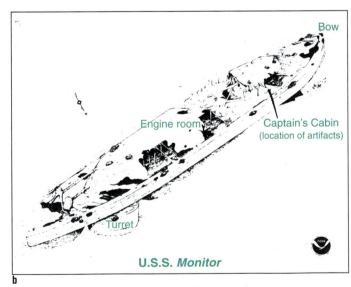

FIGURE 17-5 (a) Location of the Civil War–era ironclad U.S.S. *Monitor*, called a "cheesebox on a raft."
(b) Artist's sketch of *Monitor* wreckage, based on hundreds of underwater photographs. Much of the data was obtained during dives coordinated by NOAA. (Photographs courtesy of National Oceanic and Atmospheric Administration.)

the *Monitor,* and accurately mapping its visible parts for comparison with historical documents.

Ulu Burun Site

One of the oldest known shipwrecks lies along the Turkish coast, off the village of Ulu Burun at about 46 m (150 ft.) depth. Reported by George Bass in a 1987 issue of *National Geographic* magazine, the wreck is a trading vessel from the fourteenth-century B.C.—an amazing 34 centuries old! It was first found by a Turkish sponge diver in 1982. The 15 m (50 ft.) ship carried goods from several Mediterranean countries. Over a thousand artifacts have been recovered (see the opening figure to this chapter and Figure 17-6), including bronze swords and arrowheads, ivory, ostrich eggshells, hippopotamus teeth, copper and tin ingots, and amphorae (two-handled ancient Greek jars or vases) filled with aromatic resin.

An extremely detailed study of the site is in progress, including mapping and photography of every artifact. At the time of the *National Geographic* article, over 200 cop-

FIGURE 17-6 The Ulu Burun site off the coast of Turkey.

(a) Photographer Donald Rosencrantz uses a camera mounted on a bar suspended over the wreck area. At left is the "telephone booth" from which divers communicate with the research vessel, firmly anchored directly above the site.
(b) Ascending a staircase of copper ingots, nautical student Nicolle Hirschfeld carries an amphora from the wreck.
(c) Field director Cemal Pulak and Professor George F. Bass (right) examine a Canaanite pilgrim flask.
(d) A gold signet ring (scarab) that bears the name of the Egyptian queen Nefertiti. (Photographs by Donald A. Frey, courtesy of Institute of Nautical Archeology.)

a

b

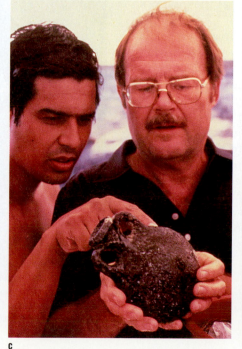

c

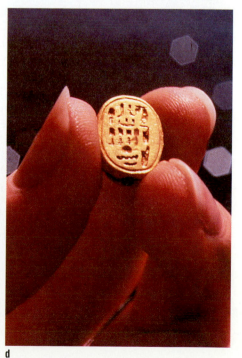

d

per ingots had been found, each weighing about 132 kg (290 lbs.). Also recovered were remarkably preserved pottery, jewelry, ivory, and daggers. Ulu Burun is a remarkable discovery that has given insight into Mediterranean life about 3,400 years ago.

S.S. *Central America**

The sinking of the S.S. (Steamship) *Central America* in the Atlantic Ocean in September 1857 was a national tragedy for the United States. According to accounts published at that time, it was one of the first major media events. The story begins with the discovery of gold in California in 1848, which created a demand for quick, safe transportation between the East Coast and West Coast. Stagecoach travel was slow and risked confrontation with bandits and Native Americans. The transcontinental railroad would not be completed for another 20 years. Thus, the best travel route was a steamship ride from the East Coast to Panama in Central America, followed by a short trek across the isthmus of Panama, and then completion of the journey by steamship up the Pacific Coast to California.

This route was to become a favorite, especially for bringing gold from California back to the East Coast. Side-wheel steamships often made these trips, carrying passengers as well as gold. One was the S.S. *Central America,* which made 43 successful trips between Panama and New York (Figure 17-7). Her 44th trip, however, ended in disaster. In early September 1857, the vessel steamed unwittingly into an Atlantic hurricane. As the storm worsened, the vessel began to take on water. Eventually its boiler fires were extinguished and the ship lost power, putting it completely at the mercy of the storm.

Crew and passengers bailed for 30 hours but could not stop the rising water. Eventually the vessel sank about 320 km (200 miles) off the South Carolina coast. Several nearby ships rescued some of the passengers and crew from lifeboats. Amazingly, another vessel saved 50 people who were floating in the water, and three days later 9 more were found alive on a lifeboat. All told, 155 of the approximately 580 people aboard the vessel survived, including all the women and children. The story and accounts of heroism made front-page news throughout the United States.

When the *Central America* sank, it took to the bottom the *largest lost treasure in U.S. history*—about 3 tons of gold in ingots and coins. The loss caused a minor financial

*Much of the information in this summary is from reports by Judy Conrad (see Further Reading) and material supplied by Charles E. Herdendorf.

FIGURE 17-7 S. S. *Central America*, a sidewheel steamship that sank in 1857 (note sidewheel in Figure 17-10a). (Photograph from Mariner's Museum, Newport, Virginia, courtesy of Charles E. Herdendorf.)

FIGURE 17-8 The research submersible *NEMO*, used to photograph and study the S. S. *Central America.* (Photograph © Columbus-American Discovery Group, courtesy of Charles E. Herdendorf.)

panic in New York and led to the failure of several banks. For nearly 130 years, the *Central America* lay undisturbed in its seawater grave.

In the 1970s, a young ocean engineer named Tom Thompson became interested in finding and recovering deep-water shipwrecks. In 1983 the *Central America* was chosen as a prime candidate, and in 1985 the Columbus-American Discovery Group was formed to find the vessel. This group of private investors initially contributed $1.4 million to finance the search. The group designed a recovery vehicle called **NEMO** (Figure 17-8), a remote-controlled device with stereo video, still cameras, and robotic arms. *NEMO* was designed to perform actual recovery as well as observation.

The first problem was to obtain as much information as possible about the *Central America* and where it probably sank. This meant months of library research, collecting the written accounts of survivors. A computer program was designed to analyze the data. Initial results suggested that the 83 m (272 ft.) ship might be found within an area of 3,626 km^2 (1,400 miles^2) about 320 km (200 miles) off the Carolinas, on what is called the Blake Ridge (see Figure 4-8, page 84). Such an undertaking exemplifies the proverbial searching for a needle in a haystack.

The first search in 1986 used a high-resolution side-scan sonar (see Box 5-1, page 103) and took about 40 days. Searchers found a very promising target (Figure 17-9), but it would be over a year before the Columbus-American Group confirmed that they had found the vessel in about 3 km (1.5 miles) of water, 257 km (160 miles) off the coast. The next two years were spent surveying and testing the recovery systems. Confirmation of their find came in the summer of 1988, with the recovery of the ship's bell (which bore its name) and with the many detailed pictures of the wreck (Figure 17-10).

This project was different from most treasure recovery efforts in several important ways. Perhaps most significant was its scientific component. The discovery/recovery system permitted studies of deep-water fauna, for which the wreck had become an oasis (Figures 17-10c and 17-10d). In addition, the Columbus-American Discovery Group meticulously documented and photographed objects from the wreck.

The group initially got legal protection from a U.S. court, ensuring their exclusive right to work on the wreck in place. This ruling was unique because the wreck had been seen only by remote-controlled devices and because it was the first private claim accepted in deep-water. Resolution of who owned recovered items would take a little longer. By the end of September 1989, the group had successfully raised about 3 tons of gold from the *Central America*. Some of the coins were in superb condition (see Figure 1 in Box 17-1).

The recovered gold has an estimated worth up to $1 billion, but the question remained: who was to own it? Eventually, 39 British and American insurance companies made claims on the gold, based on the $1.2 million these companies (or their predecessors) had paid on the loss of the vessel over 130 years ago, prior to the Civil War! The Columbus-American Discovery Group claimed that, because the insurers had not looked for the ship, they had legally abandoned it. In August 1990, a U.S. federal judge awarded the gold to the Columbus-American Discovery Group, apparently making their investors wealthy individuals. Things, however, took an unexpected turn in 1992 (see Box 17-1 for the rest of the story).

R.M.S. *Titanic*

The R.M.S. (Royal Mail Steamship) *Titanic* has truly captured the public's imagination. Her well-known story has been told in movies, books, and television specials. The 269 m (882 ft.) double-hulled luxury liner was built to be the height of technological safety. Designed to be unsinkable, people felt the *Titanic* had conquered the ocean environment—in fact, so assured was this vessel's safety that only 1,178 lifeboat spaces were provided for the 2,235 people on board.

Her maiden voyage began April 10, 1912, at Southampton, England. *Titanic* stopped in France and Ireland for mail, then headed toward the United States. On April 14, at 11:40 P.M., traveling at 22.5 knots (26 miles per hour), she hit an iceberg. Within 3 hours, the unsinkable ship slipped beneath the surface to the seafloor over

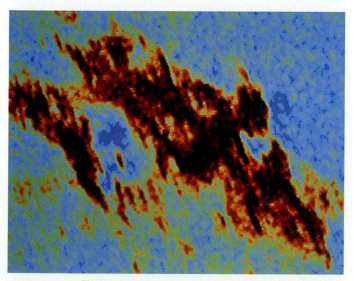

FIGURE 17-9 The key target obtained from a side-scan sonar search for the S. S. *Central America*. This picture is a color-enhanced, filtered, side-scan sonar image later found to be the lost ship. (Photograph © Columbus-American Discovery Group, courtesy of Charles E. Herdendorf.)

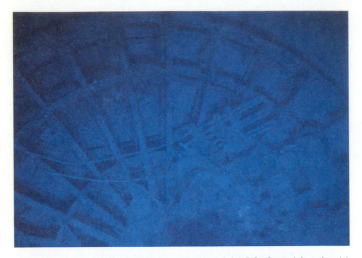

FIGURE 17-10   Some pictures from the site of the S.S. *Central America:* (a) distinctive sidewheel of the vessel;

(b) the ship's bell and steam whistle;

(c) timbers from *Central America* showing degradation by boring clams; also visible are sponges, hydrozoans, and anthozoans; the plumelike animals on the crossed timbers are glass sponges;

(d) an iron tank shows considerable deterioration and the formation of "rusticles." Animals growing on the tank include gorgonian anthozoans and sponges. (Photographs © Columbus-American Discovery Group, courtesy of Charles E. Herdendorf.)

3,650 m (12,000 ft.) below. Her resting place is about 550 km (340 miles) south of Newfoundland.

Stories of heroism emerged, of men remaining aboard to let women and children use the limited number of lifeboats; 1,522 died and 713 survived. Criticism emerged, too: Why was the ship traveling so fast in iceberg waters? Controversy appeared as well; one ship in the area apparently did not respond to calls for assistance, although recent evidence suggests otherwise.

Several groups tried to find the *Titanic,* especially during the 1980s. But it was not until 73 years after the sinking, on September 1, 1985, at 2:00 A.M., that the *Titanic* was found on the seafloor. Searching the area with the sophisticated surveying system *Argo* (see Figure 5-13, page 117), a team of U.S. and French scientists, led by Dr. Robert Ballard of the Woods Hole Oceanographic Institution, noted that pictures transmitted to the surface ship started showing debris on the seafloor. Eventually clear evidence of the *Titanic* was found (Figure 17-11).

The events and subsequent days of this expedition were observed by millions on television worldwide. The discovery focused new attention on the ocean and the

Box 17-1
Can You Make a Small Fortune from the Ocean?

Perhaps more than most environments, the sea appears to offer opportunities for making lots of money. Examples include mining mineral resources, as mentioned in Chapter 15. Other resource possibilities are aquaculture, innovative ways to catch fish and lobsters, and so forth.

How can you make a small fortune from the ocean? According to a running joke, just start with a large fortune and wait! But the recent finding of marine archeological treasures, such as those associated with the S.S *Central America*, fire the imagination (Figure 1). Certainly those in the *Central America* project may do well financially. Yet how many other expeditions have been fruitless, with little or no profit for their investors? This author knows of no useful data allowing comparison of successes versus failures in the search for marine treasure.

Investing in a company looking for marine treasures is alluring. However, before rushing to invest, one must recognize the great risk. The cost of working at sea can be immense, the probability of success is small, and there is a persistent question of ownership of anything recovered.

The investment procedure for the *Central America* was imaginative and especially rewarding to those who invested early. It went something like this:

- *Phase 1*—seed money to start the project: 20 investment units at $10,000 each were sold, raising $200,000. These investors would receive $\frac{1}{2}$ of 1 percent of the profits (for a total of 10 percent of the 100 percent available).
- *Phase 2*—research phase: 50 investment units at $28,000 each were sold, raising $1.4 million. These investors also would receive $\frac{1}{2}$ of 1 percent of the profits (total of 25 percent).
- *Phase 3*—initial recovery: 50 units at $72,000 each were sold, raising $3.6 million. This phase was to deploy equipment and start raising treasure. These individual investors would get $\frac{1}{2}$ of 1 percent of the profits (total of 25 percent). At the end of this phase, 60 percent of the total profit had been sold.
- *Phase 4*—last sale: 150 units at $50,000 each were sold, raising $7.5 million. Each investor got only $\frac{1}{10}$ of 1 percent (for a total of 15 percent).

Note how the price went up as the risk went down. Those who formed and funded the operation in its early stages were eligible for about 25 percent of the profits, while the subsequent investors were eligible for

75 percent. The money raised paid the cost of operation, legal fees, and so on. Of course, the risk always existed that nothing would be found or that insurance companies or a national or state government could successfully claim the treasure, or at least delay an ownership decision for years. This, as described below, was what actually happened.

The *Central America* was a first-class operation, with competent individuals who did their homework. (For an individual investor to confirm this before investing would be difficult.)

If the recovered treasure from the *Central America* is ultimately worth $500 million, individual investors would share a $375 million pie (75 percent of $500 million) and a share of $\frac{1}{2}$ of 1 percent would equal $1,875,000. For the first-phase investor who risked $10,000, this is a 187.5-fold return; for the second-phase investor who risked $28,000, it is a 67.5-fold return. Third-phase investors who risked $72,000 got a 26-fold return and fourth-phase investors (who by then knew that a treasure was available and risked $50,000 for $\frac{1}{10}$ of 1 percent), got a 7.5-fold return.

In this example, the return potential was good. A cynic might say that the same amount of money invested in a spectacular-growth stock might pay a 7.5-fold return in a few years. Likewise, one must note that the *Central America* was clearly an exception. Maybe one can make a small fortune from the ocean, but losing investment money is a more likely outcome.

For a while everything looked very good for investors in the *Central America* project, but soon things changed. In 1992 a Circuit Court of Appeals in Virginia reversed the prior 1990 ruling that gave the treasure to the Columbus-America Discovery Group, saying that the insurance companies were the rightful owners.

The Columbus-America Discovery Group then appealed to the U.S. Supreme Court, which in 1993 decided not to reconsider the 1992 Circuit Court decision. Even though further appeals may be possible, the present outcome is not a complete loss for the group, for they are entitled to a significant, although presently undefined, salvage award.

The marine treasure-hunting business was startled by the decision, because they want to continue operating under a simple "finders, keepers" rule. Without assured ownership of salvaged objects, it could be hard to finance the expensive expeditions necessary to find and recover treasure from the sea.

continued

Box 17-1 continued

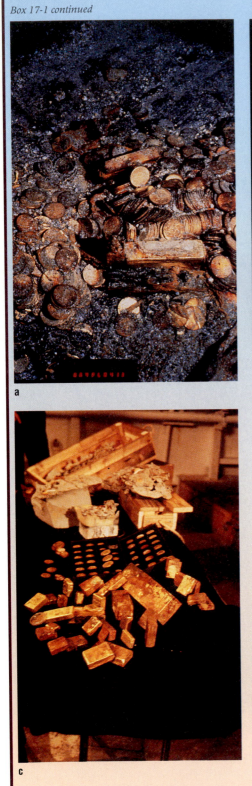

a

b

FIGURE 1 Some of the treasure recovered from the S.S. *Central America:* (a and b) gold bars and mint (uncirculated) double-eagle $20 gold coins photographed on the ocean floor; (c) some of the treasure after recovery; (d) Dr. Charles E. Herdendorf, head of the scientific team associated with the Columbus-American Discovery Group, with some of the treasure. (Photographs © Columbus-American Discovery Group, courtesy of Charles E. Herdendorf.)

c

d

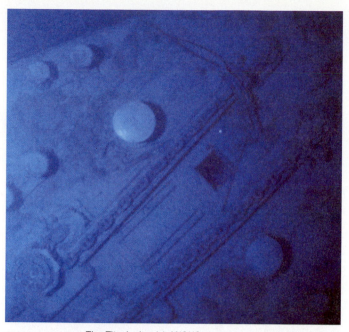

FIGURE 17-11 The *Titanic* site: (a) ANGUS picture of *Titanic*'s bow, showing anchor chain and winches;

(b) ANGUS picture showing port and starboard cranes;

(c) two bollards, used to secure mooring lines, and a railing on the starboard side of the luxury liner's bow;

(d) cranes from *Titanic* in a field of debris. (All photographs copyright © 1985 Woods Hole Oceanographic Institution, courtesy of Woods Hole Oceanographic Institution.)

new technologies used to explore its depths. The "technological hero" was **Argo,** a "swimming eyeball," a system of cameras, lights, sonar, and television. It is towed about 36.5 m (120 ft.) above the seafloor and is capable of working to depths of 7,000 m (20,000 ft.) for weeks at a time, transmitting pictures to the surface vessel by coaxial cable. Later, higher-resolution pictures were obtained by another system, **ANGUS (Acoustically Navigated Geo-logical Underwater Survey),** which has three 35 mm cameras that hold large quantities of film to photograph a swath of seafloor about 60 m (200 ft.) wide.

These technologies, combined with sophisticated navigation and sidescan sonar, make discoveries like the *Titanic* almost inevitable. Yet, without the enthusiasm of individuals and financial support for the expedition, such discoveries could not happen.

Almost immediately, speculation began about raising *Titanic*, or at least recovering artifacts from the area. Ballard had photographed wine bottles and silver trays lying intact on the seafloor, but did not disturb or move anything. The vessel lies in what most believe to be international waters and no law prohibits the recovery and sale of artifacts. However, the idea of disturbing the vessel—the grave of 1,522 people and a major historical site—was viewed by most, including Ballard, with dismay. Most, if not all, countries prefer that sunken ships remain undisturbed on the seafloor as memorials.

In 1985, hearings by a U.S. Congressional Committee attempted to designate the *Titanic* site as a maritime memorial and to prevent any country from undertaking recovery activities. The U.S. Congress, however, has no jurisdiction over international waters, nor can a country prevent one of its citizens from performing such an operation. Nevertheless, in 1986 the U.S. Congress passed a bill declaring the site an international maritime memorial.

In 1987, a French group (also financed by a Connecticut firm) recovered some items from the *Titanic* site, including wine bottles, jewelry, and pieces of clothing, and presented the story on television. In 1992, the French government offered to sell some of these items to survivors of the shipwreck and their relatives. Further attempts to salvage parts of the *Titanic* are anticipated.

In 1986, Ballard returned to the *Titanic* site with more technological tools, including the submersible *Alvin*, ANGUS, and a new ROV, **Jason Jr.** One objective of this expedition was to test *Jason Jr.*, which was deployed from *Alvin* (Figures 17-12a and 17-12e). The device produced some remarkable pictures, including one of a chandelier photographed several decks deep inside the vessel (Figure 17-12d).

Other pictures and observations from *Alvin* showed considerable destruction—twisted metal and collapsed decks. "Rusticles" (Figure 17-12b) on iron structures were common both inside and outside the ship. But some copper, brass, and silver items looked almost brand new (Figure 17-12c). Virtually all the wooden structure was missing, apparently destroyed by marine organisms. *Titanic* was in two parts on the seafloor, probably having broken in two near the surface, with the parts then rotating about 180° to each other prior to hitting the bottom.

Bismarck

Underwater discoveries need not always involve old vessels. A recent and successful search was for the World War II German Battleship *Bismarck* (Figure 17-13). The pride of the German navy, the *Bismarck* was 253 m (820 ft.) long, displaced about 50,000 tons, and had a maximum speed of 30 knots. It sailed on May 19, 1941, to intercept merchant shipping in the Atlantic. On May 24 the *Bismarck* was challenged by two British warships, the *Hood* and the *Prince of Wales*. The *Bismarck* hit the *Hood* with cannon fire, sinking it within minutes, with only 3 of its crew of 1,419 surviving. The *Hood* was the largest warship in the world prior to World War II and the major vessel in the British fleet.

The *Bismarck* was also damaged in the exchange and headed for a French port for repairs. Britain, stunned by the loss of the *Hood*, launched an all-out effort to sink the *Bismarck*. Two days later and only a few hours from safety, the *Bismarck* was found and attacked by British warplanes, damaging the ship's steering and rendering her unmaneuverable. The *Bismarck* was attacked again and sank with over 2,200 lost and only 115 survivors.

After a two-year search with the same *Argo* technology that found the *Titanic*, Dr. Robert Ballard discovered the *Bismarck* in over 4,600 m (about 15,000 ft.) of water (Figure 17-14). *Jason Jr.* could not be used because a sufficiently long fiber-optic cable was not available. The exact location of the vessel was made known only to the German government, and nothing at the site was disturbed. (Lost warships are considered the property of their country of origin.) Ballard believes that, because of the well-preserved condition of the vessel and its upright position on the seafloor, her crew may have scuttled the vessel before it could be sunk by British fire or captured.

The *Bismarck* expedition was funded by the Turner Broadcasting System, Inc., and the Quest Group, Ltd., a group of private individuals interested in deep-sea exploration technology.

The *Jason* Project

Observation and exploration for wrecks on the seafloor is compelling to the public. Dr. Ballard has combined this fact with his interest in attracting more students to science by using a program called the *Jason* Project. It uses *Argo/Jason* technology to broadcast live from the seafloor via satellite to students in science museums across the United States and Canada (Figure 17-15). The result is a "telepresence," through which students literally participate in underwater work.

This telepresence includes asking questions of Ballard and his scientific team and allowing students to control *Jason*'s motion. Ballard's point is that we can use advanced robots to explore marine archeological sites from the comfort of our homes, as if going to a museum. The view is better than if artifacts were brought to the surface to deteriorate.

The *Jason* Project has led to interesting archeological finds and indeed seems to have made science more at-

a

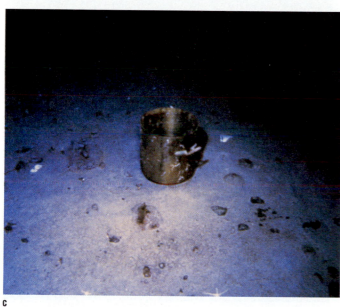

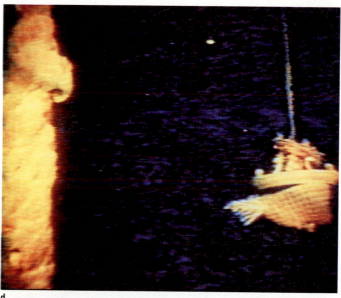
c

d

FIGURE 17-12 (a) Headlights of *Jason Jr.* shine back at *Alvin,* which is illuminating the *Titanic*'s bow.

(b) "Rusticles" ("icicles" of rust) nearly cover a porthole of the *Titanic* in this photograph, taken at the bottom of the Atlantic Ocean from the manned submersible *Alvin. Alvin*'s brilliant lights shining on the wreckage revealed rust covering vast areas of the hull.

(c) A copper kettle from *Titanic*'s galley lies in a debris field about a half a kilometer (a third of a mile) south of the vessel's bow. Strong bottom currents and particles in the water have polished the kettle so that it appears as clean as when the ship sank in 1912.

(d) A feathery sea pen on a dangling light fixture photographed by *Jason Jr.* several floors down the Grand Staircase.

(e) The robotic vehicle *Jason Jr.* leaves its "garage" aboard the manned submersible *Alvin* to photograph the remains of the luxury liner. *Jason Jr.* operates at the end of a tether 76.2 m (250 ft.) long, its movements controlled by a pilot inside *Alvin.* The *Titanic* expedition was the first deep-sea test for *Jason Jr.,* a prototype underwater camera system. (Photograph (a) by Dr. Robert Ballard and Martin Bowen; photographs (b)–(e) copyright © 1986 Woods Hole Oceanographic Institution; all photographs courtesy of Woods Hole Oceanographic Institution.)

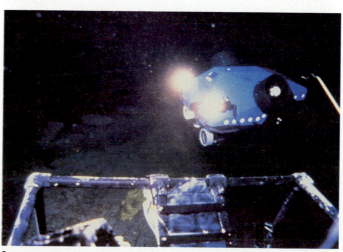
e

FIGURE 17-13   The *Bismarck* was the pride of the German navy. Adolf Hitler hoped that the vessel would match the reputation for unyielding determination of her namesake: Otto von Bismarck, founder of the German Empire, known as the "Iron Chancellor." (Photograph from the James C. Fahey Collection, courtesy of the U.S. Naval Institute.)

tractive to students. Its first phase was conducted from the Mediterranean for two weeks in 1989. Two sites—an active hydrothermal vent and an ancient ship—were explored. Several amphorae were collected from the second site and studied by archeologists (see Figure 17-3). An estimated 250,000 students observed and participated in the Mediterranean program. The National Science Teachers Association developed teaching materials for the project, focusing on the physical and social science aspects of oceanography, archeology, geography, seafaring, and trade.

The *Jason* Project in subsequent years visited the Galápagos, dived to a wreck in Lake Champlain (in the New York–Vermont region), explored a black smoker off Baja California, and the coral reefs of the Central American country of Belize. The Galápagos expedition almost ended in tragedy when considerable equipment, including *Jason Jr.,* was lost on a barge that sank off Ecuador.

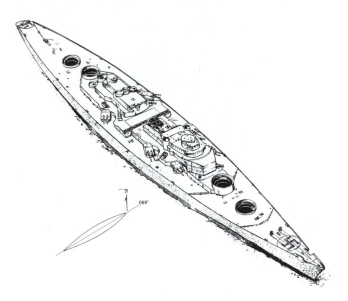

a

b

c

FIGURE 17-14   (a) Artist's representation of the *Bismarck* on the ocean bottom, based on photographs taken by *Argo*. During the battle, the four main gun turrets were blown away.
(b) An anti-aircraft gun on the *Bismarck*.
(c) Fire control station. (Photograph (a) copyright © 1989 National Geographic Society, (b) and (c) copyright © Quest Group, Ltd.; all courtesy of Woods Hole Oceanographic Institution.)

a

b

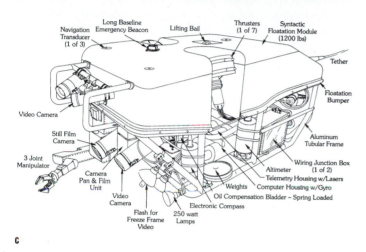

c

d

FIGURE 17-15 Pictures from the *Jason* Project: (a) The ROV *Jason* in front of its carrier *Argo;* (b) *Argo* at dockside; (c) schematic of the ROV *Jason;* (d) artist's rendition of the completed *Argo/Jason* system with its satellite transmission capability. (Photographs (a) and (b) by Tom Kleindinst; illustration (d) prepared by E. Paul Oberlander; all images courtesy of Woods Hole Oceanographic Institution.)

The Future of Marine Archeology

We have only begun the adventure of finding sunken ships, especially in deep water. Precisely how many ships have sunk through the centuries will never be known, but the number is large. Peter Throckmorton, in his excellent book on shipwrecks and archeology (see Further Reading), notes that *half* the sailing ships operating in the British Isles during the eighteenth and nineteenth centuries were lost. During the 1860s, the loss rate was about 250 ships per year.

Extending these statistics to the Mediterranean region, Throckmorton suggests that up to 15,000 merchant ships may have sunk in the first century B.C. alone. He further estimates that 5,000 military ships sank during that same century. With today's technology, we may expect new and exciting discoveries in the coming years. (See Box 17-2 for information on shipwrecks around the United States.)

From an archeological perspective, deep water is a problem because direct observation can be made only with a submersible. However, deep water has some very positive archeological aspects:

1. The sedimentation rate is less than in shallow water, so a wreck has less chance of being buried.

2. The lower temperature frequently helps preservation, because many chemical reactions are slower at lower temperatures.

3. Biological activity is less and occurs at a slower rate in deep water.

As noted, a major problem with wooden wrecks is that marine worms and mollusks bore into, eat, and destroy wood and other soft material. These creatures are absent where no dissolved oxygen exists, which is the situation below about 100 m depth (330 ft.) in the Black Sea. Unfortunately, the deep Mediterranean is well-oxygenated. Conceivably, even human remains could be preserved in an area like the Black Sea, which at this time is relatively unexplored archeologically.

This chapter has considered only shipwrecks, but other examples of human activity exist on the seafloor. Some ancient cities and habitats have been inundated by rising sea level or perhaps because tectonic movement has lowered them below the surface. Some of these sites have already been explored in the Mediterranean and the Middle East, and certainly will be the subject of more study in the coming years.

SUMMARY

As the emerging discipline of marine archeology develops and becomes more technologically advanced, there will be many new discoveries. As always, scientific attention to detail and careful documentation will be required. The competition and conflict between marine archeologists and treasure hunters will continue.

QUESTIONS

1. Why will new discoveries in marine archeology be dependent on or driven by technology?

2. What characteristics of the marine environment are favorable for preserving shipwrecks? What characteristics are not?

3. Why is a legal claim to a shipwreck site important? Why should one be granted?

4. What information about life in the past can be obtained from shipwrecks?

KEY TERMS

Bismarck

ANGUS (Acoustically Navigated Geological Underwater Survey)

Argo

Jason Jr.

marine archeology

NEMO

remotely operated vehicle (ROV)

R.M.S *Titanic*

SCUBA (Self-Contained Underwater Breathing Apparatus)

S.S. *Central America*

U.S.S. *Monitor*

FURTHER READING

Ballard, Robert D. 1990. *The Discovery of the Bismarck.* New York: Warner Books.

———. 1989. "*Bismarck,* Found." *National Geographic* 176, no. 5, pp. 622–37.

———. 1989. "*Titanic.*" *National Geographic* 168, no. 6, pp. 696–719.

Bass, George, F. 1987. "Oldest Known Shipwreck Reveals Bronze Age Splendors." *National Geographic* 172, no. 6, pp. 693–733.

———. 1980. "Marine Archeology: A Misunderstood Science." In *Ocean Yearbook 2* Chicago: University of Chicago Press, pp. 137–52.

Bass, George, F. 1988. *Ships and Shipwrecks of the Americas: A History Based on Underwater Archeology.* New York: Thames and Hudson.

Conrad, Judy. 1991. "Final Voyage" (of S.S. *Central America*). *American History* 26, no. 1, pp. 58–65, 72.

Conrad, Judy, ed. 1988. "Story of an American Tragedy: Survivors' Accounts of the Sinking of the Steamship *Central America.*" Columbus, Ohio: Columbus-American Discovery Group.

Herdendorf, Charles E. 1991. "Discovery in an Alien Environment (*S.S. Central America*)." *American History* 26, no. 1, pp 66–71.

"A Symbol of American Ingenuity: Assessing the Significance of U.S.S. *Monitor.*" Prepared for U.S. Department of Commerce, NOAA, Marine and Estuarine Management Division, National Park Service, 1988.

Throckmorton, Peter, ed. 1987. *The Sea Remembers Shipwrecks and Archeology.* New York: Weidenfeld and Nicolson.

CHAPTER 18

Future Uses of the Ocean

IN YOUR LIFETIME, STARTLING DIS-COVERIES ABOUT THE MARINE WORLD WILL OCCUR, AND INNO-VATIVE USES OF THE OCEAN'S RESOURCES WILL BE DEVELOPED. IN THIS AUTHOR'S OPINION, THE GREATEST ATTENTION WILL BE DI-RECTED TOWARD THE OCEAN'S POS-SIBILITIES AS AN ENERGY SOURCE AND TOWARD ITS POTENTIAL BIO-LOGICAL WEALTH. SO RAPID ARE NEW DEVELOPMENTS IN OCEAN RE-SEARCH AND TECHNOLOGY THAT A SIGNIFICANT DEVELOPMENT COULD OCCUR EVEN AS THIS BOOK IS PUB-LISHED. THIS CHAPTER LOOKS FOR-WARD TO THE OCEAN'S GREAT POTENTIAL.

Two small artificial islands in the Beaufort Sea off northern Alaska, less than 60 acres (0.24 km²) in area. These gravel islands were built at a cost of over $1 billion by British Petroleum to obtain oil from the offshore Endicott Field. About 100,000 barrels a day are obtained from the field. The Endicott Islands are connected to the mainland by a narrow causeway, visible in the upper left of the figure. The building of such island structures improves the safety of this operation against the storms and high waves common in this area. Without such islands, this oil field—sixth largest in the United States—might never have been developed.

(Photograph by Paul Fusco/Magnum, courtesy of British Petroleum America, Inc.)

Innovative Biological Uses

Our knowledge of the biological environment and its organisms is far from complete. Simple proof of this is the discovery of unique biological communities around deep-sea vents and the frequent discovery and identification of new marine species. This suggests an exciting future for marine biologists.

Marine Biotechnology and Genetic Engineering

Simply stated, **biotechnology** *is the application of engineering concepts to biological activities and processes.* Some biotechnology concepts have been employed for years: using greenhouses to improve and extend plant growth, breeding sterile males of certain unwanted insect species to reduce their frequency of reproduction, and selecting microorganisms to make cheese, beer, bread, and wine.

Genetic engineering *is the manipulation of the genetic code of an organism.* Genetic engineering has been used to select desired qualities of a plant or animal, but this is a long-term process. Now, key genetic materials—deoxyribonucleic acid (DNA) and ribonucleic acid (RNA)—can be manipulated and inserted into a new organism. This allows movement of genetic material that controls specific functions, from one organism to another. The applications for this process seem unlimited.

Biotechnology has been widely used in land agriculture to increase crop harvest. It offers similar possibilities in the ocean, but biotechnology using marine organisms is still in its infancy. One promising marine application is **aquaculture.** Strains of fish and shellfish could be developed by genetic manipulation to grow faster, to be meatier, and to better resist disease.

Another application of biotechnology could be marine pharmaceuticals, wherein new drugs or compounds could be developed, extracted, or produced. For example, the liver oil of some fish is already a good source of vitamins A and D, and insulin (to treat diabetes) has been extracted from tuna and some whales.

In environmental maintenance, biotechnologists have used microorganisms in the cleanup of the 1989 *Exxon Valdez* oil spill in Prince William Sound, Alaska. Some damaged beaches (Figure 18-1) were sprayed with a fertilizer that stimulated the growth of a naturally occurring oil-consuming bacteria. The results were apparently positive, although some scientists have questioned the usefulness of the technique.

Biotechnology could also be applied to marine plants to make them more tolerant of saline conditions, to increase growth, or to improve their quality or edibility. For example, growth and survival rates of marsh and coastal grasses could be improved, making them more useful in controlling erosion. Biotechnology could help solve waste-disposal problems in the coastal region, per-

FIGURE 18-1 Effects of using fertilizer to clean an oil-polluted beach in Prince William Sound, Alaska. The area on the left is the treated and apparently improved side. (Photograph courtesy of Exxon Company U.S.A.)

haps by developing materials that decay faster and more safely.

The economic potential of improved marine biological resources is considerable. For example, in the United States almost $6 billion worth of fish products were imported in 1993, consisting mainly of shrimp and salmon that could be produced by aquaculture.

Worldwide, only 15 percent of fishery products come from aquaculture. This percentage easily could improve with biotechnology, as has been done with domestic land animals. Growth hormones could be added to fish eggs, so that bigger and faster-growing fish would result. Hormones are being developed to initiate the reproductive cycle of fish at any time of the year. This would enable numerous breeding periods, most useful in an aquaculture program. Other possibilities include improving disease resistance, adjusting the sex ratio, or providing a uniform-sized final product (see Box 18-1).

The price of hard-shelled crabs is about $0.30 per pound, whereas soft-shelled crabs sell for about $1.75 per pound. Any process that can control or initiate molting (during which crabs go from their hard-shelled stage to the soft-shelled stage) has large economic potential. If molting could be restricted to times when predators are absent, it would reduce predation, because crabs are very vulnerable during their molt or when their shells are soft.

Biologically, less is known about the ocean than about the land. Therefore, biotechnology's potential is even more promising in the marine environment, and the coming years should see exciting discoveries.

Box 18-1

Aquaculture and Genetic Engineering

Aquaculture is not new. As early as 2000 B.C., the Chinese experimented with growing fish, as did the Egyptians even earlier. Today, aquaculture is big business, producing about 15 percent of the fish consumed, with sales exceeding $22 billion worldwide, according to the United Nations Food and Agricultural Organization (FAO). The future for aquaculture should become even better, for two reasons. First, the world's fish catch seems to have reached a maximum and some species may already be overfished. Second, Earth's increasing population demands more fish and shellfish.

Breeding and growing fish, however, is not an easy task because they are vulnerable to disease, sensitive to pollution and environmental change, and take considerable time to reach harvestable size. Some biotechnology procedures, such as gene-splicing, offer techniques to overcome or at least reduce some of these problems.

In the 1980s, synthetic growth hormones were fed to fish to increase their growth rate. This technique was costly and public concern grew about consuming organisms fed with such hormones. New techniques were needed. The next approach was to clone specific genes that promoted natural growth into fertilized eggs. Scientists led by Professor Thomas Chen of the University of Maryland's Center for Marine Biotechnology have succeeded in using this technique on carp and rainbow trout. (Although these are freshwater fish, the techniques will be applicable to marine organisms.) Preliminary evidence indicates a growth rate from 20 to 46 percent faster in the laboratory. The next step is to see how these fish do in an aquaculture environment, a procedure underway at Auburn University in Alabama.

Other scientists are considering the survival aspect rather than the growth component. One approach is to develop disease-fighting or disease-preventing genes and introduce them into fish embryos. Genes also could be developed to enable organisms to better withstand cold or freezing conditions (see also Box 11-2, page 253). Some success has already been achieved in giving agricultural crops this ability.

These techniques are still experimental. Scientists must take care to prevent fish with engineered genes from mixing and breeding with natural organisms, and national and local regulations surely will be written to control mixing of fish stock. Nevertheless, genetic engineering of fish is clearly a business for the future. This could be important for the United States, which recently had a $6 billion trade deficit in fishery products despite having the world's largest marine Exclusive Economic Zone.

REFERENCE
Fischetti, Mark. 1991. "A Feast of Gene-Splicing Down on the Fish Farm." *Science* 253 (August), pp. 512–13.

The following are some areas of marine biotechnology potential:

- Develop new synthetic chemicals, based on natural marine compounds that can have commercial applications. An example may come from the common marine mussel, which produces a strong adhesive protein that might be used to attach human skin grafts or to aid in replacing teeth.

- Produce compounds from marine organisms, such as beta carotene from algae, thought to be a cancer inhibitor.

- Develop bacteria to treat human or industrial wastes so that they can be discharged safely into natural waters. Some bacteria have already been used to dissolve or chemically break down oil and some other pollutants. Other possibilities are bacteria that can reduce or prevent biofouling of the hulls of ships or other objects put into the ocean.

- Produce drugs from marine organisms to combat various human and animal diseases. An example is toxic compounds from some tropical fish that could be used to treat high blood pressure.

- Develop varieties of shellfish or finfish that have increased growth rates or resistance to disease so that they can be better used in aquaculture systems.

Live Better—Eat More Fish

A common folk expression calls fish "brain food," and another offers that "fish is good for your health." Recent scientific tests and surveys seem to confirm that fish—in particular certain fish oils—can help protect the human heart and blood vessels. These fish oils contain compounds that can lower the levels of cholesterol and other fats in the blood. In so doing, they also reduce the possibility of clogged arteries. Fish oils may also increase the amounts of the so-called "good" cholesterol, HDL (high-density lipoproteins), which can remove cholesterol from the bloodstream.

More recent studies show that fish oils can lower blood pressure. The absence of these oils (or the compounds they contain) in a regular diet can affect organ development, including the brain. Another finding is that fish oil may extend pregnancy, thus offering the potential for preventing premature birth. These fish oils (called *omega-3 long-chain fatty acids*) can be purchased as a vitamin supplement, but sufficient amounts can be obtained by eating fish two or three times per week (Table 18-1).

The discovery of fish oil's value came from studying eskimos in Greenland. They consume large amounts of cholesterol-containing animal fat, but have a relatively low rate of coronary problems—the opposite of what one would expect. One group of 1,800 had only three heart attacks in 24 years, whereas in the United States,

TABLE 18-1

Seafood Grouped According to Level of Omega-3 Fatty Acids

Higher-Level Group (more than 1.0 gram omega-3 fatty acids per 3.5 oz. raw portion)	Medium-Level Group (between 0.5 to 1.0 gram omega-3 fatty acids per 3.5 oz. raw portion)	Lower-Level Group (0.5 gram or less of omega-3 fatty acids per 3.5 oz. raw portion)
FINFISH Anchovy Atlantic herring Atlantic mackerel Bluefin tuna Fish roe (mixed species) Salmon (Atlantic, king, sockeye, pink) Spanish mackerel Whitefish PRODUCTS[a] Canned sardines in oil Canned pink, chum, and sockeye salmon	FINFISH Bluefish Freshwater bass Pompano Rainbow trout Salmon (Coho and chum) Sea bass Shark Smelt Striped bass Swordfish Wolffish	FINFISH Atlantic cod Carp Catfish Croaker Eel Flatfish (e.g., sole, flounder, fluke) Grouper Haddock Mahimahi (dolphin fish) Mullet Northern pike Ocean perch Pollock Sea trout Skipjack tuna Snappers (red and other tropical) Whiting Yellowfin tuna SHELLFISH Blue crab Blue mussels Clams Crayfish Lobster Oysters Scallops Shrimp Squid PRODUCTS Gefilte fish Fish sticks and portions (frozen and reheated) Light tuna (canned in oil) Light tuna (canned in water) Canned clams Smoked whitefish

[a]Canned products made from other fish in this group are also likely to have high levels of omega-3 fatty acids.

Source: Composition of Foods: Finfish and Shellfish Products, Handbook No. 8-15, U.S. Dept. of Agriculture, Human Nutrition Information Service, September 1987.

six heart attacks per 1,000 individuals *per year* is common. A similar observation was made for Japanese coastal inhabitants. Blood studies revealed that omega-3 long-chain fatty acids in their seafood diets were the most likely explanation. Studies continue, but the evidence so far strongly suggests that eating seafood is indeed good for your health.

A very successful new seafood product, *surimi*, is a fish gel used to produce "restructured fish products." Such products look and taste like other types of fish or shellfish and have become very popular. Surimi starts as the deboned flesh of various fish and goes through several processing steps, eventually being shaped and colored to resemble, for example, crab meat. Besides being very tasty, it is relatively inexpensive and stores well. It appears that the manufacture of surimi (often sold under such names as "Snow Krab" or "Sea Legs") will lead to an increasing use of fish, especially those not now commonly consumed.

Improving the Harvest

Fishing in the ocean and hunting on land are similar because both are inefficient hit-and-miss operations. An obvious strategy is to maneuver fish into concentrations for harvest. Better techniques for tracking and herding fish would increase the catch for the same effort (see Box 18-2). To this end, the use of sound or of other fish as attractants has been attempted, with moderate success. Naturally this raises the risk of overfishing, so such schemes require careful management.

Another technique takes advantage of the life cycles of fish to maneuver them for harvest. Salmon demonstrate this principle: they are born in freshwater, travel to the salty ocean where they grow and fatten, and return to their freshwater birth site to reproduce. To do so, they follow chemicals in the water back to their original spawning ground. Thus, exposing salmon to the proper chemical at the right point in their growth will cause them to travel to the area where the chemical is released. In other words, they become *imprinted* with the smell of that chemical and will return to where they can be captured.

Weyerhauser Corporation developed a hatchery system that uses waste heat from an energy production plant to help salmon grow faster. Salmon eggs are hatched, and the young fry are fed until they grow to 10 to 13 cm (about 4 to 5 in.). This occurs over only 6 months, instead of the usual 18 months required in nature or in a cold-water hatchery. The fish are imprinted with the odors and geography of the area to which they will return as adults (Figure 18-2).

In this manner, the fish not only concentrate themselves for harvest, but also eliminate the travel normally required to bring in the catch, and the harvest occurs near the processing plant.

This salmon harvesting method works, but it has many problems. The fish become public property when released, and the company has no control over them during their 2-to-5-year development in the ocean, where anyone can catch them. Too few may return to make the operation profitable. (This eventually turned out to be the case for the Weyerhauser project.) One percent is considered to be the minimum return required, but this is rarely achieved. The reasons include predation, harvesting by fishermen, and perhaps genetic contamination (mixing of wild salmon with hatchery-bred salmon, possibly producing less hardy individuals). If these problems can be solved, hatchery systems that release fish to the "free" environment may become more successful.

Krill: Protein of the Future?

Krill is a general term for a group of crustaceans that comprises some 85 species. The most common, *Euphausia superba*, averages 2 to 4 cm (0.8 to 1.6 in.) in length (see Figure 11-19, page 266). Krill is an important food for some whales. However, overhunting has reduced the whale population, so the world krill stock has increased considerably; estimates range between 500 million and 5 billion tons. Some feel that 150 million tons a year of krill could be harvested (almost twice the present *total* world fish harvest). Krill catches can be immense, sometimes attaining 60 tons an hour. A problem with krill is processing speed, because the organisms deteriorate very quickly. However, krill contains about 20 percent protein and substantial amino acids, making it a potentially valuable food.

Krill is an extremely important organism in the Antarctic food web (see Figure 12-26, page 308). Beside whales, other animals eat krill directly, or consume it indirectly by eating other creatures that feed on it, such as seals, birds, squid, fish, and penguins. Large krill catches are possible because they congregate in large swarms. The Japanese have fished for krill since the early 1960s and make several products from it, including krill "meat" and krill protein concentrate.

Development of a krill fishery is widely anticipated, largely due to a leveling-off of the world's fish harvest to about 90 million tons per year. The present krill harvest is about 1 million tons per year, and most goes to Russia and eastern European countries to feed chickens and other livestock.

Although the krill market certainly can expand, it will probably never be the dramatic ocean product that some anticipate. The reasons are both biological and economic. Krill life span is much longer than originally estimated (up to 8 years), so their growth rate is slower than thought. Therefore, less can be harvested before seriously reducing the population. Another unknown is the effect of the harvest on other animals in the food chain. Ironically, the main whale-harvesting countries also may be affecting whale growth by harvesting the whales'

Box 18-2

Fish TV

Increasing our use of marine biological resources relies on very fundamental information: knowing what organisms are in the water. This sounds elementary, but it is not. Towing nets is one way, but it provides too small a sample of too small an area, and too many organisms, especially fish, avoid some nets. Using echo-sounders or fish-finders from a ship can help (see Figure 15-17, page 384), but this provides "returns" (echoes) only from organisms immediately below the ship. The result is a limited, "single-dimensional view" of the biological resource.

Dr. Jules Jaffe, an oceanographer at the Scripps Institution of Oceanography, is developing a technique to provide a three-dimensional view of everything in the water column that is larger than a centimeter (about half an inch). The device uses 16 high-frequency sonar units and a computer. Reflected signals from the sonar units are received, processed, and produce a computer image that can be manipulated to show a three-dimensional view of the organisms in the water, especially the fish (Figure 1).

According to Jaffe, his system (which he calls F-TV, for "fish television") will allow marine biologists to see the ocean in a manner similar to viewing creatures on land with binoculars. The F-TV system can also be used by scientists to study pollution and the dynamics of marine populations. The next step for F-TV is testing in a large fish tank, and then in the ocean.

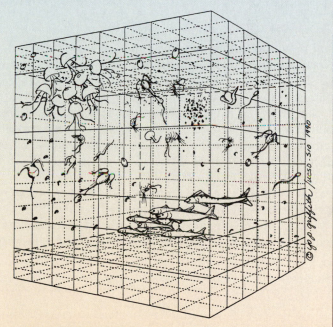

FIGURE 1 (a) Dr. Jules Jaffe adjusting controls of his system. (b) Artist's conception of the F-TV system. Illustration (b) by J. P. Griffith; both images courtesy of Dr. Jules S. Jaffe,

food. Finally, the cost of catching krill so far from where they are used, the preservation problem, and the difficulty of selling the product (it does not taste very good), all mean that the earlier enthusiasm for krill may have been overly optimistic.

Military Use of Marine Mammals

Some marine animals appear to be very intelligent, especially whales, dolphins, and porpoises. They are quick to learn and can perform complex tasks. Their use in herding harvestable fish has been suggested. Another use is recovery of devices lost on the seafloor, and some of these creatures have already been trained to dive to depths of more than 500 m (1,640 ft.) for this purpose.

The U.S. Navy has trained bottlenose dolphins for military operations (Figure 18-3). These creatures can swim up to 42 km (26 miles) an hour, and have evolved a sonar ability that allows them to detect submerged objects with great skill. This is essentially how they find a

FIGURE 18-2 (a) A fish ladder at Newport, Oregon, used by the Weyerhauser Corporation to harvest coho salmon. The salmon return to the area of their birth to breed, locating the region by its unique water chemistry. The returning fish swim up the ladder and can be captured easily, as shown in (b).

(b) Coho salmon being harvested. The man is standing atop the fish ladder in the small building shown in (a). Not all the fish are harvested, as some must be kept free to breed and to produce new young to maintain the cycle. (Photographs courtesy of Weyerhauser Corporation.)

ball in an aquarium, as seen in marine shows. Their detection ability is so sophisticated that they can distinguish sheets of different metals and locate objects less than 2.5 cm (1 in.) in diameter!

Details of the Navy's dolphin program are not public, but over 100 have been trained, apparently for surveillance and detection operations. One goal might be to protect vessels and harbor facilities from terrorist SCUBA divers. Another task is simply to recover or locate materials lost on the seafloor. In 1988, several dolphins were used in a U.S. military operation in the Persian Gulf.

Public protests against military use of marine mammals have occurred, but the U.S. Navy says these animals are not intentionally placed at risk, nor are they badly treated. The Navy explains that the animals do their "work" while swimming free and thus can swim away if they so choose. (On the other hand, these animals are trained not to swim away.) The use of nonhuman organisms in military operations is certainly not new: horses, camels, and elephants have been used to carry personnel and equipment, pigeons were used to deliver messages, and dogs and even geese have been used to guard areas. Nevertheless, the concern over potential abuse of marine animals is frequently justifiable, especially regarding the care of mammals that perform in marine shows.

Drugs from the Sea

Marine organisms may become a new drug source; several important drugs have already been obtained from the ocean. An example is Manoalide, a new anti-inflammatory drug produced from a Pacific Ocean sponge. The drug may be used for arthritis sufferers and appears to have limited—if any—side effects.

The systematic search for marine organisms that are good drug sources is relatively new, and is based in part on the toxicity of many marine organisms. They have the potential for use as medicines because a poison usually has strong potential to affect the physiological activity of another organism (such as reducing, or even stopping, its heartbeat). This effect sometimes can be incorporated into a drug, such as to treat high blood pressure. Little of the great diversity of marine life has been studied for its drug potential.

Some algae and seaweed are already used in cosmetics, drugs, and food. Brown algae, for example, are used

FIGURE 18-3 A school of spinner dolphins—a new military weapon? (Photograph courtesy of National Oceanic and Atmospheric Administration.)

to produce alginates, which can reduce bleeding and are used in some dental operations. Several new drugs have been produced from animals, such as didemnin from tunicates, which shows antiviral and anticancer potential. Another is a substance isolated from some sharks that can restrict blood flow to tumors and thus reduce their growth.

Energy from the Ocean

Any motion can be a potential source of energy: wind, people walking, animals moving, a waterfall, a running river, or the constantly moving ocean. The challenge in every case is how to *harness* the energy so we can put it to work. In the ocean, the presence of great energy is evident in waves, tides, and ocean currents. Less obvious are temperature differences between surface and deep waters, salinity differences, gas hydrates, and the large volume of biological material in the ocean, all potential energy sources.

Because the ocean covers about 71 percent of Earth's surface, it receives a major portion of the incoming solar energy. Because of the solar angle, more of this energy reaches the equatorial regions than the poles (see Figure 9-6, page 205). This large differential in incoming solar energy produces a continuing circulation that moves warm water and air toward the poles. Directly or indirectly, this circulation also causes the ocean's waves and currents—all movements that are driven by solar energy.

The potential power levels of different areas and processes of the ocean are vast and compare favorably with humanity's power appetite (Tables 18-2 and 18-3). To tap and utilize these sources of power, however, is quite another matter.

TABLE 18-2
Estimates of Power Levels in Natural Processes on the Planet Earth

Available Sources of Power	Total Power in Watts*
Direct solar power	
Where sun hits atmosphere	10^{17}
At earth's surface	10^{16}
Photosynthesis (stores sunlight in the form of chemical energy in fats, proteins, and carbohydrates, all of which are combustible)	
Marine plants	10^{14}
Arable lands, forests	10^{13}
Bioconversion of waste materials	
Plant residues and manure (can be converted by bacteria to gaseous fuels—hydrogen and methane—by storing them in airless containers at proper temperatures)	10^{12}
Garbage, sewage, and dumps (can be converted by the same process)	10^{12}
Ocean thermal power	
Solar heat absorbed by ocean water (can possibly be put to use by exploiting temperature differences between surface and depths, producing power to drive turbine)	10^{13}
Steady surface-wind power (such as that from trade winds)	10^{12}
Variable surface-wind power (in middle latitudes, where winds are unsteady)	10^{12}
Hydroelectric power (from harnessing the kinetic energy of moving waters)	
Power in rainfall (conceivably could be harnessed, but the world's total rainfall, even if the rain falling on the oceans is included, would satisfy only 10% of the world's power demand)	10^{12}
Flow of rivers (by traditional hydroelectric plants)	10^{11}

Natural evaporative exchanges between large bodies of water (The Mediterranean Sea and Red Sea are examples: evaporation is greater in them than in the ocean at large; therefore, there is a continual flow into them from the oceans to replace evaporated water. This flow can be harnessed just as in a millrace.)	10^{9}
Damming of evaporative sinks (By damming ocean openings to Red Sea and Mediterranean Sea, letting these seas evaporate until a drop of 100 m or more occurs, and then letting the ocean flow in, turning mill wheels, power might be obtained. It is not very practicable to build these dams, however.)	10^{11}
Tidal flow (This is done at such places as the Rance River, where flow can be harnessed.)	10^{9}
Power of the great ocean currents (such as the Gulf Stream and Kuroshio Current; theoretically can be harnessed the way rivers are, with some sort of "water wheel")	10^{8}
Ocean surface waves at the coastline (The power of waves is available at a potential total yield of 10^6 W/km of coastline.)	10^{10}
Geothermal power (This could be harnessed particularly at the "ring of fire" around the Pacific Ocean basin, so called because this is where tectonic plates merge and volcanoes erupt; volcanoes also exist along mid-ocean ridges.)	10^{10}

Present Power Demands	Total Power (Watts)
Worldwide power demand for all needs of civilization	10^{13}
Human metabolism (total power in terms of food needed to sustain population level)	10^{11}

*See Box 2-1, page 22, for a "refresher" on the use of exponents.

Source: From von Arx, 1979.

TABLE 18-3
Some Energy Units

1 watt*	= 3.413 Btu/hour
1 British thermal unit (Btu)**	= 252 calories
1 kilowatt-hour (kW-h)	= 3,413 Btu
1 kilowatt (kW)	= 1 thousand watts
1 megawatt (MW)	= 1 million watts
1 gallon of gasoline	= 125,000 Btu
1 barrel of oil	= 5,800,000 Btu
1 cubic foot of gas	= 1,031 Btu
1 ton of bituminous coal	= 20–30 million Btu
1 cord of wood	= 20 million Btu
Energy use in the United States per person	= 300 million Btu/yr
Energy use in India per person	= 5 million Btu/yr

*A 100-W light bulb used for 10 hours is equal to 1 kW-h.

**A Btu, or British thermal unit, is the heat needed to raise the temperature of one pound of water by 1°F.

Tidal Power

Tidal power is a form of ocean energy considered since ancient times. Although numerous tidal-energy plants have been suggested, only a few have come to fruition. The world's first major tidal-power system was built on the Rance River estuary in France in 1966, at a cost of about $100 million (Figure 18-4). Russia has an experimental tidal-power station on the Barents Sea, about 80 km (50 miles) from Finland. Neither plant is a major source of power. Russia intends to build larger plants.

The biggest problem with harnessing tidal energy is that it is not constant. Tidal currents do not flow continuously but change amplitude and direction several times a day, and vary in strength over the 2-week lunar cycle (see Figures 10-18 to 10-20, pages 238–241). This pattern is completely different from that for electrical power demand, which is fairly consistent and does not coincide with the lunar period that produces the tides. The Rance tidal station alleviates this problem somewhat by storing water in a reservoir at high tide and letting it out slowly to turn turbines.

The Rance River plant is a successful operation and its 24 turbine units can each produce 10 megawatts of energy, for a total of 240 megawatts. Nevertheless, the plant can produce power during only about half the tidal

a

b

FIGURE 18-4 (a) The Rance tidal power plant in northern France is 750 m (2,460 ft.) long. It has a reservoir of 184 million m³ (6.5 billion ft.³). Tides in this region are exceptionally high and can have an amplitude up to 13.5 m (44 ft.). The open ocean is to the left of the figure.
(b) A closer view, with a strong running tide. The tidal flow of water drives 24 turbine generating units, situated 10 m (33 ft.) below low-tide level. To generate electricity, a significant difference in seawater elevation is required, which limits energy production to only about half of each tidal cycle. (Photographs courtesy of Electricite de France.)

period, because of the up-and-down aspect of the tidal pattern.

Other areas where the tides are sufficiently high and the land would permit storage basins or reservoirs include Cook Inlet in Alaska, some areas along the Yellow Sea portion of the China coast, parts of the English Channel, and the Bay of Fundy region in the northeastern United States and southeastern Canada. Areas having potential for tidal power generation are shown in Figure 18-5.

For many years, a joint U.S.-Canadian project was considered in the Bay of Fundy and the Passamaquoddy area of Maine, but economics and politics have prevented development of this region. The Canadians did build a small tidal plant on the Annapolis River off the Bay of Fundy. This $46 million pilot project, completed in the mid-1980s, is about 213 m (700 ft.) long and is the first tidal station in North America. The Canadians plan a much larger facility that will cut across 8 km (5 miles) of the upper part of the bay.

This is a controversial project, because some feel it will affect tide levels as far south as Cape Cod, Massachusetts, with a change of up to 0.5 m (1.5 ft.) at Boston. If correct, it could dramatically impact coastal processes, causing widespread coastal erosion in the northeastern U.S. and southeastern Canada. The impact would be even greater should sea level rise due to global temperature increase.

Despite enthusiasm for tidal plants, they can produce only a small fraction of the energy of a nuclear power plant, and often cannot compete on cost. Construction of a tidal power plant involves building a dam or causeway across an estuary or bay. Besides being expensive, the construction interferes with other activities, such as shipping and movement of some marine species. A tidal power plant can affect the ecology of an area because it interferes with the normal tidal pattern within an estuary. Also, tides must have at least a 5 m (18 ft.) difference between high tide and low tide for the plant to work efficiently. On the positive side, tidal plants have a fairly long life, and tides are free and environmentally "clean."

Wave Power

One has only to watch the crashing surf to appreciate the energy available in waves, especially during storms. Wave energy is considerable, capable of moving large boulders, piers, or sea walls that weigh thousands of tons. When thinking about using waves for energy, however, one should not consider the storm situation, which is erratic, but instead should think of the average state of the sea. The energy in even moderate-sized waves can be immense. For example, a typical wave 3 m (about 10 ft.) high has the potential to transmit energy at the rate of about 100 kilowatts per meter (3.3 ft.) in its wave crest. Such power is essentially equivalent to that of a line of automobiles (along the wave and facing in its direction of motion) with their engines running at full power.

There are three basic techniques for extracting energy from waves:

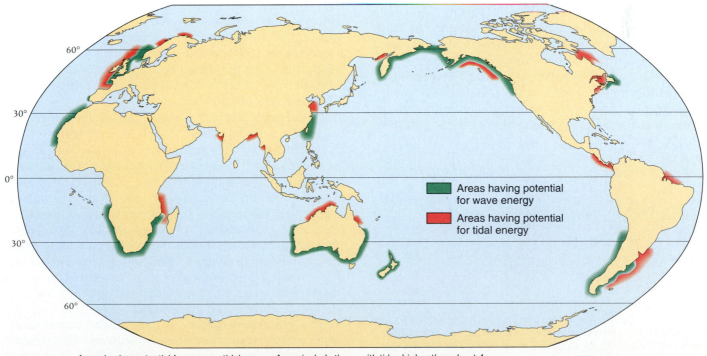

FIGURE 18-5 Areas having potential for wave or tidal energy. Areas include those with tides higher than about 4 m (13 ft.) or wave heights that exceed 3 m (10 ft.) about 30 percent of the time.

1. Harnessing the vertical rise and fall of crests and troughs of successive waves to drive an air-powered or water-powered turbine

2. Using the rolling motion of waves to move vanes or cams that turn turbines

3. Focusing waves into channels to concentrate their energy.

Energy from the rise and fall of wave crests has been used for several years in buoys, providing energy for their lights or to power a whistle. Several hundred are used in the United States and elsewhere.

Larger systems, up to several hundred feet in diameter, have been proposed in which the waves could compress air to drive a turbine. Realistically, such systems probably would generate only small amounts of power. In Japan, the experimental 500-ton *Kaimei* floating platform uses the up-and-down movements of waves to drive turbines and produces 125 kilowatts of power. The interest in such systems is considerable among island inhabitants, who must import energy (in the form of petroleum), usually at considerable cost. Nearly 100,000 such island communities exist worldwide. However, no successful and reasonably priced device has been developed.

An exciting development that concentrates wave energy is the Dam Atoll. This dome-shaped device sits just below the water surface (Figure 18-6). Waves enter at top and move through a central vortex, turning a turbine to generate electricity. According to the project designer, 1.6 km (1 mile) of waves could yield 64 megawatts of electricity, and a group of 500 to 1,000 Dam Atolls anchored to the seafloor could produce as much electricity as Hoover Dam. The device could also be used to protect beaches or harbors from wave erosion.

Even the capture of wave power can have ecological effects. Interrupting normal wave patterns may change the pattern in which they break, modifying longshore drift and the movement of sand along the shore. (These

effects are discussed in Chapters 10 and 14.) Areas that generally have high waves and thus the potential for generating energy are shown in Figure 18-5.

Power from Major Ocean Currents

Near Florida, the Gulf Stream transports up to 100 million m³ (3.5 billion ft.³) of water per second, moving at almost 3 knots (3 nautical mi./hr.). This is an impressive amount of energy. Underwater windmills or turbines could be placed in this flow so the moving water would drive them to generate electricity (Figure 18-7).

More imaginative ideas (perhaps too imaginative) include a series of parachutes attached to a continuous cable. These parachutes would catch the current and move the cable, which in turn would generate electricity.

Any of these devices would have to be large and strong to operate in a powerful current. This is not an easy feat, and none are presently in operation or being built.

Power from Thermal Differences

Recall the *thermocline,* the water zone where temperature declines rapidly with depth (see Figure 8-9, page 189). The thermocline exists between surface and deep waters, and the temperature difference can be substantial, especially in tropical regions. These temperature differences have the potential for providing energy.

The thermocline is caused by incoming solar radiation, which heats surface waters, making them considerably warmer than deeper waters. The technique that seems to have the best chance of deriving power from this temperature difference is **Ocean Thermal Energy Conversion (OTEC).** The system is fairly simple (Figure 18-8). It works somewhat like a refrigerator or air conditioner, using a fluid inside tubing to absorb heat in one place and release it in another.

Liquid ammonia is pumped through a loop of tubing. When the ammonia goes through the evaporator, which is heated by warm water drawn in at the surface, the heat evaporates the ammonia into a gas. This expands the ammonia, and the expanding gas drives a turbine, which rotates an electrical generator. (A turbine turning a generator is a standard technique used in conventional electricity generation.) The ammonia gas then passes through the condenser, which is cooled by colder seawater pumped up from depth. This cools the ammonia gas, condensing it back to a liquid state. The liquid ammonia then circulates back to the evaporator, and the cycle continues, 24 hours a day, year-round.

This is a closed system, so none of the ammonia escapes into seawater, nor does the seawater contaminate the ammonia in the tubes. Note that all of the heating and cooling take place via heat exchangers (evaporators and condensers), similar to a refrigerator.

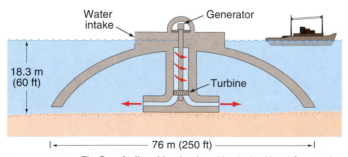

Figure 18-6 The Dam Atoll, an idea developed by the Lockheed Corporation to generate electricity from waves. These devices can be up to 76 m (250 ft.) in diameter and 18.3 m (60 ft.) high. Water enters through the inlet and spirals down the central vertical cylinder to drive an electricity-generating turbine at the bottom of the device. These Dam Atolls would be built of concrete and anchored to the seafloor.

FIGURE 18-7 An imaginative scheme for using buoyant turbines tethered in strong currents such as the Gulf Stream or Kuroshio Current.

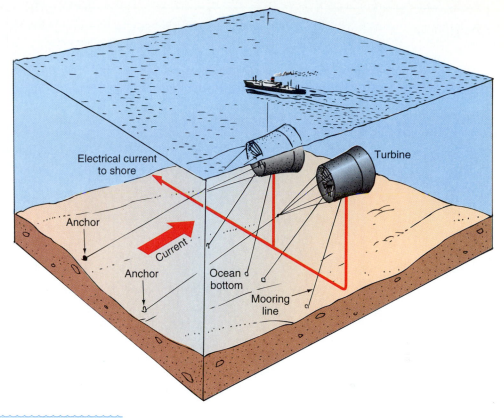

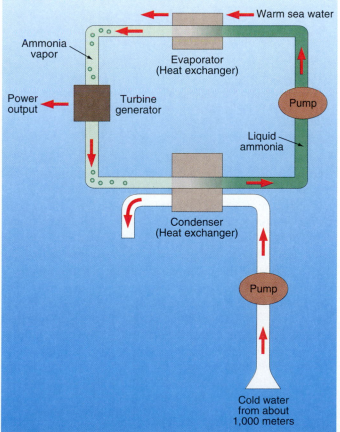

FIGURE 18-8 Basic concept of the Ocean Thermal Energy Conversion (OTEC) system. This system is not drawn to scale. (See the text and Figure 18-9 for details on its operation.)

The Ocean Thermal Energy Conversion concept is not new. It was first proposed in the 1880s by a French physicist named Jacques d'Arsonval. A small OTEC plant was built in Cuba in 1930 but lasted only a few weeks before being damaged by heavy seas. The idea has become popular in the search for alternative energy sources, and several models and working systems have been proposed (Figure 18-9).

Generally, a temperature difference of about 20°C (about 36°F) is needed for an OTEC system to work efficiently. This temperature difference in some areas would span the vertical distance from the surface to 1,000 m (3,300 ft.) depth. An OTEC site also has to be close enough to land for electrical cables to connect the OTEC's output to the land power-distribution system. Areas that are both close enough to land and have a suitable temperature difference are few, restricting the areas where OTECs can be built (Figure 18-10).

An OTEC system is very inefficient, in that only about 2 percent of the potential energy from the temperature differences is obtainable for use. In addition, a 100-megawatt plant has only about one-tenth the energy-generating potential of a modern nuclear power plant or fossil fuel plant. This means that, to bring OTEC technology on-line, it would be necessary to build many of these devices. For the United States, it has been suggested that OTEC plants cannot be more than 160 km (100 miles) offshore, because of problems in transmitting the energy, and probably would have to be located

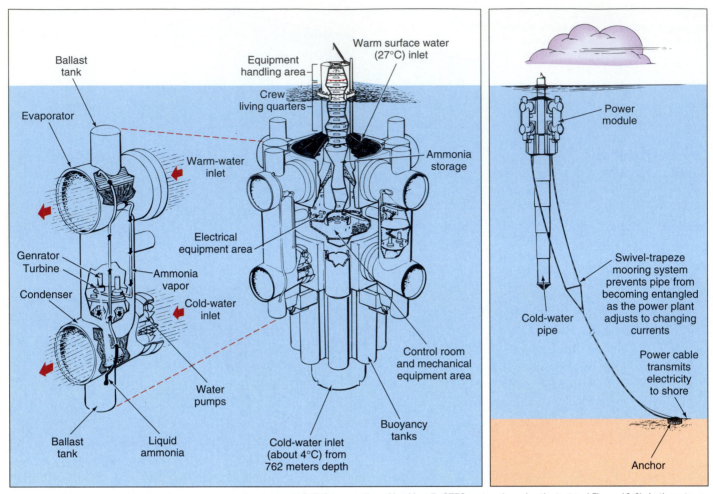

FIGURE 18-9   Artist's conception of Lockheed's OTEC system (see also the text and Figure 18-8). In the cutaway view, note the space for crew quarters. The total length of the platform is about 181 m (592 ft.) and its diameter is 75 m (246 ft.). Each module is 93 m (305 ft.) high and 22 m (72 ft.) in diameter. The cold-water pipe is 304.8 m (1,000 ft.) long. Courtesy of Lockheed Missiles and Space Company, Inc.

in suitable areas in the Gulf of Mexico or off Hawaii (see Figure 18-10). One solution to transporting the energy is to produce energy in a form that can be stored, and then transport it to the point of use.

In 1979, a Mini-OTEC test system was built off Hawaii with private funds and successfully operated for 125 hours. It produced 50 kilowatts of power, with 40 kilowatts used to operate the system and 10 kilowatts being surplus, available for distribution. Even this small device used a cold-water intake pipe that was 600 m (2,000 ft.) long and about 56 cm (22 in.) in diameter.

Some people question whether OTEC can ever work effectively. Corrosion from seawater can damage and reduce the efficiency of heat exchangers, and marine organisms can foul the intakes, pipes, and pumps. If the system is not a floating one, it would have to be anchored to the seafloor, which can be a difficult task.

Another problem is the vast quantity of water OTEC demands. A commercial 250-megawatt system would

need 1,416 m³ (50,000 ft.³) of water per second, equivalent to the flow of the Missouri River! This raises environmental questions: The system would lower the ocean surface temperature locally, so could this affect local climate? How would it affect sea ecology? A legal question exists, too: Who, if anyone, "owns" the temperature difference (energy) of ocean water outside the territory of a coastal state?

Other aspects of OTEC could help make it economically viable (Figure 18-11). The cold water from depth could be piped to air condition a hotel. The seawater that evaporates and recondenses in the process is desalinated, and could be used for agriculture or human consumption. The artificial upwelling of nutrients is another advantage, discussed in the following section.

OTEC devices should be most valuable to tropical islands (such as Hawaii or Puerto Rico) or less-developed equatorial countries that are highly dependent on imported oil. Countries involved in OTEC research and

FIGURE 18-10 Areas where the temperature difference is greater than 20°C (36°F) between the ocean surface and a depth of 1,000 m (3,281 ft.) and are close to land, have the highest potential for OTEC.

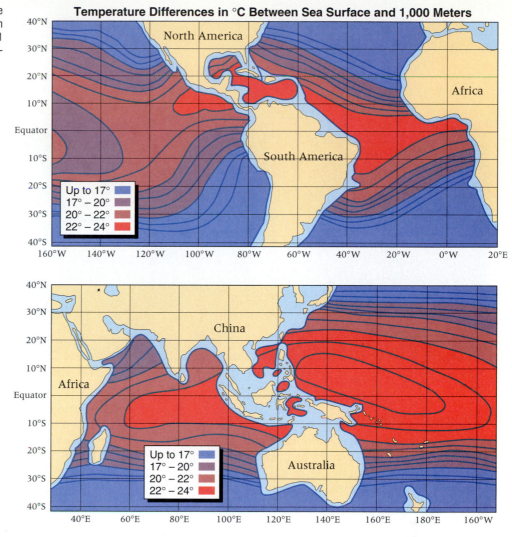

Temperature Differences in °C Between Sea Surface and 1,000 Meters

planning are Japan, France, Sweden, Holland, and the United States. Japan built several small units (less than 100 kilowatts) in the late 1980s. Recent U.S. OTEC research has emphasized the "open-system" approach, in which seawater is evaporated into steam, which drives turbines to produce power. This approach has the added advantage of producing freshwater when the steam is cooled. The open system also has less potential for fouling by organisms and thus can be easier to maintain. Although no open systems have yet been built, considerable optimism exists for their potential.

Artificial Upwelling

An interesting sidelight of the OTEC scheme could be an *artificial upwelling system* that brings nutrient-rich deeper water to the surface. This nutrient-rich water could be used for aquaculture or **mariculture.** (The terms *aquaculture* and *mariculture* are often used interchangeably;

mariculture, however, is more appropriate for the saltwater culture of marine organisms.)

Incorporating mariculture into an OTEC system would add a biological component, thus reducing the operating cost of the energy generation. Nutrient-rich water could be used to grow algae that, in turn, could nourish other organisms higher in the food chain.

The cold water could also help moderate high tropical temperatures, where most OTEC systems would be located (Figure 18-12). This moderation could be especially effective near islands situated in humid wind systems. The cold seawater could be passed through condensers in the moisture-laden winds to condense water from the air, just like setting out a huge glass of iced tea and collecting the condensation from the outside of the glass. This distilled water could be recovered for drinking. Another possibility is that the cold water from an artificial upwelling system could be used to cool an existing nuclear power plant.

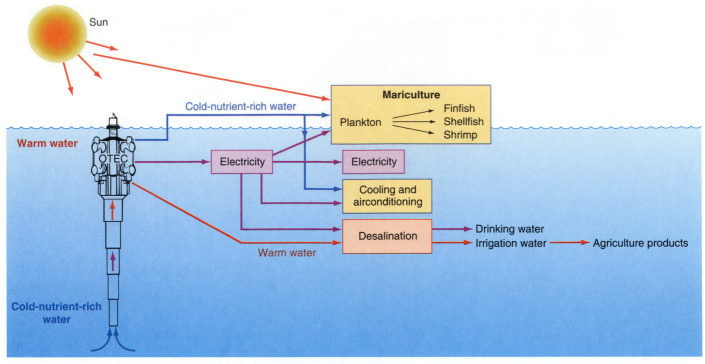

FIGURE 18-11   A potential future system that combines OTEC/mariculture/desalination. (See also Figure 18-12.)

Power from Marine Biomass

Can some organisms, especially plants and algae, be energy sources? Under the right conditions, these organisms can grow very rapidly. After harvesting and treatment, they can be converted into natural gas and other useful products.

One candidate is the giant California kelp *(Macrocystis)*. This brown alga can grow as much as *a foot a day* and is presently harvested for various biological products, such as iodine-based drugs. Natural gas, fertilizer, and livestock feed can also be obtained from the physical, chemical, and biological breakdown of the kelp. Kelp or other plants could be incorporated into an OTEC or artificial

FIGURE 18-12   A possible artificial upwelling mariculture system. The key part of the system is the availability of cold nutrient-rich deep water, which could be obtained from an OTEC system.

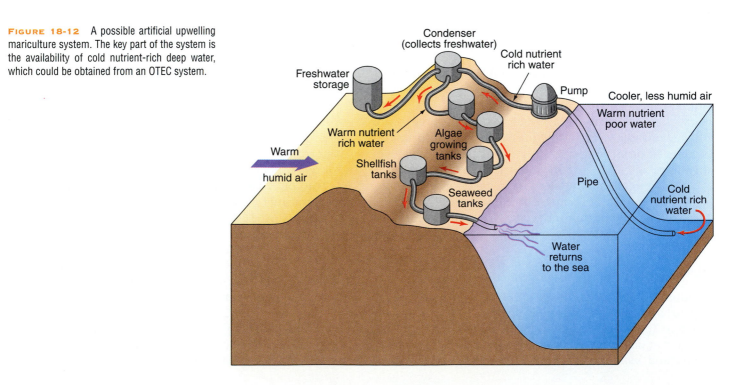

upwelling system to increase the nutrient supply, accelerating growth (Figure 18-13).

Could a *marine farm* be developed in shallow water to produce food, fuel, and other products? Kelp, or a similar plant, could be harvested, cutting the upper portions regularly. Microscopic algae could also be used. It might even be possible to develop these "farms" in ponds or small lakes, using local waste products as a source of nutrients.

If you are interested in helping the environment—and this author hopes that you are—you may become enthusiastic about the "environmentally friendly" ideas discussed here. The concept of using the products of photosynthesis (plants) is reasonable, but it is easy to become too optimistic about solar energy (the initial source of energy for photosynthesis). Certainly, the products of photosynthesis are renewable and virtually inexhaustible. However, the *efficiency* of starting with solar energy, then converting it to a biological product, and then releasing and using the energy stored in that product, is really very small. Although the ideas discussed here can work in some situations, they probably cannot become a major energy source, competitive with oil and gas.

Power from Gas Hydrates

It appears that vast volumes of natural gas, mainly methane, occur in sediment beneath some areas of the continental margin. The gas is trapped among water molecules in an icelike solid, giving rise to the name, **gas hydrate.** Gas hydrates form under specific pressure and temperature conditions that can occur in water depths below 300 m (1,000 ft.) and down to about 1,100 m (3,600 ft.) beneath the seafloor. The location of these hydrates can be determined by seismic techniques.

Gas hydrates (also called **clathrates**) may be one of the largest energy sources known: up to *700 million trillion ft.³ of gas* may be contained within them, by one esti-

mate. For comparison, 600 trillion ft.³ of natural gas is considered to be a giant conventional gas field.

Clathrates also have possible implications for the greenhouse effect because methane is a greenhouse gas (see pages 321–326 for implications). Despite these interesting possibilities, little research has been done on methods to recover the methane in clathrates, if indeed it is even possible.

Other Possible Sources of Energy

Here is yet another interesting idea for developing energy from the ocean: dam large bodies of water that connect to the ocean, especially those partially confined by land, where the evaporation rate is high. Examples would be to dam the Strait of Gibraltar, confining the Mediterranean Sea; to dam the Strait of Hormuz, confining the Persian Gulf; or to dam the southern end of the Red Sea, confining it.

The high evaporation rate will reduce the water height on the confined side of the dam, whereas the water on the other side, being connected to the large supply of water from the open ocean, remains at the original elevation. Over time, perhaps a year or so, the elevation difference would become great enough to operate hydroelectric power plants, just like those on major rivers. This intriguing idea will probably never come to fruition, however, because it would seriously disrupt ocean commerce as well as the migration of ocean organisms.

Another interesting plan to modify a small sea has been proposed by the government of Israel. The Dead Sea is the saltiest water body in the world, situated more than 400 m (1,300 ft.) below sea level. The Dead Sea is literally dying: it is drying out. A canal has been suggested to connect the Mediterranean Sea to the Dead Sea. Water moving down to the Dead Sea could provide hydroelectric power as well as recharge the Dead Sea.

FIGURE 18-13 Conceptual model of a kelp-processing plant, showing some possible byproducts. Adapted from Flowers and Bryce, 1977.

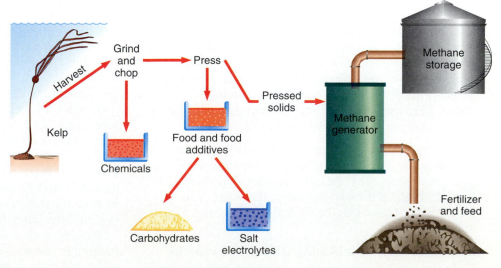

Still in the discussion phase, this program is controversial because of its ecological impact and its political ramifications (Jordan also borders on the Dead Sea).

Finally, ocean vents (see Figure 15-8, page 373) have been suggested as an energy source, because they discharge water at up to 350°C (660°F). One estimate is that the heat discharged from a single vent is equivalent to that from a small nuclear power plant. The idea is intriguing, but the technological challenge is to convert the heat being released at the seafloor into usable energy, in a water depth of several thousand meters, in the central part of the ocean, and transmitting that energy to a useful location. No workable scheme has been developed. Land-based counterparts of ocean vents, like the hot springs found in Iceland (which is located on the Mid-Atlantic Ridge) are presently used as a source of energy and hot water.

Freshwater from the Sea

Besides shortages in energy and food, many countries and regions have, or will have, shortages of freshwater. In 1960, the 160 million people then populating the United States used 16 billion gallons of water per day, or 90 gallons per person. By 1990, the greater population of 252 million used over 30 billion gallons per day, or 117 gallons per person. Similar increase in water use is common in less-developed countries, but they frequently lack a way to obtain additional water. The water needed is not all for drinking; greater quantities are needed for agricultural and industrial use, in which case the water does not have to be as pure.

Desalination

Referring back to Table 6-1 (page 133), all the water in Earth's freshwater lakes, rivers, and streams totals only about 0.01 percent of the worldwide water supply. Another 0.6 percent is available as groundwater, by drilling wells. The ocean contains about 97.2 percent of the world's water supply and is a tempting resource for freshwater. However, it is not easy to convert saltwater to potable freshwater. Desalination of seawater, although a simple process, involves a complex technology.

Three principal desalination techniques exist: membrane, distillation, and crystallization or freezing.

- In the **membrane** (or reverse osmosis) **process,** a thin sheet or membrane is used to separate saltwater from freshwater. The membrane allows freshwater to seep through, but traps the salts. (Osmosis, you may recall, is an important biological process; see Figure 6-10, page 146.)

- **Distillation** involves heating saltwater to vaporize it. The salts remain behind, and the water vapor is then cooled and condenses as distilled freshwater. This is the most common process, but it also uses considerable energy. In sunny climates solar energy can be used to reduce cost.

- **Freezing** saltwater leaves behind the salts, and the ice crystals can be separated from the salt and melted to produce freshwater.

Nearly 7,500 desalination plants are operating worldwide. About 60 percent of these plants are in the Middle East, with about 400 in the United States. The U.S. plants are used mainly for industry, not to produce drinking water.

One alternative to using desalinated or freshwater for agriculture is to develop agricultural products that can grow in salty or brackish waters. Over 500,000 acres of farmland per year become salty due to irrigation, because many of the mineral salts in the irrigation water remain in the soil. The net effect is to reduce the area available for agriculture. An innovative application of biotechnology techniques is to find plants that can tolerate salty soil or salty water. Many can do so (for example, a saltwater marsh is a highly productive area), but most of these plants are not useful food products.

The search has focused on plants that humans or livestock can eat or plants that can be used as fuel. Because many countries use wood as inexpensive fuel, a salt-tolerant crop could substitute as fuel, freeing woodlands for clearing and raising of food crops instead. At present, other than finding a few species that might be used under limited conditions, no breakthroughs have occurred. However, the new field of biotechnology should lead to exciting discoveries.

Glaciers and Ice Caps

Glaciers and ice caps include about 2.15 percent of Earth's water supply, and these of course are frozen freshwater. Thus, icebergs are yet another possible freshwater source. The largest ones come from Antarctica (Figure 18-14) and could be towed to melt where water is needed. A moderately large iceberg can contain over 1.4 billion m³ (50 billion ft.³) of water, and be worth $20 million. Such numbers compare very favorably with desalination.

Moving an iceberg involves serious technical, environmental, and legal problems, many without precedent. On the other hand, icebergs contain a fairly good quality freshwater, in some instances having fewer impurities than typical drinking water. Possible users include Saudi Arabia (which sponsored a scientific conference on the subject), California, Australia, and New Zealand; the latter two are relatively close to Antarctica.

Moving such a large piece of ice and minimizing its melting in transit is an engineering challenge. The trip itself could take over a year. Tow speeds of icebergs, re-

FIGURE 18-14 A portion of the Antarctic ice sheet, over 30 m (100 ft.) high. The largest iceberg on record was sighted in 1956 off Antarctica. It was about 31,000 km^2 (12,000 miles^2) in size, slightly larger than Maryland or about 50 percent larger than New Jersey.

gardless of method, would be less than 1.6 km per hour (1 mile per hour). The slow speed results from the iceberg's size and from friction, because about 90 percent of it is below the water surface. As an iceberg is towed through the ocean, it would produce interesting biological effects, would cool the water, and indeed could even produce local fog or rain. Mechanisms would have to be developed for breaking the iceberg apart and melting it when it arrived at its destination.

The Saudi Arabian who sponsored the scientific conference on freshwater from icebergs proposed to move an iceberg weighing 100 million tons from Antarctica to Saudi Arabia, wrapping it in cloth and plastic for towing by several ships. The trip was estimated to take 8 months and cost $100 million. It was assumed that the mile-long iceberg would lose only about 20 percent of its mass in travel. If these calculations are correct, the resulting water would be cheaper than the equivalent amount obtained by desalination. There are those, of course, who have a hard time visualizing towing an iceberg across the equator and who suspect that by the time one reached the equator, one would be towing nothing but a piece of rope. Nevertheless, it is an interesting idea.

Recently a new industry has developed in Alaska: harvesting icebergs and selling pieces of them as "gourmet ice cubes." This product fetches a nice price—for example, some of the ice is packaged in sacks of 1 kg (about 2.2 lbs.) and shipped to Japan, where it sells for $6.80, or about $3 per pound. (In fairness, "gourmet ice" is relatively pure compared to the conventional water

used to make ice cubes.) Another innovation in Japan is the establishment of "water bars," where bartenders mix exotic spring waters from various countries (glacial water probably will be next).

Disposal of Nuclear Waste in the Deep Sea

Like most things, the nuclear age has good and bad aspects. On the positive side, it has created exciting technology for manufacturing, scientific measurement, and medicine. Nuclear power plants generate about 20 percent of electrical power worldwide, without producing the greenhouse gases emitted by conventional power plants.

On the bad side are nuclear weapons and the potential for nuclear war, the risk of an accident at a nuclear power plant or weapons facility, and radioactive nuclear waste. This waste comes from nuclear power plants worldwide, both civilian and military. Nuclear waste is dangerous because it emits ionizing radiation that can damage living cells and eventually kill.

Some radioactive elements in nuclear waste give off high levels of dangerous particles and energy. Some have extremely long half-lives, radiating hazardously for 100,000 years or longer. These elements are extremely dangerous, and they make safe storage of the waste both a necessity and a challenge.

The total worldwide volume of radioactive waste is unknown, and it cannot easily be determined because of the secretive nature of nuclear programs. To appreciate the scope of this problem, it is estimated that 43,000 tons of used uranium fuel rods from U.S. power plants will have accumulated by the year 2000. These rods presently are stored in concrete or steel tanks above ground, and some have already leaked material. Such radioactive material requires secure storage for tens of thousands of years until radioactivity decays to a safe level.

The volume of this toxic material continues to increase as more countries acquire nuclear technology and as more applications are discovered. Three basic options exist for nuclear waste disposal: (1) transportation into space, (2) transmutation into safer compounds, and (3) secure storage within Earth's land or sea environment.

At this time, the technology for the first two options is unavailable, undependable, or inadequate. Shooting nuclear waste into space, besides being very expensive, may use as much energy as the nuclear material itself produced. The risk of rocket failure is fairly high, as is evident from occasional space program launches going awry, and no one would risk scattering nuclear waste from an exploding rocket. As for transmutation, no process now exists to convert radioactive elements into safe products. Thus, we are left with the option of storage in or on Earth or its oceans, and with the hornet's nest of controversy that surrounds it.

Nuclear waste can be stored in several ways, and each has pros and cons:

1. Store nuclear waste above ground in strong, guarded containers. These containers might have to be maintained for tens of thousands, or even hundreds of thousands of years. Considering that humankind's recorded history extends only about 4,000 years, and that our history is filled with war, conquest of territory, and dubious judgment, this option would be temporary at best.

2. Dispose of nuclear waste in subsurface rock or salt formations. This long has been a favored option, but recent investigations reveal that these formations are not as stable as many had hoped. Among the destabilizing factors are faulting, earthquakes, dangerous reactions with groundwater, and long-term plate tectonic effects.

3. Store canistered material in major ice sheets. However, as you saw in Chapter 13, climatic conditions can change. Melting or movement of these ice sheets is very probable.

4. Store the material below the deep ocean floor.

None of these possibilities is fully acceptable, for each has environmental risks. The deep-sea option does have interesting advantages. The deep ocean floor is some of the lowest-valued real estate on Earth, is far from the human population, and probably has no significant resources of fishing, oil, or gas. This part of the ocean also is extremely stable; some areas are thought to have remained stable for millions of years.

Do not confuse the deep-sea option with the idea of placing radioactive material into subduction-zone deep trenches such as the Java, Mindanao, or Puerto Rico trenches. These are very unstable areas with a high incidence of faulting and earthquake activity. Waste containers placed there likely would be damaged and radioactive materials could escape into the environment.

The deep-sea areas being considered are the stable central parts of major ocean plates. These areas have very slow sedimentation rates, on the order of a centimeter (about 0.4 in.) per 1,000 years. They are in areas having slow currents, and appear to be little influenced by major climatic or geologic conditions elsewhere on Earth. One proposed plan is to drill holes into the seafloor and emplace waste canisters (Figure 18-15).

However, several important questions must be answered before nuclear waste can be stored in the ocean:

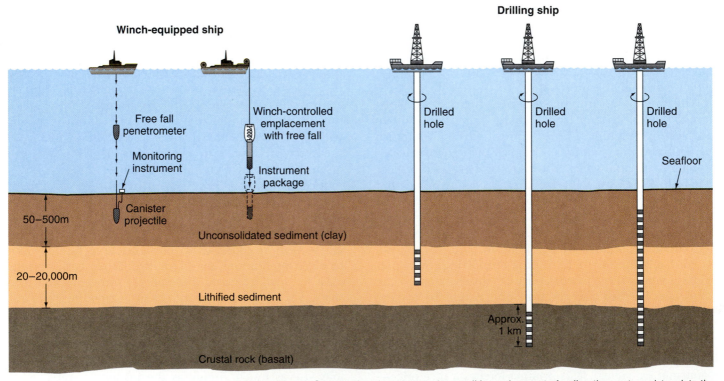

FIGURE 18-15 Some engineering concepts for possible emplacement of radioactive waste canisters into the deep seafloor. The two possibilities on the left, using a conventional ship, are either to freely drop the waste canister from the sea surface or to lower it near the bottom and then let it free-fall. In both instances the canister should bury itself into the bottom sediments. The possibilities on the right involve putting the nuclear waste canisters into holes drilled into the seafloor or underlying rock. (Adapted from Silva, 1977.)

Are seafloor sediments an effective barrier to radioactive material that might escape? How can the material be safely transported and implanted into the seafloor? Can the world's nuclear nations agree to cooperative disposal in this manner? What are the legal aspects of placing radioactive material in the seafloor? (The Law of the Sea Treaty—see Chapter 19—covers many aspects of the ocean, but surprisingly does not specifically address nuclear waste.)

If marine sediments can act as an effective barrier for radionuclides and the other issues can be solved, the deep-sea option might someday become very appealing.

Underwater Robotics

To use the ocean and seafloor intelligently, we must conduct long-term observation and monitoring. Although this is very expensive, especially when working far from shore, new robotic technology could make such long-term observation routine.

Autonomous Underwater Vehicles (AUVs) are similar to robots in that they operate without direct human guidance. AUVs are directed by onboard computers and have no connection to the land or to surface ships. They can move about on the seafloor or swim in the water, using their programmed intelligence. AUVs are released into the ocean and programmed to return after performing observations and measurements.

One such robot, called ABE (Automated Benthic Explorer), will sit dormant on the ocean bottom until a programmed time, when it will start exploring the seafloor in a predetermined pattern (Figure 18-16). ABE can measure parameters such as temperature and store images that it sees on a computer hard disk. The robot will have enough battery power to operate for about a year. ABE's first job will be near a deep sea vent.

Several U.S. institutions have led AUV development, including the University of New Hampshire, the Monterey Bay Aquarium Research Institute, Woods Hole Oceanographic Institution, and the Massachusetts Institute of Technology. Marine scientists have high hopes that AUVs will sharply refine our understanding of ocean processes, especially if they can be produced economically in large numbers.

Satellites for the Seas

Satellite technology holds great promise for oceanography. Some satellites orbit Earth constantly, transmitting images and measurements of the ocean. Other *synchronous* or geostationary satellites have orbits synchronized with Earth's rotation to maintain them over the same lo-

FIGURE 18-16 ABE (Automated Benthic Explorer), an autonomous underwater vehicle, or AUV. These robotic devices can work in the ocean, independent of a surface ship, for a year or longer. ABE was designed by Albert Bradley of Woods Hole Oceanographic Institution. (Photograph by T. Kleindinst, courtesy of Woods Hole Oceanographic Institution.)

cation, at a height of about 35,400 km (22,000 miles). These satellites monitor one area constantly, providing an excellent opportunity to observe oceanographic phenomena. Past emphasis was on outer space exploration, but studies of our home planet are the focus today.

Space technology is being developed to obtain worldwide views of the upper ocean (Figure 18-17). This information will help us better understand the ocean's interactions with the atmosphere, and vice versa (see Box 18-3). The ultimate goal is both scientific and practical: to better predict climate and oceanographic changes, and to determine whether the ocean is warming and whether polar icecaps are melting.

To achieve this goal, we must simultaneously measure currents, eddies, and ocean surface conditions and compare them with wind, solar radiation, and changes in seawater density that cause these phenomena. The measurements must be done over long periods (years) to assure that we are seeing long-term patterns, and not anomalies. The only practical way to obtain such broad information is from satellites that can rapidly circle the world, combined with detailed confirming information from surface ships or buoys.

The most effective ocean satellite at present is the European ERS-1, developed by the 13-nation European Space Agency (ESA) and launched in 1991. It has performed very well, measuring wave height, wind speed, ocean surface temperature, and ice-sheet thickness at the poles.

Another device, called *LANDSAT*, revolves around Earth 18 times each day in an almost polar orbit (that is, it passes over the North Pole, then the South Pole, and so on). It travels at about 800 km (500 miles) height,

a

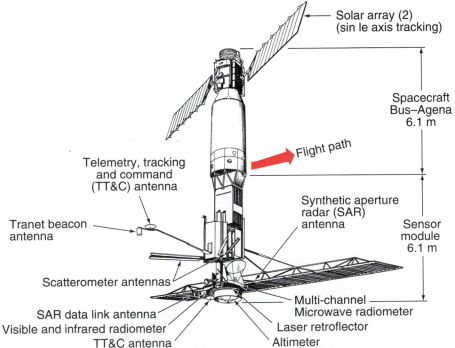

b

FIGURE 18-17 (a) Some past, present, and future ocean-sensing satellites: *TOPEX* will carry a precise altimeter to measure the ocean's surface topography and will be an important satellite for the World Ocean Circulation Experiment (WOCE) program. The data can be used to determine ocean circulation patterns. *TIROS,* a series of polar-orbiting satellites (they pass over both poles on each orbit) can measure sea surface temperature and other parameters. *NROSS* will measure surface wind over the ocean, wave height, temperature, sea ice distribution, and ocean eddies (large swirls). *GEOSAT,* a U.S. Navy satellite, uses radar altimetry to determine average ocean topography. *NIMBUS-7* monitored wind speed, sea surface temperatures, and sea ice conditions (it functioned from 1978 to 1987). *SEASAT A* was the first satellite dedicated to ocean observations. Launched in 1978, it lasted only 89 days, yet collected considerable data on wave height and sea-surface topography, surface-wind velocity, sea-surface temperature, and sea-ice distribution.

(b) *SEASAT A* transmitted only briefly, but collected copious ocean data clearly showing the value of such observations. (For some *SEASAT* results, see Figure 9-9 on page 209.)

Box 18-3
Satellite Missions for the '90s

Here are four major satellite programs with important marine components for the 1990s:

1. *NROSS (Navy Remote Ocean Sensing System):* This satellite provides basic information on ocean waves and eddies and measures winds at the ocean surface.

2. *TOPEX (The Ocean Topography Experiment):* This satellite measures the topography (surface relief, or highs and lows) of the ocean's surface to an accuracy of about 2.5 cm (1 in.) every 10 days for three years. *TOPEX* has a high-precision altimeter to measure sea topography, which is a product of the combined effects of currents, winds, and gravity. *TOPEX* data, combined with *NROSS* ocean-surface wave information, will yield a synoptic description of ocean circulation worldwide.

3. *OCI (Ocean Color Imager):* This device permits estima-

tion of phytoplankton pigment in surface waters. Combined with solar radiation data, it indicates the variability and magnitude of biological productivity. The plan is to collect data for at least 3 years, particularly during an El Niño.

4. *GRM (Geopotential Research Mission):* This mission uses two satellites traveling in similar orbits, but separated by about 100 to 560 km (60 to 350 miles). Variations in Earth's gravity field are determined by very accurate measurement of the changing distances between the two satellites and the ground. By precisely knowing these variations, the effect of gravity on Earth's topography can be studied. Combining this information with altimeter measurements from *TOPEX,* one can determine ocean current velocity with great precision.

---

and records a complete image every 103 minutes. Each pass generates a north-south strip that is contiguous to that of the previous day, with a small overlap. Therefore, every 18 days the satellite passes over every location on Earth at essentially the same time, with similar lighting.

SEASAT A, launched in 1978, was the first satellite designed for oceanography (Figure 18-17b). Unfortunately, it lasted only 89 days before failing. During its brief life *SEASAT* transmitted extremely valuable information about the ocean, including data on its waves, surface slopes, and temperature.

Satellite images have many scientific uses. They measure or monitor pollution, study ocean circulation and currents, determine areas of biological productivity by determining chlorophyll content in the surface water, assist meteorologists on predicting weather, search for minerals on land, locate schools of fish, and detect coastal zone pollution (see, for example, Figure 1-11 on page 13 and Figure 9-9 on page 209).

Underwater Habitats: Offshore Islands and Platforms

Can people live and work underwater? Seafloor habitats have been tried at shallow depths (usually less than 180 m or 600 ft.). Women and men have lived and worked in such settings for several weeks, generally for scientific study.

With the continuing exploration for oil and gas and the possible development of OTEC, offshore structures will proliferate. Offshore structures and underwater platforms could also be used for nuclear power stations, aircraft landing strips, fish-processing factories, recreational facilities, and for other purposes yet to be imagined. An unrealistic suggestion has been made that offshore habitats or seafloor habitats could provide routine living space when the land habitat becomes saturated with people, but the expense, inconvenience, and physiological stress of living underwater make this very unlikely.

The Japanese have shown imagination and enterprise in building offshore islands. Indeed, they built an offshore international airport about 480 km (300 miles) southwest of Tokyo. The Kansai International Airport is 4.8 km (3 miles) offshore, forming an island of about 500 hectares (1,200 acres). The island was created with 150 million m³ (about 5.3 billion ft.³) of fill and is connected to land by a highway. The $7 billion project required over 6 years to build.

The Ocean for Tourists

Ocean beaches always have attracted vacationers, and they always will. But new features are drawing tourists to the marine world. These include imaginative aquariums and sea parks, whale-watching tours, and even short submarine trips. Such activities are driven by the public's growing fascination with the ocean's wonders, and such interest will continue to grow.

New aquariums and marine parks are a growth industry, with over 70 new projects being considered as this is written. Several aquariums are already well-known national tourist attractions, such as the Monterey Bay Aquarium in California, the National Aquarium in Baltimore, the Aquarium of the Americas in New Orleans, and the New England Aquarium in Boston.

New aquariums, remodeled older ones, and other facilities constantly generate imaginative ways to put the visitor in closer contact with marine creatures. One uses tubes through which people walk, literally surrounded by the water and aquarium creatures. The Monterey Bay Aquarium's famous kelp beds display unique marine communities (Figure 18-18). Many marine parks are known for their trained animals, such as the famous jumping killer whales. Whale watching from small boats is very popular in some coastal areas, and has attracted so many boats that the whales literally must be protected from the tourists.

The tourist submarine industry is a more recent development. At present, about 40 such submarines around the globe afford an underwater trip. Several operate on Australia's Great Barrier Reef, where submariners can observe the beauty of this natural wonder (see Figure 12-16, page 298). Perhaps the ultimate is your own personal submarine, and several companies sell these devices.

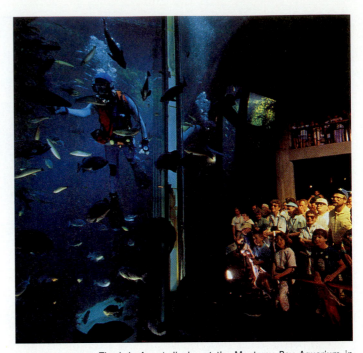

FIGURE 18-18 The kelp forest display at the Monterey Bay Aquarium in Monterey, California. This display, 28 feet high, is the first recreation of a living kelp forest community. It holds 335,000 gallons of water and is populated with kelp, sharks, and other fish. Nearly 2 million people visit this aquarium each year. (Photograph copyrighted 1992 by the Monterey Bay Aquarium, courtesy Monterey Bay Aquarium.)

SUMMARY

Future uses of the ocean are limited only by imagination, technology, legal constraints, and cost. Biotechnology and genetic engineering should lead to new and better products, including new drugs from the sea. Potential energy sources from the ocean include tides, waves, ocean currents, gas hydrates, marine biomass, and OTEC (Ocean Thermal Energy Conversion, derived from temperature differences between surface and deep waters). An OTEC device combined with an artificial upwelling system could reduce electricity cost and produce a biologically useful byproduct.

Freshwater from the ocean is now obtained by desalination, and icebergs have been proposed as another source. Nuclear waste disposal in the seafloor has been proposed, and though it is unappealing, it may be the only reasonable option for secure storage of the large volumes of the nuclear waste existing now and to be generated in the future.

The ocean's biological resources might be increased by imprinting fish to return to areas where they can be caught. Other possibilities include the expanded harvest of organisms such as krill and new methods of aquaculture.

Satellite monitoring of the ocean could monitor biological productivity, coastal changes, and pollution input and effects. Several innovative programs are being developed for the coming decade that could help answer some major questions in oceanography and climate.

QUESTIONS

1. What ocean-related endeavors can benefit from innovative ideas and technologies?

2. Why is the ocean an especially promising environment for biotechnology? Give some examples.

3. How can usable energy be obtained from the sea?

4. Discuss the advantages and disadvantages of using the tides to produce power.

5. Discuss some ways to increase ocean productivity.

6. What are some of the benefits of satellites for oceanographic research?

KEY TERMS

aquaculture

Autonomous Underwater
 Vehicle (AUV)

biotechnology

clathrate

distillation

freezing

gas hydrate (clathrate)

genetic engineering

mariculture

membrane process

Ocean Thermal Energy
Conversion (OTEC)

FURTHER READING

Baker, J. D. 1991. "Towards a Global Ocean Observing System." *Oceanus* 34, no. 1, pp. 76–83. Describes how worldwide data obtained from satellites would be used to update weather and monitor the health of the ocean.

Davis, G. R., ed. 1991. *Energy for Planet Earth.* New York: W. H. Freeman and Co. A collection of readings from *Scientific American* on various aspects of present and future energy needs.

Duedall, I. W., and M. A. Champ. 1991. "Artificial Reefs: Emerging Science and Technology." *Oceanus* 34, no.1, pp. 76–83. Shows innovative ways in which artificial reefs can be used, and the progress that Japan has made in this field.

Faulkner, D. J. 1992. "Biomedical Uses for Natural Marine Chemicals." *Oceanus* 35, no. 1, pp. 29–35. Describes how some chemicals derived from marine sources can be used in medical research and for making new drugs.

Hammer, W. M. 1984. "Krill: Untapped Bounty from the Sea." *National Geographic* 165, no. 5, pp. 626–43. Some details about and pretty pictures of krill.

Holing, Dwight. 1988. "Dolphin Defense." *Discovery* 9, no. 10, pp. 68–74. Various views about the use of marine mammals by the military.

McGoodwin, J. R. 1990. *Crisis in the World's Fisheries: People, Problems and Policies.* Stanford, Calif.: Stanford University Press.

Oceanography in the Next Decade. 1992. Washington, D.C.: National Research Council. A group of noted marine scientists give their collective view of new directions in oceanography during the next decade.

Penny, T. R., and D. Bharathan. 1987. "Power from the Sea." *Scientific American* 256, no. 1, pp. 86–92. A discussion of some ways that energy can be obtained from the ocean.

Ross, D. A. 1980. *Opportunities and Uses of the Ocean.* New York: Springer Verlag. Description of various innovative uses of the ocean.

Wunsch, Carl. 1992. "Observing Ocean Circulation from Space." *Oceanus* 35, no. 2, pp. 9–17. How satellites can provide new insights into global and regional oceanography.

CHAPTER 19

The Law of the Sea

For centuries, the ocean was considered "public," belonging to no one person or country, and available for everyone's use—except for a narrow strip along the shore, which logically belonged to the adjacent country. In recent years, traditional ocean uses (fishing) plus new uses (mining of minerals and dumping of waste) have encouraged many countries to declare much wider territorial claims on the sea. This extension of claims eventually led to an international Law of the Sea Conference, which concluded in 1982 and was ratified by the necessary number of countries in late 1993. As a result of that conference and ratified treaty, up to 42 percent of the ocean can now be controlled by adjacent countries. This is a very substantial area—larger than the land area of Earth.

U.S. Research Vessel *Knorr*, of the Woods Hole Oceanographic Institution, visits the United Nations during negotiation of the Law of the Sea Conference. The *Knorr* took some U.N. delegates on a short cruise (a few hours) to show them a little about oceanography.

(Photograph courtesy of Woods Hole Oceanographic Institution.)

Early History of the Law of the Sea

From Early History to Harry S Truman

Certain freedoms and rules of the sea have been generally accepted since people began to sail it. The first significant legal statement about the sea is credited to Dutch lawyer Hugo Grotius. In 1609, he published *Mare Liberum,* which is Latin for freedom (*liberum*) of the seas (*mare*). Grotius said that nations are essentially free to use the sea as long as their use doesn't interfere with another nation's use.

It was also generally agreed that a **coastal state** (a country having a shoreline) should have some control over the ocean immediately adjacent to its coast. This led to the definition of a zone called the **territorial sea,** over which each coastal nation had complete sovereignty. The zone was determined in a most unscientific manner, but one that was very pragmatic at the time: it was established by the distance that a cannon could fire a round from the shore out to sea. A fired cannon ball typically traveled about three nautical miles; thus the power of a 1700s cannon established the well-known "three-mile limit." Interestingly, this three-nautical-mile-wide territorial sea never has been officially ratified by an international agreement. (In marine legal terms, distances are almost always expressed in nautical miles. One nautical mile is equal to 6,072 feet, or 1.15 statute miles, or 1.852 kilometers.)

The area beyond the territorial sea, called the **high sea,** was considered *res nullius,* or "property of no one." International law generally protected the rights of all countries to use all parts of the high seas for whatever purpose they chose. This was not seriously challenged prior to World War II, because most nations had the same interests in the sea.

After World War II (1939–45), interest in the ocean increased, due partly to actions by the United States and to a new awareness that some of the sea's valuable resources—such as fish—are limited, and that valuable minerals exist on the seafloor.

The first major challenge to the three-mile territorial sea was by U.S. President Harry S Truman in 1945. In what is now called the *Truman Proclamation on the Continental Shelf,* he established a national policy with respect to a much wider area than the three-mile territorial sea. His proclamation stated:

> [T]he United States felt that the exercise of jurisdiction over the natural resources of the subsoil and seabed of the continental shelf by the contiguous nation is reasonable and just, since the effectiveness of measures to utilize or conserve these resources would be contingent upon cooperation and protection from the shore, since the continental shelf may be regarded as an extension of the landmass of the coastal nation and thus naturally appurtenant to it, since these resources frequently form a seaward extension of a pool or deposit lying within the territory, and since self-protection compels the coastal nation to keep close watch over activities off its shores which are of the nature necessary for utilization of these resources. [Thus] *the Government of the United States regards the natural resources of the subsoil and seabed of the continental shelf beneath the high seas but contiguous to the coasts of the United States as appertaining to the United States, subject to its jurisdiction and control.**

One motivation for this action was the *"long-range worldwide need for new sources of petroleum and other minerals."*[†] Even in 1945, some leaders had anticipated the energy problems to come.

President Truman basically claimed ownership for the United States of all seabed resources on or under its entire continental shelf. He suggested that an international agreement be reached for regulation of fisheries and conservation. His controversial proclamation created an unanticipated result: lost in these discussions was the fact that the United States was not changing its three-mile territorial sea position, but was claiming ownership of part of the *seafloor.* In so doing, however, the United States effectively challenged some basic freedoms of the ocean.

The Truman Proclamation caused problems with offshore mineral claims between individual U.S. coastal states and the U.S. government. Before the proclamation, individual U.S. states believed they owned the land immediately off their coasts. For example, oil exploration was conducted in the 1880s off California without federal control and, initially, even without state control. However, because the Truman Proclamation claimed *federal* ownership of seabed mineral resources, legal battles started between some states and the federal government.

In 1947, in *United States v. California,* the U.S. Supreme Court decided that the federal government had full power over the lands and minerals lying seaward of the "ordinary low-water mark" (low tide) on the coast of California and outside the inland waters. Subsequent cases between the United States and Texas and in *United States v. Louisiana* reaffirmed the federal government's ownership of the submerged land off these states.

The U.S. Congress, ever sensitive to states' interests, changed the court's actions with two congressional acts in 1953. The *Submerged Lands Act* gave states title to the lands and resources out to three miles in the Atlantic and Pacific Oceans and out to three leagues (10.5 statute miles) in the Gulf of Mexico. Within this zone, however, the federal government retained its rights to navigation, commerce, national defense, and international affairs.

Policy of the United States with Respect to the Natural Resources of the Subsoil and Seabed of the Continental Shelf, Proclamation No. 2667, 1945. Italics added for emphasis in this and other quotations.

†Ibid. (This means the same reference as previously given.)

This act defined the *inner* continental shelf as extending from the low-water line to the end of the three-mile territorial sea, and the *outer* continental shelf as the remaining continental shelf seaward of the territorial sea. The second act, the *Outer Continental Shelf Lands Act,* gave control of the seabed and subsoil of the outer continental shelf to the federal government.

Several states bordering the Gulf of Mexico challenged the three-league width, based on their earlier claims, which were in place when they attained statehood (that is, before they joined the United States). Also in question was whether their control of the sea started at their coastline or from offshore islands. The outcome: Florida (west coast only) and Texas legally won boundaries that today extend about 16.6 km (9 miles) from the coast. Louisiana, Mississippi, and Alabama received only about 5.5 km (3 miles).

The 1958 and 1960 Geneva Conventions on the Law of the Sea

The Truman Proclamation of 1945 did not define the limits of the continental shelf, but a White House media release did: the continental shelf is the submerged land that is contiguous to the continent and is covered by no more than 100 fathoms (183 m, or 600 ft.) of water. The release also claimed the Truman Proclamation did not abridge the right of free navigation of the high seas above the shelf, nor did it extend the limit of the United States.

The Truman Proclamation is extremely important in the history of sea law because some nations saw it as an opportunity to adjust inequities, real or imagined. Several Pacific-coastal South American countries were anxious to protect their few offshore resources and their share of a narrow continental shelf.

In 1952, Chile, Ecuador, and Peru extended their jurisdiction and sovereignty over the sea, seabed, and subsoil out to 200 miles (370 km) from their coastline. In other words, they declared a 200-mile territorial sea. Their move went well beyond the Truman Proclamation by including the sea surface and water as well as the ocean bottom, dramatically extending their jurisdiction. Peru based its claim on its large guano industry (bird droppings used for fertilizer) on islands off its coast.

This unilateral extension of the territorial sea by these nations (and eventually by several others) indicated that an international set of laws concerning the sea was necessary. This led to the 1958 and 1960 Geneva Conferences on the Law of the Sea, which had representatives from 86 seafaring countries. They developed **conventions,** or international agreements, on several points. These four conventions were approved by many, but not all, of the participants:

1. Convention on the Territorial Sea and the Contiguous Zone

2. Convention on the Continental Shelf

3. Convention on the High Seas

4. Convention on Fishing and the Conservation of the Living Resources of the High Seas

We will now look at each of these.

Convention on the Territorial Sea and Contiguous Zone

This agreement defined the territorial sea: "The sovereignty of the state extends beyond its land territory and its internal waters to a belt of sea adjacent to its coast, described as a *territorial sea.** A territorial sea's starting point was defined as "the low-water line along the coasts as marked on large-scale charts officially recognized by the coastal state."† Unfortunately, this convention avoided a key issue of the conference: it did not specify *how far* the territorial sea could extend seaward.

The convention also defined a new zone, the **contiguous zone,** as an area of high seas contiguous to a coastal state's territorial sea and extending not more than 12 nautical miles (22.2 km) from the baseline used to measure the territorial sea. Within this zone, the coastal state can control fishing by foreign states. This part of the convention did not specify territorial sea width, but because of the 12-nautical-mile contiguous zone limit, the territorial sea cannot exceed 12 nautical miles either. However, for many nations, a 12-nautical-mile-territorial sea covered only a portion of their continental shelf.

The coastal nation has essentially the same judicial authority over its internal waters and territorial sea as it has over its land territory. For example, it can exclude foreign ships. One exception is the right of a foreign ship to "innocent passage" within the territorial sea, meaning that it does not harm the peace, good order, or security of the coastal state.

Following the Geneva Conferences, the countries that agreed to the conventions could divide their waters into four main categories (Figure 19-1):

1. *Internal Waters:* This includes rivers, lakes, and all waters landward of the inner limit of territorial seas (low-

---

*1958 Convention on the Territorial Sea and the Contiguous Zone—Convention on the Law of the Sea Adopted by U.N. Conference in Geneva, 1958. Authors italics added for emphasis.

†*Ibid.

FIGURE 19-1   General divisions of the ocean after the 1958 Geneva Convention. The territorial sea can extend to 12 nautical miles.

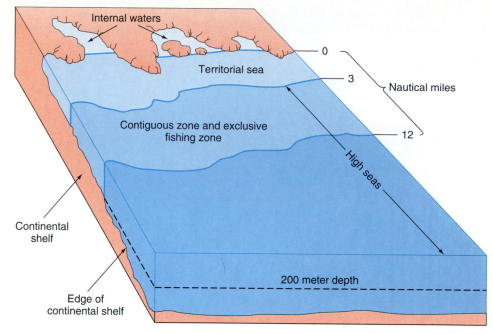

water line) as well as estuaries, harbors, and bays between the coast and offshore islands. The coastal state has complete sovereignty over these waters. (An exception: the right of innocent passage or other historical rights might be allowed where these waters once had been territorial seas or high seas.)

2. *Territorial Seas:* This is the zone of undefined width, although most states then claimed about 12 nautical miles. The rights of the coastal state are similar to those for internal waters, except that foreign states have the right of innocent passage, innocent overflight, and entry under distress conditions.

3. *Contiguous Zone and Exclusive Fishing Zone:* This zone does not extend beyond 12 nautical miles from where the territorial sea starts. The contiguous zone is part of the high sea, except in the case of fishing and a few other restrictions.

4. *High Seas:* The high seas are defined as all waters not included within territorial or internal waters. Nearly complete freedom exists within the high seas. The high seas are considered to be international waters.

In the late 1950s and early 1960s, most nations accepted a nine-nautical-mile exclusive fishery contiguous zone combined with a three-nautical-mile territorial sea, for a total of 12 nautical miles. Some countries, however, still claimed a 200-nautical-mile-wide territorial sea.

Convention on the Continental Shelf

This agreement defined **continental shelf** as "... the seabed and subsoil of the submarine areas adjacent to the coast but outside the area of the territorial sea to a depth of 200 m, or beyond that limit, to where the depth of the superadjacent [overlying] waters [allows] the exploitation of the natural resources ... [and] to the seabed and subsoil of similar submarine areas adjacent to the coasts of islands."* This critical clause, one of the most important, stipulates that the coastal state can claim and mine that part of the continental shelf beyond 200 m (656 ft.) depth. This is commonly called the *exploitability clause.* However, this convention failed to define where the continental shelf ended.

The Convention on the Continental Shelf gave coastal states the sovereign right to explore and exploit natural resources of the continental shelf. It defined *natural resources* as:

... the mineral and other nonliving resources of the seabed and subsoil together with living organisms belonging to sedentary species, which is to say, organisms which, at the harvestable stage, are immobile on or under the seabed and are unable to move except in constant physical contact with the seabed or subsoil. ... the exploitation of its natural resources must not result in any unjustifiable interference with navigation, fishing, or the conservation of the living resources of the sea, nor result in any interference

---

*1958 Convention on the Continental Shelf—Convention on the Law of the Sea Adopted by the U.N. Conference at Geneva, 1958.

with fundamental oceanographic or other scientific research carried out with the intention of open publication.*

The question of scientific research is crucial for oceanography. The Convention on the Continental Shelf stated:

> [T]he consent of the coastal state shall be obtained [for] any research concerning the continental shelf and undertaken there. Nevertheless, the coastal state shall not normally withhold its consent if the request is submitted by a qualified institution with a view to purely scientific research into the physical or biological characteristics of the continental shelf, [providing that] the coastal state shall have the right . . . to participate or to be represented in the research and that in any event the results shall be published.†

These consent conditions were to become an obstacle to some research, as discussed in a later section.

Convention on the High Seas

This agreement defines **high seas** as parts of the ocean not included in the territorial sea or in the internal waters of a state. It states that the high seas are open to all countries and can be freely used for navigation, fishing, and overflight.

Convention on Fishing and Conservation of Living Resources of the High Seas

This agreement states that all countries have the right to allow their nationals to fish on the high seas, subject to past treaties and articles adopted at the conference. It also says that the countries should cooperate in conserving the living resources of the high seas.

Shortcomings of the 1958/1960 Geneva Conventions

The 1958/1960 Conventions led to three major controversies over territorial sea width, continental shelf definition, and the exploitability of marine minerals:

Failure to define the width of the territorial sea encouraged some coastal states to establish their own width, sometimes creating claims hundreds of miles wide. And, of course, states that never ratified the Conventions were not bound by them at all.

The 200-m depth limit of the continental shelf usually did not coincide with the geologic or oceanographic definition of the shelf. The depth contour of 200 m (656 ft.) is arbitrary. In most instances, the edge of the continental

shelf is shallower; in fewer instances, it is deeper. A shelf defined by depth was unacceptable to countries having very narrow shelves, such as Peru and Chile, and to countries having shelves that extend for hundreds of miles with parts exceeding 200-m depth.

The exploitability of marine minerals was ambiguous because of the wording in the continental shelf definition, that the shelf extends "to a depth of 200 m, *or beyond that limit, to where the depth of the superadjacent [overlying] waters [allows] the exploitation of the natural resources of the said areas.** This left seafloor ownership open to various interpretations. One was that the *entire ocean* could be divided among the countries bordering it (Figure 19-2). As you study this map, note the startling expanse of undersea real estate that some nations could have, like tiny Portugal and Chile. Also observe the equally startling postage-stamp sea bottoms that could be controlled by nations like Spain and France (excluding colonies and possessions) if this interpretation were used.

Another interpretation was that the United States or any other technologically advanced country could mine its shelf, continental slope, continental rise, ocean basin, ocean ridge, and perhaps right across and up onto the other side of the ocean! In fact, the United States had accepted this interpretation, as evidenced by its leasing areas for oil exploration and exploitation in water depths greater than 200 m off California and over 100 miles into the Gulf of Mexico.

The exploitability clause was not as important to conference delegates in 1958/1960 as it became to many countries in the 1970s and 1980s, when the promise of exotic minerals and biological resources from the ocean (frequently exaggerated) caused considerable interest in their exploitation.

Despite these shortcomings, the 1958/1960 conferences were very important. This was one of the last times that the high seas would be considered as an area of unqualified freedom. Following the conferences, over 100 nations eventually indicated that they had some jurisdiction over the minerals in areas adjacent to their coasts. These claims took many forms, including individual and unilateral declarations, domestic legislation, treaties, offshore concessions, and ratification of the 1958 Convention of the Continental Shelf.

One irony was that the United States, perhaps realizing the Pandora's box it had opened with the 1945 Truman Proclamation, argued strongly in 1958 for a three-nautical-mile territorial sea. It was unsuccessful, however. In the 1960 conference, the United States supported a six-nautical-mile-wide territorial sea, but it again failed to carry—by one vote.

---

*Ibid.

†Ibid.

*1958 Convention on the Continental Shelf—Convention on the Law of the Sea Adopted by the U.N. Conference at Geneva, 1958.

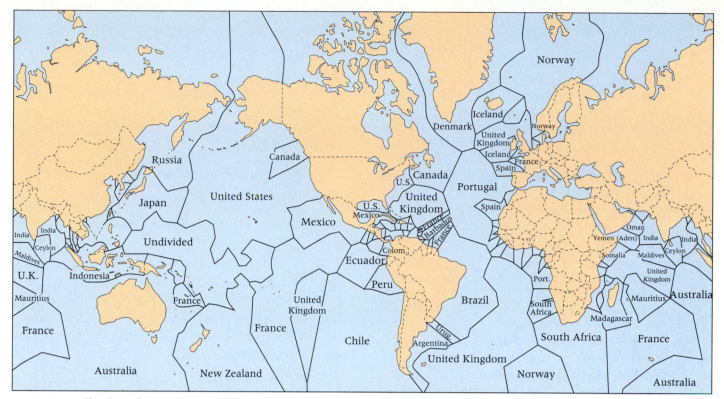

FIGURE 19-2 How the seafloor would look if divided along lines equidistant from the shorelines of adjacent and opposite coastal nations, as they existed in 1967. This shows a possible division of the seafloor, based on the exploitability concept. (Adapted from Christy and Herfindahl, 1967.)

Third Law of the Sea Conference

During the few years from the 1958 conference until the early 1960s, when the Conventions were ratified, marine technology advanced, making portions of the seabed beyond 200 m explorable for minerals. In other words, the "exploitability clause" was already changing from theoretical to real. Capability for offshore oil drilling improved (Chapter 15, opening figure) and technology for mining manganese nodules and other seafloor resources was developing. Fishing fleets also expanded and mechanized, allowing them to stay anywhere at sea for months and process their catch on large support ships. The 1958 conference on the ownership of resources beyond the 200-m depth had opened the door to the international disputes that were brewing.

From Malta, a tiny Mediterranean island just south of Italy, came the dramatic impetus for change of international ocean policy. In 1967, Malta proposed to the U.N. General Assembly a treaty to reserve that part of the ocean floor that lies beyond national jurisdictions for peaceful uses, and that its resources become *the common heritage of humankind.* Money derived from use of "The Area," as it would be called, could assist less-developed countries.

In 1969, after long opposition by the United States and the former Soviet Union, the United Nations resolved to convene a conference ". . . on the law of the sea, to review the [legal] regimes of the high seas, the continental shelf, the territorial sea and contiguous zone, fishing and conservation of the living resources of the high seas, particularly in order to arrive at a clear, precise, and internationally accepted definition of the seabed and ocean floor which lies beyond national jurisdiction."*

The United States objected, not to avoid a conference, but rather in an attempt to reach an international consensus by other means, such as by individual nations declaring their policy. The U.S. Congress felt that the Geneva Conventions had sufficiently defined U.S. resource ownership on the continental margin. As the United States was then the most advanced maritime nation, it was no surprise that Congress supported the exploitability clause.

The Third Law of the Sea Conference ran nine years, in a series of meetings from 1973 through 1982 (Table 19-1). It took almost 12 years before the resulting treaty was ratified by the 60 countries needed. At present the United States still has not signed, but it may reconsider

---

*U.N. Document A/2574A, 1969.

TABLE 19-1
Chronology of the Law of the Sea Conferences

1958	First U.N. Conference on Law of the Sea (LOS), New York Adoption of four conventions		1978	Seventh Session resumed, New York Frustration with pace of progress; concern over possibility of unilateral legislation
1960	Second U.N. Conference on LOS (UNCLOS), New York Failure to agree on width of territorial sea		1979	Eighth Session, Geneva Revision of ICNT Remaining concern over deep seabed mining, marine scientific research, continental shelf, and delimitation of offshore boundaries between adjacent or opposite nations
1970	U.N. General Assembly Declaration of the common heritage of mankind Resolutions to start the Third United Nations Conference on the Law of the Sea (UNCLOS III)			
			1979	Eighth Session resumed, New York Setting of deadlines for adoption of convention
1973	Third U.N. Conference on LOS, First Session in New York Organizational meeting, list of issues and subjects to be negotiated		1980	Ninth Session, New York Continuing consultations
1974	Second Session, Caracas, Venezuela		1980	Ninth Session resumed, Geneva Development of draft convention
1975	Third Session, Geneva, Switzerland Distribution of Informal Single Negotiating Texts covering all subjects before the conference		1981	Tenth Session, New York U.S. Review
			1981	Tenth Session resumed, Geneva
1976	Fourth Session, New York Revised Single Negotiating Text issued Deep seabed mining, area of chief disagreement		1982	Eleventh Session, New York Convention adopted in April by a vote of 130 to 4 (United States, Turkey, Venezuela, and Israel), with 19 abstentions
1976	Fifth Session, New York Issuance of Revised Text on dispute settlement Concentration of deep seabed mining agreement and economic zone delineation		1993	On 16 November, almost 12 years after the last session of the convention, the required sixtieth nation, Guyana, ratified the treaty.
1977	Sixth Session, New York Production of Informal Composite Negotiating Text (ICNT)		1994	Revision of the deep seabed mining regulations.
1978	Seventh Session, Geneva Establishment of seven negotiating groups to deal with "hard-core" issues		1994	On November 16, 1994, the Law of the Sea Treaty comes into effect.

this position in the near future. The diverse desires of conference participants makes agreement difficult. Nevertheless, many of the over 300 articles within the treaty are being adopted by coastal countries. The treaty covers almost all aspects of ocean use.

The Third Law of the Sea Conference had more than twice the participants than the 1958/1960 conferences. Many new participants were developing nations that wanted a change in the old freedom-of-the-sea concept. They felt that old customs allowed the few developed countries to dominate the ocean and its resources. Interest flourished in establishing economic resource zones, up to 200 nautical miles wide, where each coastal nation would have control over the marine resources.

Basic Issues of the Third Law of the Sea Conference

Important issues of the conference were:

- Definition of the width of the territorial sea and adjacent zones, and the degree of coastal state control in these zones

- Ownership of resources in the water, seabed, and subsoil, including (a) resources within a coastal state's jurisdiction (hence the question of the extent of this jurisdiction), and (b) resources beyond national jurisdiction, or in the high seas

- Right of overflight and navigation through what were international straits (like Gibraltar) but which might get included in someone's expanded territorial seas or exclusive economic zone

- Management of living resources in the ocean, especially species that migrate and those where many countries traditionally have fished, but which now might come under a single country's jurisdiction (such as Georges Bank; see Box 19-1)

- Protection from and reduction of pollution in the ocean

- Freedom of scientific research in the ocean

- A regime for control or management of the high seas.

Box 19-1

Good Marine Boundaries Make Good Neighbors

In "Mending Wall," poet Robert Frost writes, "good fences make good neighbors." This also applies to the ocean. With the acceptance of marine jurisdictions from the Third United Nations Conference on the Law of the Sea, especially EEZs, several hundred marine boundaries among coastal nations remain undecided. These boundaries can be very important because of the large territory size involved, mineral and biological resources, pollution, military use, and shipping.

Boundary disputes will probably exist as long as people do. Here are some examples:

1. In the South Pacific, 13 island nations that have a land area of only 522,000 km² (about 200,000 miles²) can control more than 19,000,000 km² (about 7,300,000 miles²) of marine territory. In this region, at least 20 marine boundaries need resolution.

2. In the Caribbean Sea, boundary disputes over fish and energy resources continue between Venezuela and some of the island nations.

3. Conflicting U.S./Canadian claims east of Massachusetts began in the 1960s when Canada issued oil-exploration permits for the eastern part of Georges Bank. The United States protested. In 1977 both countries established EEZs of 200 nautical miles and wanted to control foreign fishing access to the region. Georges Bank has been a rich fishing area, yielding large amounts of cod, haddock, flounder, and scallops. The United States claimed the entire bank, whereas Canada claimed the eastern third, an especially good fishing area. In 1981, the two countries agreed to resolve the dispute by binding arbitration in the International Court of Justice at the Hague, in the Netherlands. In 1984, after 20 years of dispute, the court awarded about half of the disputed eastern part to each country (Figure 1). At this writing, the fishermen of both countries are still unhappy with the judgment and no significant hydrocarbon deposits have been found in the disputed region.

4. Several other major boundaries between the United States and Canada remain in dispute, including the line between Alaska and British Columbia and the Juan de Fuca Strait boundary between Vancouver Island and the State of Washington.

Boundaries seem easy to establish: simply draw a line equidistant between the two countries. But it is never this simple, for reasons including lack of precise bathymetric or other geographic data; presence of offshore islands; past agreements (formal or informal), such as traditional fishing grounds; and disagreement between the parties on where to start the line and in what direction to extend it. A slight difference in direction can create large territorial differences when the line extends 200 nautical miles out to sea. Maps can also introduce an error, since they are two-dimensional representations of a three-dimensional world.

In most instances, questionable boundaries are negotiated with give and take. If unsuccessful, the dispute ends up in court. Inevitably and unfortunately, force has been used to settle some disputes. An example not long ago was the near war between Iceland and Great Britain over fishing rights near Iceland. Good marine boundaries can make good neighbors.

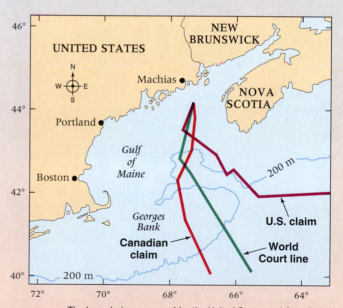

FIGURE 1 The boundaries proposed by the United States and Canada for Georges Bank and the settlement decided by the World Court. (Adapted from Richard Kelly, Maine State Planning Office.)

The U.S. position on these issues has varied over time. However, the United States would accept extended jurisdiction over resources under certain considerations, including prevention of unreasonable interference with other uses of the ocean (such as navigation rights), mechanisms for mining marine resources and sharing revenues with other nations, and compulsory settling of disputes.

The United States was also willing to adopt a 12-nautical-mile territorial sea if free passage were guaranteed through international straits. This is important because 116 straits worldwide now have a passageway to the high seas (Figure 19-3; see also Figure 19-6). Because these straits are more than 6 nautical miles wide but less than 24 nautical miles wide, they would become part of someone's territorial sea if the limit changed from 3 miles to 12 miles. These include important straits, such as Gibraltar (Mediterranean Sea passage to the Atlantic), Malacca (Pacific Ocean passage to the Indian Ocean in southeast Asia), and Bab el Mandeb (Mediterranean passage to the Indian Ocean, via the Suez Canal and the southern end of the Red Sea).

If these straits were to become territorial seas with no guarantee of passage, countries would have only the "right of innocent passage" to allow them through. This would require submarines to travel on the surface through them and would not allow military planes to fly over them, seriously restricting U.S., Russian, and other military powers around the world. The issue is also important to oil-importing countries, such as Japan, whose maritime trade could be seriously affected by unreasonable transit regulations that could result from these straits becoming entirely territorial (Figure 16-13, page 409).

Results of the Third Law of the Sea Conference

The conference and subsequent actions by many countries have effectively divided the oceans into four main zones (Figure 19-4):

1. A 12-nautical-mile-wide territorial sea

2. A 200-nautical-mile-wide exclusive economic zone (EEZ) that includes the territorial sea

3. A "continental shelf" that may extend to 350 nautical miles from land

4. A zone enclosing the rest of the ocean, called "The Area."

These zones were unofficial until the treaty was ratified by the necessary 60 nations in late 1993. Countries that have not approved the treaty are of course not required to recognize these zones, although they may choose to do so. Many countries that have accepted the zones have done so with reservations.

Exclusive Economic Zones

Since 1982, many countries, including the United States, have declared 12-nautical-mile-wide territorial seas and 200-nautical-mile-wide EEZs (Table 19-2). The United States, although presently unwilling to sign the treaty, declared its own EEZ in 1983 and 12-nautical-mile-wide territorial sea in 1988. It was the 60th nation to declare an EEZ.

Ironically, the U.S. EEZ is the largest in the world (3,362,600 nautical miles^2 or about 11,500,000 km^2), and is 1.21 times the size of the U.S. land area (Figure 19-5). Our EEZ is so large because inhabitable islands, such as Hawaii, Guam, and Samoa, have their own 200-nautical-mile-wide EEZ.

The concept of "inhabitable island" can be carried to extremes. In 1987, Japan planned to spend over $200 million to protect two small rocks that barely remain above water at low tide. If these rocks erode away, or if sea level rises a bit, Japan could lose a claim to nearly 300,000 square miles of seafloor.

Within its EEZ, a coastal country has complete control over marine resources, can control access for scientific

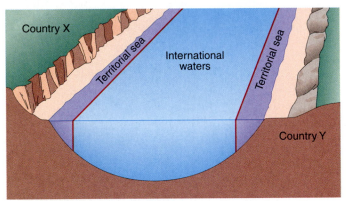

a 3 nautical-mile wide territorial sea

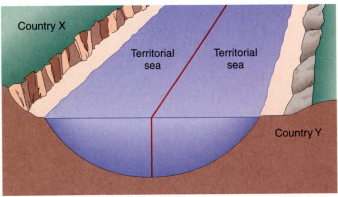

b 12 nautical-mile wide territorial sea

FIGURE 19-3 Effects on a strait that is 24 miles or less in width with (a) a three-nautical-mile-wide territorial sea and (b) if the territorial sea was extended to 12 nautical miles.

FIGURE 19-4 Main divisions of the seafloor resulting from the Third Law of the Sea Conference.

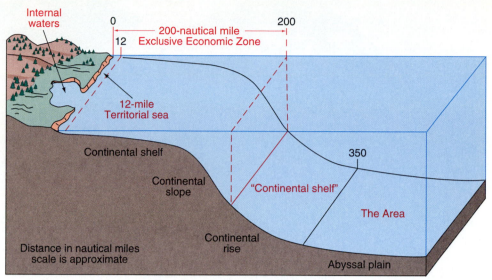

research, and can make rules concerning other ocean uses and pollution. In certain situations, the coastal nation can also control marine resources and scientific research on its continental margin out to 350 nautical miles (648 km). With establishment of a 200-nautical-mile EEZ, 37,745,000 nautical miles² (about 128,000,000 km²) of ocean will come under the jurisdiction of coastal countries. These numbers represent about 35.8 percent of the ocean and exceed Earth's total land surface (Figure 19-6). If the boundary were to be the 350 nauti-

cal miles allowed under certain conditions (see the following section), about 42 percent of the ocean would be controlled by coastal nations.

Continental Margins

If a country's true continental margin in the geologic sense—continental shelf, plus continental slope, plus continental rise—extends beyond 200 nautical miles, the coastal country has the right to claim even more territory, following a complicated set of options (Figure 19-7). This new zone is called the *continental shelf* (a bad choice of terms). This ruling probably will be as controversial as the exploitability clause of the 1958 convention.

Note that each of these depths or distances has no real significant oceanographic or geologic meaning associated with it. Many countries have continental shelves shallower than 200 m, and several are deeper. Likewise, a distance criterion of 200 nautical miles would not extend over all parts of the continental shelves of some nations, such as Argentina and Russia.

Developed vs. Developing Nations

Most of the winners in acquiring seafloor are developed countries. This is ironic, for a key issue of the Law of the Sea Conference was the "common heritage concept" and the sharing of resources. However, as shown in Chapter 15, almost all marine resources (save some migratory fish and manganese nodules) occur in what will be coastal nations' EEZs and the biggest ones are off developed countries. There is no provision in the treaty to share the resources in the EEZ.

Developing countries pushed very hard for the 200-nautical-mile EEZ, for the ocean presents many opportunities. However, most lack even basic knowledge of their

TABLE 19-2

Summary of Territorial Sea and Exclusive Economic Zone Claims (for 143 Coastal Countries as of February 1993)

Territorial Sea

Breadth in Nautical Miles	Countries	Number of Countries
3	Bahamas, Bahrain, Denmark, Germany, Jordan, Singapore, United Arab Emirates	7
4	Finland, Norway	2
6	Dominican Republic, Greece, Turkey	3
12	Most of the remaining countries	115
20	Angola	1
30	Nigeria, Togo	2
35	Syria	1
50	Cameroon	1
200	Benin, Congo, Ecuador, El Salvador, Liberia, Nicaragua, Panama, Peru, Sierra Leone, Somalia, Uruguay	11

Exclusive Economic Zones

86 countries have declared 200-nautical-mile-wide EEZs

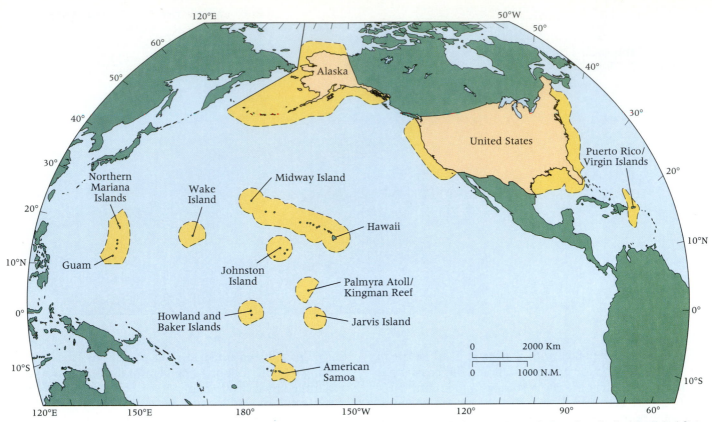

FIGURE 19-5 Map showing the 200-nautical-mile Exclusive Economic Zone (in yellow) of the United States, Commonwealth of Puerto Rico, Commonwealth of Northern Mariana Islands, and overseas possessions. The total U.S. EEZ area is about 1.21 times larger than the land area of the United States and its possessions. (From B.A. McGregor and M. Lockwood, *Mapping and Research in the Exclusive Economic Zone* (Reston, Va.: U.S. Geological Survey Publication, 1985).)

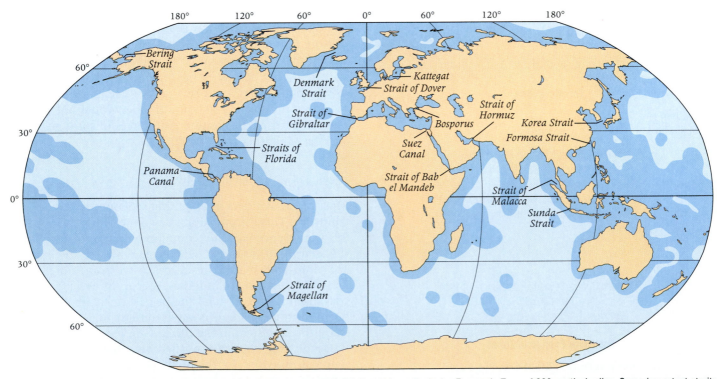

FIGURE 19-6 Area that falls within an Exclusive Economic Zone of 200 nautical miles. Some important straits are indicated.

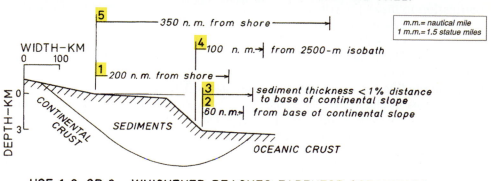

LEGAL DEFINITION OF OUTER LIMIT OF "CONTINENTAL SHELF"

USE 1, 2, OR 3– WHICHEVER REACHES FARTHEST OCEANWARD;
BUT NOT FARTHER THAN EITHER 4 OR 5.

offshore geology, biology, and resource potential, due to a lack of expertise in the marine sciences. Likewise, developing nations realize that total freedom of the seas offers the greatest advantage to countries sufficiently advanced to exploit resources oceanwide. Based on these two points, it is logical that developing countries want to obtain an extensive EEZ over which they would have control.

Developing nations also generally place higher priority on economic growth than on environmental quality and pollution. They view pollution as a problem of developed countries that have caused it through their own industrial growth and are the only ones with resources to clean it up. At this writing, most nations that ratified the treaty are developing countries and *no major maritime nation has yet approved it.*

On November 16, 1993, Guyana ratified the treaty, becoming the 60th nation to do so. According to its rules the treaty went into effect one year after the 60 nations have ratified it, on November 16, 1994.

Deep-Sea Mining

The potential mining of deep-sea nodules has been controversial since early in the treaty negotiations. Issues include the rate of mining, the sharing of mining technologies, and the international organization called the *Enterprise* that would control many activities in the deep sea. Without a treaty, there is the possibility that one or more nations will start mining the seafloor.

With a treaty, manganese nodules might be mined for their copper, manganese, nickel, and cobalt, but several factors would restrict the quantity mined in a year (see Chapter 15, page 376), limiting the financial return. If the world's nations shared resources within their own 200-mile zone, then the "common heritage concept" would have meaning. There is no indication that this will happen.

The Third Law of the Sea Conference, which progressed slowly (see Table 19-1 for chronology of the

conference), was probably the most complex series of negotiations ever undertaken. Upon completion of the negotiations in 1982, the United States was disenchanted with the treaty due to its impact on U.S. deep-sea mining interests. United States representatives indicated that they would not participate further in treaty negotiations, nor sign the agreement. Later, the United States indicated that it would follow most aspects of the treaty, but still would not consider ratifying it. In 1994 there were a series of modifications to the deep-sea mining provisions in the treaty, which were acceptable to the United States and which may eventually change the U.S. position toward signing the treaty.

Freedom of Scientific Research

Freedom of scientific research was an especially complex issue in the Law of the Sea negotiations. Research is essential for developing basic oceanographic knowledge and for learning about marine resources. With growing concern about global climatic change, it is critical for basic oceanographic information to be collected from all parts of the ocean.

The concept of **freedom of scientific research** for the ocean means being able to conduct research that causes no environmental damage in essentially any part of the ocean. This concept was assumed by marine scientists almost from the beginning of oceanographic study. The only exception was in a country's inland waters and in its three-mile territorial sea, where the coastal country has complete sovereignty. However, a coastal nation could and frequently did give permission for scientific research there.

Later, as countries extended their marine claims, little effect on scientific research was felt initially. If a country did restrict research in their waters, the restrictions could usually be reduced or removed by contacting a scientific colleague in that country. The visiting scientist could then be invited to join a cruise and through this gain permission to make the study.

Another reason that marine scientists did not worry about restrictions in the 1950s and 1960s was that so many areas of the ocean still had not been studied. If permission was not given, the scientific program could easily be modified to avoid entering those areas. The scientist just went elsewhere.

Recently, restrictions on scientific investigation have increased. Although the U.S. State Department has become more active in helping to obtain permission for scientific work, it cannot help in an area of disputed control. For example, until 1988, when the United States declared a 12-nautical-mile territorial sea, the State Department did not recognize territorial seas greater than three miles in width. It was only a few years prior that we recognized foreign EEZs and the right of countries to control research in them. When the United States declared its own EEZ, it did not claim control over research and hoped that other countries would do the same, but so far, none have. Increasing control over marine science has followed both extended jurisdiction prior to the Third Law of the Sea Conference and establishment of EEZs afterward.

The Law of the Sea Conference discussed freedom of scientific research. The main result was establishment of a *consent regime* (meaning that the coastal country must give permission for research) within the 200-nautical-mile EEZs and on the "legal" continental shelves of a coastal nation. This permission can still be hard to get and oceanographers are concerned about access to scientific research within these regions.

The problem is critical because, to understand ocean processes, scientists must be able to study conditions where land and water meet, both along the coastline and the continental shelf. This region is the source of sediments for the deep sea. It is where waves and many currents end, where most upwelling of ocean water occurs, and where most pollutants and other chemicals first enter the ocean. Many ocean phenomena are global in scale and certainly none are affected by, nor do they respect, political boundaries. If oceanographers are limited to studying the deep sea or that area outside the EEZ, oceanography could become a sterile and esoteric science.

Permission to perform scientific research can be granted with specific conditions. These requirements may not be unreasonable, but the *initial decision to give permission* is often based on political rather than scientific reasons. Some conditions can restrict the progress of an expedition. For example, the coastal nation might require being informed of the exact geographic area of work six months in advance. This sounds reasonable until you consider the exploratory nature of oceanography and the possibility that an important discovery could be made that would change the original cruise track. When at sea, this decision must often be made within hours.

The procedure is so bureaucratic that to get quick permission for such a change could be almost impossible.

The world's EEZ encloses approximately 35.8 percent of the ocean, and if the region around Antarctica were considered (Antarctica is excluded from the treaty), the percentage would be even higher. Some countries did suggest regulating research in the deep sea and having it controlled by an international authority, but this view did not prevail.

Concerns of Developing Nations

Consider the position of a country that has limited oceanographic capabilities. Its concerns about foreign scientists' working within its territorial waters or exclusive economic zone are understandable. Some countries' administrators neither understand nor appreciate the difference between *basic* and *applied* scientific research. They might feel that basic marine science, which attempts to understand the principles and processes of the ocean, might somehow affect their control over their resources. They might feel that, because they don't have scientific expertise to evaluate scientific findings, they could be exploited and not even know it.

Another concern of some coastal nations is the belief that marine research can be a form of espionage and could jeopardize their national security. Many scientists and administrators in the United States are aware of these problems and try to help developing countries acquire a marine science capability of their own. This would give them an understanding of marine research and allow for their meaningful participation in marine research programs.

During the Law of the Sea Conference, several marine scientists met with delegates, formally and informally, to discuss mutual advantages of marine science research. In one instance, a research vessel was brought to the United Nations to demonstrate its use in marine science (see opening figure to this chapter). Dr. John Knauss, then Provost of the Graduate School of Oceanography of the University of Rhode Island and later to be head of NOAA, explained the importance of science to society and how most technological and economic advances of modern times have directly resulted from scientific research. He also noted the irony of the fact that were it not for the technological advancements generated by science, there would be no *need* for the Law of the Sea Conference.

It is difficult to predict the long-term effects of restrictions on marine scientific research. Advantages to the restrictions could even emerge if they encourage oceanographers to create cooperative research programs with developing countries and to choose research objectives with value for all nations. It is also true that nations that elect to impose severe restrictions on basic marine scientific research in their waters will probably hurt only themselves in the long run.

SUMMARY

Until recently, the ocean was considered "public," belonging to no one country and available for everyone's use, except for the narrow territorial sea immediately adjacent to each coastal country. The Truman Proclamation of 1945 claimed the natural resources of the U.S. continental shelf. It was likely this action that encouraged other coastal nations to make claims and extensions to their territorial seas.

These events led to the First and Second Law of the Sea Conferences held in Geneva in 1958 and 1960. Four conventions were adopted: Convention on the Territorial Sea and the Contiguous Zone; Convention on the Continental Shelf; Convention on the High Seas; and Convention on Fishing and the Conservation of the Living Resources of the High Sea. These conventions, although codifying (documenting) many rules of ocean use, actually failed in part. They did not define the width of the territorial sea and gave an arbitrary definition to the edge of the continental shelf (a depth of 200 m, or 656 ft.) and the distance to which it could be exploited.

Countries continued to extend their claims seaward, largely due to gaps in the conventions, the development of marine technology, the awareness of marine resources, and political reasons. These actions and a desire by some to consider the resources of the ocean as a common heritage of humankind, to be shared by all, led to the Third Law of the Sea Conference, 1973–82, which considered almost all aspects of the ocean and resulted in confrontations between developed and developing countries. The resulting treaty was finally ratified by the necessary 60 countries in late 1993, but so far none of the ratifiers is a major maritime nation.

Nevertheless, many aspects of the treaty have been adopted by various countries; these include a 12-nautical-mile territorial sea and a 200-nautical-mile Exclusive Economic Zone (EEZ). In the EEZ, the adjacent coastal state has control over many activities, including exploiting its resources and marine scientific research. An international regime, details of which are still unresolved, may manage exploitation of deep sea resources, mainly manganese nodules.

The desire to share the resources of the ocean following the common heritage concept has been lost because almost all resources (in particular oil and gas and biological resources) lie within the EEZ and will be claimed by the adjacent nation. The control over marine science within the EEZ could have a damaging effect in developing our scientific understanding of the ocean, a point that has become more urgent as global-change problems have become more evident.

QUESTIONS

1. Summarize some of the important events leading to the Third Law of the Sea Conference that started in 1973.

2. What were the shortcomings of the 1958 and 1960 Geneva Conventions?

3. What were some of the key results of the Third Law of the Sea Conference?

4. What is an EEZ and what are the advantages and disadvantages of the establishment of EEZs around the world?

5. How might marine scientific research be affected by the results of the Third Law of the Sea Conference?

6. Why is a worldwide knowledge of oceanographic processes important?

KEY TERMS

coastal state

contiguous zone

continental shelf

convention

freedom of scientific research

high sea

high seas

Law of the Sea Conference

territorial sea

FURTHER READING

Borgese, E. M. 1982. "The Law of the Sea." *Scientific American* 247, no. 3, pp. 42–49. Implications of the Law of the Sea from a keen observer of marine matters.

Brittin, B. H. 1986. *International Law for Seagoing Officers.* Annapolis: Naval Institute Press. A well-written and interesting text on marine law for those who go to sea.

Champ, M. A., and N. A. Ostenso. 1984–85. "Future Uses and Research Needs in the EEZ." *Oceanus* 27, no. 4, pp. 62–69.

Discussion about some of the opportunities for marine science in the EEZs of the world.

Mangone, G. J. 1988. *Marine Policy for America*. Bristol, Pa.: Taylor and Francis. A broad and comprehensive treatment of marine policy by one of its better-known practitioners.

Nandan, Satya. 1989. "A Constitution for the Ocean: The 1982 U.N. Law of the Sea Convention." *Marine Policy Reports* 1, pp. 1–12. The head of the U.N. Law of the Sea Office discusses the Law of the Sea Convention.

Ross, D. A., and J. A. Knauss. 1982. "How the Law of the Sea Treaty Will Affect U.S. Marine Science." *Science* 217, pp. 1003–8. A detailed account of the possible effects on marine science by the Law of the Sea Treaty.

The United Nations publishes a Law of the Sea Bulletin that provides timely information concerning Law of the Sea issues. It is available from their New York City Office of the United Nations at their Ocean Affairs and the Law of the Sea.

Terms and Statistics

Oceanographers, for various reasons, use a confusing mixture of terms when discussing the ocean. The metric system, adopted by most scientists, is also generally used in oceanography. This system is based on multiples of 10 (see also Box 2-1, page 22). The smallest unit with which we shall be concerned is a micro ([μ]); 1,000 [μ] equal 1 millimeter (mm), 10 mm equal 1 centimeter (cm), 100 cm equal 1 meter (m), and 1,000 m equal 1 kilometer (km). A kilometer is about 0.6 miles (Tables 1 and 2).

Depth is measured either in fathoms or in meters. A fathom is 6 feet (ft.), or approximately the length of a line a person can hold between outspread hands; 100 fathoms equal 183 m.

Velocity is usually measured in knots (kn.); 1 kn. equals 1 nautical mile (6,080 ft.) per hour. The commonly used metric equivalent of 1 kn. is approximately 50 cm per second. The term *knot* was derived from a device called a Dutchman's log, which is a piece of wood or log attached to a string with knots tied in it at equal units. The log was thrown overboard and as the ship moved away, the string paid out. The number of knots in the line passing overboard in a certain duration of time were counted, and thus the speed of the ship was measured in knots.

Temperature is measured in degrees Celsius (°C); 0°C equals 32°F (Fahrenheit) (the freezing point of water), 20°C equals 68°F (about room temperature), and 100°C equals 212°F (the boiling point of water). Conversion of temperature from Fahrenheit to Celsius (or vice versa) can be done using the following equations:

$$°F = (1.8 \times °C) + 32$$

$$°C = (°F - 32)/1.8$$

TABLE 1
Metric-English Equivalents

	Metric			English		
Unit	Centimeters	Meters	Kilometers	Inches	Feet	Miles
cm	1	1/100	1/100,000	0.3937	——	——
m	100	1	1/1,000	39.37	3.28	——
km	100,000	1,000	1	——	3,280	0.6214
in.	2.54	——	——	1	1/12	——
ft.	30.48	0.3048	——	12	1	1/5,280
mi.	——	1,609	1.609	——	5,280	1

	Grams	Kilograms	Ounces	Pounds
g	1	1/1,000	0.035	——
kg	1,000	1	——	2.20
oz.	28.35	——	1	1/16
lb.	453.54	0.453	16	1

TABLE 2
Conversion of Various Units Used in Oceanography

To convert	Into	Multiply by
centimeters	inches	0.3937
meters	feet	3.28
meters	centimeters	100.0
meters	fathoms	0.546
kilometers	miles	0.624
kilometers	meters	1000.0
grams	ounces (avdp)	0.035
kilograms	pounds	2.2

Glossary

Absorption The conversion of light or sound energy into heat.

Abyssal That part of the ocean between a depth of about 2,000 to 6,000 m (about 6,550 to 19,700 ft.).

Abyssal hills Small irregular hills, rising to a height of 30 to 1,000 m (100 to 3,300 ft.), that cover large areas of the ocean floor; especially common in the Pacific Ocean.

Abyssal plain A very flat portion of the ocean floor underlain by sediments; usually found between 3,000 to 6,000 m in depth. The slope of abyssal plains are less than 1:1,000.

Acidic solution A liquid whose hydrogen ion concentration (H^+) is greater than its hydroxyl ion concentration (OH^-) or whose pH is less than 7.0.

Aerobic A condition where oxygen is present in the environment.

Albedo The ratio of the solar radiation reflected by a body or surface to the amount incident or received.

Alkaline solution A liquid whose hydroxyl ion concentration (OH^-) is greater than its hydrogen ion concentration (H^+) or whose pH is greater than 7.0. Seawater is slightly alkaline, having a pH between 7.5 and 8.4.

Anadromous Animals that migrate from the sea to freshwater (usually a river) to spawn; salmon are an example.

Anaerobic A condition in which oxygen is absent. The deeper parts of the Black Sea are an example of an anaerobic environment.

Anaerobic organisms Organisms that do not need free oxygen for their life processes.

Anion A negatively charged ion; examples are chloride (Cl^-) and oxygen (O^{2-}).

Anoxic Devoid of free oxygen.

Aphotic zone That part of the ocean where insufficient light is present for photosynthesis by plants.

Aquaculture Farming of the ocean, whereby organisms, such as fish, algae, and shellfish, are grown under controlled conditions. At present this technique is used only in nearshore areas.

Asthenosphere The upper part of the earth's mantle, reaching from the base of the lithosphere to about 250 km below the ocean basin and continents. The plates of the lithosphere move on its surface.

Atmospheric pressure The pressure exerted by the atmosphere. At sea level atmospheric pressure is equal to 14.7 pounds per square inch.

Atoll A circular-shaped series of coral reefs surrounding a central lagoon.

Atom The smallest component of an element that has all the properties of the element. An atom consists of protons, electrons, and neutrons.

Atomic weight The relative weight of an atom on the basis that the most abundant isotope of carbon (^{12}C) has a weight of 12. The atomic weight is essentially equal to the number of protons in the atom.

Autotroph An organism, such as a photosynthetic plant, that produces its own food from inorganic material using light or chemical energy.

Autotrophic bacteria Bacteria that produce their own food from inorganic compounds.

Bar An offshore accumulation of sand generally forming a long, low ridge paralleling the shore.

Barrier island A long, generally narrow island, peninsula, or bar parallel to the coast but separated from it by a lagoon or marsh.

Barrier reef A reef mainly composed of coral that parallels land but is separated from it by a lagoon.

Basalt A basic igneous rock, commonly found on the sea floor, composed mainly of feldspar and pyroxene minerals. Basalt rocks, which are dark and fine grained, are thought to underlie most of the ocean basin.

Bathyal That part of the ocean between a depth of about 200 to 2,000 m (about 650 to 6,600 ft.).

Bathymetry The measurement of the depth of the ocean.

Bathyscaphe An essentially immobile diving craft similar to a balloon, with an enclosed passenger compartment, but it sinks rather than floats.

Bathythermograph An instrument used to measure temperature with depth in the ocean.

Beaches Unconsolidated sediments (generally sand or gravel) that cover parts or all of the shore.

Benthic or benthonic The area of the ocean bottom inhabited by marine organisms.

Benthos Organisms—such as certain fish—that live on or in close contact with the ocean bottom.

Biological magnification An increase in concentration of certain compounds, including pollutants, in successively higher levels in the food web.

Bioluminescence The production of visible light by living organisms as a result of a chemical reaction.

Biomass The amount of living organisms in a certain area or volume at a certain time, which is usually expressed as grams per unit area or volume.

Biosphere A collective term for the area of habitat or the organisms of the earth.

Biotope An area where the principal habitat conditions and the living forms adapted to the conditions are uniform.

Black Smoker A type of oceanic vent where high-temperature, mineralized water is rapidly discharged.

Bloom A dense growth of phytoplankton occurring because of optimum growth conditions, frequently an excess of nutrients.

Boiling point The temperature at which a liquid starts to boil. For pure water this temperature is 100°C (212°F) at normal pressure.

Brackish water Water with a salinity less than seawater; generally formed by the mixing of seawater and freshwater.

Buffer A compound that can minimize changes in the pH by reversibly releasing or taking up H⁺ ions.

Calorie A measure or unit of heat generally defined as the amount of heat needed to raise the temperature of 1 g of water by 1°C; abbreviated cal.

Carbon-14 A radioactive isotope that can be used for dating. This isotope is especially useful in dating material that was once alive, since all living matter contains carbon. The half-life of carbon-14 is 5,680 years.

Carnivore An animal that eats other animals.

Catadromous Animals that migrate from rivers to the sea to breed; some eels are examples.

Catastrophic waves Large waves, resulting from intense storms or submarine slumping, that can cause immense damage and loss of life.

Cation A positively charged ion; examples are hydrogen (H⁺), and sodium (Na⁺).

Celsius A temperature scale in which water freezes at 0° and boils at 100° (at standard atmospheric pressure). One degree Celsius equals 1.8 degrees Fahrenheit.

Centigrade A term frequently used as an alternate term for Celsius.

Centrifugal force A force due to rotation that causes motion away or out from the rotating object.

Chemical bond The force of attraction that occurs between adjacent atoms that holds them together in a molecule.

Chemosynthesis The production of organic matter in the absence of sunlight by certain bacteria using chemical compounds such as sulfur to produce hydrogen sulfide. This type of process occurs at active hydrothermal vents.

Chlorophyll A group of green pigments found in plants that are essential for photosynthesis.

Cohesion The tendency of molecules to hold together.

Colligative property Those properties that vary with the number of chemical elements in the solution and not with their composition. In seawater boiling point and osmotic pressure increase with increasing salinity and freezing point and vapor pressure decrease.

Colloidal particles Very small particles, usually smaller than 0.00024 mm.

Commensalism A type of symbiotic relationship where one organism benefits without harm to the other.

Community An integrated group of organisms inhabiting a common area. These organisms may be dependent on each other or possibly on the environment. The community may be defined by their habitat or by the composition of the organisms.

Compensation depth The depth at which the primary production by the plant equals the amount of organic material that the plant needs for its own respiration during a 24-hour period.

Compensation depth oxygen The depth at which the oxygen produced by a plant during photosynthesis equals the amount the plant needs for respiration during a 24-hour period.

Compound A substance containing two or more elements in fixed proportions.

Conduction The transfer of heat by the collision of one atom with another atom.

Conservative elements Elements in seawater whose ratio to other conservative elements remains constant; examples are chlorine, sodium, and magnesium.

Continental crust The part of the crust underlying the continents and composed mainly of granitic rocks, generally 35 km (22 miles) thick.

Continental drift The concept that continents can drift or move about on the surface of the earth.

Continental margin That portion of the ocean adjacent to the continent and separating it from the deep sea; includes the continental shelf, continental slope, and continental rise.

Continental rise An area of gentle slope (usually less than half a degree, or 1:100) at the base of the continental slope.

Continental shelf The shallow part of the seafloor immediately adjacent to the continent; generally has a smooth seaward slope that leads to the continental slope.

Continental slope A declivity, averaging about 4°, that extends from the seaward edge of the continental shelf down to the continental rise or deep-sea floor.

Continuous seismic profile A record produced by using a high-energy acoustic device that shows the thickness and structure of the upper layers of the crust.

Convection Motion within a fluid due to differences in density or temperature.

Convergence zone Area of the earth's crust where crustal material is lost due to the collision of two or more plates. One plate may be subducted under another one or combined with another to thicken the crust or form mountains.

Coriolis effect A deflection due to the earth's rotation. This effect causes moving objects to turn to the right in the

Northern Hemisphere and to the left in the Southern Hemisphere.

Covalent bond The bond or linkage between two atoms in a molecule, formed by the sharing of electrons.

Current meter A device to measure the speed and direction of moving water.

Dead reckoning A type of navigation that mainly uses the speed and direction of the ship to estimate position.

Decibar A measure of pressure equal to one-tenth normal atmospheric pressure and approximately equal to the pressure change of 1 m depth in seawater.

Deep A relatively deep portion of the ocean floor.

Deep-scattering layer A sound-reflecting layer caused by the presence of certain organisms in the water. The layer, or layers, which may be 100 m (330 ft.) thick, usually rises toward the surface at night and descends when the sun rises.

Deep Sea Drilling Project A large-scale scientific project whose main aim was to drill numerous deep holes into the sediments on the ocean floor. See also JOIDES.

Deep-water wave A surface wave in water deeper than half its wavelength.

Delta A low-lying sediment deposit found at the mouth of a river.

Density The ratio of the mass of a substance to its volume.

Density current A current that is driven by gravity, where denser water tends to flow or sink downhill.

Desalination A variety of processes whereby the salts are removed from seawater, resulting in water that can be used for human consumption or for agriculture.

Desiccation Drying up, or loss of moisture.

Detrital deposits Sedimentary deposits resulting from the erosion and weathering of rocks.

Diagenesis Chemical and physical changes that sediments experience after their deposition.

Diapir A geological structure resulting from the upward movement of less dense rock (frequently salt) into more dense rock. See also Salt dome.

Diffusion The movement or transfer of a substance along a gradient; movement will be from areas of high concentrations to areas of low concentrations.

Dipole An object that has opposite charges at two points, usually refers to a water molecule.

Discontinuity An abrupt change in a physical property.

Diurnal Occurring daily. Referring to tides, one low and one high tide within one lunar day (about 24 hours and 50 minutes).

Divergence The flow of water in different directions away from a particular area or zone; often associated with areas of upwelling.

Earthquake A sudden motion of the earth caused by faulting or volcanic activity. Earthquakes can occur in the near-surface rocks or down to as deep as 700 km (440 miles) below the surface. The actual area of the earthquake is called the focus; the point on the earth's surface above the focus is called the epicenter.

Ebb tide The portion of the tidal cycle when the water is flowing from high to low tide. Thus the water level will drop in an estuary or marsh.

Echo sounding A method of determining the depth of the ocean by measuring the time interval between the emission of an acoustic signal and its return or echo from the seafloor. The returning signal is usually printed to give a visual picture of the topography of the seafloor. The instrument used in this method is called an echo sounder.

Ecology The study of the interactions of organisms with each other and with their physical, chemical, and biological environments.

Ecosystem An ecological unit including the environment and the organisms, each interacting with the other.

Eddy A generally circular flow of water (or air) often on the side or edge of a main current.

EEZ See Exclusive Economic Zone.

Electrical conductivity A measure of the ability of a material to conduct electricity.

Electron A negatively charged particle orbiting the nucleus of an atom.

Electrical conductivity A measure of the ability of a material to conduct electricity.

El Niño Warm current that occassionly forms, around Christmas time, off the coasts of Ecuador, Peru, and Chile. The current can cause large-scale death of fish and plankton.

Epicenter See Earthquake.

Erosion The physical and chemical breakdown of a rock and the movement of these broken or dissolved particles from one place to another.

Estuary A semi-enclosed coastal body of water having a free connection with the open sea and within which seawater is diluted by freshwater derived from land drainage.

Eustatic Worldwide.

Eutrophication A process involving the rapid growth of organisms due to excess nutrients that leads to the eventual oxidation of organic matter, which in turn will consume all the dissolved oxygen in the water.

Evaporation The physical process of a liquid becoming a gas.

Exclusive Economic Zone (EEZ) An offshore area extending 200 nautical miles (370 km) from the shoreline in which the adjacent country has extensive rights.

Fahrenheit A temperature scale on which the freezing point of water is 32° and the boiling point is 212°.

Fathom A common unit of measure of depth equal to 1.83 m (6 ft.).

Fault A fracture of rock along which the opposite sides have been relatively displaced.

Fauna The animal life of a particular time or area.

Fecal pellets Feces of marine organisms that may sink through the ocean to eventually become food for bottom living organisms.

Flood tide The portion of a tidal cycle when the water is flowing from low to high tide.

Flora The plant life of a particular time or area.

Focus See Earthquake.

Food chain A sequence of organisms in which each feeds on the next member of the sequence.

Food web A complex set of food chain interactions involving all the food relationships of a biological community.

Gabbro A category of igneous rocks that are granular, dark-colored, and commonly found in the oceanic crust.

Geostrophic current A current that is the result of the balance between the Coriolis effect and gravitational forces.

Glaciation The change or alteration of land or ocean floor caused by the movement of a glacier over it.

Global change A series of worldwide changes that may result from an increase in the greenhouse effect. The global changes could include a rise in sea level, increased frequency of hurricanes, and decreased supplies of fresh water.

Glucose A carbohydrate ($C_6H_{12}O_6$) produced by plants during photosynthesis, a basic food substance.

Gradient Rate of increase or decrease of one factor with respect to another, such as temperature with depth.

Granite A light-colored, coarse-grained igneous rock consisting mainly of quartz and alkali feldspar.

Gravimeter A device used to measure differences in the earth's gravitational field.

Gravity The force of attraction that causes objects on earth to fall toward the center of the earth. The universal law of gravitation as first given by Newton states that every particle in the universe attracts every other particle with a force that is proportional to the product of their masses and inversely proportional to the square of the distances between the particles.

Grazing The feeding by zooplankton upon phytoplankton.

Greenhouse effect The warming of the earth's atmosphere because of its transparency to incoming sunlight but relative opacity to radiation of heat from the earth. The opacity (and hence heat) is increased by increasing amounts of carbon dioxide, water vapor, methane, and dust in the atmosphere.

Gross production The total amount of organic matter photosynthesized by plants over a certain period and within a certain area or volume.

Guano The accumulated excrement of seabirds, often used as a fertilizer.

Guyot A flat-topped seamount.

Gyre A large, essentially closed circulation system.

Habitat Location where a specific plant or animal lives.

Hadal The deepest parts of the ocean; that part below a depth of 6,000 m (about 20,000 ft.).

Half-life The time it takes for one-half the atoms of a radioactive isotope to decay into another isotope. The half-life is different for each radioactive element.

Halocline A zone, usually 50 to 100 m (160 to 330 ft.) below the surface and extending to perhaps 1,000 m (3,300 ft.), where the salinity changes rapidly. The salinity change is greater in the halocline than in the water above or below it.

Heat capacity A ratio of the amount of heat absorbed or released by an object to the change in temperature of the object; in other words, the amount of heat necessary to raise the temperature of the object.

Herbivores Organisms that eat plants.

Heterotroph An organism that requires preformed organic compounds for food and energy.

Heterotrophic bacteria Bacteria that use organic material, produced by other organisms, for their food.

Holocene See Pleistocene.

Holoplankton Organisms that spend their entire life as plankton.

Hydrocarbon Organic compounds composed of carbon and hydrogen. Petroleum, methane and natural gas are hydrocarbons.

Hydrogen bond The relatively weak attraction that allows one water molecule to attract and form a bond with another water molecule.

Hydrologic cycle System whereby the water is removed from the ocean by evaporation into the atmosphere and eventually returns to the ocean either directly as precipitation or indirectly by rivers.

Hydrothermal deposits Deposits resulting from high-temperature water, either by altering existing rocks or by forming their own precipitates.

Ice Age Period of time in the earth's history when temperatures were below normal and ice sheets or glaciers covered part of the continents.

Iceberg Large piece of a glacier that has broken off and is floating in the ocean or stranded on a shoal.

Igneous rocks Rocks formed by the solidification of molten magma. Magma is composed of numerous minerals (mainly silicates) and gases derived from the earth's crust and mantle and is in a molten state.

Indicator species An organism that can give a clear indication of some natural aspect of the environment it was or is living in.

Inorganic compound A compound that does not contain both hydrogen and carbon, and therefore not of biological origin.

In situ In place.

Interface A surface between two media that have a discontinuity in some property, such as their sound velocities. The major interfaces in the ocean are the water-atmosphere, water-biosphere, and water-sediment.

Internal wave A wave occurring at the boundary between liquids of different density.

Ion An atom or group of atoms having an electrical charge. Most of the atoms in seawater are in the ionic form. An ion having a positive charge is called a cation; one having a negative charge is called an anion.

Isotope Different forms of the same element, differing mainly in their atomic weights.

Isostacy The balance of large portions of the Earth's crust as though they are floating in a denser medium.

JOIDES Joint Oceanographic Institutions for Deep Earth Sampling, a program started in 1964, that initially drilled six holes into the sediments on the continental shelf and slope off eastern Florida. The second phase of the program, the Deep Sea Drilling Program (DSDP), has resulted in over 800 holes drilled into the Atlantic, Pacific, and Indian Oceans and several marginal seas. The third and present phase is the Ocean Drilling Program (ODP).

Kilometer A metric measure of distance equal to 1,000 m, 0.62 statute miles, or 0.54 nautical miles; abbreviated km.

Kingdom The highest catagory in the biological classification of organisms.

Knot A unit of velocity equal to 1 nautical mile (6,080 ft.) per hour; approximately equal to 50 cm per second, or 1.69 ft. per second; abbreviated kn.

Krill A small shrimplike animal, frequently very abundant in polar waters and a favorite food of whales and other carnivores.

Lagoon A shallow pond or lake partially or completely separated from the sea by a shallow bar or bank.

Langmuir circulation A cellular type of surface-water circulation having alternating vortices with its axes in the direction of the wind.

Latent heat The amount of heat absorbed or released by a substance during a change of state, under conditions of constant temperature and pressure.

Latitude Location on Earth's surface defined by its angular distance north or south of the equator (0° latitude).

Lava Liquid rock that comes to the surface, usually from a volcano or a fissure.

Leeward The direction toward which the wind is blowing; the opposite is windward.

Lithosphere The outer rigid shell of the earth which includes the crust and the upper part of the mantle. The lithosphere, which is above the less rigid asthenosphere, is divided into numerous slowly moving plates.

Longitude Location on the earth's surface defined by its angular distance east or west of the prime meridian (0° longitude).

Magma Molten rock, called lava when it reaches the surface of the earth. When magma cools it forms igneous rock.

Manganese nodules Concretionary lumps of the oxides of iron, manganese, nickel and copper found over large areas of the seafloor. Some nodules may have economic potential.

Mantle The layer of the Earth between the outer crust and the innermost core.

Mariculture The culture of marine organisms in seawater. The terms *mariculture* and *aquaculture* are often used interchangeably; however, mariculture is more appropriate for the saltwater culture of marine organisms.

Marine Science See Oceanography.

Microtemperature structure Small-scale (in terms of vertical or horizontal dimensions) changes in temperature.

Mixing processes Any process or condition that causes mixing of the seawater.

Mohorovičić discontinuity The sharp change in seismic velocity occurring at a depth of about 11 km (7 miles) in the ocean and 35 km (21 miles) under land that defines the top of the earth's mantle. This discontinuity, commonly called the Moho, may represent either a chemical or a phase change in the layering of the earth.

Monsoon Seasonal winds, usually applied to the changing wind patterns in the Indian Ocean.

Mutualism A symbiotic relationship between two organisms that is beneficial to both organisms.

Nansen bottle A common device used by oceanographers to obtain subsurface water samples.

Neap tide Weak tides that occur about every 2 weeks when the moon is in its quarter position.

Nekton Animals that are able to swim independently of current action.

Neritic The part of the pelagic environment that extends from the nearshore zone out to a depth of about 200 m (650 ft.); in other words, the waters overlying the continental shelf.

Niche The function or role of a particular species in a biological community.

NOAA National Oceanic and Atmospheric Administration.

Node A portion of a stationary wave where the vertical motion is at its minimum but where horizontal velocity is highest.

Nonrenewable resource A resource that is present in a fixed amount and cannot be replenished in a reasonable time period.

Nutrients Compounds or ions that plants need for the production of organic matter.

Ocean basin That portion of the ocean seaward of the continental margin that includes the deep-sea floor.

Oceanographic station A position at sea where oceanographic observations are made.

Oceanography The application of all science to the phenomenon of the ocean.

Ooze Marine sediment that contains more than 30 percent of various microorganism shells.

Organic compound A compound that contains both carbon and hydrogen, it is of biological origin.

Osmosis The movement of dissolved ions or molecules through a semipermeable membrane. An osmotic pressure results when a difference in concentration exists on either side of the membrane. The greater the difference, the higher the osmotic pressure; the flow is toward the more concentrated solution.

Outcrop An exposure of rocks at the earth's surface.

Oxidation The process whereby an element or compound combines with oxygen or whereby electrons are removed from an ion or atom; opposite of respiration.

Oxygen-minimum zone A layer below the surface where the oxygen content is very low or zero.

Paleomagnetism The study of variations in the earth's magnetic field as recorded in ancient rocks.

Pangaea Large continent that split apart about 200 million years ago to form the present continents.

Parasitism A symbiotic relationship in which one organism (the parasite) lives on or in another organism (the host) and benefits from the relationship at the expense of the host organism.

Peat A brownish sediment composed of partially decomposed plant tissue.

Pelagic A division of the marine environment, including the entire mass of water. The pelagic environment can be divided into a neritic (water that overlies the continental shelf) province and an oceanic (the water of the deep sea) province.

Pelagic sediments Sediments deposited in the deep sea that have little or no coarse-grained terrigenous material.

pH A measure of the alkalinity or acidity of a solution. pH is the logarithm of the reciprocal of the hydrogen ion concentration, or $pH = \log 1/H^+$. A pH value of 7.0 indicates a neutral solution; lower than 7.0 is acidic; higher than 7.0 is alkaline.

Photic zone That part of the water that receives sufficient sunlight for plants to be able to photosynthesize.

Photosynthesis The production of organic matter by plants using water and carbon dioxide in the presence of chlorophyll and light; oxygen is released in the reaction.

Photosynthetic zone That part of the ocean where photosynthesis is possible, usually defined by the availability of light. In the open ocean this zone usually extends from the surface to about 200 m (660 ft.).

Phytoplankton Those plankton that are plants.

Phylum The highest division (or grouping) used in the classification of organisms in the various kingdoms; the plural is phyla.

Pinger An acoustic device used to determine distance of instruments above the bottom.

Placers Mineral deposits resulting from the reworking effect of waves or currents that remove the lighter density material, leaving the heavier (and often valuable) minerals behind.

Plankton Floating or weakly swimming organisms that are carried by the ocean currents. Plankton range in size from microscopic plants to large jellyfish.

Plankton bloom A large concentration of plankton within an area due to a rapid growth of the organisms. The large numbers of plankton can color the water, causing a red tide in some instances.

Plate tectonics The concept of plate movement caused by seafloor spreading and mantle convection.

Pleistocene A geological epoch that ended about 10,000 years ago and lasted about 1 to 2 million years. This epoch has been subdivided into four glacial states and three interglacial stages. The last of the Pleistocene glacial stages is the Wisconsin stage. The period we are now in, the Holocene Epoch, is not part of the Pleistocene and may be an interglacial stage.

Polymorphism Organisms that can occur in several different forms, independent of sexual differences.

Population A group of individuals from one species in a particular area.

Pressure The force per unit area upon an object.

Primary production The rate of carbon fixed into organic matter by autotrophs using energy from either solar or chemical reactions. Primary production is expressed for a given volume in a given period of time; also called gross primary production.

Primordial Pertaining to the beginning or initial times of the earth's history.

Productivity The production of organic material.

Protoplanets The early planets that preceded and developed (according to the condensation hypothesis of the origin of the planets) into the present planets.

Pycnocline A zone where the water density changes rapidly. The increase is greater than that in the water above or below it. The density change, or pycnocline, is due to changes in temperature and salinity.

Radioactive elements Those elements capable of changing into other elements by the emission of charged particles from their nuclei.

Rare gases Those gases—such as krypton, xenon, and argon—that are present in the earth's atmosphere in very small quantities.

Red tide A brownish or reddish coloration of surface waters caused by large concentrations of microscopic dinoflagellates (sometimes no color is evident). Toxins produced by the dinoflagellates can cause large fish kills and be harmful to humans who eat shellfish infected with these dinoflagellates.

Reef A solid marine structure composed of shells, skeletons, and secretions of marine organisms.

Refraction See Wave refraction.

Relict sediments Sediments whose character does not represent present-day conditions but rather a past environment.

Renewable resource A resource that is replaced by natural processes, usually on a daily or seasonal basis.

Residence time The residence time of an element in seawater is defined as the total amount of the element in the ocean divided by the rate of introduction of the element or the rate of its precipitation to the sediments.

Respiration An oxidation process whereby organic matter is used by plants and animals and converted to energy. Oxygen is used in this process and carbon dioxide and water are liberated; the opposite of respiration.

Rift valley Narrow trough or valley found along an area of seafloor spreading.

Rip current A narrow seaward-flowing current that results from breaking of waves and subsequent accumulation of water in the nearshore zone.

Rock weathering The chemical and physical processes that cause rocks to decay and eventually form soil.

Salinity The total amount of dissolved material in seawater; measured in parts per thousand by weight in 1 kg of seawater.

Salt dome A large cylindrical mass of salt that has risen through the surrounding sediment. It can form a trap for oil or gas. See also Diapir.

Sand A particle having a grain size between $1/16$ and 2 mm.

SCUBA Self-Contained Underwater Breathing Apparatus.

Seamounts Isolated elevations, usually higher than 1,000 m (3,000 ft.), on the seafloor. They usually resemble an inverted cone in shape, and frequently are of volcanic origin.

Sea waves Waves that, within their area of generation, are usually irregular without a definite pattern.

Secondary shoreline Shoreline where the coastal region has been formed mainly by marine or biological agents; examples are coral reefs, barrier beaches, and marshes.

Sedimentary rocks Rocks that have formed from the accumulation of particles (sediments) in water or from the air.

Seiche See Stationary waves.

Semidiurnal tides Tides that have two highs and two lows within one lunar day (24 hours and 50 minutes).

Seismicity Related to earth movements or earthquakes; an area of high seismicity has numerous earthquakes.

Sextant An instrument for measuring angles.

Shelf break The sharp break in slope that marks the edge of the continental shelf and beginning of the continental slope.

Shoal A submerged but shallow bank or ridge that can be dangerous to ships.

Shoreline The place where land and water meet.

Sill A ridge separating one partially closed ocean basin from the ocean or another basin.

Silt A particle having a grain size between $1/128$ and $1/16$ mm. Silt is intermediate in size to sand and clay (less than $1/128$ mm).

Slumping The sliding or moving of sediments down a submarine slope.

Solubility The degree to which a substance mixes with another substance.

Solvent A liquid that can dissolve other substances in itself.

Sound channel A zone where the sound velocity reaches a minimum value. Sound in this zone is reflected upward or downward back into the zone, with little energy loss. Thus the sound traveling in this channel can be transmitted over distances of many thousands of kilometers.

Sounding The determination of the depth of the ocean either by lowering a line to the bottom or electronically by noting how long sound takes to travel to the bottom and return.

Species Similar individuals that can breed and produce fertile offspring with each other, but not with other organisms.

Specific heat The number of calories, or amount of energy, needed to raise the temperature of 1g of a substance by 1°C.

Spontaneous liquefaction The movement or flow of an entire sediment mass, similar to an avalanche, that occurs when water-saturated sediments are subjected to a sudden shock, shear, or increase in pore-water pressure, and the internal grain-to-grain contacts within the sediment change.

Spring tides Strong tides that occur about every 2 weeks, when the moon is full or new.

Standing crop See Biomass.

Stationary waves A type of wave in which the waveform does not move forward; however, the surface moves up and down. At certain fixed points, called nodes, the water surface remains stationary.

Storm surges Abnormally high water levels due to strong winds blowing on the water surface.

Stratigraphy The study of layered rocks in particular their sequence and their correlation with similar rocks in other areas.

Subduction zone The area where an oceanic plate plunges beneath a crustal plate into the asthenosphere; often the surface expression is an oceanic trench.

Submarine canyon A steep, often V-shaped canyon generally cut into the outer continental shelf or continental slope.

Surf Breaking waves in a coastal area.

Surface tension A phenomenon peculiar to liquids in which the molecules on the surface of the liquid pull on each other and form a thin but elastic film.

Swell Waves that have traveled away from their generating area. These waves have a more regular pattern than sea waves.

Symbiosis The close relationship existing between two organisms, the relationship can be commensalism, mutualism, or parasitism

Synoptic measurements Numerous measurements taken simultaneously over a large area.

Taxonomic classification A systematic method of classifying animals and plants.

Tectonics The study of the deformation and construction of the earth structure.

Telemetry A method, usually electronic, of measuring something and then transmitting the measurement to a receiving station.

Terrigenous sediments Sediments composed of material derived from the land. Usually these deposits are found close to land.

Territorial sea A zone extending seaward from the shore in which the adjacent coastal nation has essentially full authority. The recent Law of the Sea Convention, or UNCLOS III, established 12 nautical miles as the maximum width of the territorial sea.

Thermistor A heat-sensitive device that can be used to measure temperature.

Thermocline A zone where the water temperature changes more rapidly than the water above or below it. This zone usually starts from 10 to 500 m (30 to 1,600 ft.) below the surface and can extend to over 1,500 m (5,000 ft.) in depth.

Thermohaline circulation A vertical circulation of seawater caused by differences in surface temperature and salinity that causes density differences and sinking of water.

Tidal bore A large wave of tidal origin that travels up some rivers and estuaries.

Tidal currents Currents due to tides.

Tidal period Time between successive high or low waters.

Tidal range The difference between successive high and low waters. The period of comparison can be different, such as the range over a week, month, or year.

Tides The regular rising and falling of sea level, caused mainly by the gravitational attraction of the moon and the sun on the earth.

Topography The study or description of the physical features of the earth's surface.

Transform fault A type of fault characteristic of areas of seafloor spreading (ocean ridges) where there is a horizontal offset of the ridge.

Trench A long, narrow, and deep depression with steep sides on the seafloor.

Trophic level The position of an organism in a food chain.

Tsunami A long-period wave caused by a submarine earthquake, slumping, or volcanic eruptions. Tsunamis, sometimes called tidal waves, have heights of only a few centimeters in deep water but can reach several tens of meters by the time they break on the beach.

Turbidite A generally coarse-grained sediment deposited from a turbidity current; generally found interbedded with the fine-grained muds typical of the deep sea.

Turbidity current A turbid, relatively dense current composed of water and sediment that flows downslope through less dense seawater. The sediment eventually settles out, forming a turbidite.

Turbulence A flow of water in which the motion of individual particles appears irregular and confused.

UNCLOS III The Third United Nations Conference on the Law of the Sea.

Upwelling The movement of water from depth to the surface.

Velocity The rate of change of position (distance) in a given time, such as 50 km per hour.

Viscosity A property (sometimes called internal friction) of a substance to resist flow.

Volcanic rocks Rocks that have resulted from volcanic eruptions.

Water mass A body of water identifiable by its temperature and salinity or by some other distinct marker.

Wave A disturbance that moves through the ocean or the earth. Its speed depends on the characteristics of the medium and the wave itself.

Wave attenuation The decrease in the waveform or height with distance from a wave's origin.

Wave crest The highest part of a wave.

Wavelength The horizontal distance between two wave crests (or similar points on the waveform) measured parallel to the direction of travel of the wave.

Wave period The time required for successive wave crests to pass by a fixed point.

Wave refraction The change in direction of waves that occurs when one portion of the wave reaches shallow water and is slowed down while the other portion is in deep water and moving relatively fast.

Wave trough The lowest part of a wave between two successive crests.

Wave velocity The speed with which the waveform proceeds; equal to the wavelength divided by the wave period.

Weathering The destruction (complete or partial) of a rock by mechanical or chemical processes.

Windward Direction from which the wind is blowing; opposite is leeward.

Zooplankton Those plankton that are of the animal kingdom.

References

Atkins, W.R.G. 1926. "The Phosphate Content of Sea Water in Relation to the Growth of the Algal Plankton, Part III." *Journal of the Marine Biological Association* 14, pp. 447–67.

Beck, W. S., K. F. Liem, and G. G. Simpson. 1991. *Life: An Introduction to Biology,* 3d ed., New York: HarperCollins.

Bigelow, H. B. 1931. *Oceanography: Its Scope, Problems, and Economic Importance.* Boston: Houghton Mifflin.

Brewer, P. G., W. S. Broecker, W. J. Jenkins, P. B. Rhines, C. G. Rooth, J. S. Swift, T. Takahashi, and R. T. Williams. 1983. "A Climatic Freshening of the Deep Atlantic North of 50°N Over the Past 20 Years." *Science* 222, pp. 1237–39.

Broecker, W. S. 1974. *Chemical Oceanography.* New York: Harcourt Brace Janovich.

Budyko, M. I. 1974. *Climate and Life.* New York: Academic Press.

Christy, F. T., Jr., and H. Herfindahl. 1967. "A Hypothetical Division of the Sea Floor: A Chart Prepared for the Law of the Sea Institute." University of Rhode Island.

Cronan, D. S., and J. S. Tooms 1969. "The Geochemistry of Manganese Nodules and Associated Pelagic Deposits from the Pacific and Indian Ocean." *Deep-Sea Research* 16, pp. 335–61.

Deacon, M. 1971. *Scientists and the Sea, 1650–1900: A Study of Marine Science.* London: Academic Press.

Edmond, J. M. 1980. "GEOSECS Is Like the Yankees: Everybody Hates It and It Always Wins . . . K. K. Turekian (1978)." *Oceanus* 23, no. 1, pp. 33–39.

Flipse, J. E. 1980. "The Potential Cost of Deep-Ocean Mining Environmental Regulation." Texas A&M Sea Grant Publication 80–205.

Fuglister, F. C. 1960. "Temperature and Salinity Profiles and Data from the International Geophysical Year of 1957–1958." Atlantic Ocean Atlas, Woods Hole Oceanographic Institution Atlas Series, 1.

Goldberg, E. D. 1963. "*The Oceans as a Chemical System,*" In *The Sea,* vol. 2, ed. M. N. Hill. New York: Interscience.

Gross, M. G. 1990. *Oceanography: A View of the Earth,* 5th ed. Englewood Cliffs, N.J.: Prentice Hall.

Hamblin, K. W. 1992. *Earth's Dynamic Systems,* 6th ed. New York: Macmillan Publishing Co.

Hamilton, E. L. 1959. "Thickness and Consolidation of Deep-Sea Sediments," *Bulletin of the Geological Society of America* 70, pp. 1399–1424.

Harris, C. L. 1992. *Concepts in Zoology.* New York: HarperCollins.

Hartline, B. K. 1980. "Coastal Upwelling: Physical Factors Feed Fish." *Science* 208, pp. 38–40.

Harvey, H. W. 1950. "On the Production of Living Matter in the Sea Off Plymouth." *Journal of the Marine Biological Association* 29, pp. 97–137.

Heezen, B. C., and M. Ewing. 1963. "The Mid-Oceanic Ridge." In *The Sea,* vol. 3, ed. M. N. Hill. New York: Interscience.

———. 1952. "Turbidity Currents and Submarine Slumps, and the 1929 Grand Banks Earthquake." *American Journal of Science* 250, pp. 849–73.

Heezen, B. C., M. Tharp, and M. Ewing. 1959. "The Floors of the Ocean, North Atlantic." *Geological Society of America,* Special Paper no. 65.

Hill, M. N., 1963. *The Sea: Composition of Seawater.* New York: Wiley Interscience.

Hood, D.W. 1966. "Seawater: Chemistry." In *Encyclopedia of Oceanography,* ed. R. W. Fairbridge. New York: Reinhold.

———. 1963. "Chemical Oceanography." *Oceanographic and Marine Biology Annual Review* 1, pp. 129–55.

Horn, D. R., B. M. Horn, and M. N. Delach. 1972. "Distribution of Ferromanganese Deposits in the World Ocean." In *Ferromanganese Deposits on the Ocean Floor,* ed. D. R. Horn. New York: Arden House and Lamont–Doherty Geological Observatory.

Isaacs, J. D., D. Castel, and G. L. Wick. 1976. "Utilization of the Energy in Ocean Waves." *Ocean Engineering* 3, pp. 175–87.

Jerlov, N. G. 1951. "Optical Studies of Ocean Waters." *Reports of the Swedish Deep Sea Expedition, Physics and Chemistry* 3, no. 1, pp. 1–59.

Kaufman D. G., and C. M. Franz. 1993. *Biosphere 2000: Protecting Our Global Environment.* New York: HarperCollins.

Kaufman, W., and O. Pilkey. 1979. *The Beaches Are Moving.* Garden City, N.Y.: Anchor Press/Doubleday.

Kennett, James. 1982. *Marine Geology.* Englewood Cliffs, N.J.: Prentice-Hall.

Leopold, L. P., and K. S. Davis. 1966. *Water.* New York: Time Incorporated.

Menard, H. W., and S. M. Smith. 1966. "Hypsometry of Ocean Basin Provinces." *Journal of Geophysical Research* 71, pp. 4305–25.

Milliman, J. D., and K. O. Emery. 1968. "Sea Levels During the Past 35,000 Years." *Science* 162, pp. 1121–23.

Mix, M. C., P. Farber, and K. I. King. 1992. *Biology: The Network of Life*. New York: HarperCollins.

National Geographic Society. 1990. *Atlas of the World*, 6th ed. Washington, D.C.

Nybakken, J. W. 1993. *Marine Biology: An Ecological Approach*, 3d ed. New York: HarperCollins.

Peterson, S. 1980. "The Common Heritage Of Mankind?" *Environment* 22, no. 1, pp. 6–11.

Picard, G. L. 1964. *Descriptive Physical Oceanography*, Oxford: Pergamon Press Ltd.

Richardson, P. L. 1980. "Benjamin Franklin and Timothy Folger's First Printed Chart of the Gulf Stream." *Science* 207, pp. 643–45.

Riley, G. A. 1946. "Factors Controlling Phytoplankton Populations on Georges Bank." *Journal of Marine Research* 6, pp. 54–78.

Ross, D. A., and J. A. Knauss. 1982. "How the Law of the Sea Treaty Will Affect U.S. Marine Science." *Science* 217, pp. 1003–8.

Ryther, J. H. 1969. "Photosynthesis and Fish Production in the Sea." *Science* 166, pp. 72–76.

Ryther, J. H., and C. S. Yentsch. 1958. "Primary Production of Continental Shelf Waters off New York." *Limnology and Oceanography* 3, pp. 327–35.

Spaeth, M. G., and S. C. Berkman. 1967. "The Tsunami of March 28, 1964, as Recorded at Tide Stations." *Coast and Geodetic Survey Technical Bulletin* 1, no. 33.

Stommel, H. 1965. "The Anatomy of the Atlantic." *Scientific American* 192 (January), pp. 30–35.

Valiela, I., and S. Vince. 1976. "Green Borders of the Sea." *Oceanus* 19, no. 5, pp. 10–17.

Van Dover, C. L., S. Kaartvedt, S. M. Bollens, P. H. Wiebe, J.W. Martin, and S. C. France. 1992. "Deep-Sea Amphipod Swarms." *Nature* 358 (July), pp. 25–26.

Von Arx, W. S. 1979. "Prospects: A Social Context for Natural Science." *Oceanus* 22, no. 4, pp. 3–11.

Weihaupt, J. G. 1979. *Exploration of the Oceans*. New York: Macmillan.

Wentworth, C. K. 1992. "A Scale of Grade and Class Terms for Clastic Sediments." *Journal of Geology* 30, pp. 377–92.

Williams, S. J. 1986. *Sand and Gravel Deposits Within the United States Exclusive Economic Zone: Resource Assessment and Uses*. Houston: 18th Annual Offshore Technology Conference, pp. 377–86.

Wright, L. D., and J. M. Coleman. 1973. "Variations in Morphology of Major River Deltas as Functions of Ocean Wave and River Discharge Regimes." *American Association of Petroleum Geologists* 57, no. 2, pp. 370–98.

Wust, G., W. Brogmus, and E. N. Noodt. 1954. "Die Zonale Verteilung von Salzgehalt, Neiderschlag, Verdungstung, Temperatur and Dichte an der Oberflache der Ozeane." *Kieler Meeresforsch* 10, pp. 137–61.

Zarudski, E.F.K. 1967. "Swordfish Rams the Alvin." *Oceanus* 4, pp. 14–18.

Index

Abyssal hills, 96
Abyssal plains, 91, 92, 96
Acanthaster (Crown of Thorns), 268, 269
Acidity, 154, 155
Acid rain, 154, 400, 401
Accoustically Navigated Geological Underwater Survey (ANGUS), 105, 431
Acoustic tomography, 186, 187
Afar triangle, 59
Agar, 259, 391
Agnatha, 62
Air-sea interaction (see Ocean-atmosphere interactions)
Albatross, 9, 372
Aldrin (see Pesticides)
Alexander's Acres, 294, 295
Algae
  blue-green, 256
  brown, 258
  green, 258
  red, 259
  use of, 444, 445
  yellow-green, 259
Algin, 258, 391
Alkalinity, 154, 155
Alpha Ridge, 98, 99
Alps, 52
Altimeter, 76
Aluminaut, 115
Alvarez, Luis, 32
Alvin, 112, 114–116, 168, 432
Amino acids, 29
Amoco Cadiz oil spill, 411
Amphipoda, 272
Anadromous fish, 272
Angiospermophyta, 257
Animalia Kingdom, 261–277
Animals in the ocean, 293–296
  adaptation to environment, 293
  color, 294
  distribution, 293
  food and effects on distribution of organisms, 294
  salinity effects, 294
  sampling problems, 293

temperature effects on organisms, 293
Animals of the sea, 261–277
  annelida, 261, 262
  arthropoda, 266–266
  chaetognatha, 261
  chordata, 268–277
  cnidaria, 261
  echinodermata, 266–268
  mollusca, 266
  nematoda, 261
  platyhelminthes, 261
  porifera, 261
Anions, 27
  source, 145
Antarctica, 149
  climate, 324
  ozone problem, 330
Antarctic bottom water, 217
Antarctic Circumpolar Current, 208, 214, 331
Antarctic ice shelf, 150
Antarctic intermediate water, 216, 217
Aphotic, 245
Appalachian Mountains, 52, 56
Aquaculture, 390, 391, 439, 440
Aquariums, 459, 460
Arabian Plate, 57–60
Arabian Sea, 321
Aragonite, mining, 368
Archeology (see Marine archeology)
Arctic Ocean, 37, 98, 99
Areas of nondeposition, 125
ARGO/JASON, 116, 117, 428, 431, 432
Argo Merchant oil spill, 411
Arthropoda (see Animals of the sea)
Artificial upwelling, 451, 452
Asthenosphere, 51
Atlantic Ocean, 36, 92–94, 148,149
Atlantis, lost continent, 451, 452
Atlantis II, 102
Atlantis II Deep, 374

Atmosphere, origin, 26,27
Atmosphere, pressure, 140
Atmospheric circulation, 206–211
Atolls, 297, 298
Atoms, 134
Attenuation, 141
Automated Benthic Explorer ABE, 457
Autonomous Underwater Vehicles (AUVs), 456, 457
Autotrophs, 256, 284
Aves 272
Axial tilt (see Milankovitch hypothesis)
Azoic zone, 6

Bacteria, 254, 256
  bacterioplankton, 249, 250
Baird, Spencer, 10
Baker, Edward, T., 218
Ballard, Robert, 116, 428, 432
Baltic Sea, 36
Barite, 368
Barnacles, 263, 264
Barrier beaches, 343–347
Barrier islands, 343–347
Basaltic rock, 49
Bass, George, 423, 425
Bathyal (see Marine environment)
Bathymetric map, 7
Bathymetry, 75, 76
Bathyshere (see *Trieste*)
Bathythermograph, 182
  expendable, 183
Bay of Fundy, 239
Beaches
  effect of waves, 226, 227, 339–344
  erosion, 226, 340–343, 355, 356
  human effects, 341–343, 345
  sea level, 342, 343
  seasonal effects, 342, 343
Beagle, 6
Beaufort Wind Scale, 224, 225
Benthic environments (see Marine environment)

Benthos, 249, 252, 253 (see also Population of the ocean)
Bergman, Olaf, 143
Bering Strait, 331
Beta carotene, 440
Beta Pictoris, 24, 25
Biochemical cycle, 170
Biochemical reactions in the ocean, 170–171
Biofouling, 203, 204, 440
Biogenic sediments, 122–125 (see see also Oozes)
  for dating, 128,129
Biological community, 284–286
Biological consequences of seawater, 286, 287
Biological dredge, 284
Biological, innovative uses of the ocean, 439–445
Biological magnification, 404, 405
Biological oceanography, 14, 280–283
  instruments, 282–284
  modern studies, 281
  sampling devices, 282–284
Biological pump, 165
Biological resources of the ocean, 380–391
  animal protein, 380
  efficiency of conversion of organic matter, 381, 382
  fish harvest, 381–385 (see also Fishing)
  organisms used to extract elements from seawater, 391
  phytoplankton production, 380, 381
Bioluminescence, 294, 295
Biomass, 246, 248
Biotechnology, 439, 440
Biotope, 245, 284, 285
Birds, 272
Bismarck, 432, 434
Black Sea, 162, 175, 436
Black smokers, 67, 168, 169, 373
Blake Plateau, 379

Bloom, 259
Blue water diving, 282
Blue whale (see Whales)
Bottom photographs (see
    Underwater photography)
Boyle, Robert, 143
Box corer, 107, 110
Breaching, 34
Breakwater (see Coastal struc-
    tures)
Brine, 147
Broecker, Wallace, 334
Brooke, Midshipman, 6
Brown clay (see Oozes)
Buffered solution, 154, 155
    biological effects, 287
By-catch (see Fishing)
Buoyancy, 287
Buoys, 182, 185 (see also Current
    measurements)
Buzzards Bay oil spill, 415

Calcium carbonate, 122
California current, 215
California kelp (see Kelp)
Calorie, 139
Cameras, underwater (see
    Underwater photography)
Cape Cod, beach erosion, 346,
    347
Capillary waves, 224
Cap rock, 370 (see also Sulfur de-
    posits)
Carbon, 164, 165, 303
Carbon dioxide, 162, 163
    anthropogenic, 323
    in atmosphere, 164, 322–324
    and climate, 165, 320–324
    pathways, 164
    in seawater, 160–163
    sources, 161–165
Carbon-14, 161, 174, 303
Carbonate compensation depth,
    126
Caribbean Sea, 36
Carlton, J. T., 400
Carrageenan, 391
Cartilaginous fish (see
    Chondrichthyes)
Catadromous fish, 272
Catastrophic waves, 230–236
    landslide surges, 231
    stationary waves, 235, 236
    storm surges, 230
    tsunami, 231–236
Cations, 27
    source, 145
Celsius, 34
Center for Maritime Studies,
    Israel, 423
Central America, S.S., 426–430
    ownership, 427,429
    value, 429–430
Centrifugal force, 238
Cephalopoda 266
Cetacea, 272–277 (see also Whales)

Chaetognatha, 261 (see also
    Animals of the Sea)
Challenger, 8
Challenger Deep, 9, 96
Challenger expedition, 8, 9, 143,
    281, 374
Chemical bond, 134
Chemical oceanography, 14
    present objectives, 154, 156
    sampling devices, 156–158
Chemical reactions in the ocean,
    166–174
Chemosynthesis, 253, 300, 301
Chernobyl nuclear power plant,
    175
Chesapeake Bay, 348, 351
China Seas, 34
Chlorinated hydrocarbons (see
    Synthetic organic com-
    pounds)
Chlorinity, 158
Chlorofluorocarbons (CFCs), 321,
    330, 405 (see also Synthetic
    organic compounds)
Chondrichthyes, 268–271
Chlorophyll, 259, 304, 321
Cholesterol, 440
Chordata, 268–277
Christmas Island, 376
Circulation (see Ocean circulation)
Climate, 311–336
    atmosphere and ocean interac-
        tions with, 320–326
    modeling, 324
    and phytoplankton, 321
    trends, 334
Coastal dunes, 356
Coastal environments, 337–361
Coastal erosion, 346–347, 354
Coastal problems, 354–358
    global change, 354, 355
    sea level, 355
Coastal regions, 81
Coastal structures, 356–358
    artificial fill, 356
    breakwater, 356
    effects of, 341–343
    floating breakwater, 358
    groins, 340, 356
    jetties, 341, 356
    revetments, 356
    seawalls, 342, 356
Coastal Upwelling Ecosystems
    Analysis, 215
Coastal zone, 338
    extent, 338
    human activities and develop-
        ment, 356–360
    importance to ocean, 338
    management, 338, 359, 360
    problems, 354–358
    use, 358, 359
Coastal Zone Management Act,
    U.S., 359
Cobalt crusts, 377, 379 (see also
    Resources of the ocean)

Coccolithophoridae 259–261 (see
    also Algae, yellow-green)
Coccoliths (see Oozes)
Coelenterata (see Animals of the
    Sea)
Cohesion, 141
Columbus, Christopher, 3, 5
Columbus-America Discovery
    Group, 427–430
Commensalistic relationships, 285
Common heritage of mankind,
    467, 471, 473
Communities, 245
Compensation depth, 288, 289
Compounds, 134
Condensation hypothesis, 23, 24
Conductivity of seawater, 158
Conservative elements, 158
Contiguous zone and exclusive
    fishing zone, 465
Continental borderland, 86
Continental drift, 41–46
    difficulties, 45, 46
Continental margin, 80–92
    general characteristics, 82
    legal sense, 471
Continental rise, 54, 87, 88
Continental shelf, 81–87
    defined, international conven-
        tion, 465, 471
    glaciated shelves, 83, 85
    resources, 2
    shelf break, 82
    unglaciated shelves, 83, 85
    United States, 84–87
Continental slope, 87
Continuous seismic profiling, 76,
    78, 79
Convection, 43
Convection cells, 45
Convection currents, 43, 45
Convention on the Continental
    Shelf (see Law of the Sea)
Convergence, of waves, 227
Conveyor belt, seafloor spreading,
    66
Cook, Captain James, 4
Copepoda, 264, 265
Coral atoll, 87
Coral bleaching, 296, 297
Coral reefs 2296–299
    community, 296–298
    origin, 298, 299
    types, 296, 297
Coralline algae, 296
Core, of Earth, 41
Coring devices, 107, 108
Coriolis effect, 206, 207
Corrosion, 392
Corwith Cramer, 16
Cosmogenous sediments, 125
Cousteau, Jacques, 423
Covalent bond, 134, 135
Craig, Harmon, 218
Cretaceous mass extincton, 32
Cromwell Current, 211

Crown of Thorns, 268
Crustacea (see Animals of the sea,
    arthropoda)
Crustal layer, 26
Crustal structure, 41, 79, 81 (see
    also Seafloor spreading)
    basalt, 80
    gabbro, 80
    granite, 80
    Mohorovičić discontinuity, 80
    oceanic layer (third layer), 80
    second layer, 79
CTD(conductivity-temperature-
    depth) probe, 144, 158
Current measurements
    buoys, 200, 202
    current meters, 200, 203
    derelict vessels, 202
    drift bottles, 200, 201
    drogues, 200
    free-fall devices, 200
    indirect methods, 200, 204
    remote sensing, 200, 204
    ship drift, 200, 201
    swallow float, 200, 201
    telemetering buoys, 200
    vertical current meter, 204
Currents (see Ocean currents)
Cyclones (see Hurricanes)

Dam Atoll, 448
d'Arsonval, Jacques, 449
Darwin, Charles, 6
    coral reef hypothesis, 298,
        299
Dating methods (see Isotopes)
DDT, 154 (see also Pesticides)
Dead Sea, 60
Deep Rover, 422
Deep-scattering layer, 294–296
Deep Sea Drilling Project, 11, 108
Deep-sea fauna, origin, 294
Deep-sea muds and oozes, min-
    eral deposits, 377
Deep-sea mining, 473
Deep-sea sediments, 122–125
Deep-sea system (see Marine
    environment)
Deep-sea vents (see Vents)
Delaware Bay, 348
Delaware, University of, 17
Deltas, 85, 352, 353
Denitrifying bacteria, 290
Density, 138 (see also Seawater,
    density)
    crustal, 80
    factors influencing, 138, 190,
        192
Desalination, 368 (see also
    Freshwater from the sea)
Dessication, 286
Deuterium, 135
Diamonds, mining, 368
Diapir, 369–371
Diatom ooze 259 (see Oozes)
Diatoms, 163, 259, 260

Dieldrin (see Pesticides)
Dietz, Robert, 46
Differentiation, of magma, 26
Dimethyl sulfide (DMS), 321
Dinoflagellate, 259, 260 (see also Algae, yellow-green)
Dinosaurs, 31 (see also Extinctions)
Discovery, R.R.S., 373
Disposal of nuclear waste, 455, 456
Dissolved gases, 160–163
Dissolved material in the ocean, 145
Dissolved inorganic matter, 159, 160
Dissolved organic matter, 163
Distichlis, 351
Distributaries, 352
Dittmar, William, 143, 144, 158
Diurnal tides (see Tides)
Divergence, 46, 48
    of waves, 227
Dolphins (see Cetacea)
Downwelling (see Subduction)
Dredge, 108, 110
Drift bottle, 200 (see also Current measurements)
Drilling platforms, 363, 369
Drilling ships, 108, 109, 113
Drogues, 200 (see also Current measurements)
Drugs from the sea, 444, 445
Duke University, 424

Earle, Sylvia, 119, 422
Early earth, 25
Earth internal structure, 26
Earthquakes, 41, 42
    and plate tectonics, 55
Earth's orbit (see Milankovitch hypothesis)
Eastern boundary currents, 208
East Pacific Rise, 37, 67, 95, 96, 169
Eastward, 424
Echinodermata, 266–268
Echo sounders, 9, 105 (see also Sounding methods)
Ecology, 284
Ecosystem, 284
Eddies, 211–214
Eelgrass, 257, 285
EEZ (see Exclusive Economic Zone)
Ekman spiral and transport, 211, 212
Elections, 34
Electromagnetic spectrum, 142
El Chicon, 325, 329
El Niño, 199, 325–329, 388
    causes, 327, 329
    effects, 329
    prediction, 327, 328
    southern oscillation, 327
Embley, Robert, 218
Emperor Seamount Chain, 55
Endeavour, 4

Endangered Species Act, 195
Energy from the ocean 445–454
    (see also Future uses of the ocean)
    artificial upwelling, 451, 452
    damming large bodies of water, 453
    gas hydrates, 453, 454
    marine biomass, 452, 453
    ocean currents, 448
    thermal differences, 448–452
    tides, 446, 447
    vents, 454
    waves, 447, 448
Environmental Protection Agency, 16, 404, 406
Epicenter, 42
Equatorial current, 208
Equatorial undercurrent, 211
Eratosthenes, 3
Estuaries, 338, 344, 345, 348–351
    definition, 344
    impact of human development on, 344, 356
    life, 350, 351
    water mixing and flow, 349, 350
Ethogram, 76
Eukorytic, 255
Euphausia superba (see Krill)
Euphausids, 264–266
Euphotic zone (see Marine environment)
Eurasian Plate, 60
Evaporite deposits, 145
Evaporation, 139
Exclusive Economic Zone (EEZ)
    extent, 472
    law of the sea, 470, 471
    resources, 377–379
Exploitability clause, 465, 466
Extinctions
    and the greenhouse effect, 31–33
    hypotheses, 32
    periodicity, 31
Exxon Valdez oil spill, 412–415, 439

Factory ships, 384
Fahrenheit, 34
Falmouth oil spill (see Buzzards Bay oil spill)
Faulting, 50
Fecal pellets, 122, 124, 145
Fecal matter, 247
Federal Flood Plain Insurance Program, 360
Ferrel, William, 7
Ferromanganese oxides (see Manganese nodules)
Fetch, 224
Fish (Pisces), 268–272
    antifreeze, 253
    bony, 269–272
    deep sea, 248
    improving the harvest, 442

meal, 380
oils, 440–442
surimi, 442
tagging, 283
Fishery Conservation and Management Act, 360, 388
Fishery Management Councils, 388
Fishing, 380–390
    by-catch, 384, 285
    conflicts, 388–390
    legal problems, 388, 390
    major areas, 381
    maximum sustainable yield, 384
    methods, 382–388
    Peru, fisheries, 385, 388
    overfishing, 382
    pollution, 389, 390
    resources, 381–385
    technology, 384–388
    U.S. fish catch, 382, 383
    U.S. fishing industry, 382, 383,
    U.S. recreational fisherman, 382
    world fish production, 389
Fish kills, 405
Fish TV, 443
Fjords, 348, 349
Fletcher's Ice Island, 184
Floating Instrument Platform (FLIP), 185
Flood-plain deposits, 352
Florida oil spill, 415
Flotation of plants, 288
Focus, 42
Food chain, 14, 286, 305, 306
Food cycle, 305–308
    implications, 307
Food web, 305, 307, 308
Foraminifera, 129, 251
Forbes, Edward, 6, 281
Forchhammer, Georg, 143
Forchhammer's Principle, 143, 158
Fossil fuel, contribution to carbon dioxide problem, 323
Fracture zones, 50
Fram 9, 149, 150, 184
Franklin, Benjamin, 6, 7, 199, 213
Frasch process, 370
Freedom of high seas (see High seas)
Freedom of scientific research (see Marine scientific research)
Freshwater, density, 138 (see also Water properties)
Freshwater from the sea, 454–455
    desalination, 454
    glaciers and icebergs as source, 454, 455
Frost wedging, 138
Fucus, 258 (see also Algae, brown)
Future uses of marine organisms, 439–445
    drugs/pharmaceuticals, 444, 445
    hatcheries, 442, 444

krill, 442, 443
porpoises, 443, 444
Future uses of the ocean, 438–456

Gabbro (see Crustal structure)
Gagnan, Emile, 423
Galápagos Islands, 96, 168, 269, 273, 434
    rift, 67, 127
    rocks, 127
    vents, 67, 372
Gas, solubility, 161
    bladders, 252, 286, 288
    carbon dioxide, 162, 163
    inert gases, 161
    nitrogen, 160
    methane, 169
    oxygen, 161, 162
Gas giants, 23
Gas hydrates, 452, 457
Gastropoda, 266 (see also Mollusca)
Geller, J. B., 400
Genetic engineering, 439, 440
    and aquaculture, 440
Geneva conferences (see Law of the Sea)
Geochemical cycle of the ocean, 166–172
Geochemical Ocean Sections Study (GEOSECS), 11, 155, 218
Geologic time scale, 28, 30
Geomagnetic time scale, 50
Georges Bank, 303
    boundary dispute, 469
Geostrophic currents (see Ocean currents)
Gill net, 388 (see also Fishing)
Glaciers and glaciation, 44, 147, 149, 150
    deposits, 47
    rebound, 44
    sea level and beaches, 342, 343
    sediments, 121
Glauconite, 368
Global Ocean Ecosystems Dynamics (GLOBEC), 281
Global change
    and climate, 311–336
    and coral reefs, 297
    and hurricanes, 318
Global positioning system, 77, 112
Glomar Challenger, 11, 12, 108, 117
Glucose, 161, 288
Gold, 365, 371, 379 (see also Seawater, recoverable elements)
Gorda Ridge, 379
Grab samplers, 107
Graben, 50
Grain size, 119, 120
Grand Banks earthquake, 90, 91
Granite (see Crustal structure)
Gravity, 23, 71, 237
Gravity corer, 107, 108

Gravity measurements, 75, 76
Grazers (*see* Zooplankton)
Grazing, 304, 305
Great Barrier Reef, 268, 297–299, 460
Great Lakes, pollution, 397
Great Ocean Conveyor, 334, 335
Greenhouse effect, 30, 156, 320–326
  and circulation, 320
  and fossil fuels, 323
  gases, 321–323
  and global warming, 321–326
  sea level rise, 321
Greenland ice sheets, 334
Groins (*see* Coastal structures)
Grotius, Hugo, 463
Gross production (*see* Organic production)
Guano, 329
Gulfs
  Aden, 59, 60
  Aqaba (Elat), 60, 61
  California, 86
  Maine, 469
  Mexico, 36, 169
  Oman, 60, 64
Gulf Stream, 6, 205, 334, 448
Gulf Stream rings, 213, 214
Guyots, 96
Gyres, 208

Hadal (*see* Marine environment)
Half life, 172
Halobates, 263
Halocline, 188, 189
Hawaii, 66
Hawaii, University of, 10
Hawaiian Islands Chain, 55, 57
Heard Island, 194, 195, 274
Heat, 139, 313–316
Heat budget, Earth, 204–206
Heating of land and water, 313–315
Heavy-metal muds, 372–372
Herbivores (*see* Food chain; Food cycle; Zooplankton)
Herdendorf, Charles E., 426, 427, 430
Hess, Harry, 46
Heterotrophs, 256, 284
High resolution profiler, 183
High seas, 463, 465, 466
Himalayan Mountains, 52
Hoffman, David, 325
Holdfast, 258
Holocene epoch, 28
Holothurian, 267, 270
Holoplankton, 250
Hot spots, 55–57, 67
Hot vents (*see* Vents)
Hudson River, 348
Hughes, Howard, 375
Hughes Glomar Explorer, 375–377
Human input into the ocean, 172
  (*see also* Pollution, marine)

Humboldt, Alexander von, 43
Hurricane Andrew, 319
Hurricane Frederic, 317
Hurricane Hugo, 317, 319
Hurricanes, 315–320
  and global change 318
  origin, 315
  Saffir-Simpson scale, 318
  strong, U.S., 316
Huxley, Thomas, 6
Hydraulic piston corer, 109
Hydrocarbons (*see* Oil and gas; Petroleum; Resources of the ocean)
Hydroelectric power (*see* Tidal energy)
Hydrogen bonds, 135, 136
Hydrogenous sediments, 125
Hydrogen sulfide, 138, 301
Hydrographic station, 181
Hydrologic cycle, 133, 134, 167
  (*see also* Water, properties)
Hydrophones, 78
Hydrothermal deposits, 156
Hydrothermal fluids, 371
Hydrozoa (*see* Animals of the sea, coelenterata)
Hypothesis, 23
Hypsographic curve, 33, 35

Ice, 147–150
  climate, 147
  cores, 334
  station, 184
Icebergs, 147, 149, 150 (*see also* Freshwater from the ocean)
Iceland, 66
Igneous rocks, 49
Indian Ocean, 37, 96–99
Indian Ocean Ridge, 37, 97, 99
Indicator species, 251
Inertia, 237, 238
Infrared Astronomical Satellite (IRAS), 24
Infrared radiation, 141
Innocent passage, 464
Institute for Maritime Archeology, Scotland, 424
Institute of Nautical Archaeology, 422, 423
Institute of Oceanographic Sciences, 158
Interfaces, 28
Internal waters, 464, 465
Internal waves, 230
International Court of Justice, 469
International straits issue, 470
International Whaling Commission, 277
Invertebrates (*see* Animals of the sea)
Ions, 134
Isaacs, John, 358
Isostasy, 43, 44, 80
Isotopes, 134, 135, 172–175 (*see also* Radioactive gases)

argon-40, 172
  artificial radioactive, 174–175
  dating, 172, 173
  half-life, 172
  nuclear explosions as source, 174
  parent-daughter, 172
  radioactive, 174–175
  stable, 173
  use, 172–175
Ixtoc I oil spill, 412

Janssen, John, 285
Japan, Sea of, 34
Japan current, 214
Japan trench, 169
Jason, Jr., 423, 432, 433
Jason Project, 432, 434
Java trench, 97
Jetties (*see* Coastal structures)
Johnson-Sea Link II, 112, 115
JOIDES Resolution, 11, 108, 109, 113
Joint Oceanographic Institutions Deep Earth Sampling (JOIDES), 108
Joint Global Ocean Flux Study (JGOFS), 13, 165
Juan de Fuca ridge, 62, 65, 66, 218, 372

Kaimei, 448
Kelp, 258 (*see also* Algae, brown)
  processing plant, 453
Kharg-5 oil spill, 415
Kingdom, 254 (*see also* Taxonomic classification)
  animalia, 254, 261–277
  fungi, 254
  monera, 254, 256
  plantae, 254, 257, 258
  protista, 254, 258–261
Knauss, John, 474
Knot, 34
Krakatoa, 97, 122, 233
Krill, 274, 442, 443
  role in food web, 307
Kuroshio Current, 214
Kwajalein, 297

Lagoons, 344, 345, 348
Lake Baikal, 169
Lamont-Doherty Earth Observatory, 10, 334
La Niña, 328
Land breeze, 331, 332
Land bridge, 332
LANDSAT, 181, 457
Landslide surge, 231
Langmuir cells, 227, 230
Latent heat, 137–139, 205
  of condensation, 205
  of evaporation, 137–139, 205
  of melting, 137
  of vaporization 137–139
Lava, 49,

Lavoisier, Antoine, 143
Law of the Sea, 462–476
  conflict between states and U.S. federal government, 463
  developing countries and, 471–473
  early history, 463, 464
  impacts on scientific research, 473, 474
  1958 and 1960 Geneva Conventions, 464–466
    Convention on the Continental Shelf, 465, 466
    Convention on Fishing and Conservation of Living Resources of the High Seas, 466
    Convention on the High Seas, 466
    Convention on the Territorial Sea and Contiguous Zone, 464
  scientific research, 466
  shortcomings of 1958 and 1960 Geneva Conventions, 466
  Third Law of the Sea Conference, 467–474
    basic issues, 568, 470
    chronology, 468
    common heritage, 467
    continental margin, 471
    deep-sea mining, 473
    developed versus developing nations, 471, 473, 474
    Exclusive Economic Zone, 470, 471–473
    fishing, 389
    manganese nodules, 376, 377
    results of, 470–474
    scientific research, 473, 474
    straits issue, 470
    territorial sea, 471
Lead (*see* Pollution, marine)
Life, favorable conditions, 29, 30
Light in the ocean, 141, 286
  effect on photosynthesis, 292
Light-year, 22
Limestone, 123
Line Islands, 55
Lithosphere, 51
Lithothamnion, 259 (*see also* Algae, red)
Little ice age, 334
Littoral system (*see* Marine environment)
Lituya Bay, Alaska, 231, 232
Lobsters, 265
Loihi Seamount, 67
Longshore current, 228, 339–342
Lupton, John, 218
L wave, 42

Macroystis (*see* Kelp)
Magellan, Ferdinand, 4
Magma, 26, 48

Magnetic field, 48, 71, 76
Magnetic reversals, 48, 50, 128
Magnetic stratigraphy, 50, 128
Malcolm Baldrige, 16
Major currents, characteristics, 210
Mammalia, 272–277
Manganese nodules, 156, 374–377, 379 (*see also* Resources of the ocean)
  composition, 375,376
  distribution, 374, 375
  mining pollution, 416, 417
  origin, 374
  recovery, 376, 378
  value, 374, 375
Mangroves, 258, 350
Mantle, 26, 41 (*see also* Crustal structure)
Marcet, Alexander, 143
Mare Liberum, 463
Marginal seas, 34, 92
Marianas Trench, 9, 96
Mariculture, 451–453 (*see also* Aquaculture)
Marine archeology, 15, 420–437
  competition, 421
  discipline of, 421–424
  future, 435, 436
  history, 423, 424
Marine Biological Laboratory, 10
Marine environment, divisions, 245–249
  abyssal, 245, 246
  aphotic zone, 245
  bathyal, 245, 246
  benthic, 245
  benthic realm, 245
  deep-sea system, 245, 246
  hadal, 249
  intertidal, 246
  littoral system, 245, 246
  neritic system, 245–247
  oceanic system, 245, 246, 248
  pelagic realm, 245–249
  supratidal, 246
  subtidal, 246
Marine geology and geophysics, 13, 14
  equipment and techniques, 71–78, 103–116
Marine Mammal Protection Act, 195
Marine pharmaceuticals (*see* Future uses of marine organisms)
Marine policy, 15
Marine pollution (*see* Pollution, marine)
Marine Protection Research and Sanctuaries Act , 360
Marine scientific research (*see* Law of the Sea)
  consent regime for marine scientific research, 474
  freedom of, 473
  restrictions on research, 474

Marine sediments, 102–131
  missing, 116, 117
  pelagic, 122–125
  terrigenous, 119–122
  U.S. continental margin, 128
Marine snow, 163, 165, 247
Marshes, 350, 351
  human development, 356
  productivity, 350
Martin, John, 326
Massachusetts Lobster Hatchery, 391
Mauna Loa Observatory, 322
Maximum sustainable yield, 384 (*see also* Fishing)
Maury, Matthew Fontaine, 6, 7, 199
McClintock, James, 285
Mediterranean Sea, 36, 212, 435
Megaplume, 66, 169, 218, 219
Mercury (*see* Pollution, heavy-metal)
Meroplankton, 250
Merrimac, 424
Metamorphic rocks, 50
Metalliferous muds, 372–374 (*see also* Red Sea brine deposits)
Meteor expedition, 9, 71, 199
Methane, 169
Miami, University of, 10
Microplankton, 251
Microns, 34
Mid-Atlantic Ridge, 37, 66, 67, 74, 92, 93, 169, 217
Mid-Indian Ocean Ridge, 37
Mid-ocean ridges, 37, 145
Milankovitch hypothesis, 332–334
Military uses of mammals, 443, 444
Milky Way Galaxy, 22
Minamata disease, 406
Mineral resources (*see* Resources of the ocean)
Mining the seafloor, 364–368
Mir, 112, 115
Mississippi River and delta, 352, 353
Moho (Mohorovičić discontinuity), 41, 79, 80
Molecule, 134
Mollusca, 266
Monera, 254, 256
Monitor, U.S.S., 424
Monsoons, 3, 208
Monterey Bay Aquarium, 459, 460
Montreal Protocol, 331
Moon, 29, 118
Moraine, 83
Moss Landing Marine Observatory, 326
Mount Pinatubo, 122, 324, 325, 329
Mount Saint Helens, 62, 65, 122

Muds, deep-sea, 120 (*see also* Oozes)
Multiple Opening/Closing Net-Environmental Sensing System (MOCNESS), 283
Mussel watch, 398
Mutualism, 285

Nanoplankton, 251
Nansen, Fridtjof, 9, 157
Nansen bottle, 9, 157, 179, 181
National Academy of Sciences, 409, 410, 415
National Coastal Zone Management Act, 359
National Estuarine Research Reserve System, 360
National Marine Fisheries Service, 10, 195
National Marine Sanctuary Program, 360
National Ocean and Atmospheric Administration (NOAA), 10, 15
National Science Foundation, 16
National Science Teachers Association, 434
National Status and Trends Program, 398
Nauset Beach, 346, 347
Nautilus, 10
Navigation, 11
  dead reckoning, 11a
  Loran, 111
  global positioning system, 111
  set, 111
  sextant, 111
Neap tides (*see* Tides)
Nearshore region, 338
Nekton, 249, 252 (*see also* Population of the ocean)
Nematoda, 261
NEMO (submarine), 427
Nereocystis, 258 (*see also* Algae, brown)
Neritic system (*see* Marine environment)
Neutrons, 134
Niche (*see* Biotope)
Nile Delta, 352
Nitrogen, 160, 161, 289–291
  fixing bacteria, 290, 291
Nitrous oxides, 322
NOAA (*see* National Oceanic and Atmospheric Administration)
Nodes, 235, 237
Nonconservative elements, 158
Nondeposition areas, 125
Nonrenewable resources (*see* Resources of the ocean)
North American Plate, 62, 65
North Atlantic Deep Water, 216, 217
North Sea, 36
Norwegian Sea, 367
Notochord, 268

Nuclear explosions 173 (*see also* Radioactive Isotopes)
Nuclear waste disposal, 455, 456
Nucleus, of atom, 134
Nutrients, 170–171
  cycles, 289–291
  distribution, 170–171
  seasonal changes, 292

Ocean
  density, 190, 192
  dimensions, 37
  general characteristics, 33–37, 186–196
  heat regulator, 314
  heat storing capacity, 313, 314
  interfaces, 28
  origin, 27, 28
  salinity, 187, 189
  temperature and climate, 190, 191, 320, 321
Ocean-atmosphere interactions, 313–315
Ocean basins, 33, 92–99
  and mountains, 56
Ocean circulation, 208–219
Ocean-climate interactions, 311–316 (*see also* Ice and climate)
Ocean crust (*see* Crustal structure)
Ocean currents, 200–219
  boundary, 208, 210, 211
  countercurrents, 211
  energy from, 448
  geostrophic, 210, 211
  thermohaline, 216–218
  undercurrents, 211
Ocean Drilling Program, 108
Ocean dumping, 407
  burning of wastes at sea, 408, 409
  Ocean Dumping Act, 408
Ocean engineering, 14, 15
Ocean floor origin (*see* Seafloor spreading)
Ocean tides (*see* Tides)
Oceanic environment (*see* Marine Environment)
Oceanic layer (*see* Crustal structure)
Oceanic ridge, 37, 46 (*see also* Mid-ocean ridges)
Ocean Minerals Company, 376
Oceanography
  career, 15–17
  components, 14–15
  definition, 13
  early history, 3–9
  global environment, 2
  its resources, 2
  modern history, 9–13
  for the nonoceanographer, 18
  unified view, 15
Ocean Thermal Energy Conversion System, 448–452

Ocean Topography Experiment (TOPEX), 458, 459
Oceanus, 198
Ocean vents (*see* vents)
Octopus, 266, 268
Office of Naval Research, 16
Offshore islands and platforms, 459
Oil and gas, 368–371 (*see also* Petroleum; Pollution, oil)
  drilling for, 369
  exploration for, 369
  origin of, 368, 369
  potential, 369, 370
  sources of, 370
  sulfur deposits with, 370, 371
  value of, 368
Okhotsk, Sea of, 34
Omega-3 fatty acids, 441
Oozes, 6, 122–125
  calcareous, 122, 123
  coccoliths, 122, 124
  diatoms, 122
  foraminifera, 122–124
  radiolarian, 123
  resource, 377
  siliceous, 122
  types, 122
Orbital motion of waves, 222, 223
Oregon, 62
Oregon State University, 10
Organic cycle, 288–290
Organic matter, 163, 170
Organic production, 302–305
  factors, 304, 305
  measurements, 303, 304
  standing crop, 304
Osmosis, 287
Osmotic pressure, 146, 287
Osteichthyes, 269–272
OTEC (*see* Ocean Thermal Energy Conversion)
Outer Continental Shelf Lands Act, 464
Owen Fracture Zone, 60
Oxygen, in seawater, 161, 162
Oxygen isotopes, 173
Oxygen minimum zone, 161
Oyster farming, 390, 391
Ozone, 322, 330
  depletion, 330
  hole, 330
  layer, 330
  problem, 329–331
  solving it, 331

Pacific Ocean, 94–96
Pacific Plate, 65
Padre Island, 343, 353
Pack ice, 147
Paleoceanography 128–130
Pangaea, 28, 31, 45, 66, 331
  supporting data, 47
Parasitic relationships, 285

Parent/daughter, 172 (*see also* Radioactive dating)
Particulate matter, 163
Past climates, 331–335
PCB (*see also* Synthetic organic compounds)
  and fish, 390
Peat, 351
Pelagic clays, 125
Pelagic environment (*see* Marine environment)
Pelagic sediments (*see* Marine sediments)
Pelecypoda (*see* Mollusca)
Persian Gulf, 37, 60, 64, 444
  oil spills, 412, 413
Peru (*see* Fishing)
Peru-Chile Trench, 169
Pesticides, 404
  Aldrin, 404
  DDT, 154, 404
  Dieldrin, 404
Petroleum, 368–370 (*see also* Oil, and gas)
  conditions necessary for, 359
  origin of, 368, 369
pH, 154, 155, 162
Phosphate deposits, 156, 367, 368, 379
Phosphorite, 125
Phosphorus, 289–291
Photic zone, 142, 245
Photosynthesis, 27, 142, 161, 165, 288–293, 301–305
Phylum (*see* Taxonomic classification)
Physical oceanography, 14
Physical processes in the ocean, 171, 172, 178–197
Physical resources of the ocean, 392
Phytoplankton, 249, 258–261 (*see also* Food chain; Population of ocean)
  and climate, 320
  experiment, 326
Piccard, Auguste, 110
Pingers, 103, 104, 107
Pisces (*see* Fish)
Piston corer, 107–109
Placer, 365, 366
Planetesimals, 23, 24
Plankton, 249 (*see also* Population of the ocean)
Plankton net, 282
Plantae (*see* Kingdom)
Plants in the ocean, 256–261, 288–293
  biological processes, 288–293
  light effects, 286, 287, 292
  salinity, influences by, 290
  temperature, influences by, 290, 291
Plastics (*see* Pollution, plastics)
Plate tectonics, 46–57 (*see also* Seafloor spreading)

earthquakes, 54, 55
mineral deposits, 61
motions involved, 54
ocean chemistry, 61
types of plates, 52, 54
Platforms, floating and fixed, 184, 185
Pleistocene epoch, 28, 30
  glaciation, 333, 334, 355
Popup profiler, 186
Polar regions, 147 (*see also* Arctic Ocean; Antarctica)
Pollution, biological magnification, 404
Pollution, from domestic, industrial, and agricultural sources, 403–407
  pesticides, 404, 405
  plant nutrients, 403
  sediments, 403
  synthetic organic compounds, 405
Pollution, from mineral resource exploration and exploitation, 416, 417
Pollution, heavy-metal, 405–407
Pollution, marine, 154 (*see also* Ocean dumping; Sewage)
  categories, 399
  control, 402, 403
  definition, 396, 397
  effects, 396, 397
  Great Lakes, 397, 399
  human activity and impact, 399–402
  lead distribution, 402
  monitoring, 398, 402
  natural causes, 399
  pathways to the sea, 402
Pollution, oil, 409–416
  acute and chronic effects, 415
  behavior in marine environment, 415, 416
  detergent damage to environment, 411
  natural seeps, 411
  offshore drilling and platforms, 412, 415
  oil in marine environment, 410
  shipping, 409, 410
  spills, 410–415
  tanker accidents, 410–415
Pollution, plastics, 408
Pollution, radioactive waste, 417
Pollution, thermal, 417
Polychlorinated biphenyls (PCB) (*see* Synthetic organic compounds)
Polymetallic sulfides (*see* Mineral resources)
Pompeii worm, 302
Population of the ocean
  benthos, 252, 253
  classification, 249–277
  nekton, 252
  plankton, 249–252

Porifera, 261 (*see also* Animals of the Sea)
Porpoises (*see* Cetacea)
Positioning, 77
Potassium-argon dating method, 172
Precambrian, 28
Pressure, 140
  biological effects, 286
Primary production, 284, 303, 305
Prince William Sound
  earthquake, 237
  oil spill, 414, 414
Prokaryotic, 255
Protein, 380
Protista (*see* Kingdom)
Protosun, 23, 24
Protozoa (*see* Animals of the Sea)
Ptolemy, 3
Puerto Rico trench, 94
Purse seine, 385, 386 (*see also* Fishing)
P wave, 42
Pycnocline, 188, 189

Quaternary Period, 28 (*see also* Climate, recent changes)
Quest Group, Ltd., 432

Radioactivity, 134, 135
Radioactive dating, 172, 173
Radioactive decay, 66, 161
Radioactive gases
  argon-39, 161
  carbon-14, 161, 174
  tritium 161, 175
Radioactive isotopes 172–175 (*see also* Isotopes)
Radioactive waste, 417
Radiolaria (*see* Oozes)
Radiotelemetry, 182
Rance River, 241, 446
Red Sea, 57–60, 137, 168, 256
  brine deposits, 60, 372–374 (*see also* Resources of the ocean)
Red tide, 259, 405, 407
Reefs (*see* Coral reefs)
Refraction, 227, 228
RELAYS System, 186
Relict sediments, 128
Remotely operated vehicles (ROVs), 12, 116, 169, 424
Remote sensing (*see* Satellites)
Renewable resources (*see* Resources of the Ocean)
Reptilia, 272
Reservoir beds, 369 (*see also* Oil and gas)
Residence time, 166, 167
Res nullius, 463
Resources of the ocean, 363–394 (*see also* Cobalt crusts; Manganese nodules)
  continental margin, 364–371
  deep sea, 371–377
  difficulty in assessing, 364, 365

nonrenewable, 364
other minerals, 368
petroleum (see Oil and gas)
phosphates, 367, 368
Red Sea metalliferous muds, 372–374
sand and gravel 366, 367
seawater, 368
sediment deposits, 365–368
renewable, 364
value, 366
Remotely Operated Platform for Ocean Science (ROPOS), 116
Respiration, 161, 288, 289
Resting spores, 259
Rhode Island, University of, 10
Richardson, Philip, L., 200
Ridge Interdisciplinary Global Experiments (RIDGE), 13, 63–66
Ridges (see Mid-ocean ridges)
Riftia (see Tube worms)
Ring of fire, 96
Rings, 211–214
Rip current, 227–230, 339, 340
Robison, Bruce, 119
Rocks, on seafloor, 125
Rogue waves, 224
Ross, Sir John, 6
ROVs (see Remotely operated vehicles)
Rule of Constant Proportions (see Forchhammer's Principle)
Rusticles, 432, 433

Salicornia, 257 (see also Marshes)
Saffir-Simpson hurricane scale, 318
Salinity, 158
definition, 158, 187
with depth, 188, 189
determination, 157, 158
electrical conductivity, 187
factors influencing, 187
and geography, 187, 188
plants, 146
Salt dome, 85, 369, 370
Salt marsh, 350, 351
plants, 257, 351
Salt-wedge estuary, 349 (see also Estuaries)
San Andreas Fault, 45,62
Sand and gravel, resources, 366, 367, 379
Sandwich Trench, 94
San Salvador, 5
Santa Barbara oil spill, 411
Santorini, 234
Sargassum, 249, 258 (see also Algae, brown)
Sashimi, 389
Satellites, 12, 180, 181, 457–459
advantages, 180
arctic images, 148
bathymetry and gravity, 75, 76
LANDSAT, 181

predicting El Niños, 328
SEASAT A, 181
varieties, 457–459
Saudi Arabia, 374
Scientific law, 23
Scientific research (see Law of the Sea)
Scorpion, 110
Scripps Institution of Oceanography, 10,11, 358
Scuba, 116, 118, 282, 423
Sea Beam, 70, 71, 73, 74
Sea breeze, 314,315
Sea cucumbers, 267,270
Sea Education Association, 16
Seafloor rocks, 125
Seafloor spreading, 28, 31, 46–57, 168 (see also Crustal structure; Plate tectonics)
age of seafloor, 53
conveyor belt example, 50,53
and deep-sea sedimentation, 126
mechanisms, 63,66
and mineral resources, 371–374
unanswered questions, 63, 66
Sea Grant, 17
Sea ice, 147,149
Sea level, 31 (see also Climate; Glaciers and glaciation)
changes due to glaciation, 147
changes due to seafloor spreading, 62
effect on beaches, 344
rising, 355
Seamounts, 87, 96
SEASAT A, 12, 209, 459, 458
Seasons, effect on plant growth, 292
Sea star, 267,268
Seawalls (see Coastal structures)
Seawater (see Water, properties)
biologically important aspects, 159
buffered solution, 154,155
carbon dioxide, 162–165
chemistry of, 158–175
composition, 158–166
density, 140, 146, 190, 192
dissolved gases, 160–163
dissolved organic matter, 163
freezing point, 137,138,146
heat capacity, 146
inorganic matter, 159, 160
major elements, 158,159
minor elements, 158,159
organic matter, 159
osmotic pressure, 146
oxygen concentration,161,162
particulate matter, 163
pressure, 140
properties, 146, 147
salinity, 146, 147, 187–189
specific heat, 138
standard, 158
temperature, 188–191

trace elements, 159
vapor pressure, 141,173
Secchi disks, 143
Sediment reworking, 366
Sediment sampling, 107–110
Sediment traps, 163,247
Sedimentary rocks, 50
Seiche, 235, 237
Seismic activity, 55 (see also Earthquakes)
Seismic reflection and refraction, 57, 76, 77
Semidiurnal tides (see Tides)
Semipermeable membrane, 287
Sensible heat, 139
Settling velocity, 125
Shadow zone, 195, 196
Sharks 268–271 (see also Vertebrates)
Shepard, Francis, P., 82, 84
Shinkai 6500, 112
Shipping, 392
Ship wrecks, U.S., 436
Shoreline, 338 (see also Beaches)
Shore zone, 338
Side-scan sonar, 105, 106
Slump deposits, 121
Slumping, 79, 87
Sorting, 119
SOFAR channel, 195, 198, 200
Solar energy, 204–206
Solar radiation, 204
Solar system, 21, 22
origin and evolution, 23–25
Solomon, Susan, 325
Sonar, 72
Sound channel, 196
Sounding methods, 61, 72–74, 104, 105
echo sounding, 71, 72
Sound 274 (see also Underwater sound)
Sound velocity, 193
Southern Ocean, 99
Southern Oscillation, 327
Spain, mussels, 391
Spartina, 257 (see also Marshes)
Spat, 390
Species, 254 (see also Taxonomic classification)
Spermatophytes (see Plant kingdom)
Sponges (see Animals of the sea, porifera)
Spreading rates, 50
Spring tide (see Tides)
Squid, 266,268
Standard seawater, 158
Standing crop, 304
Standing wave, 235–237
Stationary wave, 235–237
Stipe, 258
Stommel, Henry, 334
Storm surge, 230
Strabo, 3
Stratigraphy, 128

Strike slip, 54
Subduction, 51, 54, 169
Submarine canyons, 87–91
Corsair Canyon, 90
La Jolla and Scripps Canyons, 87, 88
Lydonia Canyon, 88, 89, 91
Submarines (see also *Scorpion*, *Thresher*)
submarine warfare, 10
Submerged Lands Act, 463
Submersibles, 11, 109, 110, 112 (see also Aluminaut; Alvin; Johnson Sea Link II; Shinkai 6500; Trieste)
pros and cons, 115, 116
Sulfur deposits, 370, 371 (see also Oil and gas)
Surface tension, 135,141
Sushi, 389
Swallow Float, 198, 200 (see also Current measurements)
S wave, 42
Symbiotic relationships, 285
Synoptic measurements, 179
Synthetic organic compounds, 398, 405
chlorinated and halogenated hydrocarbons, 398
polychlorinated biphenyls (PCBs), 398, 405

Taxonomic classification, 253–256
Tectonics (see Plate tectonics)
Temperature, 139
affect on plants, 290
conversion, 135
factors influencing, 189, 190
methods of measurement, 179–186
Nansen bottle, 179, 181
pattern in the ocean, 189, 190
satellite measurements, 180
Temperature stratification (see Thermocline)
Teredo, 266
Terrestrial planets, 23
Terrigenous sediments (see Marine sediments)
Territorial sea, 463, 470, 471
Texaco Italia, 392
Texas A&M University, 10, 422
Thermal differences, energy from (see OTEC)
Thermal pollution, 417
Thermocline, 188, 189
Thermohaline, circulation, 208, 216–218
Thermometers, 157
Theory, 23
Thira, 234
Third Law of the Sea Conference (see Law of the Sea)
Thompson, Tom, 427
Thomson, Sir Charles Wyville, 8
Thresher, 110

Three-mile limit, 463
Throckmorton, Peter, 435
Thunder Bay, Ontario, 371
Tidal bore, 239
Tidal energy, 392, 446, 447
Tidal waves (see Tsunami)
Tides, 236–242
    causes, 237–240
    diurnal, 239, 241
    ebb, 239
    flood, 239
    importance, 239
    mixed, 239, 241
    neap tide, 239
    semidiurnal, 239, 241
    spring tide, 239
    tidal currents, 239
    tidal friction, 241, 242
    tidal range, 239
    tide gauge records, 235
Titanic, 11, 150, 427, 428, 431, 432
TOGA (see Tropical Ocean Global
    Atmospheric Experiment)
TOPEX, (see Ocean Topography
    Experiment)
Topography, 71
Torrey Canyon, 410, 411
Trade winds, westerlies, 206
Transform faults, 51, 54
Transparency of water, 287
Transpiration, 133
Transient Tracers in the Ocean
    (TTO), 155, 175, 320
Trawling, 386 (see also Fishing)
Trenches, 94
Trench-island-arc systems, 52
Trichinella, 261
Trichodesmium, 256 (see also Algae,
    blue-green)
Trieste, 10, 110
Tritium, 135, 161 (see also
    Radioactive gases)
Tropical Ocean Global
    Atmospheric Experiment
    (TOGA), 13, 199, 313, 328
Truman, Harry S, 463
Truman Proclamation, 463, 464
Tsunami, 231–236
    warning system, 234–236

TTO (see Transient Tracers in the
    Ocean)
Tube worms, 300, 301
Tunicates, 268
Turbidites, 90
Turbidity current, 89–91
Turner Broadcasting System, Inc.,
    432
Turtles, 273, 274
Typhoons (see Hurricanes)

Ulu Burun, 420, 425
Ulva, 258 (see also Algae, green)
Ultraplankton, 250
UNCLOS (see Law of the Sea
    Conference)
Undercurrents (see Ocean currents)
Underwater eruption, 67
Underwater habitats, 459
Underwater photography,
    103–105
Underwater robotics, 457
Underwater sound,
    193–196, 274
    channel, 195
    losses, 196
    refraction, 195
    velocity, 193
U.S.–Canada boundary disputes,
    469
U.S. coastal zone, 353, 354
U.S. Coast Guard, 16
U.S. Congress, Titanic, 432
U.S. Exclusive Economic Zone,
    470, 472
U. S. Geological Survey, 16
U.S. mineral resources,
    377–379
U.S. Supreme Court, S.S. Central
    America, 429
United States v. California, 463
United States v. Louisiana, 463
Universe, 22
Upwelling, 214, 215, 290, 293

Vapor pressure, 141
Vega, 24
Vents, 67, 155, 156, 168–170
    ancient, 301, 302

biological communities, 67,
    298, 300–302
and chemosynthesis, 300, 301
fish, 302
mineral deposits, 372, 373
mining, 375
Vertebrates, 268–277
Vikings, 4
Viscosity, 141
Vitamins, 163
Volcanic deposits, 49, 122
Vulcanus II, 409

Washington, state of, 62
Washington, University of, 10
WASP (diving suit), 119
Water, origin (see Origin of ocean
    water)
Water, properties
    boiling point, 137
    dissolving power, 136, 137
    distribution, 133
    freezing point, 137, 138
    heat capacity, 137–140, 205
    hydrologic cycle, 133, 134, 167
    ice, 137
    phases, 137, 138
    solvent, 136
    structure, 133–136
    surface tension, 135, 141
    vapor, 322
    vapor pressure, 141
Water masses, 216, 217
Water molecule, 135, 136
Watling's island, 5
Waves (see also Catastrophic waves;
    Wind-generated waves)
    in coastal region 226, 227
    convergence, 227
    divergence, 227
    energy, 222
    erosion by, 226, 227
    height, 222
    internal, 230
    orbital motion, 222, 223
    period, 222
    refraction, 227, 228, 339
    sea, 225, 226
    surf, 225, 226

swell, 225, 226
velocity, 222
wavelength, 222
wave power pump, 447, 448
Weather, 311, 312
    effect on productivity, 304
    forecasting, 313
    satellite, 312
Weathering, of volcanic rock, 27
Wegener, Alfred, 43
Wentworth scale of sediment
    grain scale, 120
Western boundary current, 208
West Wind Drift (see Antarctic
    Circumpolar Current)
Weyerhaeuser Corporation, 442,
    444
Whales, 272–277 (see also Cetacea)
    behavior, 276
    ethograms, 274, 276
    songs, 274
    types of, 275
Wind-driven circulation, 208–216
Wind-generated waves, 224–230
    capillary waves, 224
    fetch, 224
    important factors, 224
    shallow water waves, 226, 227
    types, 225, 226
    wave convergence, 227
    wave refraction, 227, 228
Winds, 207–211
    easterlies, 207
    trade, 207
    westerlies, 207
Woods Hole Oceanographic
    Institution, 5, 10
World Ocean Circulation
    Experiment (WOCE), 13,
    199, 313, 328, 373
Wyrtki, Klaus, 327

Yellow Sea, 34

Zagros Mountains, 60
Zebra mussels, 399–401
Zooplankton, 249–252 (see also
    Population of the ocean)
Zostera, 257, 258 (see also Eelgrass;
    Marshes)